K. Wildermuth
Y. C. Tang

A Unified Theory of the Nucleus

Clustering Phenomena in Nuclei

Edited by
K. Wildermuth
P. Kramer

Volume 1

Volume 1
Wildermuth/Tang
A Unified Theory of the Nucleus

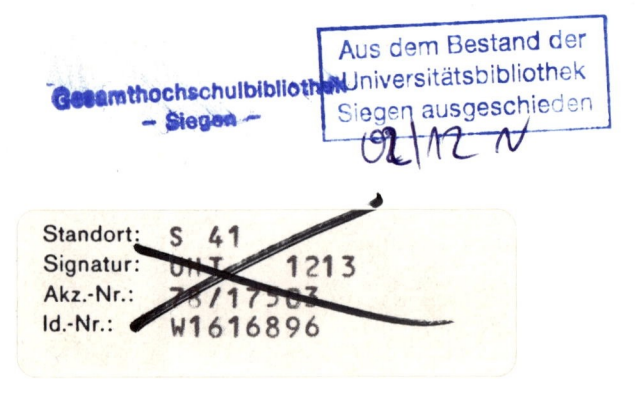

K. Wildermuth
Y. C. Tang

A Unified Theory of the Nucleus

With 77 Figures

Vieweg

1977

Set by Vieweg, Braunschweig
Printed by E. Hunold, Braunschweig
Bookbinder: W. Langelüddecke, Braunschweig
Cover design: Peter Morys, Wolfenbüttel
Printed in Germany-West

ISBN 3 528 08373 5

Preface

The purpose of this monograph is to describe a microscopic nuclear theory which can be used to consider all low-energy nuclear phenomena from a unified viewpoint. In this theory, the Pauli principle is completely taken into account and translationally invariant wave functions are always employed. Also, this theory is quite flexible; it can be utilized to study reactions initiated not only by nucleons but also by arbitrary composite particles.

Throughout this monograph, we have endeavoured to keep the underlying physical ideas as easily comprehensible as possible. Consequently, it becomes frequently necessary to sacrifice mathematical rigour in favour of clarity in presenting these ideas. In this way, it is our hope that this monograph could be useful to many research physicists in the nuclear field, experimentalists and theorists alike.

In chapters 1 through 4, the formulation of this theory is presented. Numerical examples concerning bound-state, scattering, and reaction calculations are mainly described in chapters 5 through 7. In chapters 8 through 15 we discuss, within the framework of this theory, general properties of nuclear systems. Finally, in chapters 16 and 17, we show in specific cases how one can achieve, without carrying out explicit calculations, a qualitative or even semi-quantitative understanding of these cases by applying the general physical concepts contained inherently in this theory.

Many of our colleagues have offered valuable advice and constructive criticism. These people are: *R. E. Brown, D. Clement, F. Gönnenwein, E. Kanellopoulos, P. Kramer, E. Schmid, H. Schultheis, R. Schultheis, G. Staudt, W. Sünkel,* and *D. R. Thompson.* To them we express our most sincere gratitude.

It is also our pleasure to acknowledge *W. Pilf, S. Smith,* and *G. Tollefson* for their assistance in the preparation of the manuscript. Finally, we are grateful to our wives, *Erika Wildermuth* and *Helen Tang,* for their patience and encouragement throughout the period in which this monograph was written.

K. Wildermuth
Y. C. Tang

Tübingen, Germany
Minneapolis, Minnesota, U.S.A.

December, 1975

Contents

1. Introduction

1.1. General Remarks

Numerous experimental and theoretical investigations have confirmed the proposal that nuclei are built up of protons and neutrons [HE 32, IV 32]. Therefore it should be possible to derive all the properties of a nucleus consisting of A nucleons, or of a nuclear reaction in which A nucleons participate, from the Schrödinger equation for this A-body system

$$\begin{aligned} H\Psi(\tilde{r}_1,\ldots,\tilde{r}_A;t) &= \left[-\frac{\hbar^2}{2M}\sum_{i=1}^{A}\nabla_i^2 + V(\tilde{r}_1,\ldots,\tilde{r}_A)\right]\Psi(\tilde{r}_1,\ldots,\tilde{r}_A;t) \\ &= i\hbar\frac{\partial}{\partial t}\Psi(\tilde{r}_1,\ldots,\tilde{r}_A;t), \end{aligned} \tag{1.1}$$

where $\tilde{r}_i$ denotes the space, spin, and isobaric-spin coordinates of the ith nucleon.

To carry out this program we must overcome two problems:

(i) We must know the specific form of the nuclear interaction potential energy $V(\tilde{r}_1,\ldots,\tilde{r}_A)$.

(ii) We must solve, at least approximately, the A-body Schrödinger equation.

The form of the nuclear potential is in fact not completely known, although it has become increasingly evident recently that this potential can be considered as primarily a superposition of two-body potentials between all pairs of nucleons [BE 71]. The problem with the two-body potentials proposed earlier was that they could not reproduce the nuclear saturation character, i. e., the property that the nuclear volume must increase proportionally with the nucleon number A. This difficulty has led to speculation [DR 63] that the nuclear interaction might include many-body forces in addition to the two-body terms. Thus, for example, in the case of a three-body problem with an additional three-body force [BR 73], one would have to write the nuclear interaction as

$$V = V_{12} + V_{13} + V_{23} + V_{123}. \tag{1.2}$$

However, over the past twenty years, increasingly reliable two-nucleon potentials have been developed describing the two-nucleon scattering data up to several hundred MeV [BE 71, BR 60, BR 62, HU 62]. In contrast with the earlier two-body interactions, these potentials contain not only strong tensor and Majorana (space-exchange) components, but also velocity-dependent terms or a repulsive core which prevents any two nucleons from approaching each other too closely. Such improved two-body potentials have been found capable of producing saturation [BE 71]. Hence, it has become gradually clear that potentials without large many-body terms, like V_{123}, are adequate to fit the nuclear data, including the saturation character.†

† Most of the general considerations which we bring into this monograph remain valid even if the nuclear interaction should contain large many-body terms of short-range character.

In our choice of a nuclear potential, we shall adopt a phenomenological point of view; that is, we shall assume a nuclear potential composed only of two-body forces, and we require only that our two-body potentials be reasonably consistent with the two-nucleon scattering data, give the correct deuteron binding energy, and have the proper nuclear saturation character [WU 62]. The presently unresolved question as to which of the possible potentials meeting these criteria is the correct one will not be considered in this monograph, since it is more a question for relativistic quantum theory. Hence, for our A-nucleon Schrödinger equation, we write

$$\begin{aligned} H\Psi(\tilde{r}_1, \ldots, \tilde{r}_A; t) &= \left[-\frac{\hbar^2}{2M}\sum_{i=1}^{A}\nabla_i^2 + \sum_{i<j}^{A}\sum_{j=1}^{A} V_{ij}\right]\Psi(\tilde{r}_1, \ldots, \tilde{r}_A; t) \\ &= i\hbar\frac{\partial}{\partial t}\Psi(\tilde{r}_1, \ldots, \tilde{r}_A; t). \end{aligned} \tag{1.3}$$

For quantitative calculations typical two-nucleon potentials meeting our criteria will be given in Chapter 5.

Our adoption of a potential corresponds to forces which act without propagation delays and hence are not relativistically correct. This neglect of relativistic effects and our consequent use of the non-relativistic Schrödinger equation are reasonable as long as the kinetic energy per nucleon is much less than the nucleon rest mass of about 940 MeV. With this restriction on the energy, we may consider the problem of the nuclear potential as more or less settled, at least for our purposes; for, whatever the outcome of the relativistic theory, it must in the low-energy limit yield a potential in reasonable accord with the phenomenological one.

We shall now turn our attention to the second problem – we wish to develop a flexible and consistent approximation method for solving the A-nucleon Schrödinger equation. The need for flexibility arises, because the properties of nuclei vary considerably from nucleus to nucleus and even from level to level. Hence it is necessary to have a method of great generality such that for each individual nuclear state we can, first, systematically use our physical intuition to include every physical effect that might help to determine the character of the wave function, and then, quantitatively test and improve this approximate wave function to which our intuition has led us.

As will be seen, the method to be discussed in this monograph does meet this requirement. It is in fact a unified theory of the nucleus, since it considers nuclear reactions and nuclear structure from a unified point of view. In addition, with the help of this method, we shall be able to understand the relations between the various nuclear models currently in use. These models have been developed to explain specific nuclear phenomena and very often seem to contradict one another. For the resolution of these contradictions, it will be shown that the Pauli exclusion principle is of fundamental importance, because it reduces to a large extent the differences between different conceptions.

To obtain this method we shall rewrite the many-body Schrödinger equation as a projection equation. But before we do this, we shall discuss why with the usual consideration of the many-body Schrödinger equation, it is in practice not possible to derive such a unified theory of the nucleus.

1.2. Difficulties of Some Reaction Theories

Instead of using the time-dependent Schrödinger equation (1.1) as the starting equation for the treatment of reactions, we can also use the time-independent Schrödinger equation

$$H\psi = E\psi, \tag{1.4}$$

because any time-dependent solution Ψ of eq. (1.1) can be expressed as a linear superposition of the solutions ψ of eq. (1.4).

As is well known, it is possible only in very special cases to solve exactly eq. (1.4) for $A > 2$. Therefore, one is generally forced to use approximation methods which allow one to approximate the exact solutions in successive steps. For the derivation of these methods one commonly formulates the Schrödinger equation (1.4) as an integral equation. The advantage of this is that from the beginning the boundary conditions for a given reaction problem can be incorporated into the solution.

To obtain this integral equation (Lippmann-Schwinger equation) one divides the Hamiltonian H into two parts,

$$H = H_0 + H', \tag{1.5}$$

where H_0 is the Hamiltonian for the target and the bombarding particle without mutual interaction, and H' describes their mutual interaction. Substituting eq. (1.5) into eq. (1.4) yields

$$(H_0 - E)\,\psi = -H'\psi, \tag{1.6}$$

from which one obtains immediately the following Lippmann-Schwinger equation:

$$\psi = \phi_0 - (H_0 - E - i\epsilon)^{-1} H'\psi. \qquad \epsilon > 0 \tag{1.7}$$

In eq. (1.7), ϕ_0 represents a solution of the homogeneous equation

$$(H_0 - E)\,\phi_0 = 0 \tag{1.8}$$

and describes the target and the incoming particle without interaction. The term $-(H_0 - E - i\epsilon)^{-1} H'\psi$ is responsible for the ensuing scattering and reactions. The infinitesimal part $-i\epsilon$ guarantees that only outgoing waves are produced by this term, as is required by physical argument.[†]

Without proceeding any further, one can already recognize the difficulties inherent in eq. (1.7). For this we consider eq. (1.7) in the position representation; that is,

$$\begin{aligned}\psi = \phi_0(\tilde{r}_1, \ldots, \tilde{r}_A) - \underset{\tilde{r}', \tilde{r}''}{S}\; & G(\tilde{r}_1, \ldots, \tilde{r}_A; \tilde{r}'_1, \ldots, \tilde{r}'_A) \\ & \times \langle \tilde{r}'_1, \ldots, \tilde{r}'_A | H' | \tilde{r}''_1, \ldots, \tilde{r}''_A \rangle\, \psi(\tilde{r}''_1, \ldots, \tilde{r}''_A),\end{aligned} \tag{1.9}$$

† Our emphasis in this section is to point out the essential difficulties associated with reaction theories formulated in a straightforward way from eq. (1.7). Hence, mathematical problems connected with three-body breakup and so on will not be dealt with here.

where the sign S indicates summation over discrete variables and integration over continuous variables. The function $G(\tilde{r}_1, \ldots, \tilde{r}_A; \tilde{r}'_1, \ldots, \tilde{r}'_A)$ represents the Green's function for the resolvent operator $-(H_0 - E - i\epsilon)^{-1}$ and has the following form:

$$G(\tilde{r}_1, \ldots, \tilde{r}_A; \tilde{r}'_1, \ldots, \tilde{r}'_A) = \underset{E_0^\alpha, \lambda_\alpha}{S} \langle \tilde{r}_1, \ldots, \tilde{r}_A | E_0^\alpha, \lambda_\alpha \rangle \frac{1}{E_0^\alpha - E - i\epsilon} \langle E_0^\alpha, \lambda_\alpha | \tilde{r}'_1, \ldots, \tilde{r}'_A \rangle \tag{1.10}$$

In the above equation, the functions $\langle \tilde{r}_1, \ldots, \tilde{r}_A | E_0^\alpha, \lambda_\alpha \rangle$ are orthonormalized solutions of the homogeneous equation (1.8), with λ_α denoting collectively the eigenvalues of all those operators which can be diagonalized simultaneously with H_0.

The representation of $G(\tilde{r}_1, \ldots, \tilde{r}_A; \tilde{r}'_1, \ldots, \tilde{r}'_A)$ by means of the orthonormalized eigenfunctions of H_0 has the important consequence that in all approximation methods which one uses to solve eq. (1.9), the wave function ψ has to be expanded in terms of the orthonormalized eigenfunction set of H_0. From this it follows that only transitions to those final states which are eigenstates of H_0 can be calculated in a relatively simple manner. For example, let us consider the interaction of two hydrogen atoms and assume for the moment that both the protons and the electrons involved are distinguishable. Then one can calculate the transition probabilities to all those states where the hydrogen atoms are in their ground or energetically allowed excited states and where the two electrons are not exchanged. On the other hand, transitions to final states which are not eigenstates of H_0 cannot be described in a useful way, because these states have to be described in general by a very complicated superposition of eigenstates of H_0, including highly excited continuum states. In our example here, these are for instance transitions where the reaction products are a H^- ion plus a H^+ ion, i.e., both electrons belong to one hydrogen nucleus. In nuclear physics, the $^6Li(p, {}^3He)\,{}^4He$ rearrangement reaction is an example of such a transition.

Final states where nothing else occurs except identical particles being exchanged are also not eigenstates of H_0. Therefore, such exchange transitions also cannot be described in a simple way by a reaction theory which is based on the Lippmann-Schwinger equation (1.7). This means that with such a reaction theory, even a proper consideration of the indistinguishability of identical particles cannot be simply carried out.

The underlying reason for the above-mentioned difficulty is that the splitting of H into H_0 and H' is not symmetrical in all particle coordinates. This has the consequence that the Green's function $G(\tilde{r}_1, \ldots, \tilde{r}_A; \tilde{r}'_1, \ldots, \tilde{r}'_A)$ is also not symmetrical in these coordinates. For the case of two hydrogen atoms, the Hamiltonian operators H, H_0, and H' have the following forms:

$$\begin{aligned} H = &-\frac{\hbar^2}{2M_P}\nabla^2_{\mathbf{R}_1} - \frac{\hbar^2}{2M_P}\nabla^2_{\mathbf{R}_2} - \frac{\hbar^2}{2m_e}\nabla^2_{\mathbf{r}_1} - \frac{\hbar^2}{2m_e}\nabla^2_{\mathbf{r}_2} \\ &+ \frac{e^2}{|\mathbf{R}_1 - \mathbf{R}_2|} - \frac{e^2}{|\mathbf{R}_1 - \mathbf{r}_1|} - \frac{e^2}{|\mathbf{R}_1 - \mathbf{r}_2|} - \frac{e^2}{|\mathbf{R}_2 - \mathbf{r}_1|} - \frac{e^2}{|\mathbf{R}_2 - \mathbf{r}_2|} + \frac{e^2}{|\mathbf{r}_1 - \mathbf{r}_2|}, \end{aligned} \tag{1.11}$$

$$H_0 = -\frac{\hbar^2}{2M_p}\nabla^2_{R_1} - \frac{\hbar^2}{2m_e}\nabla^2_{r_1} - \frac{e^2}{|\mathbf{R}_1 - \mathbf{r}_1|} - \frac{\hbar^2}{2M_p}\nabla^2_{R_2} - \frac{\hbar^2}{2m_e}\nabla^2_{r_2} - \frac{e^2}{|\mathbf{R}_2 - \mathbf{r}_2|} \quad (1.12)$$

$$H' = \frac{e^2}{|\mathbf{R}_1 - \mathbf{R}_2|} - \frac{e^2}{|\mathbf{R}_1 - \mathbf{r}_2|} - \frac{e^2}{|\mathbf{R}_2 - \mathbf{r}_1|} + \frac{e^2}{|\mathbf{r}_1 - \mathbf{r}_2|}\,. \quad (1.13)$$

As is easily seen, H is invariant against a permutation of the proton or electron coordinates, but not H_0 and H'.

One can certainly split H into H_0 and H' in such a way that H_0 describes the reaction products in the final state (outgoing channel) without interaction. But now the description of the particles in the initial state (incoming channel) becomes very complicated. This means that, by means of the Lippmann-Schwinger equation discussed here, the boundary conditions for either the incoming channel or the considered outgoing channel can be fulfilled, but not both together.

For the scattering of nucleons by nuclei, the difficulty associated with the indistinguishability of the nucleons can be removed in the framework of a shell-model reaction theory [MA 69] by truncating the Hilbert space in such a way that the resultant approximate Hamiltonian can be split into H_0 and H' which are both symmetrical with respect to all nucleon coordinates. As a consequence, the corresponding Lippmann-Schwinger equation also becomes symmetrical in these coordinates and it is possible to define a complete set of antisymmetrized orthonormal functions in terms of which the wave function of the system can be expanded. It should be emphasized, however, that this particular reaction theory has a rather limited domain of applicability. It can be used to study elastic, inelastic, and charge-exchange scattering of nucleons, but not processes in which two or more composite particles are involved in either the initial or the final state.

In summary, we wish to make the following remarks. In the usual reaction theories where one splits H into H_0 and H', an orthonormal set of basis functions is defined, in which the wave function of the system has to be expanded. By this procedure the theory is made so inflexible from the beginning that it is in general not possible to introduce the incoming and reaction channels in a symmetrical manner into the integral equation which describes the dynamical behaviour of the system. This has the consequence that one can conveniently introduce only the boundary conditions belonging either to the incoming channel or to the outgoing channel, but not both together. A practical limitation resulting from this is that rearrangement processes cannot be considered. For the treatment of such processes, one has to find a method which allows the handling of the incoming and outgoing channels in a symmetrical way.

2. Reformulation of the Schrödinger Equation

A symmetrical treatment of the incoming and outgoing channels can be done in a basically very simple way by formulating the Schrödinger equation (1.1) in the form of a projection equation (see, e.g., ref. [WI 72])

$$\left\langle \delta \Psi | H + \frac{\hbar}{i} \frac{\partial}{\partial t} | \Psi \right\rangle = 0. \qquad (2.1)$$

As usual, the Dirac bracket denotes the integration over spatial coordinates and the summation over spin and isospin coordinates. If $\delta\Psi$ represents a completely arbitrary variation of Ψ at a given instant t in the complete Hilbert space, then eq. (2.1) implies that the vector

$$\left(H + \frac{\hbar}{i} \frac{\partial}{\partial t} \right) | \Psi \rangle$$

must be orthogonal to any arbitrary vector in this space. This evidently will be the case only if Ψ obeys eq. (1.1). Therefore, eq. (2.1) is merely another formulation of the time-dependent Schrödinger equation. However, as we shall see later, eq. (2.1) does allow us to treat the incoming and outgoing channels in a symmetrical way.

If we write Ψ as

$$\Psi = \psi \exp(-iEt/\hbar) \qquad (2.2)$$

and insert it into eq. (2.1), we obtain the time-independent Schrödinger equation formulated as a projection equation

$$\langle \delta \psi | H - E | \psi \rangle = 0. \qquad (2.3)$$

As has been mentioned previously, we can also use this simpler equation as the starting point for our future considerations.

We shall now briefly discuss some general properties of eq. (2.3), which we shall need at a later stage. Let us make for ψ the ansatz

$$\psi = \sum_r a_r \phi_r + \int a_p \phi_p \, dp \equiv \mathrm{S}\, a_k \phi_k , \qquad (2.4)$$

where the coefficients a_r and a_p are the discrete and continuous linear variational amplitudes for the trial functions ϕ_r and ϕ_p, respectively. By substituting eq. (2.4) into eq. (2.3) and using the fact that $\delta\psi$ is obtained by an arbitrary variation of the discrete and continuous amplitudes a_k, i.e.,

$$\delta \psi = \underset{k}{\mathrm{S}}\, \delta a_k \phi_k , \qquad (2.5)$$

we obtain the following set of coupled equations:

$$\left\langle \phi_n \left| H - E \right| \mathop{S}_{k} a_k \phi_k \right\rangle = 0, \tag{2.6}$$

where the subscript n takes on both discrete and continuous values. If the trial functions ϕ_k form a complete set, then the solutions of the coupled equations (2.6) are identical to the solutions of the time-independent Schrödinger equation. We should emphasize that the functions ϕ_k must be linearly independent, but need not be orthogonal to each other. This point is of great importance for our future considerations, because only by choosing in general a nonorthogonal set of functions can one expect to introduce the incoming and outgoing channels symmetrically into the theory. At the same time, the relaxation of the orthogonality requirement also gives us some extra flexibility to choose the set of functions ϕ_k in the most appropriate manner according to the problem under consideration.

If we substitute eq. (2.4) into the time-dependent projection equation (2.1), then we obtain a set of coupled equations similar to eq. (2.6) in the stationary case. The only modification is that the quantity E is replaced by the operator $-\frac{\hbar}{i}\frac{\partial}{\partial t}$ which acts on the now time-dependent amplitudes a_k.

In contrast to the trial functions ϕ_k, the eigensolutions of eq. (2.6) are always mutually orthogonal, if all degeneracies are removed. This will be the case even if we restrict the number of variational amplitudes in eq. (2.4). In order to prove this orthogonality property, we consider the following normalized solutions:

$$\psi_l = \mathop{S}_{k} a_k^l \phi_k, \tag{2.7a}$$

$$\psi_m = \mathop{S}_{k} a_k^m \phi_k, \tag{2.7b}$$

which belong to the sets of equations

$$\langle \phi_n | H - E_l | \psi_l \rangle = 0, \tag{2.8a}$$

$$\langle \phi_n | H - E_m | \psi_m \rangle = 0. \tag{2.8b}$$

Upon multiplying eq. (2.8a) by a_n^{m*}, eq. (2.8b) by a_n^{l*} and summing over all possible values which the index n can assume, we obtain

$$\langle \psi_m | H - E_l | \psi_l \rangle = 0, \tag{2.9a}$$

$$\langle \psi_l | H - E_m | \psi_m \rangle = 0. \tag{2.9b}$$

If we now subtract the complex conjugate of eq. (2.9b) from eq. (2.9a), then, due to the hermiticity of H, we obtain the following relations for properly normalized functions ψ_l and ψ_m:

$$\langle \psi_l | \psi_m \rangle = \delta(l, m), \tag{2.10}$$

$$\langle \psi_1 | H | \psi_m \rangle = \langle \psi_m | H | \psi_l \rangle = E_l \, \delta(l, m), \tag{2.11}$$

where the symbol $\delta(l, m)$ represents the Kronecker symbol for discrete values of l and the Dirac δ-function $\delta(l-m)$ for continuous values of l.

Equations (2.10) and (2.11) show that even when we restrict the number of linear variational parameters, i.e., we only work in a subspace of the Hilbert space, we still obtain the result† that (i) any two solutions ψ_l and ψ_m are orthonormalized, and (ii) in this subspace the Hamiltonian H can be represented by a real diagonal matrix. These results have the consequence that the normalization of the time-dependent solution of eq. (2.1) remains constant in time. From this it follows that the conservation law of probability current, or in other words the unitarity of the S matrix, is still satisfied, even if we work with a set of functions which does not span the entire Hilbert space.

For numerical calculations, one has to limit the number of linear variational parameters, i.e., the number of trial functions ϕ_k in the ansatz (2.4). It is reasonable to neglect all those terms in eq. (2.4) which can be expected to have a very small amplitude in the final solution of eq. (2.3).†† Of course, the decision as to which terms will be unimportant is not easy to make and will have to rely on one's experience and physical intuition about the system under investigation (see Chapter 3). Indeed, we know at present of no rigorous rule which could help us in predicting the relative magnitudes of the various variational amplitudes. Therefore, we shall give here only a possibly useful hint which results from the following crude considerations. Let us assume that by some means, we have arrived at a normalized trial function ϕ_a, yielding an energy expectation value fairly close to the eigenvalue of interest. To this function ϕ_a we now add another normalized trial function ϕ_b which, for simplicity, will be assumed to be orthogonal to ϕ_a. A simple calculation then shows that the variational amplitudes a and b satisfy the following relation:

$$\frac{b}{a} \approx -\frac{\langle \phi_b | H | \phi_a \rangle}{\langle \phi_b | H | \phi_b \rangle - \langle \phi_a | H | \phi_a \rangle} . \tag{2.12}$$

From this one sees that $|b/a|$ will be very small if the energy overlapping integral $|\langle \phi_b | H | \phi_a \rangle|$ is much smaller than the energy difference $[\langle \phi_b | H | \phi_b \rangle - \langle \phi_a | H | \phi_a \rangle]$. In the case where ϕ_a and ϕ_b are not entirely orthogonal, a similar consideration will be harder to make, but we do not expect this result to change in any essential way.

---

† Strictly speaking, the proof given here is valid only if all functions ϕ_k are finite-normalizable and hence, belong to a mathematically exact Hilbert space. We shall not discuss the mathematical complications which will appear in our proof if the functions belong partially to a more extended function space (e.g., Banach space), because for all physical situations which we shall describe, we can in principle always work with functions which are normalizable. For practical considerations it is often more convenient to use wave functions ϕ_k which belong to a function space more extended than the Hilbert space.

†† It should be mentioned that one has to be very careful in omitting such terms, because it can happen that the sum of many ϕ_k with small amplitudes a_k could contribute significantly to the wave function ψ. For instance, it is well known that the addition of many small terms does play an important roie in connection with the saturation character of the nuclear forces.

3. Discussion of the Basis Wave Functions for Nuclear Systems

3.1. General Remarks

The considerations given so far have been quite general. They can be applied not only to nuclear physics, but also to atomic physics, solid-state physics, and other fields of quantum mechanics. The distinction between these different fields lies in the choice of the basis wave functions in terms of which the wave function of the considered physical system is most conveniently expanded. In this chapter, we shall discuss extensively how to choose the proper basis-wave-function set for use in nuclear-physics problems.

The guiding idea for the choice of a proper basis-wave-function set is that the wave function of the system or process which one wishes to describe can be represented by a linear superposition of a small number of terms in this basis set. In this way one obtains a satisfactory description of the physical situation, which is still amenable to quantitative study with present-day computational facilities. More importantly, as we shall see later, a good choice of the basis wave functions is also indispensable for the purpose of obtaining a physical insight into the bound-state structure and reaction mechanisms in a many-particle system.

A proper choice of the basis wave functions depends strongly on the general characteristics of the interaction forces between the particles of which the considered system is composed. In atomic physics, for instance, the interacting atoms can be polarized over large distances due to the long-range nature of the Coulomb interaction. The basic-wave-function sets for atomic systems have therefore to be chosen such that these polarization effects can be included in the description of these systems by a relatively small number of basis functions. In nuclear physics, the situation is quite different. Here the long-range polarization effects on the nuclear participants by the Coulomb forces can be neglected in a good approximation, because the nucleons in a nucleus are tightly bound as a consequence of the great strength of the short-ranged nuclear forces. Therefore, in the choice of basis wave functions for nuclear problems, the emphasis should be on the short-range character of the nuclear interaction and other relevant factors to be discussed in section 3.2, rather than on the presence of the Coulomb repulsion.

3.2. Qualitative Discussion of Cluster Correlations

For nuclear systems two factors predominate in determining the basis wave functions: the character of the nuclear forces and the influence of the Pauli exclusion principle. The important facts about the phenomenological nuclear forces are that they are short-ranged (≈ 2 fm), are strongly attractive over most of this range, but at short distances ($\lesssim 0.5$ fm) become strongly repulsive. The effect of the Pauli principle in a system of nuclear dimension is to allow low-energy nucleons to move relatively undisturbed throughout the nuclear volume, because these nucleons may not be scattered into other already occupied energy levels.

These factors severely restrict the types of correlation which the nucleons may have, especially when the nucleus is in its ground or low excited states. In order to minimize the total energy, they must take maximum advantage of the attractive potential energy compatible with these restrictions. If the nucleons are too far apart, they lose attractive potential energy and the total energy of the nucleus rises. If the nucleons are too close together, repulsive-core and kinetic-energy (from both the uncertainty principle and the Pauli principle) effects again raise the total energy. Hence the nucleons must spend a great deal of time (more accurately, there must be a high probability) in a configuration where their separation distances lie within a certain, narrow energetically favoured range. But again the Pauli principle forbids any very stringent correlations to achieve this.

There remains a type of loose but effective correlation by which the total energy can be brought to a minimum. We can begin to understand this correlation with a simplified physical argument which we subsequently refine. Consider four nucleons moving in the volume of a large nucleus. Let us for the moment assume that the Pauli principle is inoperative and the sole effect of all other nucleons in this nucleus is to confine these four to the nuclear volume. Now if we suppose that the four nucleons move completely uncorrelated from one another, then most of the time they will be outside the range of each other's attractive forces and there will be little mutual potential energy among them. On the other hand, suppose we introduce a correlation such that there exists the probability that they will be found, more often than not, close to one another. As this probability is increased, within the above-mentioned limits, the attractive potential energy among the four nucleons can be greatly increased without a particularly large compensating increase in kinetic energy, and thereby the total energy of the nucleus is reduced.

This simple density correlation is greatly modified by the Pauli principle, however, which we must now add into our picture. First, because nucleons are indistinguishable, any of these four nucleons can exchange places with any other nucleon in the nucleus. Thus any individual nucleon can move throughout the nuclear volume, while at the same time, somewhere in the nucleus four of the nucleons are participating in the correlation. Second, unless the spins and isospins of these four nucleons are anti-alligned, the Pauli principle forbids them to be brought very closely together. Third and more drastically, the presence of the other nucleons means that many single-particle levels of the nucleus are already occupied. Hence, there is much less freedom to correlate the low-energy nucleons (which requires a superposition of single-particle levels). Consequently, these nucleons spread out more in the nuclear volume and can participate only weakly in the correlation.[†]

This type of correlation, where within the above restrictions of the Pauli principle there exists an enhanced probability to find four nucleons, with properly alligned spins and isospins, more often than not close together, we call an α-cluster correlation. The four nucleons which participate in this correlation are, therefore, referred to as forming an α cluster.

---

† This means that the weakly correlated substructure described above cannot be naively considered as an α particle within the nucleus. A partial exception occurs in configurations where the four nucleons are somewhat separated from the others, in which case they can act somewhat more like a free α particle. Hence, we expect the correlation to be strongest in the nuclear surface.

We can immediately generalize the idea. Consider ^{6}Li as an example. In the ground and low excited states, we should expect four of the nucleons to correlate as an α cluster. However, because the Pauli principle prevents more than four nucleons from entering into the mutual short-ranged interaction region, we expect the remaining two to correlate most strongly with each other, forming a deuteron cluster. The structure of ^{6}Li is then mainly described by these two clusters, roughly speaking, moving about each other. In heavier nuclei, correlations can exist between small clusters, so that even larger clusters can be built up.

The cluster structure of a given nucleus can change with increasing excitation energy due to the changed energetical and orthogonality requirements. Because the Pauli principle severely limits the variety of structures possible in the nucleus, these changes in cluster structure are not liable to be too drastic, although some sensitive quantities, such as gamma transition probabilities, level widths, etc., may be much altered thereby.

The theoretical justification of the above picture rests in formulating wave functions embodying cluster correlations and demonstrating with the projection equation (2.1) or (2.3) that these "cluster functions" or a linear superposition of a small number of them provide good approximate descriptions of the systems under consideration.[†] We now turn to the problem of formulating such cluster functions quantitatively.

3.3. Construction of Oscillator Cluster Wave Functions

It is well known that nuclei exhibit different kinds of behaviour. Some of these are due to single-particle features, while others are connected with collective motions of the nucleons. The relative importance of the various types of behaviour can change significantly from nucleus to nucleus and even from one level to the next. Thus, there exists very often a particular set of single-particle or collective coordinates, which is most appropriate to be used in a description of a given nuclear level.

It is very useful to find basis systems, in which the nuclear wave function which employs this most appropriate coordinate set can be expanded. One such system is, for example, that generated from an oscillator potential, the common harmonic-oscillator shell-model-eigenfunction system (see for instance [MO 57]). But the shell-model system is only of limited practical value in expanding trial functions, because it is restricted to single-particle coordinates. We now wish to investigate the consequences of introducing various sets of collective coordinates into the oscillator Hamiltonian in order to generate new, more practical basis systems for solving the nuclear Schrödinger equation, formulated in the form of a projection equation.

There are several reasons why we start with the oscillator potential. First, this potential reflects many broad features of the actual averaged nuclear potential, yet even upon introducing collective coordinates, the resultant Schrödinger equation is simple enough to

---

† Since a cluster correlation represents only a probability statement about certain nuclear configurations, there is no experimental way to determine the presence of a cluster any more than there is an experimental way to measure the wave function itself. However, we can certainly predict experimental quantities with these cluster functions.

solve exactly. Second, in order to see how antisymmetrization serves to remove the contradictions among the various collective and single-particle viewpoints, it is important that we be able to compare our new basis systems with each other and with the shell-model system; this is particularly easy if we use the oscillator potential to generate all the systems. Finally, the functions in these new oscillator basis systems can be generalized in a very natural way to obtain good trial functions to be used in the projection equation (2.1) or (2.3) which contains the actual Hamiltonian with two-nucleon interactions (see section 3.7).

We shall first solve the oscillator-Hamiltonian problem using single-particle coordinates, and then show how to introduce collective-coordinate sets to find new basis systems [WI 58]. Consider the motion of A nucleons in an oscillator potential. The Schrödinger equation is

$$\frac{1}{2M}\left[\sum_{i=1}^{A} p_i^2 + a^2\hbar^2 \sum_{i=1}^{A} r_i^2\right] \phi_n(\mathbf{r}_1, \ldots, \mathbf{r}_A) = E_n\, \phi_n(\mathbf{r}_1, \ldots, \mathbf{r}_A), \tag{3.1}$$

where $\mathbf{p}_i$ and $\mathbf{r}_i$ denote the momentum and position vectors of the ith nucleon, respectively. The quantity $a = M\omega/\hbar$ is the width parameter of the oscillator potential, with M being the mass of a nucleon and ω being the angular frequency of the oscillator potential.

The eigenfunctions $\phi_n(\mathbf{r}_1, \ldots, \mathbf{r}_A)$ of eq. (3.1) are products of single-particle harmonic-oscillator wave functions and form a complete set of orthogonal functions for A variables $\mathbf{r}_i$. We can express the antisymmetrized wave function of any state of a nucleus composed of A nucleons as a superposition of these eigenfunctions, if in addition we introduce the spin and isobaric-spin functions of the nucleons. As has been indicated previously, this single-particle basis system usually has limited utility for trial functions however, because many terms are required to adequately represent any collective behaviour of the nucleons.

Now, collective motion occurs when a certain number of nucleons are energetically favoured to move more or less in a coherent manner. In general, there may be one or more such subgroups within a nucleus. Therefore, to introduce more effective coordinates, let us divide the A nucleons into N groups, or clusters† as we shall call them, with the kth cluster consisting of n_k nucleons such that

$$\sum_{k=1}^{N} n_k = A. \tag{3.2}$$

The set of indices $\{n_1, n_2, \ldots, n_N\}$ denotes what we call a cluster representation of the nucleus. If we now introduce the center-of-mass coordinates

$$\mathbf{R}_k = \frac{1}{n_k}\sum_{i=1}^{n_k} \mathbf{r}_i \tag{3.3a}$$

† We must emphasize again that one must be careful not to think of such substructures too literally as "real composite particles"; exchange effects due to the Pauli principle complicate any such simple, intuitive picture.

and the center-of-mass momenta

$$\mathbf{P}_k = \sum_{i=1}^{n_k} \mathbf{p}_i \tag{3.3b}$$

of these clusters, the oscillator-model Schrödinger equation (3.1) becomes

$$\left[\sum_{k=1}^{N} \left(H_k + \frac{1}{2\,M n_k} \mathbf{P}_k^2 + \frac{a^2 \hbar^2 n_k}{2\,M} \mathbf{R}_k^2\right)\right] \phi_n = E_n\,\phi_n, \tag{3.4}$$

where the Hamiltonian H_k depends only on the relative coordinates $(\mathbf{r}_i - \mathbf{R}_k)$, and their derivatives, of the nucleons in the kth cluster. The eigenfunctions of eq. (3.4) are therefore products of functions, in which every function either depends only on the relative coordinates of one cluster or depends only on a single center-of-mass coordinate $\mathbf{R}_k$.

After introducing the spin and isobaric-spin functions and antisymmetrizing, all these eigenfunction systems are equivalent to the antisymmetrized single-particle wave-function system – the usual oscillator shell-model system – in the sense that we can expand all the states of a nucleus in terms of any one of these eigenfunction systems. Thus, we see that to each cluster representation there belongs an antisymmetrized complete oscillator eigenfunction system, which we can characterize by the representation indices $\{n_1, n_2, \ldots n_N\}$.

We should also note another important point that, because we have only performed a mathematical transformation on our original oscillator Hamiltonian, all these eigenfunction systems must have identically the same energy spectrum.† This means that any eigenfunction corresponding to a given energy eigenvalue in one eigenfunction system can be expanded in any other eigenfunction system as a linear superposition of just those degenerate eigenfunctions corresponding to the same energy eigenvalue. This makes the comparison between eigenfunctions of different oscillator eigenfunction systems especially simple. We shall see the utility of such a comparison later.

It now depends on the nuclear forces in which eigenfunction system the different states of the nuclei are most simply represented, i. e., which kind of correlation among the nucleons is particularly favoured in the different nuclear states. This favoured representation, as we have emphasized, can change from nucleus to nucleus and from level to level. One practical criterion for what constitutes a simple description of a state in a certain eigenfunction system is that in this basis system, the state is described essentially by one or by a superposition of only a small number of antisymmetrized eigenfunctions.

Antisymmetrized eigenfunctions of the kind discussed above are called oscillator cluster functions. It must be emphasized that the antisymmetrization very often influences the physical properties of a wave function profoundly. Therefore one has always to investigate in principle how much of the physical properties of unantisymmetrized cluster wave functions remain after antisymmetrization. This important point will be discussed later in detail. Special oscillator cluster functions have been given by other authors [BR 57, EL 55].

† This means that, for the oscillator cluster wave functions as defined here, the frequencies of the internal, relative, and total center-of-mass motions of the clusters are all the same. This can be easily seen from eqs. (3.4), (3.3a), and (3.3b).

3.4. Discussion of ^{8}Be as an Illustrative Example

In this section we shall use the general considerations discussed above to study the ground and low excited states of the nucleus ^{8}Be. The purpose is to illustrate that very often, suitably chosen oscillator cluster functions can already give us a good insight into the structure of the nucleus in its various states of excitation. We begin by using our physical knowledge to decide what representation we shall use.

A cluster consisting of two protons and two neutrons and corresponding before antisymmetrization of the nuclear wave function to an α particle in the ground state is called an α cluster. Because of the large binding energy of a free α particle, we can presume that the α cluster is a rather stable substructure particularly in light nuclei where, aside from exchange distortion effects, little further distortion should occur [NI 71]. Hence we shall assume that the ^{8}Be nucleus can be represented to a good approximation by the motion of two α clusters, with the low excited states corresponding to excitations in the relative motion of the two clusters, but with no internal excitation of the clusters themselves.†

We must now formulate this description quantitatively. As has been mentioned in section 3.3, we generate suitable wave functions by transforming the following single-particle oscillator Schrödinger equation:

$$\left[\frac{1}{2M}\sum_{i=1}^{8}(\mathbf{p}_i^2 + a^2\hbar^2\mathbf{r}_i^2)\right]\tilde{\phi}_N(\tilde{r}_1,\ldots,\tilde{r}_8) = E_N\,\tilde{\phi}_N(\tilde{r}_1,\ldots,\tilde{r}_8), \tag{3.5}$$

where, as before, $\tilde{r}_i$ denotes the space $\mathbf{r}_i$, spin s_i, and isobaric-spin t_i coordinates of the ith nucleon.

Because they will be useful for comparative purposes, let us first explicitly write down the single-particle solutions of eq. (3.5):

$$\tilde{\phi}_N^s(\tilde{r}_1,\ldots,\tilde{r}_8) = \prod_{i=1}^{8}\tilde{\varphi}_{k_i}(\tilde{r}_i), \tag{3.6}$$

where each $\tilde{\varphi}_{k_i}(\tilde{r}_i)$ is an oscillator function multiplied by the appropriate spin and isobaric-spin functions and the superscript s indicates that this function is a member of the single-particle basis system. The solutions $\tilde{\phi}_N^s(\tilde{r}_1,\ldots,\tilde{r}_8)$ are complete and give rise to a complete set of antisymmetrized functions, in terms of which any antisymmetrized 8-nucleon wave function can be expressed. We write for the antisymmetrized basis wave functions

$$A\,\tilde{\phi}_N^s(\tilde{r}_1,\ldots,\tilde{r}_8) = \sum_P(-1)^p\,P\tilde{\phi}_N^s(\tilde{r}_1,\ldots,\tilde{r}_8). \tag{3.7}$$

In the above equation, the sum extends over all permutations P which can be carried out on the coordinates $\tilde{r}_1,\ldots,\tilde{r}_8$, and p is the number of interchanges that makes up the permutation P.

---

† Such states we call α-particle states. However, see our cautionary remarks in the footnote of p. 12.

We now introduce the following new coordinates appropriate to the {4,4} cluster representation:

$$\begin{aligned}
\mathbf{R}_C &= \tfrac{1}{2}(\mathbf{R}_1 + \mathbf{R}_2), & \bar{\mathbf{r}}_4 &= \mathbf{r}_4 - \mathbf{R}_1 = -(\bar{\mathbf{r}}_1 + \bar{\mathbf{r}}_2 + \bar{\mathbf{r}}_3),\\
\mathbf{R} &= \mathbf{R}_1 - \mathbf{R}_2, & \bar{\mathbf{r}}_5 &= \mathbf{r}_5 - \mathbf{R}_2,\\
\bar{\mathbf{r}}_1 &= \mathbf{r}_1 - \mathbf{R}_1, & \bar{\mathbf{r}}_6 &= \mathbf{r}_6 - \mathbf{R}_2,\\
\bar{\mathbf{r}}_2 &= \mathbf{r}_2 - \mathbf{R}_1, & \bar{\mathbf{r}}_7 &= \mathbf{r}_7 - \mathbf{R}_2,\\
\bar{\mathbf{r}}_3 &= \mathbf{r}_3 - \mathbf{R}_1, & \bar{\mathbf{r}}_8 &= \mathbf{r}_8 - \mathbf{R}_2 = -(\bar{\mathbf{r}}_5 + \bar{\mathbf{r}}_6 + \bar{\mathbf{r}}_7),
\end{aligned} \tag{3.8}$$

with

$$\mathbf{R}_1 = \frac{1}{4}\sum_{i=1}^{4} \mathbf{r}_i, \quad \mathbf{R}_2 = \frac{1}{4}\sum_{i=5}^{8} \mathbf{r}_i. \tag{3.9}$$

The coordinates $\mathbf{R}_1$ and $\mathbf{R}_2$ represent the center-of-mass coordinates of the clusters 1 and 2, respectively, while $\mathbf{R}_C$ represents the center-of-mass coordinate of the total 8-nucleon system and $\mathbf{R}$ represents the relative coordinate of the two clusters. The $\bar{\mathbf{r}}_i$ in eq. (3.8) are the respective internal coordinates of the α clusters. Note that not all the internal coordinates are independent, since they must satisfy the relations

$$\sum_{i=1}^{4} \bar{\mathbf{r}}_i = 0 \tag{3.10a}$$

and

$$\sum_{i=5}^{8} \bar{\mathbf{r}}_i = 0. \tag{3.10b}$$

For definiteness we shall say that $\bar{\mathbf{r}}_4$ and $\bar{\mathbf{r}}_8$ are dependent and the rest independent as already indicated in eq. (3.8). Introducing these new coordinates into the Hamiltonian of eq. (3.5) yields

$$H = H_1 + H_2 + \frac{1}{2}\frac{1}{8M}(\mathbf{P}_C^2 + 8^2 a^2 \hbar^2 \mathbf{R}_C^2) + \frac{1}{2}\frac{1}{2M}(\mathbf{P}^2 + 2^2 a^2 \hbar^2 \mathbf{R}^2), \tag{3.11}$$

where $\mathbf{P}_C$ and $\mathbf{P}$ are the canonically conjugate momenta of the coordinates $\mathbf{R}_C$ and $\mathbf{R}$, respectively. The operators H_1 and H_2 are Hamiltonians having as variables the internal coordinates of cluster 1 and cluster 2, respectively; explicitly, H_1 is given by

$$H_1 = \frac{\hbar^2}{2M}\sum_{j=1}^{3}\sum_{k=1}^{3}\left[-\left(\delta_{jk} - \frac{1}{4}\right)\nabla_{\bar{\mathbf{r}}_j}\cdot\nabla_{\bar{\mathbf{r}}_k} + a^2(\delta_{jk}+1)\,\bar{\mathbf{r}}_j\cdot\bar{\mathbf{r}}_k\right], \tag{3.12}$$

and similarly for H_2. Both H_1 and H_2 can be shown to have a harmonic-oscillator form with frequency ω when expressed in terms of a properly chosen set of three other independent coordinates. This procedure is explicitly carried out in appendix A.

Because of the separability of the Hamiltonian (3.11), its eigenfunction may be expressed in the following product form:

$$\begin{aligned}
&\tilde{\phi}_N^C(\mathbf{R}_C, \mathbf{R}, \bar{\mathbf{r}}_1, \ldots, \bar{\mathbf{r}}_8, s_1, \ldots, t_8)\\
&= \phi_i(\bar{\mathbf{r}}_1, \ldots, \bar{\mathbf{r}}_4)\,\phi_j(\bar{\mathbf{r}}_5, \ldots, \bar{\mathbf{r}}_8)\,\chi_k(\mathbf{R})\,\xi_m(s_1, \ldots, t_8)\,Z_M(\mathbf{R}_C),
\end{aligned} \tag{3.13}$$

with the superscript C indicating its being a member of the {4,4} cluster basis system. In eq. (3.13), the functions ϕ_i and ϕ_j are the internal functions of cluster 1 and cluster 2, respectively, and the function $\chi_k(\mathbf{R})$ is the relative-motion function. The spin and isobaric-spin dependence of $\tilde{\phi}_N^C$ is contained in $\xi_m(s_1, \ldots, t_8)$. Also, the last factor $Z_M(\mathbf{R}_C)$, giving an oscillatory motion of the center of mass of ^{8}Be, arises because the harmonic-oscillator potential is fixed in space and not invariant with respect to center-of-mass translation.† As is clear, the $\tilde{\phi}_N^C$ form a complete set of wave functions and the $A\,\tilde{\phi}_N^C$ form a complete set of antisymmetrized functions (oscillator cluster functions). Now, since all we have done so far is to change coordinates, the Hamiltonians of eqs. (3.5) and (3.11) are physically identical and must therefore have identical energy eigenvalues. Hence, as has been mentioned previously, each oscillator cluster function of a given energy can be expressed as a superposition of degenerate single-particle solutions (3.7) of the same energy; that is,

$$A\,\tilde{\phi}_N^C(\mathbf{R}_C, \mathbf{R}, \bar{\mathbf{r}}_1, \ldots, \bar{\mathbf{r}}_8, s_1, \ldots, t_8; E_N) = \sum_L C_{NL}\, A\,\tilde{\phi}_L^s(\bar{r}_1, \ldots, \bar{r}_8; E_N), \quad (3.14)$$

where the C_{NL} are expansion coefficients. From this relation we see that the oscillator cluster function must describe a state of at least four quanta of energy, i. e., $4\hbar\omega$ of energy above the zero point; otherwise, it will vanish upon antisymmetrization. This is so, since in the single-particle picture, if we are not to violate the Pauli principle, the configuration with the lowest energy is $(1\,s)^4\,(1\,p)^4$ which has four oscillator quanta of excitation energy.††

In the description of the lowest states in ^{8}Be we use the restriction that in the cluster function $\tilde{\phi}_N^C$ of eq. (3.13), $\phi_i(\bar{\mathbf{r}}_1, \ldots, \bar{\mathbf{r}}_4)$ and $\phi_j(\bar{\mathbf{r}}_5, \ldots \bar{\mathbf{r}}_8)$ should assume their ground-state forms $\phi_0(\bar{\mathbf{r}}_1, \ldots, \bar{\mathbf{r}}_4)$ and $\phi_0(\bar{\mathbf{r}}_5, \ldots, \bar{\mathbf{r}}_8)$, and the spin and isobaric-spin functions of the nucleons in each α cluster should couple to yield the value zero for both the total spin quantum number S_α and the total isobaric-spin quantum number T_α. This restriction follows from our physical knowledge that, due to the attractive nature of the nuclear forces, the breakup of an α-cluster substructure inside a nucleus is unfavourable energetically.††† Thus, the four quanta of excitation energy must be associated with the relative-motion part of the cluster wave function. Therefore, the orbital angular-momentum quantum number l of $\chi_k(\mathbf{R})$ must be either 0, 2, or 4. The energetical sequence of the levels with these l values is expected to be in the order of increasing l, because the nuclear interaction is attractive on the average and thus the energy will be the smallest for the state with the largest spatial overlap of the α clusters, i. e., for the state having the smallest

---

† The possibility of unambiguously identifying a center-of-mass part of the motion arises because the oscillator Hamiltonian has the property of separating into a part involving only the center-of-mass coordinates and a part involving only the relative coordinates. This is not true for other Hamiltonians with a space-fixed potential. It is an important reason for choosing an oscillator potential instead of some other form, such as a square-well potential, as a basis for shell-model calculations (see for instance [EL 55]).

†† Note that in the $(1\,s)^4\,(1\,p)^4$ configuration, the total center-of-mass motion has no excitation.

††† The breakup energy of an α cluster inside a nucleus is of the order of the $n + He^3$ separation energy which is about 20 MeV.

relative orbital angular momentum (penetrating-orbit argument). This predicted level sequence is in agreement with the experimental finding.

It should be noted that the oscillator cluster function $\tilde{\phi}_N^C$ has the important feature that the wave function for the center-of-mass motion occurs explicitly as a multiplicative factor, in contrast to the wave function in the single-particle basis system. Therefore, since the actual nuclear Hamiltonian is translationally invariant, one can usually drop it from $\tilde{\phi}_N^C$ and thus avoid all difficulties associated with spurious center-of-mass excitations [EL 55, EL 56]. For example, for the α-cluster or α-particle states discussed above which involve four quanta (n = 4) of excitation energy, the wave functions can be simply writen as

$$\begin{aligned} &A\,\tilde{\phi}_N^C(\mathbf{R}, \bar{\mathbf{r}}_1, \ldots, \bar{\mathbf{r}}_8, s_1, \ldots, t_8) \\ &= A\,\{\phi_0(\bar{\mathbf{r}}_1, \ldots, \bar{\mathbf{r}}_4)\,\phi_0(\bar{\mathbf{r}}_5, \ldots, \bar{\mathbf{r}}_8)\,\chi_{nlm_l}(\mathbf{R})\,\zeta(s_1, \ldots, t_8; S = 0, T = 0)\}, \end{aligned} \tag{3.15}$$

which are 15-fold degenerate with respect to the oscillator Hamiltonian, with $l = 0, 2, 4$ and $m_l = -l$ to l.

The next higher states of the α-cluster type have six quanta of excitation in the relative-motion part. The wave functions with five quanta of excitation in the relative-motion part vanish upon antisymmetrization because the α clusters obey Bose statistics.† Other states of higher excitation are produced by internally exciting the α clusters. These states are expected to lie in the ^{8}Be excitation-energy region above 20 MeV.

Similar consideration as in the case of ^{8}Be can also be applied to other nuclei. For instance, for the description of the lowest T = 0 states in ^{6}Li one uses, besides the oscillator shell-model representation, both the α + d and the t + ^{3}He representations. By employing an equivalence relation similar to eq. (3.14), one can conclude that the α + d cluster wave functions for these T = 0 states must have relative-motion parts which involve two oscillator quanta of excitation, in accordance with the Pauli principle. The resultant relative orbital angular-momentum quantum numbers for these states are therefore equal to 0 and 2. Due to the internal spin of the deuteron cluster (S = 1) and the noncentral part of the nuclear forces, the $l = 2$ states will be split in energy. Thus the expected T = 0 level spectrum of ^{6}Li in the low-energy region is a J = 1 ground state followed by a triplet of excited states with J = 3, 2, and 1. This is in complete agreement with experiment.††

---

† This can be seen easily. If one exchanges the four nucleons of one α cluster with the four nucleons of the other α cluster in every term of the antisymmetrized wave function, then due to the exchange of an even number of nucleons, the relative-motion parts of the wave function must not change their signs. On the other hand, if l is odd, these parts will change their signs; thus the odd-l wave functions must vanish when antisymmetrized.

†† The nucleus ^{6}Li is very interesting from both theoretical and experimental viewpoints. It is a stable nucleus (as opposed to the lighter nuclei ^{5}He and ^{5}Li) containing, on the one hand, sufficiently many nucleons to exhibit many important general features of nuclear phenomena and, on the other hand, sufficiently few nucleons such that detailed calculations can be made even with nuclear forces containing a strong repulsive core. Therefore, ^{6}Li is a test nucleus in nuclear physics, in much the same way as the hydrogen and helium atoms are in atomic physics. We shall return again and again to this nucleus to see how refinements in our considerations could improve its description.

We wish to mention that the method we have just used, of comparing nuclear wave functions in different basis systems to understand antisymmetrization effects, is of quite general utility, and we shall use it frequently.

3.5. Effects of Antisymmetrization

We have already mentioned that as a result of antisymmetrization, seemingly quite different wave functions can become very similar or even equivalent to each other. This reduction in the differences between different structures by antisymmetrization is a very general feature and not restricted to the oscillator representations of sections 3.3 and 3.4; it is only that with these representations the effects of antisymmetrization can be demonstrated in a particularly clear manner.

This effect of antisymmetrization is an important key to the resolution of apparently contradictory descriptions of the nucleus by different models, all of which have had some success in predicting nuclear characteristics. Here we shall illustrate the influence of antisymmetrization in three specific examples.

3.5a. Fermions without Mutual Interaction in a Square-Well Potential

As first example let us consider a large number of fermions without mutual interaction in their ground state in a square-well potential [WI 58]; this would approximately describe the conducting electrons in a conductor, for example. The energy differences among the particles in the well arise from their kinetic energies, because the energies of the states increase with the square of their momenta. In the ground state, all single-particle states are filled inside a momentum sphere known as the Fermi sphere (see fig. 1a). If this system as a whole is now given a small velocity $\Delta\mathbf{v}$, then the Fermi sphere will be slightly shifted so that its center is no longer at the origin (see fig. 1b). The change relative to the situation shown in fig. 1a is a collective excitation in which each fermion receives a small change in momentum $m\,\Delta\mathbf{v}$. Now let us instead start with the Fermi sphere at the origin (as in fig. 1a) and impart various large amounts of momenta to a few of the fermions (all those in states in shaded region 1 of fig. 1c) at the left of the sphere so as to excite them into states just to the right of the sphere (filling the states in shaded region 2 of fig. 1c). Due to the indistinguishability of the fermions, expressed through the antisymmetrization of

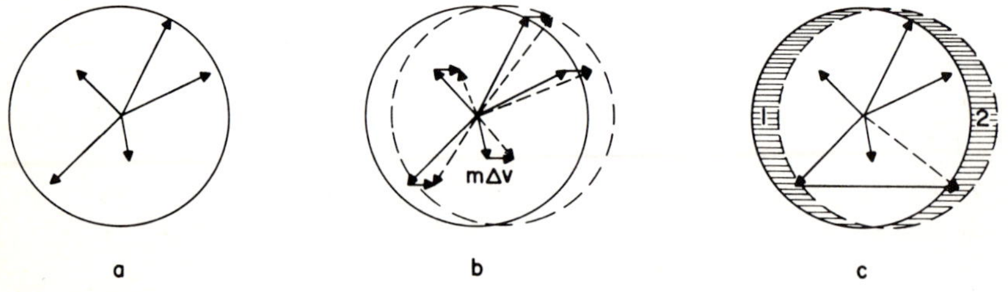

Fig. 1. Example of equivalence under antisymmetrization: (a) The arrows denote representative states of the filled Fermi sphere. (b) All momentum vectors are shifted by an amount $\Delta\mathbf{p} = m\,\Delta\mathbf{v}$. (c) Just those vectors in shaded region 1 are shifted to shaded region 2.

the wave function, the situation in fig. 1c is completely equivalent to that in fig. 1b. This shows that under antisymmetrization, a large excitation imparted to a few fermions can be equivalent to a collective excitation of all the fermions as a whole.

To see this mathematically, consider, for example, N fermions in a well with periodic boundary conditions and nearest-neighbour level spacing corresponding to a wave-number difference Δk. Then the ground-state wave function is

$$\psi = A \exp\left(\sum_{n=-f}^{f} in\,\Delta k\, x_n \right), \quad \left(f = \frac{N-1}{2} \right) \tag{3.16}$$

where, for simplicity and clarity, we consider only the case in one dimension. In the above equation, the operator A is the antisymmetrization operator as defined in eq. (3.7). In one dimension the Fermi sphere reduces to a line of points in momentum space, centered about the origin and filled up to some number $\pm f\hbar\,\Delta k$.

Now let us give each fermion an additional momentum of $\Delta p = \hbar\,\Delta k$ so that the total momentum added to the system is $N\,\Delta p$. The new wave function becomes

$$\psi = A \exp\left[\sum_{n=-f}^{f} i(n\Delta k + \Delta k)\, x_n \right]. \tag{3.17a}$$

The center of the "Fermi-line" is shifted Δp to the right of the origin. But we can write the new wave function as

$$\psi = A \exp\left(\sum_{n=-f+1}^{f+1} in\,\Delta k\, x_n \right). \tag{3.17b}$$

Because the labeling of the particles is irrelevant in an antisymmetrized wave function,† this is the same as giving the left-most particle all the momentum $N\Delta p$ and exciting it from the state with momentum $-f\hbar\Delta k$ to the state with momentum $(f+1)\,\hbar\Delta k$. Thus what we originally conceive as a collective excitation of all the fermions moving together to the right, and as a single-particle excitation of one fermion strongly excited to the right, become equivalent under antisymmetrization.

3.5b. The Lowest 4^+ α-Cluster State of ^{8}Be

As second example, we shall explicitly carry out the antisymmetrization in the relatively simple case of the $n = 4$, $l = 4$, $m_l = 4$, α-cluster state of ^{8}Be discussed in section 3.4 and show how the wave function looks when expressed as a superposition of single-particle shell-model wave functions [WI 59, KA 59].

The cluster wave function for this state is the eigenfunction of the Hamiltonian (3.11); it is given by

$$\psi = A\,\{\tilde{\phi}_0(\tilde{\alpha}_1)\,\tilde{\phi}_0(\tilde{\alpha}_2)\,\chi_{444}(\mathbf{R})\,Z_0(\mathbf{R}_C)\}, \tag{3.18}$$

† Except for an unimportant phase fector $(-1)^p$, where p is the number of interchanges that makes up the permutation producing the new labeling.

where $\tilde{\alpha}_1$ denotes the internal spatial, spin, and isobaric-spin coordinates of all the nucleons in cluster 1, and similarly for $\tilde{\alpha}_2$. The physically irrelevant center-of-mass wave function $Z_0(\mathbf{R}_C)$ is included here in order to make the transition to the single-particle wave functions.

Because in the ^{8}Be state discussed here, the α clusters have no internal excitation, their wave functions are of the form

$$\tilde{\phi}_0(\tilde{\alpha}_1) = \exp\left(-\frac{1}{2}\,a\sum_{i=1}^{4}\bar{\mathbf{r}}_i^2\right)\xi_1(s_1,\ldots,t_4), \tag{3.19}$$

with

$$\xi_1(s_1,\ldots,t_4) = \alpha_1\nu_1\alpha_2\pi_2\beta_3\nu_3\beta_4\pi_4, \tag{3.20}$$

and a similar expression for $\tilde{\phi}_0(\tilde{\alpha}_2)$. In eq. (3.20), α_i is a shorthand notation for $\alpha(s_i)$ and denotes the spin-up state for nucleon i. Similarly, β_i, ν_i, and π_i denote the spin-down state, isospin-up (neutron) state, and isospin-down (proton) state for nucleon i, respectively. Note that the wave function of eq. (3.19) is not an eigenfunction of the α-cluster spin and isospin operators $\mathbf{S}_\alpha^2$ and $\mathbf{T}_\alpha^2$, but will become an eigenfunction with eigenvalues $S_\alpha = 0$ and $T_\alpha = 0$ after antisymmetrization.

The relative-motion function $\chi_{444}(\mathbf{R})$ is a lg oscillator wave function (see appendix B) which has four quanta of excitation; it has the form

$$\chi_{444}(\mathbf{R}) = R^4\,e^{-aR^2}\,Y_{44}(\theta,\varphi), \tag{3.21}$$

where $Y_{44}(\theta,\varphi)$ is the spherical harmonic with the indices $l = 4$ and $m_l = 4$. Finally, $Z_0(\mathbf{R}_C)$ corresponds to a zeroth-order oscillation of the center of mass of ^{8}Be. It is given by

$$Z_0(\mathbf{R}_C) = \exp(-4\,a\mathbf{R}_C^2). \tag{3.22}$$

With eqs. (3.19)–(3.22) we obtain the following form for the cluster wave function ψ of the Be^8 state considered here:

$$\begin{aligned}\psi = \mathcal{A}\Bigg\{&\left[\exp\left(-\frac{1}{2}\,a\sum_{i=1}^{4}\bar{\mathbf{r}}_i^2\right)\alpha_1\nu_1\alpha_2\pi_2\beta_3\nu_3\beta_4\pi_4\right]\\ &\times\left[\exp\left(-\frac{1}{2}\,a\sum_{i=5}^{8}\bar{\mathbf{r}}_i^2\right)\alpha_5\nu_5\alpha_6\pi_6\beta_7\nu_7\beta_8\pi_8\right]\\ &\times\left[R^4\,e^{-aR^2}\,Y_{44}(\theta,\varphi)\right]\times e^{-4aR_C^2}\Bigg\}.\end{aligned} \tag{3.23}$$

Next, let us express this cluster wave function in terms of particle coordinates $\bar{\mathbf{r}}_i$ so that we may carry out the antisymmetrization explicitly. By using eq. (3.8), we obtain for the exponents in eq. (3.23) the following simplification:

$$\begin{aligned}&\frac{a}{2}\sum_{i=1}^{4}\bar{\mathbf{r}}_i^2 + \frac{a}{2}\sum_{i=5}^{8}\bar{\mathbf{r}}_i^2 + a\mathbf{R}^2 + 4a\mathbf{R}_C^2 = \frac{a}{2}\Bigg[\sum_{i=1}^{4}(\mathbf{r}_i-\mathbf{R}_1)^2 + \sum_{i=5}^{8}(\mathbf{r}_i-\mathbf{R}_2)^2\\ &+ 2(\mathbf{R}_1-\mathbf{R}_2)^2 + 2(\mathbf{R}_1+\mathbf{R}_2)^2\Bigg] = \frac{a}{2}\sum_{i=1}^{8}\mathbf{r}_i^2,\end{aligned} \tag{3.24}$$

which is a totally symmetric function of the eight spatial coordinates $\mathbf{r}_i$. Note that we obtain such a simple form for the exponent because we have retained in ψ the wave function for the center-of-mass motion. With eq. (3.24), the cluster function becomes

$$
\begin{aligned}
\psi &= \mathcal{A}\left\{ R^4\, Y_{44}(\theta,\varphi) \exp\left(-\frac{a}{2}\sum_{i=1}^{8} r_i^2\right) \alpha_1 \ldots \pi_8 \right\} \\
&= \frac{3}{16}\sqrt{\frac{35}{2\pi}}\,\mathcal{A}\left\{ R^4 (\sin\theta\, e^{i\varphi})^4 \exp\left(-\frac{a}{2}\sum_{i=1}^{8} r_i^2\right) \alpha_1 \ldots \pi_8 \right\} \\
&= \frac{3}{16}\sqrt{\frac{35}{2\pi}}\,\mathcal{A}\left\{ (X+iY)^4 \exp\left(-\frac{a}{2}\sum_{i=1}^{8} r_i^2\right) \alpha_1 \ldots \pi_8 \right\} \qquad (3.25) \\
&= \frac{3}{16}\sqrt{\frac{35}{2\pi}}\left(\frac{1}{4}\right)^4 \mathcal{A}\left\{ [(x_1+iy_1)+(x_2+iy_2)+\ldots-(x_5+iy_5)-\ldots-(x_8+iy_8)]^4 \right. \\
&\quad \left. \times \exp\left(-\frac{a}{2}\sum_{i=1}^{8} r_i^2\right)\alpha_1 \ldots \pi_8 \right\}.
\end{aligned}
$$

The term $[(x_1+iy_1)+\ldots]^4$ in eq. (3.25) is a sum of many terms of the form

$$(x_1+iy_1)^{n_1}(x_2+iy_2)^{n_2}\ldots(x_8+iy_8)^{n_8},$$

where $n_1, \ldots, n_8$ take on integral values from 0 to 4, subject to the condition that $n_1+n_2+\ldots+n_8=4$.

In accordance with the Pauli principle, only those terms which correspond to four nucleons in the 1p shell (whose wave functions, besides their exponential factors, are proportional to x_i+iy_i) and four nucleons in the 1s shell (whose wave functions, besides their exponential factors, are constants) with the proper spin and isospin configuration are different from zero. Thus, even though there is a large number of terms in ψ of eq. (3.25), most of them vanish because of antisymmetrization. In fact, it is easy to see that the terms which vanish are of the following types:

(i) Any term with one or more of the n_j larger than 1. An example of this is

$$\mathcal{A}\left\{ (x_1+iy_1)^2(x_2+iy_2)(x_3+iy_3)\exp\left(-\frac{a}{2}\sum_{i=1}^{8} r_i^2\right)\alpha_1\ldots\pi_8 \right\}.$$

which describes a system with five nucleons in the 1s shell.

(ii) Any term with two of the non-zero n_j being (n_1, n_5), (n_2, n_6), (n_3, n_7), or (n_4, n_8). An example of such a term is

$$\mathcal{A}\left\{ -(x_1+iy_1)(x_2+iy_2)(x_3+iy_3)(x_6+iy_6)\exp\left(-\frac{a}{2}\sum_{i=1}^{8} r_i^2\right)\alpha_1\ldots\pi_8 \right\},$$

which describes a system with two 1s nucleons in the spin-down, isospin-up state.

Again because of antisymmetrization, the remaining non-vanishing terms are all equal to each other, except for multiplicative constants. Therefore, when expressed in particle coordinates $\tilde{r}_i$, the function ψ can be written as

$$\psi = N_p A \left\{ (x_5 + iy_5)(x_6 + iy_6)(x_7 + iy_7)(x_8 + iy_8) \exp\left(-\frac{a}{2} \sum_{i=1}^{8} r_i^2\right) \alpha_1 \ldots \pi_8 \right\}, \tag{3.26}$$

with N_p being a constant factor. We see that this wave function corresponds to four 1 p nucleons with parallel orbital angular momenta and four 1 s nucleons. This is the way it has to be, since we have started with a state of total orbital angular momentum $l = 4$.

We have thus shown that the antisymmetrized cluster function of eq. (3.18) is completely identical to the antisymmetrized shell-model function of eq. (3.26). This gives us an important clue as to how one could proceed to resolve apparently contradictory physical descriptions of the same nuclear state. We shall return to this point again and again.

We now wish to present a frequently useful method for finding the effect of antisymmetrization on the spin-isospin wave function of a nuclear system. We shall introduce this method using our ^{8}Be wave function of eq. (3.26) as a specific example.

We shall show that the antisymmetrized wave function of eq. (3.26) has total spin $S = 0$ and total isospin $T = 0$, notwithstanding that the unantisymmetrized wave function inside the curly brackets is only an eigenfunction of S_z and T_3, but not an eigenfunction of the operators $\mathbf{S}^2$ and $\mathbf{T}^2$. This is easily proven using the following property of the antisymmetrization operator A:

$$A P_0 = [\sum_P (-1)^p P] P_0 = (-1)^{p_0} \sum_P (-1)^{pp_0} PP_0 = (-1)^{p_0} A, \tag{3.27}$$

where P_0 is any fixed permutation. In deriving the formula (3.27) we have used the fact that PP_0 extends over all permutations when P extends over all permutations.

The utilization of eq. (3.27) permits us to write

$$A \bar{\psi} \sim A(\bar{\psi} + c P_0 \bar{\psi}), \tag{3.28}$$

where $\bar{\psi}$ denotes the unantisymmetrized wave function inside the curly brackets of eq. (3.26). The constant c can be chosen arbitrarily, provided that its choice does not make the right side vanish identically. Now the spin and isobaric-spin part of the function $\bar{\psi}$, upon which the permutation act, may be written out in full as

$$\alpha_1 \nu_1 \alpha_2 \pi_2 \beta_3 \nu_3 \beta_4 \pi_4 \alpha_5 \nu_5 \alpha_6 \pi_6 \beta_7 \nu_7 \beta_8 \pi_8$$

and we note that the spatial part is unchanged by permutations either among the first four nucleons or among the last four nucleons. Thus by choosing P_0 as P_{12}, the permutation of nucleons 1 and 2, and c as -1, we obtain

$$A \bar{\psi} \sim A \left\{ (x_5 + iy_5)(x_6 + iy_6)(x_7 + iy_7)(x_8 + iy_8) \right. \\ \left. \times \exp\left(-\frac{a}{2} \sum_{i=1}^{8} r_i^2\right) (\nu_1 \pi_2 - \nu_2 \pi_1) \alpha_1 \alpha_2 \beta_3 \nu_3 \ldots \pi_8 \right\}, \tag{3.29}$$

which corresponds to a coupling of nucleons 1 and 2 into an eigenfunction of total isobaric spin zero. Proceeding further, we may write

$$A\bar{\psi} \sim A\{(1-P_{78})(1-P_{56})(1-P_{34})(1-P_{12})\bar{\psi}\}, \quad (3.30)$$

in which the right-side bracket has nucleon pairs (12), (34), (56), and (78) all coupled into an eigenfunction of isobaric-spin zero. This shows that the wave function $A\bar{\psi}$ or ψ must be a T = 0 eigenstate.† Now one can start again and couple pairs (13), (24), (57) and (68) into spin-zero eigenfunctions, thus showing that $A\bar{\psi}$ or ψ is also an eigenstate of the operator $\mathbf{S}^2$ with eigenvalue zero.

This type of argument is useful in many problems, because very often one can start with unantisymmetrized wave functions which are not eigenfunctions of total spin and isobaric spin, but become eigenfunctions automatically after antisymmetrization. Note that the behaviour of the spatial part of the wave function is very important. If the spatial part would have changed its sign in the permutation P_{12}, then this permutation would have created an isospin-one state. If the spatial part of the wave function is not an eigenfunction of such permutations, then the method does not work and we have to start from the beginning using eigenfunctions of the spin and isobaric-spin operators for the function $\bar{\psi}$.

3.5c. Mathematical Equivalence of the Lowest ^{6}Li States Described in the t + ^{3}He and the d + α Oscillator Cluster Representations

As third and last example to demonstrate the influence of antisymmetrization, we shall show explicitly, in the oscillator picture, the mathematical equivalence of the wave functions for the lowest T = 0 ^{6}Li states expressed as a triton cluster plus a ^{3}He cluster, and as an α cluster plus a deuteron cluster [TA 62]. To show this, we shall expand the wave functions in both cluster representations as a superposition of single-particle shell-model wave functions.

In the t + ^{3}He picture the wave function†† may be expressed as

$$\psi_I = A\{\tilde{\phi}_0(\tilde{t})\,\tilde{\phi}_0(\tilde{h})\,\chi_I(\mathbf{R}_t - \mathbf{R}_h)\}, \quad (3.31)$$

where $\tilde{\phi}_0(\tilde{t})$ and $\tilde{\phi}_0(\tilde{h})$ denote the ground-state internal wave functions of the triton and the ^{3}He clusters, respectively, and $\chi_I(\mathbf{R}_t - \mathbf{R}_h)$ denotes the function describing the oscillatory relative motion of the two clusters. Similarly, in the α + d picture, the wave function†† may be expressed as

$$\psi_{II} = A\{\tilde{\phi}_0(\tilde{\alpha})\,\tilde{\phi}_0(\tilde{d})\,\chi_{II}(\mathbf{R}_\alpha - \mathbf{R}_d)\} \quad (3.32)$$

with analogous meanings for the functions $\tilde{\phi}_0(\tilde{\alpha})$, $\tilde{\phi}_0(\tilde{d})$ and $\chi_{II}(\mathbf{R}_\alpha - \mathbf{R}_d)$.

---

† Since $\mathbf{T}^2 = (\mathbf{t}_1 + \mathbf{t}_2 + \ldots + \mathbf{t}_8)^2$ is a totally symmetric operator, it commutes with P_{ij} and hence with A. If for an arbitrary function $\bar{\psi}$, $\mathbf{T}^2\bar{\psi} = T(T+1)\bar{\psi}$, then

$$T^2 A\bar{\psi} = A T^2 \bar{\psi} = T(T+1) A\bar{\psi}.$$

That is, if $\bar{\psi}$ is an eigenfunction of T^2 with eigenvalue $T(T+1)$, so is $A\bar{\psi}$.

†† For simplicity, the center-of-mass function is not explicitly written out.

Using an argument similar to that given in section 3.4, we can easily show that for the lowest-lying states of ^{6}Li, the relative-motion functions χ_I and χ_{II} must have oscillator quantum number n = 2. For this oscillator quantum number, there exist four T = 0 states with total angular momentum and parity $J^\pi = 1^+, 3^+, 2^+, 1^+$ (see section 3.4). For simplicity, we select for our further considerations the $J^\pi = 3^+$ state, where the orbital angular momentum $l = 2$ and the spin angular momentum S = 1 are parallel to each other.

By using oscillator wave functions of quantum numbers n = 2, $l = 2$ and $m_l = 2$ for χ_I and χ_{II}, the two wave functions ψ_I and ψ_{II} become

$$\psi_I = A\left\{\exp\left(-\frac{a}{2}\sum_{i=1}^{6} r_i^2\right)[(x_t - x_h) + i(y_t - y_h)]^2 \times \xi(s_1, \ldots, t_6)\right\}, \tag{3.33}$$

$$\psi_{II} = A\left\{\exp\left(-\frac{a}{2}\sum_{i=1}^{6} r_i^2\right)[(x_\alpha - x_d) + i(y_\alpha - y_d)]^2 \times \xi(s_1, \ldots, t_6)\right\}. \tag{3.34}$$

The spin-isospin function ξ is taken as

$$\xi = \alpha_1 \nu_1 \alpha_2 \pi_2 \beta_3 \nu_3 \beta_4 \pi_4 \alpha_5 \nu_5 \alpha_6 \pi_6 . \tag{3.35}$$

Upon carrying out the antisymmetrization, both ψ_I and ψ_{II} become eigenfunctions of the total spin and isospin operators with eigenvalues S = 1 and T = 0.

The relative-motion terms in eqs. (3.33) and (3.34) can be written as

$$\begin{aligned}[(x_t - x_h) + i(y_t - y_h)]^2 = [&\tfrac{1}{3}\{(x_1 + iy_1) + (x_2 + iy_2) + (x_3 + iy_3)\} \\ &- \tfrac{1}{3}\{(x_4 + iy_4) + (x_5 + iy_5) + (x_6 + iy_6)\}]^2 ,\end{aligned} \tag{3.36}$$

$$\begin{aligned}[(x_\alpha - x_d) + i(y_\alpha - y_d)]^2 = [&\tfrac{1}{4}\{(x_1 + iy_1) + (x_2 + iy_2) + (x_3 + iy_3) + (x_4 + iy_4)\} \\ &- \tfrac{1}{2}\{(x_5 + iy_5) + (x_6 + iy_6)\}]^2 .\end{aligned} \tag{3.37}$$

As in the case of the antisymmetrization of the $l = 4$, $m_l = 4$ ^{8}Be wave function in section 3.5b, only such terms in eqs. (3.36) and (3.37) will remain after antisymmetrization which describe the configuration of four nucleons in the 1s shell and a neutron and a proton with parallel spins in the 1p shell. In addition, all these terms when multiplied by the exponential factor $\exp\left(-\frac{a}{2}\sum_{i=1}^{6} r_i^2\right)$ and the spin-isospin function ξ are proportional to each other after antisymmetrization. Therefore, we have shown that ψ_I and ψ_{II} are indeed equal to each other, except for a constant multiplicative factor.

3.5d. Summary

Let us now summarize our discussions of the effects of the Pauli principle. The requirement that nuclear wave functions must be totally antisymmetric is the quantitative expression of the fact that nucleons are indistinguishable fermions. Because of this requirement, one must beware of too naive conceptions of single-particle and collective motions; such conceptions often arise from a too literal interpretation of unantisymmetrized wave functions. Only what remains of such conceptions after antisymmetrization constitutes the actual behaviour of the system and may not resemble one's pictorial ideas at all. Indeed, as we have seen, seemingly different starting points can lead upon antisymmetrization to identical or nearly identical wave functions, which is a realization that may help us to resolve the apparent differences among the many existing nuclear models. For instance, the equivalence of the ^{6}Li cluster wave functions will play a very important role in the interpretation of phenomenological direct-reaction mechanisms and will make it clear that these phenomenological mechanisms are partially equivalent to each other. This will be discussed in detail in Chapter 12. Also, it may be interesting to mention that the reduction in differences among various cluster wave functions has one practical advantage. Because of this reduction effect, one may usually solve the Schrödinger equation (2.1) or (2.3) to a good approximation by employing a trial function which consists of a relatively small number of wave functions in different cluster representations.

The oscillator cluster functions which we have so far used in our discussions have the advantage of providing several different viewpoints, mathematically equivalent after antisymmetrization, from which to regard each nuclear phenomenon. By extracting noncontradictory features from the various viewpoints (see especially section 3.6c), we are often guided toward a comprehensive understanding of the correct physical behaviour without carrying out lengthy antisymmetrized calculations. Further effects of antisymmetrization, especially in regard to its physical consequences, will be discussed frequently in the following chapters.

3.6. Application of Oscillator Cluster Representations to a Qualitative Description of Low-lying Levels in Light Nuclei

In this section we shall discuss qualitatively the level schemes of some other nuclei from the point of view of oscillator cluster representations. We shall show that by a consideration of these nuclei in both the oscillator single-particle and cluster representations, we can correctly predict not only the total angular momenta and parities but also the ordering of the low-lying levels. In addition, and perhaps more importantly, we can obtain indications from these discussions as to the way in which one must generalize the oscillator cluster functions so that a superposition of a small number of these generalized functions could be used to yield results in quantitative agreement with experiment.

3.6a. ^{7}Li and ^{7}Be

As first example, we consider the ground state and the low excited states of ^{7}Li and ^{7}Be. From our general considerations given in section 3.2, it is appropriate to choose an $\alpha + t$ cluster representation for the description of low-lying ^{7}Li states [HO 61, HO 61a, IN 39, PH 60, PH 61, SP 60, TA 61, WI 59]. In this representation the wave functions have the form

$$\psi = A\,\{\tilde{\phi}_i(\tilde{\alpha})\,\tilde{\phi}_j(\tilde{t})\,\chi_{nlm}(\mathbf{R}_\alpha - \mathbf{R}_t)\}, \tag{3.38}$$

where $\tilde{\phi}_i(\tilde{\alpha})$ and $\tilde{\phi}_j(\tilde{t})$ are the internal wave functions of the alpha and the triton clusters, respectively, and $\chi_{nlm}(\mathbf{R}_\alpha - \mathbf{R}_t)$ is the relative-motion function of the clusters. Using the information that the separation energy of a nucleon in a free triton is quite large (≈ 6 MeV), we can presume that the ^{7}Li states lying below about 5 MeV can be described, in a first approximation, by a single cluster function in which neither the α cluster nor the triton cluster is excited. Since in the single-particle shell-model picture, three nucleons are in the 1 p shell, the relative motion between the clusters must be an oscillation of third order (n = 3) to comply with the Pauli principle.

Relative-motion oscillation with three quanta of excitation can be either 2p or 1f oscillation (see appendix B) with orbital angular momentum $l = 1$ or 3. By using an argument (penetrating-orbit argument) similar to that used in the case of ^{8}Be, we expect the state with $l = 1$ to be lower in energy than the state with $l = 3$.† In actuality, the states are further split in energy by the spin-orbit coupling. The triton spin being 1/2, the total angular momentum for the $l = 1$ states can be either $\frac{1}{2}$ or $\frac{3}{2}$. The spin-orbit coupling demands that the lowest state has J = 3/2. The order of the $l = 3$ states with the sequence J = 7/2, J = 5/2 is similarly established. The experimental sequence of the low-lying levels in ^{7}Li, shown in fig. 2 [LA 66, SP 67], is in agreement with this qualitative description. All these levels have negative parity due to the odd value of the relative orbital angular momentum and the even internal parities of the α and the triton clusters.

The nucleus ^{7}Be possesses a low-energy spectrum similar to that of ^{7}Li, only that all levels are somewhat shifted due to the larger Coulomb energy and the neutron-proton mass difference.

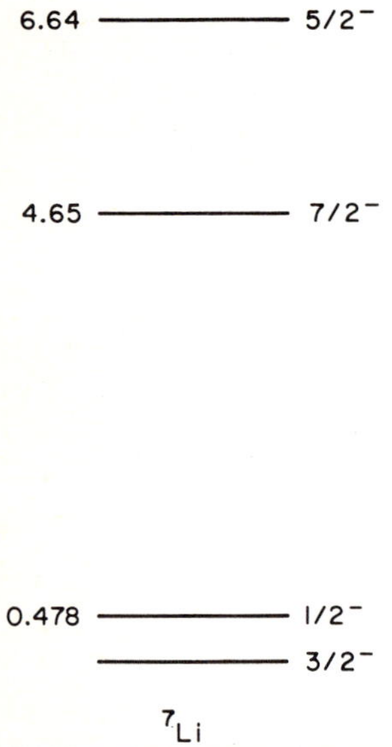

Fig. 2. Energy levels of ^{7}Li.

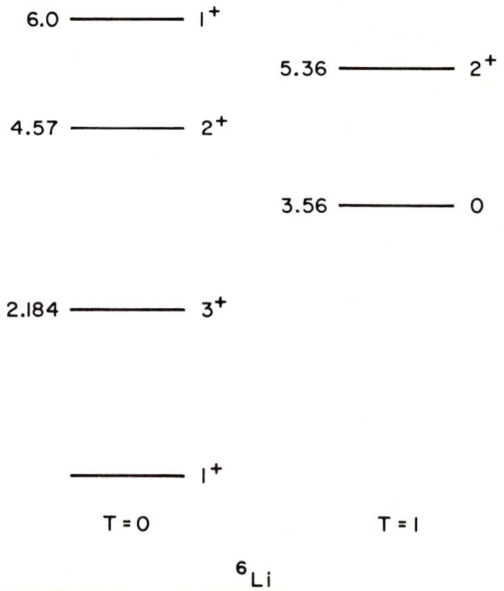

Fig. 3. Energy levels of ^{6}Li.

† In the oscillator model, the $l = 1$ and 3 states are degenerate. However, the desired approximate energy of the corresponding nuclear level is obtained by using the oscillator cluster function of eq. (3.38) as a trial function in a variational calculation with the microscopic nuclear Hamiltonian.

3.6b. ^{6}He, ^{6}Li, and ^{6}Be

We now return briefly to the case of low-lying levels in ^{6}Li. As already mentioned in section 3.4, the lowest states of ^{6}Li can be described by assuming an α + d cluster structure. For these states we shall assume again that the α cluster has no internal excitation. But it is not immediately obvious that a deuteron cluster in a larger nucleus will not be essentially broken up, because a free deuteron has a binding energy of only 2.2 MeV in the S = 1, T = 0 configuration and exists only as a virtual state in the S = 0, T = 1 configuration.

We can make the following qualitative argument in favour of the stability of a deuteron cluster in a larger nucleus [WI 62, SH 60]. Because of the strong short-range character of the nuclear forces, the potential energy of interaction of the two nucleons in a free deuteron is quite large (≈ -15 MeV in the S = 0 state and ≈ -25 MeV in the S = 1 state). But due to the finite volume of the deuteron, the uncertainty principle demands a correspondingly large internal kinetic energy of the two nucleons, so that the binding energy of the free deuteron is quite small. Now, inside a nucleus, the effect of the d-cluster structure is to introduce a correlation amongst the nucleons such that two nucleons constituting the deuteron cluster, whichever they are at any moment,† have a large probability of being close to each other.†† Hence, the deuteron cluster should retain the high internal potential and kinetic energies characteristic of the free deuteron. If, however, the deuteron cluster is broken up, then the potential energy between the two now uncorrelated nucleons will become much smaller in magnitude. But because the presence of the other cluster (in ^{6}Li the α-cluster) confines the two uncorrelated nucleons within the nuclear volume, the uncertainty principle does not permit a correspondingly large decrease in their kinetic energies††† as in the case of the free deuteron. Therefore, we conclude that it is energetically unfavourable to destroy the deuteron-cluster correlation and a deuteron cluster could be considered as a relatively stable substructure inside a larger nucleus.

Based on the qualitative argument given above, we can presume that T = 0 ^{6}Li states with excitation energies less than about 5 MeV can be described, in a first approximation, by α + d cluster wave functions with unexcited clusters. The T = 0 energy spectrum of ^{6}Li, therefore, consists of a 1^+ ground state followed by a triplet of excited states with $J^\pi = 3^+, 2^+$, and 1^+ (see section 3.4). This is in agreement with the experimentally determined level sequence as shown in fig. 3 [LA 66].

For the description of the low-energy T = 1 states in ^{6}Li, we use a cluster representation based on an unexcited α cluster and a deuteron cluster excited to its lowest T = 1, S = 0

---

† It should be mentioned again that, because nucleons are indistinguishable, exactly which nucleons participate in the deuteron-cluster correlation can change from instant to instant.

†† With an oscillator cluster wave function this correlation is not particularly strong as yet. If one strengthens the d-cluster correlation by adopting generalized cluster wave functions to be discussed in the next section, then the ^{6}Li description will be improved, as have been shown by variational calculations using eq. (2.3) (see section 5.3a).

††† In the oscillator cluster model, the total kinetic energy (internal kinetic energies of the α and the d clusters plus the kinetic energy of the relative motion of the clusters) cannot change, unless we permit excitations outside of the 1p shell.

configuration. In the single-particle picture, recoupling the spins to yield an S = 0 state still leaves two nucleons in the 1 p shell. Hence, in the oscillator cluster model, the relative motion must still be a second-order oscillation with orbital angular momentum l equal to 0 or 2. The total angular momenta of the T = 1 states are therefore also 0 and 2, since both clusters involved have zero spin. The J = 0 state will have a lower energy according to penetrating-orbit arguments; it will lie several MeV above the J = 1, T = 0 ground state, since the singlet state of the deuteron has a higher energy than the triplet state. Also, using arguments similar to those applied to the states in ^{7}Li, one sees immediately that the two T = 1 ^{6}Li states considered here have even parity. As is seen from fig. 3, these qualitative conclusions do agree with the experimental level scheme.

Level spectra of ^{6}He and ^{6}Be in the low-excitation region are similar to the T = 1 level spectrum of ^{6}Li as shown in fig. 3. The ground states of both these nuclei have $J^\pi = 0^+$. In ^{6}Be, the states are not stable with respect to particle decay, due to the large number of protons present.

3.6c. α-Cluster States of ^{16}O

As third example, we consider the low-lying levels in ^{16}O. We shall discuss this example in greater detail because of two reasons. The first reason is that one can learn a great deal about the low excited states of this nucleus without carrying out explicit calculations, but by making rather simple qualitative studies in different oscillator cluster representations and extracting nonconflicting features. The second reason is that the level structure of ^{16}O indicates in an especially clear way that one needs to generalize the oscillator cluster functions such that a quantitative description of the behaviour of a nuclear system can be obtained by using just a small number of these generalized functions.

In order to begin our description of the low-lying ^{16}O states, we assume an α + ^{12}C cluster structure. In the oscillator cluster picture, this is completely equivalent to a representation in terms of four α clusters, with the nucleus ^{12}C considered as consisting of three α clusters in relative oscillations [WI 62].

In the ground state of ^{16}O, the relative motion between the unexcited α cluster and the unexcited ^{12}C cluster has four oscillator quanta of energy and zero orbital angular momentum. This is so, because the resultant antisymmetrized cluster wave function must be mathematically equivalent to the oscillator shell-model wave function describing the configuration in which the 1 s and 1 p shells are completely filled.

We now consider the lowest negative-parity excited states. In these states where the α and ^{12}C clusters are not internally excited, the relative motion between the two clusters is an oscillation of fifth order with orbital angular momentum l = 1, 3, or 5. We obtain further negative-parity levels when the ^{12}C cluster is internally excited to its first (2^+) or second (4^+) excited state†; the spins of the internally excited ^{12}C cluster will then be coupled to the above-mentioned orbital angular-momentum values. Further, it should be

† In these two ^{12}C states, no α cluster is internally excited and, in terms of the oscillator shell model, all nucleons remain in the lowest single-particle states compatible with the Pauli principle (i. e., the $(1\,s)^4$ $(1\,p)^8$ configuration). This is quite analogous to the corresponding 2^+ and 4^+ states of ^{8}Be discussed in section 3.4.

noted that, in the oscillator model, the energy eigenvalues for all these negative-parity cluster wave functions are just one oscillator quantum larger than the energy eigenvalue for the ground-state cluster wave function.

After antisymmetrization, most of the negative-parity wave functions for ^{16}O described above (the coupling possibilities range from $J^\pi = 1^-$ to $J^\pi = 9^-$) either vanish or become identical to one another. To see this we employ our usual method of considering the levels simultaneously in different representations: the oscillator shell-model representation and the oscillator $\alpha + {}^{12}C$ cluster representation. In the shell-model representation, all these negative-parity levels of ^{16}O, having one oscillator quantum of excitation energy and in which none of the four α clusters are internally excited, correspond to one-particle excitations to the 2s-1d shell (see appendix B). In addition, one can have only such one-particle excitations in which no spin flip and no isobaric-spin flip occur (i.e., excitation to states with T = 0 and S = 0), because otherwise an α cluster would be broken up. But this means that the total angular momenta of these states must come from the orbital angular momenta of the nucleons. By lifting up one nucleon from the 1p shell to the 2s–1d shell, the orbital angular momenta of the nucleons can couple to yield only total angular-momentum values J = 3, 2, and 1. Further, one sees that for J = 3 and for J = 2, just one coupling possibility exists. Therefore, ^{16}O has one 3^- and one 2^- state which are one-particle-excitation states to the next higher oscillator shell and in which simultaneously no α cluster is broken up. For the J = 1 case, there are two coupling possibilities. But one of these couplings must describe a pure center-of-mass excitation of the ^{16}O nucleus in the space-fixed oscillator potential, which is irrelevant for us. The orbital angular momentum of this excited center-of-mass motion is L = 1. This is easily seen by briefly referring back to the oscillator cluster picture. There we can obtain from the ground state a negative-parity state with one oscillator quantum of excitation by changing the center-of-mass function from a 1s to a 1p oscillation. Because the center-of-mass coordinate is symmetric in the particle coordinates, this spurious state cannot vanish upon antisymmetrization. Hence, there exists also only one J = 1 negative-parity ^{16}O state in which a single particle is excited to the next oscillator shell and no α cluster is broken up. One sees from this that from all coupling possibilities which one expects in the $\alpha + {}^{12}C$ oscillator cluster representation ($J^\pi = 1^-, \ldots, 9^-$), only three possibilities remain at the most. This reduction is due to the Pauli principle.

As far as we know, this is the most illuminating example of how it is sometimes possible to take into account the effect of the Pauli principle without explicit calculations by considering the states of a nucleus simultaneously in different oscillator representations and retaining only those states which are compatible in all these representations. In addition, these kinds of considerations are also very helpful in constructing the appropriate nuclear oscillator cluster and shell-model wave functions explicitly.† For instance, in the case of ^{16}O, as soon as one has constructed one 1^- wave function for a negative-parity state with one oscillator quantum of excitation and no broken-up α cluster, one has found already all possible wave functions of this kind due to our discussion above. The same is true also for the corresponding 2^- and 3^- states.

---

† For the explicit construction of these wave functions, see reference [KA 59].

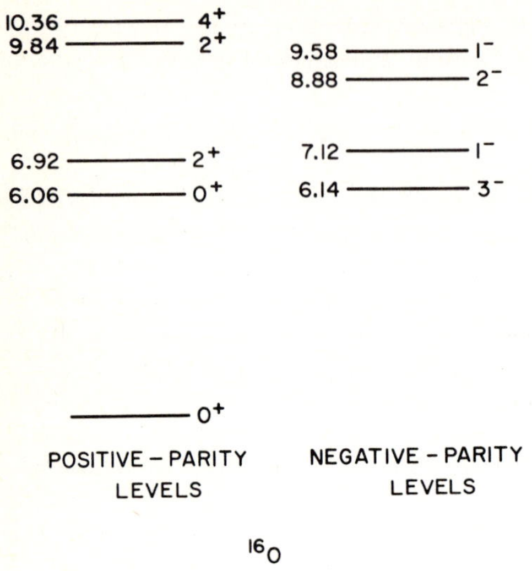

Fig. 4. Energy levels of ^{16}O.

Our considerations show that the lowest measured negative-parity levels of ^{16}O should have $J^\pi = 3^-, 1^-$, and 2^- (see fig. 4). As one might expect, the 2^- level lies above the 1^- and 3^- levels, because there the ^{12}C cluster has to be in its first excited state. The penetrating-orbit argument no longer suffices to give ordering of the 1^- and 3^- levels, since there is large cluster overlapping in both these states and other more subtle and complicated effects connected with the Pauli principle come into play. In fact, if one calculates the spacing between the 1^- and 3^- levels using oscillator cluster trial functions and a simple Serber ansatz for the nuclear forces, then one finds that these levels coincide†. The noncentral part of the nuclear forces introduces a small probability that an α cluster will breakup, and thereby depresses the state with the larger angular momentum; therefore, the first 3^- state in ^{16}O has a little smaller excitation energy than the first 1^- state.

The first calculations of the negative-parity states of ^{16}O with $T = 0$ were made in the frame of the oscillator shell model (one oscillator quantum excitation) by *Elliot* and *Flowers* [EL 57]. These calculations yielded the correct energy sequence of the three lowest states practically independent of the forces employed, whereas the other $T = 0$ negative-parity states do not fit in with the experimentally measured level scheme. The reason for this follows immediately from our discussions given above. The first three negative-parity states are the three α-cluster states we have considered.†† Due to our considerations, in the other negative-parity states of *Elliot* and *Flowers*, at least one of the α clusters must always be broken up. Therefore, their excitation energies should be at least 12 MeV or higher, instead of, for instance, 9.58 MeV as was found experimentally for the fourth negative-parity state of ^{16}O (see fig. 4). It is energetically much easier to

† Unpublished calculation of *Th. Kanellopoulos.*

†† It is interesting to mention that these states also coincide with the corresponding zeroth-approximation particle-hole states (see, for instance, [BR 64]). This is again due to the indistinguishability of the nucleons.

excite a higher-order relative motion of the α cluster against the unexcited ^{12}C cluster than to break up an α cluster. Thus, one has to assume that the fourth negative-parity state of ^{16}O is again approximately an α-particle state, but with a relative cluster oscillation of seventh order. For the lowest α-particle state with seven oscillator quanta of excitation, one expects a 4p oscillation between the clusters. The total angular momentum and parity of this state must therefore be 1^-, which agrees with the experimental finding (see fig. 4). Similar considerations are also valid for the succeeding negative-parity states.

We shall now discuss the low-excitation α-particle levels of ^{16}O with positive parity. In the frame of the oscillator cluster model, the lowest excited positive-parity level with $J^\pi = 0^+$ is described by a 4s oscillation (six oscillator quanta of excitation) of the α cluster relative to the unexcited ^{12}C cluster. This 0^+ state at 6.06 MeV is the lowest level of a rotational band whose other members are the 2^+ level at 6.92 MeV and the 4^+ level at 10.36 MeV (see fig. 4). Above the rotational 2^+ state lies a second 2^+ state at 9.84 MeV; for this state, the relative motion of the clusters is again a 4s oscillation, but with the ^{12}C cluster internally excited to its first excited state with $J^\pi = 2^+$. This description of the 9.84 MeV level is supported by the fact that its energy distance from the first excited 0^+ level is 3.8 MeV, which is approximately the excitation energy (4.4 MeV) of the first excited ^{12}C level.

3.6d. Brief Remarks

Our discussion indicates that the oscillator cluster model can very often give us good insight into the qualitative structure of nuclear spectra. To a large extent, this is due to the fact that in the oscillator model, the relation between wave functions in different representations can be seen most clearly. Indeed, our examples here have shown that, even without any explicit calculation, rather simple considerations can already lead to successful predictions not only about the spins and parities of nuclear levels, but also about the level sequence in the low-excitation region.

On the other hand, if one calculates the expectation values of the microscopic nuclear Hamiltonian with one or a very small number of oscillator cluster wave functions, then one obtains, in general, rather poor results. For instance, with the nucleon-nucleon potential given in ref. [PE 60], the calculated excitation energies of the negative-parity states in ^{16}O are 15 MeV or more and of the positive-parity states 20 MeV or more. This is not in agreement with experimental values. As is shown in fig. 4, the lowest excited state has an excitation energy of only 6.06 MeV. Furthermore, we note in fig. 4 an interesting feature which is rather puzzling from the point of view of the oscillator cluster model. This feature is that the lowest excited positive-parity level has a smaller excitation energy than the lowest negative-parity level. In the frame of the oscillator model, one should expect just the opposite, because the lowest excited positive-parity level has six oscillator quanta of excitation, while the lowest negative-parity level has only five oscillator quanta of excitation.

Therefore, one has to ask: what causes the low excited states in ^{16}O have much smaller excitation energies than those calculated in the oscillator cluster model? The main reason is that the correlation nature of the nucleons cannot be adequately described with just one or two oscillator cluster wave functions. Of course, since the oscillator cluster wave functions do form a complete set, one can certainly obtain a good, quantitative description of the behaviour of the system by using a large number of these functions. But this procedure will not be a desirable one, because in so doing, one will not only be faced with computational difficulties but also lose all hope of gaining any physical insight into the nature of the system.

Finally, before leaving the oscillator cluster model, we would like to make some additional remarks as to the question of choosing the most suitable oscillator cluster basis system. We have already emphasized that all the cluster basis systems are complete, so that our choice is a matter of simplifying the calculations and improving the physical insight. The important point we wish to make here is that our choice depends not only on the nuclear level we are considering, but also on the kind of nuclear reaction process we wish to describe. For instance, let us consider the case of γ transition from the ground state of ^{16}O. If the resultant excited state lies in the low-excitation region where ^{16}O can be expected to consist of non-broken-up α clusters, then the $\{4444\}$ α-cluster description is the most appropriate for the ^{16}O ground state. On the other hand, if the excited state happens to be the giant dipole resonance state, then the α-cluster representation will not be too convenient. As is known, an approximate description of ^{16}O in this giant resonance state is that the protons as a whole oscillate against the neutrons as a whole. Thus, for this state the most appropriate representation is the $\{88\}$ cluster representation [BR 57], in which the cluster wave function is given by

$$\psi = A\,\{\phi_0\,(8\,n)\,\phi_0\,(8\,p)\,\chi_k\,(\mathbf{R}_{8n} - \mathbf{R}_{8p})\,\xi_k\,(s, t)\}. \tag{3.39}$$

The ground state (T = 0, S = 0) and the giant resonance states (T = 1, S = 0) are then obtained with k = 0 and 1, respectively,† with k denoting the number of oscillator quanta in the relative motion between the two large clusters. Thus, in this example, we see that the most convenient representation for the ground state of ^{16}O can be either the α-cluster representation or the 8n + 8p representation, depending on the nature of the final state to which ^{16}O is excited.

3.7. Construction of Generalized Cluster Wave Functions

In the previous section, we have mentioned that it is, in general, not possible to obtain satisfactory quantitative descriptions of nuclear energy spectra with the use of one or a very small number of oscillator cluster wave functions. To see why this is so, let us consider again the example concerning the ground state and the first excited positive-parity state of ^{16}O. In the oscillator cluster model, the wave functions for these states may be written as

$$\psi = A\,\{\phi_0\,(^{12}C)\,\phi_0\,(\alpha)\,\chi_{nlm}\,(\mathbf{R})\,\xi\,(s, t)\}, \tag{3.40}$$

† In the oscillator cluster model the antisymmetrized 8n + 8p cluster wave function with k = 0 is identical to the antisymmetrized shell-model (and α-cluster-model) ground-state wave function. Similarly, the antisymmetrized 8n + 8p cluster function with k = 1 is equivalent to the antisymmetrized shell-model wave function describing the excitation of one nucleon into the 2s–1d shell [BR 57].

where A is the antisymmetrization operator and $\xi(s, t)$ is an appropriate spin-isospin function for S = 0 and T = 0. The function $\phi_0(^{12}C)$ describes the spatial behaviour of the unexcited ^{12}C cluster; it has a 3α cluster form, corresponding in the shell-model picture to a configuration of $(1s)^4\ (1p)^8$ in a harmonic well of width parameter a. The function $\phi_0(\alpha)$ describes the spatial behaviour of the unexcited α cluster and has the form

$$\phi_0(\alpha) = \exp\left(-\frac{1}{2} a \sum_{i=1}^{4} \bar{\mathbf{r}}_i^2\right), \tag{3.41}$$

with the internal coordinates $\bar{\mathbf{r}}_i$ defined as in eq. (3.8). Finally, the relative-motion function $\chi_{nlm}(\mathbf{R})$ with $l = 0$ and $m = 0$ is given by

$$\chi_{n00}(\mathbf{R}) = L_n(R^2) \exp\left(-\tfrac{3}{2} aR^2\right), \tag{3.42}$$

where $L_n(R^2)$ is a polynomial function, with n denoting the number of oscillator quanta of excitation (n = 4 and 6 for the ground and the first excited positive-parity states, respectively). Now, the important point to note is that the internal width parameters of the ^{12}C and the α clusters and the relative-motion width parameter are all the same. This means that, in calculating the expectation value of the microscopic nuclear Hamiltonian with this wave function, the only flexibility is contained in the variation of the common width parameter a. In fact, even this variation has to be severely limited. In order to obtain a fair agreement with the experimentally measured value of the rms charge radius of ^{16}O [EH 59], one should choose a around 0.32 fm^{-2} [EL 61]. Therefore, with this added restriction, the wave function has rather limited flexibility and, in general, cannot correctly account for the correlation behaviour of the nucleons.

To remedy the above situation, one could, of course, use a large number of oscillator cluster functions in calculating the energy expectation values. However, as has been mentioned several times before, this would cause us to lose physical insight into the structure of the system under investigation. A better way is to generalize the wave function of eq. (3.40) such that one or a few of the resultant generalized cluster wave functions can be used to yield satisfactory quantitative results. To see how this generalization can be done, we shall first make a qualitative discussion of the roles played by the internal and relative-motion width parameters.

Inspection of the functions in eqs. (3.41) and (3.42) shows that the internal width parameter of a cluster determines the strength of that cluster correlation. Increasing the internal width parameter increases the probability that the nucleons in the cluster must be found closer together. More roughly, one can say that the internal width parameter determines the mean radius of the cluster (insofar as we can speak of a cluster having a radius); as the internal width parameter becomes larger, the mean radius of the cluster becomes smaller. Similarly, the relative-motion width parameter determines, in a rough sense, the mean separation of the cluster centroids.

Now, let us go back to the ^{16}O problem. If one calculates the expectation values of the microscopic nuclear Hamiltonian for the ground and the first excited positive-parity states using wave functions of eq. (3.40) with a = 0.32 fm^{-2}, then one obtains an excitation energy of about 20 MeV, which is much larger than the value of 6.06 MeV found experimentally. To reduce this excitation energy we mention two important effects. The first effect (radius-change effect [SH 60, WI 59]) is associated with the fact that the width parameter ($\approx 0.52\ \text{fm}^{-2}$) of a free α particle is larger than the width parameter ($\approx 0.32\ \text{fm}^{-2}$) appropriate to ^{16}O in its ground state. This means that a free α particle has a smaller mean radius than an α cluster in the ground state of ^{16}O. Since in the first excited positive-parity state at 6.06 MeV, the α and the ^{12}C clusters are bound by only 1.09 MeV, one should expect that at least some of the α clusters will spend an appreciable amount of time outside of the nucleus and, therefore, behave more like free α particles. Thus, in the calculation of the energy expectation value for this state, it will be more appropriate to choose internal α-cluster width parameters larger than $0.32\ \text{fm}^{-2}$. If this effect is incorporated into the calculation, one would expect the excitation energy to decrease. The second effect (anharmonicity effect) follows from the observation that the relative-motion width parameter for this excited state should be smaller than that for the ground state. This must be so, because in the limit case where one α cluster becomes free, this parameter must go to zero. At present, we know of no detailed calculations in ^{16}O which take these two effects into account. But, variational calculations in analogous cases (such as ^{12}C) did indicate that these effects, if properly considered, could decrease the excitation energy of this positive-parity excited state in ^{16}O by 10 MeV or more.

Our discussion shows that the generalized cluster wave function, obtained from the oscillator cluster wave function by letting the internal and relative-motion width parameters be independent of each other, could yield a better description of the nucleon correlation behaviour. To proceed further, one can easily see that there is really no compelling reason in staying with internal and relative-motion functions of oscillator form. Thus, for example, a deuteron cluster, with its diffuse nature, is liable to be better represented by a superposition of Hulthén functions containing variational parameters. Similarly, in scattering and reaction problems, the relative-motion functions can be chosen more appropriately to satisfy the asymptotic boundary conditions.

In short, the guiding principle for the construction of generalized cluster wave functions is that one or a small number of them should suffice to yield a satisfactory description of the behaviour of the physical system. Therefore, in general, they should satisfy the following requirements:

(i) They should contain enough parameters such that they can be adjusted to give a proper account of the favoured nucleon correlation behaviour.

(ii) The center-of-mass wave function should appear as a multiplicative factor, so that there will be no trouble with unphysical center-of-mass excitations.

(iii) They should be eigenfunctions of the square of the total angular-momentum operator $\mathbf{J}^2$; at least, one should be able to obtain eigenfunctions of $\mathbf{J}^2$ by linearly superposing a small number of these wave functions.

To meet the first requirement, one should demand that the generalized cluster wave function contain an anti-correlation or Jastrow factor [JA 55]. This is necessary, since phenomenological nucleon-nucleon potentials commonly used possess a repulsive core which prevents two nucleons from approaching each other too closely.

In the following subsections, we shall give a description of the Jastrow factor, together with a discussion of Jacobi and parameter coordinates which are useful in connection with generalized cluster wave functions.

3.7a. Introduction of Jacobi Coordinates

To satisfy requirement (ii) mentioned above, it is necessary to have the center-of-mass coordinates clearly separated out. For this purpose, we shall introduce the so-called Jacobi coordinate system. In terms of a set of A single-particle coordinates $\mathbf{r}_i$, the Jacobi coordinates $\hat{\mathbf{r}}_i$ are defined as follows:

$$\begin{aligned}
\hat{\mathbf{r}}_1 &= \mathbf{r}_1 - \mathbf{r}_2, \\
\hat{\mathbf{r}}_2 &= \tfrac{1}{2}(\mathbf{r}_1 + \mathbf{r}_2) - \mathbf{r}_3, \\
\hat{\mathbf{r}}_3 &= \tfrac{1}{3}(\mathbf{r}_1 + \mathbf{r}_2 + \mathbf{r}_3) - \mathbf{r}_4, \\
&\ldots\ldots\ldots \\
\hat{\mathbf{r}}_A &= \mathbf{R}_C = \tfrac{1}{A}(\mathbf{r}_1 + \mathbf{r}_2 + \ldots + \mathbf{r}_A).
\end{aligned} \tag{3.43}$$

If we further introduce the canonically conjugate momenta

$$\hat{\mathbf{p}}_k = \frac{\hbar}{i} \nabla_{\hat{\mathbf{r}}_k}, \tag{3.44}$$

then the total orbital angular-momentum operator $\mathbf{L}$ has the form

$$\mathbf{L} = \sum_{k=1}^{A} (\hat{\mathbf{r}}_k \times \hat{\mathbf{p}}_k). \tag{3.45}$$

Thus, the total orbital angular momentum operator expressed in Jacobi coordinates has the same form as that expressed in single-particle coordinates (see appendix A for a simple proof of this).

Next, we consider a system of A nucleons forming N clusters. In this case, we first define the center-of-mass coordinates for each of these clusters

$$\mathbf{R}_k = \frac{1}{n_k}(\mathbf{r}_1^k + \ldots + \mathbf{r}_{n_k}^k), \qquad (k = 1, \ldots, N) \tag{3.46}$$

with n_k being the number of nucleons in the kth cluster. Then, we define either internal cluster coordinates $\bar{\mathbf{r}}_i$ as in eq. (3.8) or internal Jacobi coordinates $\hat{\mathbf{r}}_i$ as in eq. (3.43). In addition, the following Jacobi coordinates appropriate for the relative motions between the clusters can be defined:

$$\begin{aligned}
\hat{\mathbf{R}}_1 &= \mathbf{R}_1 - \mathbf{R}_2, \\
\hat{\mathbf{R}}_2 &= \frac{1}{n_1 + n_2}(n_1 \mathbf{R}_1 + n_2 \mathbf{R}_2) - \mathbf{R}_3, \\
&\ldots\ldots\ldots \\
\hat{\mathbf{R}}_N &= \mathbf{R}_C = \frac{1}{A}(n_1 \mathbf{R}_1 + n_2 \mathbf{R}_2 + \ldots + n_N \mathbf{R}_N).
\end{aligned} \tag{3.47}$$

With these coordinates and the canonically conjugate momenta

$$\hat{\mathbf{P}}_k = \frac{\hbar}{i} \nabla \hat{\mathbf{R}}_k, \tag{3.48}$$

one can show by using a procedure similar to that given in Appendix A that the orbital angular-momentum operator associated with the relative motions of the clusters has the form

$$\mathbf{L}_R = \sum_{k=1}^{N-1} (\hat{\mathbf{R}}_k \times \hat{\mathbf{P}}_k). \tag{3.49}$$

Now, in the N-cluster system, if we describe the relative motions by orbital angular-momentum eigenfunctions (spherical harmonics) with Jacobi coordinates as variables, then it follows from eq. (3.49) that we can construct eigenfunctions of the operator $\mathbf{L}_R^2$ by simply employing the quantum-mechanical vector-coupling theorem using Clebsch-Gordan coefficients. This is completely analogous to the case where single-particle coordinates are used.

Proceeding further, we can construct the total angular-momentum operator of the N-cluster system

$$\mathbf{J} = \mathbf{L}_R + \sum_{k=1}^{N} \boldsymbol{l}_k + \sum_{k=1}^{N} \mathbf{S}_k, \tag{3.50}$$

with $\boldsymbol{l}_k$ and $\mathbf{S}_k$ being the internal orbital angular-momentum operator and spin angular-momentum operator of the kth cluster, respectively. By using again the vector-coupling theorem, we can easily obtain the eigenfunctions of $\mathbf{J}^2$ from the various orbital and spin angular-momentum eigenfunctions. The important point to note from the viewpoint of constructing generalized cluster wave functions is that the radial parts of the relative-motion wave functions can be chosen completely arbitrarily without affecting the eigenvalue of the operator $\mathbf{J}^2$.

Another important point we should mention here is that, if one has constructed an unantisymmetrized wave function which is an eigenfunction of $\mathbf{J}^2$ and in which the total center-of-mass wave function appears as a multiplicative factor, then the wave function obtained after antisymmetrization will also have such properties. The reason for this is that the total angular-momentum operator and the center-of-mass wave function are symmetric in all the nucleon coordinates.

Special applications of this procedure to obtain angular-momentum eigenfunctions with the help of Jacobi coordinates have been made in sections 3.4 and 3.5 in the construction of ^{8}Be and ^{6}Li wave functions.

3.7b. Introduction of Parameter Coordinates

For a proper description of the collective behaviour of the nucleus, it is necessary to introduce cluster coordinates, such as $\hat{\mathbf{R}}_i$, into the formulation. These coordinates are, however, not symmetric functions of the nucleon coordinates and, hence, acquire obscure meaning when the wave function is antisymmetrized. On the other hand, there is the definite need, especially for discussions of principal nature, of a set of collective coordinates which can be defined in a mathematically unambiguous way and yet give no preferential labels to any of the nucleons. In this subsection, we shall show how this set of coordinates can be constructed.

To begin our discussion as to the construction of this set of coordinates, the so-called parameter coordinates [WI 59a], we shall illustrate with a very simple example. In this example, we have two nucleons in the same spin and isospin state and located at the points $\mathbf{r}_1'$ and $\mathbf{r}_2'$. If the nucleons were distinguishable, with nucleon 1 at $\mathbf{r}_1'$ and nucleon 2 at $\mathbf{r}_2'$, the spatial part of the wave function would be

$$\phi = \delta(\mathbf{r}_1 - \mathbf{r}_1')\,\delta(\mathbf{r}_2 - \mathbf{r}_2'), \tag{3.51}$$

where $\mathbf{r}_1$ and $\mathbf{r}_2$ are the nucleon coordinates, and $\mathbf{r}_1'$ and $\mathbf{r}_2'$ are the parameter coordinates. Since the nucleons obey the Pauli principle, we must antisymmetrize and obtain the spatial part as

$$\hat{\phi} = \delta(\mathbf{r}_1 - \mathbf{r}_1')\,\delta(\mathbf{r}_2 - \mathbf{r}_2') - \delta(\mathbf{r}_2 - \mathbf{r}_1')\,\delta(\mathbf{r}_1 - \mathbf{r}_2'). \tag{3.52}$$

This wave function does not specify which nucleon is at $\mathbf{r}_1'$ or $\mathbf{r}_2'$; this means that the parameter coordinates which specify the positions of the nucleons are completely separated from the nucleon coordinates. Therefore, we can now introduce the center-of-mass and the relative parameter coordinates of the two nucleons in the following way:

$$\mathbf{R}' = \frac{1}{2}(\mathbf{r}_1' + \mathbf{r}_2'), \qquad \mathbf{r}' = \mathbf{r}_1' - \mathbf{r}_2'. \tag{3.53}$$

With these coordinates we obtain

$$\hat{\phi} = \delta[\mathbf{r}_1 - (\mathbf{R}' + \tfrac{1}{2}\mathbf{r}')]\,\delta[\mathbf{r}_2 - (\mathbf{R}' - \tfrac{1}{2}\mathbf{r}')] - \delta[\mathbf{r}_2 - (\mathbf{R}' + \tfrac{1}{2}\mathbf{r}')]\,\delta[\mathbf{r}_1 - (\mathbf{R}' - \tfrac{1}{2}\mathbf{r}')] \tag{3.54}$$

If we now give $\mathbf{R}'$ and $\mathbf{r}'$ fixed values, we have in no way committed any particular nucleon to be preferentially fixed to one of the occupied points and our goal is achieved.

Let us now apply this method to the generalized ^{8}Be cluster wave function

$$\psi(\tilde{r}_1, \ldots, \tilde{r}_8) = A\,\{\tilde{\phi}_0(\tilde{\alpha}_1)\,\tilde{\phi}_0(\tilde{\alpha}_2)\,\chi(\mathbf{R})\,Z(\mathbf{R}_C)\}. \tag{3.55}$$

As is easily seen, this wave function can be written as

$$\begin{aligned}\psi(\tilde{r}_1, \ldots, \tilde{r}_8) = A \int &\{\tilde{\phi}_0(\tilde{\alpha}_1')\,\tilde{\phi}_0(\tilde{\alpha}_2')\,\chi(\mathbf{R}')\,Z(\mathbf{R}_C')\\ &\times [\delta(\tilde{r}_1 - \tilde{r}_1')\,\delta(\tilde{r}_2 - \tilde{r}_2') \ldots \delta(\tilde{r}_8 - \tilde{r}_8')]\}\, d\tilde{\tau}',\end{aligned} \tag{3.56}$$

where the delta function $\delta(\tilde{r}_i - \tilde{r}_i')$ includes the spin and isobaric-spin coordinates of the nucleon i as well as its spatial coordinates, i.e.,

$$\delta(\tilde{r}_i - \tilde{r}_i') = \delta(\mathbf{r}_i - \mathbf{r}_i')\,\delta_{s_i s_i'}\,\delta_{t_i t_i'} \tag{3.57}$$

and $d\tilde{\tau}' = d\tilde{r}_1'\, d\tilde{r}_2' \ldots d\tilde{r}_8'$ is a differential volume element in the combined space, spin, and isospin space. The antisymmetrization is carried out on the nucleon coordinates $\tilde{r}_1, \ldots, \tilde{r}_8$ and not on the primed parameter coordinates which are distinct from the nucleon coordinates.

We can introduce for the primed (parameter) coordinates now the internal coordinates and the center-of-mass coordinates of the clusters, which is just a change of dummy variables of integration. Using the definitions given by eqs. (3.8) and (3.9), but with all coordinates there primed to denote that the transformations are carried out on the parameter coordinates and not on the nucleon coordinates, we obtain

$$\begin{aligned}\psi(\tilde{r}_1,\ldots,\tilde{r}_8) = \int & \tilde{\phi}_0(\tilde{\alpha}_1')\,\tilde{\phi}_0(\tilde{\alpha}_2')\,\chi(\mathbf{R}')\,Z(\mathbf{R}_C') \\ \times\, & A\,\{\delta\,[\mathbf{r}_1-(\mathbf{R}_C'+\tfrac{1}{2}\,\mathbf{R}'+\bar{\mathbf{r}}_1')]\,\delta_{s_1 s_1'}\delta_{t_1 t_1'} \\ \times\, & \delta\,[\mathbf{r}_2-(\mathbf{R}_C'+\tfrac{1}{2}\,\mathbf{R}'+\bar{\mathbf{r}}_2')]\,\delta_{s_2 s_2'}\,\delta_{t_2 t_2'} \\ \times\, & \ldots\ldots \\ \times\, & \delta\,[\mathbf{r}_8-(\mathbf{R}_C'-\tfrac{1}{2}\,\mathbf{R}'+\bar{\mathbf{r}}_8')]\,\delta_{s_8 s_8'}\delta_{t_8 t_8'}\}\,d\tilde{\tau}'\end{aligned} \tag{3.58}$$

In this integral representation of the ^{8}Be wave function, the cluster (parameter) coordinates are now completely distinct from the nucleon coordinates and may be directly manipulated without treating any nucleon as distinguishable.

Very often it is useful to introduce only a smaller number of parameter coordinates. For instance, in the case of ^{8}Be, we may wish to introduce only the parameter coordinates $\mathbf{R}'$ by writing the wave function as

$$\begin{aligned}\psi(\tilde{r}_1,\ldots,\tilde{r}_8) &= A\int\{\tilde{\phi}_0(\tilde{\alpha}_1)\,\tilde{\phi}_0(\tilde{\alpha}_2)\,\delta(\mathbf{R}-\mathbf{R}')\,\chi(\mathbf{R}')\}\,d\mathbf{R}' \\ &= \int A\,\{\tilde{\phi}_0(\tilde{\alpha}_1)\,\tilde{\phi}_0(\tilde{\alpha}_2)\,\delta(\mathbf{R}-\mathbf{R}')\}\,\chi(\mathbf{R}')\,d\mathbf{R}',\end{aligned} \tag{3.59}$$

where, for simplicity, we have left out the irrelevant center-of-mass wave function. In this way, we have expressed ψ as a linear superposition of antisymmetrized wave functions $A\,\{\tilde{\phi}_0(\tilde{\alpha}_1)\,\tilde{\phi}_0(\tilde{\alpha}_2)\,\delta(\mathbf{R}-\mathbf{R}')\}$. It should be emphasized again, however, that in spite of its simple appearance, the wave function $A\,\{\tilde{\phi}_0(\tilde{\alpha}_1)\,\tilde{\phi}_0(\tilde{\alpha}_2)\,\delta(\mathbf{R}-\mathbf{R}')\}$ should be interpreted carefully. It is only when the two α clusters have very little spatial overlap that we can approximately speak of this wave function as describing a state where two α particles are found at a vector distance $\mathbf{R}'$ from each other.

3.7c. Introduction of Jastrow Factors

Since phenomenological nucleon-nucleon potentials become strongly repulsive at small internucleon distances ($\lesssim 0.5$ fm), it is necessary that generalized cluster wave functions-contain a multiplicative anti-correlation or Jastrow factor

$$\phi_J = \prod_{\substack{\text{all}\\ \text{pairs}}} f_{ik}(|\mathbf{r}_i-\mathbf{r}_k|), \tag{3.60}$$

which has the property that for values of $|\mathbf{r}_i-\mathbf{r}_k|$ smaller than the range of the repulsive core, the factor $f_{ik}(|\mathbf{r}_i-\mathbf{r}_k|)$ becomes rather small, and for values of $|\mathbf{r}_i-\mathbf{r}_k|$ much larger than the range of the repulsive core, the factor $f_{ik}(|\mathbf{r}_i-\mathbf{r}_k|)$ approaches unity. In fig. 5 we show the qualitative behaviour of such a factor, which is appropriate when the nucleon-nucleon potential has a soft repulsive core.

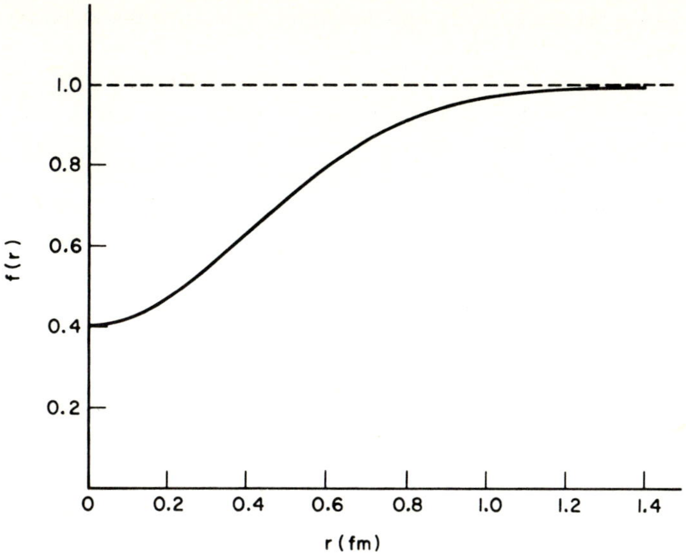

Fig. 5. Behaviour of f(r) as a function of the internucleon distance. The curve shown is for $f(r) = 1 + c \times \exp(-dr^2)$, with $c = -0.6$ and $d = 3.0\ \mathrm{fm}^{-2}$ [HE 71].

The anti-correlation factor ϕ_J is rotationally and translationally invariant. Therefore, its introduction does not disturb the geometrical properties of the generalized cluster wave function. That means if the considered cluster function is an eigenfunction of the operator $\mathbf{J}^2$, the parity operator, and so on, before the introduction of this anti-correlation factor, it will remain so after the introduction of such a factor.

If $f_{ik}(|\mathbf{r}_i - \mathbf{r}_k|)$ is chosen to have the same form for all nucleon pairs, then ϕ_J of eq. (3.60) is symmetric in all nucleon coordinates and does not have to be included in the antisymmetrization process. One consequence of this is that our discussions about equivalences between different cluster functions remain valid even after the introduction of this factor. For example, in the oscillator model, the complete equivalence of the ground-state ^{6}Li description in either the α + d cluster picture or the ^{3}He + t cluster picture will not be altered by the addition of this anti-correlation factor.

The calculations can usually be refined further by choosing different functions $f_{ik}(|\mathbf{r}_i - \mathbf{r}_k|)$ for different nucleon pairs. In this case, ϕ_J is no longer symmetric in all nucleon coordinates and, therefore, one has to include it in the antisymmetrization process.

From these remarks we see that the presence of an anti-correlation factor of the form (3.60) does not change in any fundamental way the considerations we have developed so far. Therefore, unless this factor is needed explicitly in our subsequent discussions, we shall abbreviate the formulae by not writing it out in the wave function and simply assuming its presence understood. In particular, in some of the following chapters where discussions of more general character are made, we shall adopt this practice as a simplifying convention.

Certainly the introduction of such a factor will complicate the calculation to a very large extent. This is not only because ϕ_J is a product of many functions, but also because one has to include in the function $f_{ik}(|\mathbf{r}_i - \mathbf{r}_k|)$ enough variational parameters such that the anti-correlation behaviour of the nucleons can be described properly. As has been shown in the example of an α-particle calculation with a nucleon-nucleon potential containing a hard core [SC 63], a proper description of this anti-correlation behaviour is essential for a satisfactory determination of the α-particle binding energy.

We wish to mention that, in principle, it is not necessary to introduce a Jastrow factor into the wave function to take into account the repulsive-core character of the nucleon-nucleon potential. Theoretically, one can always use a superposition of a large number of cluster wave functions such that the interference among these wave functions will yield a satisfactory description of the nucleon anti-correlation behaviour. In practice, however, this will not work since, using simple uncertainty-principle arguments, one can already see that with such a procedure one will have to include even cluster wave functions corresponding to excitation energies of as high as several hundred MeV.

3.7d. Summary

In this section it is shown that the method of generalized cluster wave functions allows us to construct angular-momentum eigenfunctions of general type for a many-nucleon system. By using Jacobi coordinates as variables, we can, for instance, choose the radial dependence of the spatial parts of the wave functions describing the internal and relative motions of the clusters in a very flexible way without affecting the angular-momentum eigenvalues for the many-nucleon wave functions. In addition, we have shown that the total center-of-mass motion can always be separated off in a clear-cut manner and consequently, there is no need for concern about spurious center-of-mass excitations.

For practical calculations it is often convenient to choose functions of Gaussian dependence for both the spatial parts of the wave functions and the spatial parts of the nuclear forces, because integrations involved in the evaluation of matrix elements can then be carried out analytically. This is, of course, not a principal limitation, since, as is known, every normalizable function can be approximated to any degree of accuracy by a linear superposition of Gaussian functions.

Finally, it should be mentioned that the characteristics of generalized cluster wave functions have been discussed in detail in ref. [WI 66]. Here we wish only to point out that with generalized cluster wave functions simple mathematical relations which exist in the oscillator model, such as that given by eq. (3.14), become now only approximately valid and one must be more careful in discussing equivalences between wave functions in different cluster representations.

4. Formulation of a Unified Microscopich Nuclear Structure and Reaction Theory

4.1. General Remarks

We apply now the basis wave functions discussed in Chapter 3, in particular the generalized cluster wave functions, to the time-dependent and time-independent Schrödinger equations, written in the form of projection equations (see eqs. (2.1) and (2.3)), to formulate a unified microscopic nuclear structure and reaction theory. As has been mentioned already in Chapter 2, the most important feature of this theory is that all open and closed channels will be treated on an equal footing and, consequently, the asymptotic boundary conditions in all the channels can be fulfilled in a very natural manner.

In order to make clear the underlying idea, we shall make the formulation as simple as possible. For this purpose, we shall not write out explicitly the Jastrow factor in the cluster wave function and simply assume its presence understood.† As was mentioned in section 3.7c, this is a convenient way to abbreviate the formulae, because the Jastrow factor does not influence our considerations in any principal manner. In addition, we shall present the formulation in a form most appropriate for the case where the nucleon-nucleon potential is purely central. This does not imply, of course, that our formulation is only applicable to those cases where central nucleon-nucleon potentials are employed; in fact, many calculations performed with this method have been done with potentials containing noncentral components. However, the mathematical modifications required are well-known and straightforward, and in no way affect the basic ideas behind our formulation. Hence, we opt for clarity and discuss the formulation as if the nucleon-nucleon potential were purely central.

Before we discuss the general formulation of this theory, we shall first present two specific examples. From these examples one can see in an especially clear way the essential features involved and the manner in which one's physical intuition is systematically employed for a successful application of this theory.

---

† For certain calculations it is permissible in good approximation to use a simple nucleon-nucleon potential containing no repulsive core, in which case the Jastrow factor may be approximated by a factor of unity.

4.2. Specific Examples

4.2a. n + α Scattering

As first example, we consider the scattering of neutrons by α particles. In the low-energy region where no reaction channels are open, it is appropriate to use the following ansatz for the wave function of the n + α system:†

$$\psi = A\,\{\phi(\alpha)\,F(\mathbf{R}_n - \mathbf{R}_\alpha)\,\xi(s,t)\}, \tag{4.1}$$

where the function $\phi(\alpha)$ describes the spatial behaviour of the α cluster and $\xi(s, t)$ is a suitable spin-isospin function.†† For $\phi(\alpha)$ one should preferably use a flexible function which is obtained from a variational calculation based on the α particle Hamiltonian, since in the asymptotic region the α cluster must behave like a free α particle. In practice, however, this is neither simple nor computationally desirable. Thus, what one normally does is to use an approximate wave function consisting of one or more Gaussian functions. In this way, the many-dimensional integrals involved in computing with the wave function ψ can usually be carried out analytically.

Our first task is to discuss the justification in using the relatively simple wave function ψ of eq. (4.1) to describe the n + α scattering problem. For this it is necessary to give a qualitative discussion of what is meant by the distortion of the clusters in a many-body problem. There are in general two types of cluster distortion present. The first type is commonly referred to as the Pauli or exchange distortion, which is connected with the fact that nucleons are indistinguishable and hence, one cannot label the nucleons which participate in the cluster correlation. This type of distortion has always been found to have a profound influence upon the behaviour of the system and is properly taken into consideration by the antisymmetrization of the wave function (see section 6.3). The second type has been called the specific distortion, which arises because the nucleon-nucleon potential is attractive in its longer-range part and, therefore, two clusters when approaching each other will tend to deform themselves for the purpose of increasing their spatial overlap to take maximum advantage of this attractive interaction. Thus, in a system where a highly deformable cluster, such as a deuteron cluster, is present, this distortion effect must be carefully taken into account in order to obtain quantitatively satisfactory results. In our present case of n + α scattering, however, the specific distortion effect can be neglected to a good approximation, because a free α particle is known to have a very stable structure. Therefore, we feel that as long as we consider only scattering in the low-energy region ($\lesssim$ 20 MeV), the wave function ψ of eq. (4.1) should give a good approximate description of the n + α system.

† A normalized center-of-mass function $Z(\mathbf{R}_C)$ is left off for convenience. This is a practice which we shall usually follow in our future discussions.

†† By writing $F(\mathbf{R})$ as

$$F(\mathbf{R}) = \sum_{lm} \frac{1}{R} f_l(R)\, Y_{lm}(\theta, \varphi),$$

one can easily see that the wave function ψ represents a sum of generalized cluster wave functions, with each of them being an eigenfunction of the operators $\mathbf{L}^2$, L_z, $\mathbf{S}^2$, S_z, $\mathbf{T}^2$ and T_3.

We should further mention that it is a common practice to refer to a calculation where the specific distortion effect is not considered but the wave function is totally antisymmetrized as a calculation in the "no-distortion" approximation. It must be emphasized that in such a calculation, the more important exchange distortion effect is always taken into account properly. Thus, as long as one is concerned only with clusters which are relatively incompressible, the "no-distortion" approximation is expected to yield a rather satisfactory description of the behaviour of the system.

Next, we discuss the significance of the relative-motion function $F(\mathbf{R}_n - \mathbf{R}_\alpha)$. In our formulation, this function is considered as a linear variational function which is arbitrarily varied to satisfy the projection equation (2.3). To make this more clear, let us write eq. (4.1) in the following parameter representation:

$$\psi = \int A \{\phi(\alpha)\, F(\mathbf{R}')\, \xi(s,t)\, \delta(\mathbf{R} - \mathbf{R}')\}\, d\mathbf{R}', \tag{4.2}$$

with

$$\begin{aligned} \mathbf{R} &= \mathbf{R}_n - \mathbf{R}_\alpha, \\ \mathbf{R}' &= \mathbf{R}_n' - \mathbf{R}_\alpha'. \end{aligned} \tag{4.3}$$

As has been mentioned previously, the antisymmetrization operator A operates only on the unprimed nucleon coordinates, but not on the primed parameter coordinates $\mathbf{R}'$. From eq. (4.2) one sees that the wave function ψ represents a continuous linear superposition of trial functions (or basis wave functions as we shall often call them) of the form

$$A \{\phi(\alpha)\, \delta(\mathbf{R} - \mathbf{R}')\, \xi(s,t)\},$$

with variational amplitudes given by $F(\mathbf{R}')$.

The variation of ψ is given by

$$\delta\psi = \int \delta F(\mathbf{R}')\, A \{\phi(\alpha)\, \delta(\mathbf{R} - \mathbf{R}')\, \xi(s,t)\}\, d\mathbf{R}'. \tag{4.4}$$

Upon substituting eqs. (4.4) and (4.1) into the projection equation (2.3), we obtain, because of the arbitrary variation of $F(\mathbf{R}')$,

$$\langle A \{\phi(\alpha)\, \delta(\mathbf{R} - \mathbf{R}')\, \xi(s,t)\} | H - E | A \{\phi(\alpha)\, F(\mathbf{R})\, \xi(s,t)\}\rangle = 0, \tag{4.5}$$

where the Hamiltonian H is given by

$$H = \sum_{i=1}^{5} \frac{1}{2M} p_i^2 + \sum_{i<j=1}^{5} V_{ij} - T_C, \tag{4.6}$$

with T_C being the kinetic-energy operator of the total center of mass. Next, we perform the integration over all the nucleon spatial coordinates and the summation over all the nucleon spin and isospin coordinates, and obtain the following integrodifferential equation satisfied by the relative-motion function $F(\mathbf{R}')$:

$$\left[-\frac{\hbar^2}{2\mu} \nabla_{\mathbf{R}'}^2 + V_D(\mathbf{R}')\right] F(\mathbf{R}') + \int K(\mathbf{R}', \mathbf{R}'')\, F(\mathbf{R}'')\, d\mathbf{R}'' = E_R\, F(\mathbf{R}'), \tag{4.7}$$

where μ denotes the reduced mass and E_R denotes the total kinetic energy of the neutron and the α cluster at large separation in the center-of-mass system. Also, in eq. (4.7), $V_D(\mathbf{R}')$ is the direct nuclear potential and $K(\mathbf{R}', \mathbf{R}'')$ is a hermitian kernel function† for the nonlocal interaction between the neutron and the α cluster, arising from the exchange character in the nucleon-nucleon potential and the antisymmetrization procedure.

Once eq. (4.7) is derived, one can then proceed in the usual manner to study the $n + \alpha$ bound and elastic-scattering states. Thus, by imposing the boundary condition that $F(\mathbf{R}')$ must approach zero asymptotically, one obtains bound-state eigenvalues for this system. For scattering states, E_R takes on positive values and the scattering amplitude can be found using the boundary condition that there are incoming plane wave and outgoing spherical wave in the asymptotic region.

For scattering energies above 17.6 MeV, it is energetically possible to break up the α particle. At these energies, one should therefore improve the description of the $n + \alpha$ system by taking into consideration at least the next energetically favoured cluster configuration, which is the deuteron plus triton (d + t) configuration, with the deuteron being in its triplet ground state. Thus, we write the wave function ψ as a superposition of two channel functions, i.e.,

$$\psi = A\,\{\phi(\alpha)\, F_1(\hat{\mathbf{R}}_1)\, \xi_1(s,t)\} + A\,\{\phi(d)\, \phi(t)\, F_2(\hat{\mathbf{R}}_2)\, \xi_2(s,t)\}, \tag{4.8}$$

where $\hat{\mathbf{R}}_1$ and $\hat{\mathbf{R}}_2$ are channel radii, given by

$$\begin{aligned} \hat{\mathbf{R}}_1 &= \mathbf{R}_n - \mathbf{R}_\alpha, \\ \hat{\mathbf{R}}_2 &= \mathbf{R}_d - \mathbf{R}_t, \end{aligned} \tag{4.9}$$

and ξ_1 und ξ_2 are appropriate spin-isospin functions for $S = \frac{1}{2}$ and $T = \frac{1}{2}$. The functions F_1 and F_2 describe the relative motions of the $n + \alpha$ and $d + t$ clusters, respectively, and are considered as linear variational functions in our formulation. Now if one substitutes eq. (4.8) into eq. (2.3), then one obtains the following coupled integrodifferential equations for F_1 and F_2:

$$\begin{aligned} &\left[-\frac{\hbar^2}{2\mu_1}\nabla_1'^2 + V_{D1}(\hat{\mathbf{R}}_1')\right] F_1(\hat{\mathbf{R}}_1') + \int K_{11}(\hat{\mathbf{R}}_1', \hat{\mathbf{R}}_1'')\, F_1(\hat{\mathbf{R}}_1'')\, d\hat{\mathbf{R}}_1'' \\ &+ \int K_{12}(\hat{\mathbf{R}}_1', \hat{\mathbf{R}}_2'')\, F_2(\hat{\mathbf{R}}_2'')\, d\hat{\mathbf{R}}_2'' = E_{R1}\, F_1(\hat{\mathbf{R}}_1'), \\ &\left[-\frac{\hbar^2}{2\mu_2}\nabla_2'^2 + V_{D2}(\hat{\mathbf{R}}_2')\right] F_2(\hat{\mathbf{R}}_2') + \int K_{22}(\hat{\mathbf{R}}_2', \hat{\mathbf{R}}_2'')\, F_2(\hat{\mathbf{R}}_2'')\, d\hat{\mathbf{R}}_2'' \\ &+ \int K_{21}(\hat{\mathbf{R}}_2', \hat{\mathbf{R}}_1'')\, F_1(\hat{\mathbf{R}}_1'')\, d\hat{\mathbf{R}}_1'' = E_{R2}\, F_2(\hat{\mathbf{R}}_2'). \end{aligned} \tag{4.10}$$

---

† It should be mentioned that $K(\mathbf{R}', \mathbf{R}'')$ depends explicitly on the total energy E. The coordinate $\mathbf{R}''$ is a parameter coordinate; its occurrence in the kernel function can be easily seen by writing the wave function on the ket side of eq. (4.5) in the form of eq. (4.2).

In the above equations, the quantities $\mu_1, \mu_2, V_{D1}, V_{D2}, K_{11}$, and K_{22} have meanings analogous to those corresponding quantities appearing in eq. (4.7), and E_{R1} and E_{R2} are relative energies of the clusters in the center-of-mass system, given by

$$
\begin{aligned}
E_{R1} &= E - E_\alpha, \\
E_{R2} &= E - (E_d + E_t),
\end{aligned}
\tag{4.11}
$$

with E_α, E_d, and E_t being the internal energies of the various clusters. The coupling between the two channels is represented by the kernel functions K_{12} and K_{21}. Because of the hermiticity of the Hamiltonian operator, these coupling kernels are adjoint to each other, i.e.,

$$K_{12}(\hat{\mathbf{R}}_1, \hat{\mathbf{R}}_2) = K_{21}^*(\hat{\mathbf{R}}_2, \hat{\mathbf{R}}_1). \tag{4.12}$$

Equation (4.10) can be used to study various bound-state and reaction problems simply by choosing appropriate boundary conditions. These problems are:

(i) Elastic scattering of neutrons by α particles, together with transitions to the d + t channel. Here one has incoming and outgoing waves in the n + α channel and only outgoing wave in the d + t channel. In the case where the d + t channel is closed, its relative-motion function must tend asymptotically to zero.

(ii) Deuteron-triton elastic scattering, together with transitions to the n + α channel. In this case, there are incoming and outgoing waves in the d + t channel, but only outgoing wave in the n + α channel.

(iii) Bound states (if present) having a mixture of d + t and n + α cluster structure. The boundary conditions which must be fulfilled are that, in the asymptotic regions, both relative-motion functions must tend exponentially to zero.

To improve further the description of the n + α system, one could for example augment the wave function of eq. (4.8) by a sum of square-integrable or bound-state-type ^{5}He wave functions with linear variational amplitudes. In this way, effects due to the specific distortion of the deuteron and triton clusters can be taken into account. We shall discuss such distortion effects in the next example where the scattering of deuterons by α particles is considered.

4.2b. d + α Scattering

Similar to the n + α case, the simplest ansatz for the d + α wave function is

$$\psi_0 = A\,\{\tilde{\phi}(\tilde{d})\,\tilde{\phi}(\tilde{\alpha})\,F(\mathbf{R}_d - \mathbf{R}_\alpha)\}, \tag{4.13}$$

where it should be noted that for simplicity in writing, we have absorbed the spin-isospin function into the deuteron and α-cluster internal wave functions.† From our discussion

† A remark on notation may be useful here. The function $\tilde{\phi}(\tilde{\alpha})$ with a tilde mark means the internal wave function of an α cluster, depending on the internal spatial, spin, and isospin coordinates of all the nucleons forming the α cluster, while the function $\phi(\alpha)$ without a tilde mark means only the spatial part of the α-cluster wave function.

in section 4.2a, it should be clear, however, that this wave function where the internal functions $\tilde{\phi}(\tilde{d})$ and $\tilde{\phi}(\tilde{\alpha})$ are not varied in the calculation will not yield a satisfactory description of the behaviour of the $d + \alpha$ system. In this case, effects due to the specific distortion of the easily deformable deuteron cluster should be quite important. Therefore, we improve the wave function ψ_0 of eq. (4.13) in the strong-interaction region by adding to it a number of square-integrable or distortion functions with linear variational amplitudes. Thus, we write

$$\psi = \psi_0 + \sum_{i=1}^{N} \sum_{l=0}^{\infty} A_{li} \hat{\varphi}_{li}, \tag{4.14}$$

where N is the number of distortion functions and l denotes the relative orbital angular-momentum quantum number between the two clusters.† The distortion functions $\hat{\varphi}_{li}$ may be chosen in various ways, depending upon one's physical intuition. One common way is to choose these functions in the form of a $d + \alpha$ bound-state cluster function, i.e.,

$$\hat{\varphi}_{li} = A \left\{ \tilde{\phi}(\tilde{\alpha}) \tilde{\phi}_i(\tilde{d}) \frac{1}{R} g_{li}(R) P_l(\cos\theta) \right\}, \tag{4.15}$$

with $\tilde{\phi}_i(\tilde{d})$ describing the behaviour of a distorted deuteron cluster, having a rms radius either larger or smaller than that of a free deuteron. It should be noted that the distortion functions are not varied at each energy in the calculation [TH 73] and the freedom in the wave function ψ is contained in the energy-dependent linear variational function $F(\mathbf{R})$ and the linear variational parameters A_{li}.

If we now substitute eq. (4.14) into eq. (2.3) and carry out the variation of the function $F(\mathbf{R})$ and the amplitudes A_{li}, then we obtain the following set of coupled integrodifferential and integral equations:

$$\langle A \{\tilde{\phi}(\tilde{d}) \tilde{\phi}(\tilde{\alpha}) \delta(\mathbf{R}-\mathbf{R}')\} | H - E | \psi_0 \rangle$$
$$+ \sum_{i=1}^{N} \sum_{l=0}^{\infty} A_{li} \langle A \{\tilde{\phi}(\tilde{d}) \tilde{\phi}(\tilde{\alpha}) \delta(\mathbf{R}-\mathbf{R}')\} | H - E | \hat{\varphi}_{li} \rangle = 0, \tag{4.16a}$$

$$\langle \hat{\varphi}_{l'j} | H - E | \psi_0 \rangle + \sum_{i=1}^{N} \sum_{l=0}^{\infty} A_{li} \langle \hat{\varphi}_{l'j} | H - E | \hat{\varphi}_{li} \rangle = 0, \tag{4.16b}$$

where the subscripts l' and j in eq. (4.16b) take on integral values from 0 to ∞ and from 1 to N, respectively. These equations can be solved by using standard techniques to obtain the function $F(\mathbf{R})$ and the amplitudes A_{li}, and consequently, the phase shifts in the various partial waves.

In the energy region where the reaction cross section is small, the wave function ψ of eq. (4.14) has been found to yield an adequate description of the $d + \alpha$ elastic-scattering process [TH 73, JA 69]. As the energy increases, the reaction cross section will become

† It should be noted that we choose the z-axis to be in the direction of the incident plane wave.

467

larger and one will then have to use more channel wave functions as in the case of n + α scattering discussed in section 4.2a.

4.2c. Discussion

With these two examples, we have demonstrated how one can use physical intuition and energetical considerations to obtain the simplest trial function necessary for a satisfactory description of the system. Indeed, the prescription is rather simple; at a given energy, one constructs a trial function consisting of all open-channel functions which may be significantly excited and a set of distortion functions which is specifically designed to improve the description of the system in the region of strong interaction. As the energy becomes high, the number of required channel and distortion functions may be quite large and the calculation becomes very complicated, but the procedure is always straightforward and readily understood.

In our examples, we have assumed that the nucleon-nucleon potential is purely central. As has already been mentioned, this simplification is desirable, because we can then bring out the basic ideas in a particularly clear manner. In actual calculations, it is of course not allowed to use such a simplified potential in many cases. For instance, to describe the resonance behaviour in the n + α scattering at around 18 MeV, one needs to employ a tensor interaction, since only this interaction can couple the n + α channel to the ^{5}He compound-nucleus level at 16.7 MeV excitation.

We wish to stress the following two important points which will remain valid when we generalize the above examples to obtain the equations for a unified theory of the nucleus which can be used to study both reactions and bound states in a many-body system:

(i) Due to the hermiticity of the potentials and kernels in eqs. (4.7) and (4.10), which follows from the hermiticity of the many-particle Hamiltonian, the law of current conservation (i.e., the unitarity of the S matrix) is always satisfied exactly.

(ii) The incoming and outgoing channels are always introduced into the theory in a symmetrical manner; consequently, the boundary conditions in all the channels can be easily fulfilled. As a price for this, we have to use channel functions and bound-state-type distortion functions which are not orthogonal to each other.† It should be pointed out, however, that this does not disturb our considerations, because, as was shown in chapter 2, the wave functions ψ which are the solutions of the projection equation (2.3) and for which all degeneracies are removed are always mutually orthogonal. Of course, because of this nonorthogonality associated with the channel and distortion functions, one does have to be careful in the interpretation of the wave function ψ with regard to the cluster structure of the nucleus under consideration. Especially in the compound-nucleus region where all the clusters have a strong spatial overlap with one another, a straightforward interpretation, based solely on the amplitudes of the channel and distortion functions, will usually lead to erroneous conclusions.

---

† It should be noted that even wave functions $A\ \{\phi(\alpha)\ \delta(\mathbf{R}-\mathbf{R}')\ \xi(s,t)\}$ (see eq. (4.4)) with different values of $\mathbf{R}'$ are not orthogonal to each other.

4.3. Extension to General Systems

From the discussion given in section 4.2, it is clear that for a proper description of nuclear bound states and reactions, one should introduce into the calculation enough channel wave functions with their relative-motion parts treated as linear variational functions and enough bound-state-type distortion functions with linear variational amplitudes.† As is quite evident, the most general ansatz for a trial function comprising linear variational functions and amplitudes is as follows [BE 69, WI 69]:

$$\psi = \sum_i A'\{\hat{\phi}(\widetilde{A}_i)\,\hat{\phi}(\widetilde{B}_i)\,F_i(\hat{\mathbf{R}}_i)\} + \sum_j A'\{\hat{\phi}(\widetilde{A}_j)\,\hat{\phi}(\widetilde{B}_j)\,\hat{\phi}(\widetilde{C}_j)\,F_j(\hat{\mathbf{R}}_{j1},\hat{\mathbf{R}}_{j2})\}$$

$$+ \sum_k A'\{\hat{\phi}(\widetilde{A}_k)\,\hat{\phi}(\widetilde{B}_k)\,\hat{\phi}(\widetilde{C}_k)\,\hat{\phi}(\widetilde{D}_k)\,F_k(\hat{\mathbf{R}}_{k1},\hat{\mathbf{R}}_{k2},\hat{\mathbf{R}}_{k3})\} \qquad (4.17a)$$

$$+ \ldots + \sum_m a_m\,\hat{\varphi}_m ,$$

where A' is an antisymmetrization operator which interchanges nucleons in different clusters but not within the individual clusters themselves, and the $\hat{\mathbf{R}}$'s are Jacobi coordinates defined in eq. (3.47). Also, in eq. (4.17a), the functions $\hat{\phi}$ are antisymmetric eigenfunctions of the cluster Hamiltonians. The terms with relative-motion functions $F_j(\hat{\mathbf{R}}_{j1},\hat{\mathbf{R}}_{j2})$, $F_k(\hat{\mathbf{R}}_{k1},\hat{\mathbf{R}}_{k2},\hat{\mathbf{R}}_{k3})$, and so on are responsible for three-, four-, and more-particle decays. The distortion functions $\hat{\varphi}_m$ are chosen to improve the wave function in the strong-interaction region; they vanish for large internucleon and intercluster distances.

It should be noted that in eq. (4.17a) antisymmetric eigenfunctions $\hat{\phi}$ of the cluster Hamiltonians are used as internal wave functions. In principle, this is necessary because the clusters must assume the behaviour of the corresponding free particles in the asymptotic region. In practical calculations, it is of course unavoidable to approximate the functions $\hat{\phi}$ by $A\tilde{\phi}$, with the approximate functions $\tilde{\phi}$ chosen to yield a good description of the internal behaviour of the clusters and, at the same time, for ease in computation. With the adoption of approximate internal functions, eq. (4.17a) is then written as

$$\psi = A\Big\{\sum_i \tilde{\phi}(\widetilde{A}_i)\,\tilde{\phi}(\widetilde{B}_i)\,F_i(\hat{\mathbf{R}}_i) + \sum_j \tilde{\phi}(\widetilde{A}_j)\,\tilde{\phi}(\widetilde{B}_j)\,\tilde{\phi}(\widetilde{C}_j)\,F_j(\hat{\mathbf{R}}_{j1},\hat{\mathbf{R}}_{j2})$$

$$+ \sum_j \tilde{\phi}(\widetilde{A}_k)\,\tilde{\phi}(\widetilde{B}_k)\,\tilde{\phi}(\widetilde{C}_k)\,\tilde{\phi}(\widetilde{D}_k)\,F_k(\hat{\mathbf{R}}_{k1},\hat{\mathbf{R}}_{k2},\hat{\mathbf{R}}_{k3}) + \ldots + \sum_m a_m\,\varphi_m\Big\} . \qquad (4.17b)$$

We should mention that the examples given in subsections 4.2a and 4.2b are formulated in the form of eq. (4.17b).

† A similar approach, but in more specialized form, has been proposed by Wheeler [WH 37] and is called the method of resonating-group structure or the resonating-group method. Already in the years before 1960, calculations using this approach have been performed on light nuclear systems by Massey and others ([MA 60] and references contained therein).

In the following chapters, we shall always use eq. (4.17a) for discussions of principal character, such as those given in chapter 8 and subsequent chapters. On the other hand, especially for illustrative examples of numerical nature, eq. (4.17b) which contain approximate cluster internal functions will necessarily have to be adopted.

To proceed, let us for simplicity confine ourselves at the moment to such reactions where three and many-particle decays do not occur. In such cases, one can then include these terms in the sum for the distortion functions and the wave function ψ of eq. (4.17b) may be written in the simpler form

$$\psi = A\left\{\sum_i \tilde{\phi}(\widetilde{A}_i)\,\tilde{\phi}(\widetilde{B}_i)\,F_i(\hat{\mathbf{R}}_i) + \sum_m a_m\,\varphi_m\right\}. \qquad (4.18)$$

By substituting eq. (4.18) into the projection equation (2.3) and carrying out the variation of the functions F_i and the amplitudes a_m we obtain the following set of coupled equations†:

$$\sum_i \langle \tilde{\phi}(\widetilde{A}_p)\,\tilde{\phi}(\widetilde{B}_p)\,\delta(\hat{\mathbf{R}}_p - \hat{\mathbf{R}}_p')|H - E|\,A\{\tilde{\phi}(\widetilde{A}_i)\,\tilde{\phi}(\widetilde{B}_i)\,F_i(\hat{\mathbf{R}}_i)\}\rangle$$

$$+ \sum_m a_m\,\langle \tilde{\phi}(\widetilde{A}_p)\,\tilde{\phi}(\widetilde{B}_p)\,\delta(\hat{\mathbf{R}}_p - \hat{\mathbf{R}}_p')|H - E|\,A\,\varphi_m\rangle = 0, \qquad (4.19)$$

$$\sum_i \langle \varphi_q|H - E|\,A\{\tilde{\phi}(\widetilde{A}_i)\,\tilde{\phi}(\widetilde{B}_i)\,F_i(\hat{\mathbf{R}}_i)\}\rangle + \sum_m a_m\,\langle \varphi_q|H - E|\,A\,\varphi_m\rangle = 0,$$

where the indices p and q take on values $1, 2, \ldots$, and $\hat{\mathbf{R}}_p'$ are parameter coordinates. In eq. (4.19), it should be noted that the antisymmetrization is carried out only on the ket side of the Dirac bracket; this is permissible, since the antisymmetrization operator is a hermitian operator which commutes with the Hamiltonian operator H and satisfies the relation

$$A^2 = A!\,A, \qquad (4.20)$$

with A being the total number of nucleons.

It should be mentioned that in solving eq. (4.19), difficulties will arise if some of the terms in eq. (4.17a) or eq. (4.17b) are linearly dependent of one another. As to how one can avoid these difficulties in a basically simple way, we shall discuss in chapter 8.

Due to the rotational and reflectional invariance of the Hamiltonian H, the wave function ψ can be expanded into partial waves of given total angular momentum and parity. By so doing, eq. (4.19) can be further divided into sets of coupled integro-differential and integral equations, from which one can determine by numerical techniques the S matrix and bound-state eigenenergies. As to how this is done explicitly, we shall demonstrate later in chapters 5 and 7 by means of specific examples.

† In the remainder of this monograph, the symbol H denotes the Hamiltonian in the center-of-mass system; that is, it does not include the kinetic-energy operator of the center of mass (see eq. (4.6)). Also, we wish to remind the reader that in eqs. (4.17a) and (4.17b) we have left out, for convenience, a normalizable center-of-mass wave function $Z(R_C)$.

If we take into consideration three- and more-particle-decay terms in eq. (4.17b), then we have to evaluate also matrix elements such as

$$\langle \tilde{\phi}(\widetilde{A}_r)\, \tilde{\phi}(\widetilde{B}_r)\, \tilde{\phi}(\widetilde{C}_r)\, \delta(\hat{\mathbf{R}}_{r1} - \hat{\mathbf{R}}'_{r1})\, \delta(\hat{\mathbf{R}}_{r2} - \hat{\mathbf{R}}'_{r2}) | H - E | A \{\tilde{\phi}(\widetilde{A}_j)\, \tilde{\phi}(\widetilde{B}_j)\, \tilde{\phi}(\widetilde{C}_j)\, F_j(\hat{\mathbf{R}}_{j1}, \hat{\mathbf{R}}_{j2})\} \rangle,$$

$$\langle \tilde{\phi}(\widetilde{A}_r)\, \tilde{\phi}(\widetilde{B}_r)\, \tilde{\phi}(\widetilde{C}_r)\, \delta(\hat{\mathbf{R}}_{r1} - \hat{\mathbf{R}}'_{r1})\, \delta(\hat{\mathbf{R}}_{r2} - \hat{\mathbf{R}}'_{r2}) | H - E | A \{\tilde{\phi}(\widetilde{A}_i)\, \tilde{\phi}(\widetilde{B}_i)\, F_i(\hat{\mathbf{R}}_i)\} \rangle,$$

$$\langle \tilde{\phi}(\widetilde{A}_r)\, \tilde{\phi}(\widetilde{B}_r)\, \tilde{\phi}(\widetilde{C}_r)\, \delta(\hat{\mathbf{R}}_{r1} - \hat{\mathbf{R}}'_{r1})\, \delta(\hat{\mathbf{R}}_{r2} - \hat{\mathbf{R}}'_{r2}) | H - E | A\, \varphi_m \rangle,$$

and so on. Obviously, the calculation now will become very complicated, but the procedure is straightforward and there are, in principle, no insurmountable difficulties.

In practical calculations, it is clearly necessary to restrict the number of channel and distortion functions in eq. (4.17a). This means that, by our variational procedure, we project out of the complete Hilbert space a certain subspace in which the coupled equations are defined. As was pointed out in chapter 2, neither the general structure of the set of coupled equations nor the general quality of the pertinent solutions depends on how extended this subspace is. For instance, as has been mentioned several times already, the law of current conservation is always fulfilled exactly, independent of the size of this subspace. Therefore, because of these properties possessed by the solutions of the coupled equations, we can study general features of the unified theory proposed here even with simple examples, in which the Hilbert space is rather severely truncated.

As to the selection of terms to be included in the wave function for the considered system, one will normally have to resort to energetical considerations and arguments based on spatial overlap of the clusters. Thus, one neglects terms whose energy expectation values are far from the energy region under consideration and terms which do not have large spatial overlap with the main part of the wave function. In addition, it should be pointed out that because the Pauli principle serves to reduce the differences between different structures, one can usually obtain fairly accurate results even by adopting a relatively small number of terms in the wave function for the system.

The present formulation of a unified nuclear theory allows us to combine different types of model wave functions in a very natural way for the description of nuclear bound states and reactions. For instance, it is especially useful in heavier systems to describe in a nuclear reaction the compound-nucleus region by translationally invariant shell-model wave functions, including Hill-Wheeler type functions (see 14.7–14.9) and the region in which the reaction participants are separated by the corresponding channel wave functions [FE 70]. This means that for the functions φ_m in the trial function of eq. (4.17b), one chooses shell-model wave functions, rather than cluster-type wave functions such as that given by eq. (4.15). If one inserts now the wave function so chosen into eq. (4.19), then one obtains automatically equations which describe the coupling of the shell-model wave functions with the channel wave functions. In subsequent chapters, we shall discuss calculations of this kind to show the flexibility of the unified theory in this formulation.

So far, we have considered stationary solutions in our discussions. Now, we wish to add a few remarks about time-dependent processes, such as the time development of nuclear reactions. For the description of such processes, one makes for the wave function Ψ again an ansatz of the form (4.17b), but considers the variational functions and discrete variational amplitudes as time dependent. By substituting now the wave function Ψ into the projection equation (2.1), one obtains a set of coupled integrodifferential and integral equations satisfied by the time-dependent variational functions and amplitudes. This set of coupled equations is very similar to eq. (4.19), except that the quantity E is replaced by the operator $-\frac{\hbar}{i}\frac{\partial}{\partial t}$.

We emphasize again that with the coupled equations derived in this section, one can in principle carry out accurate calculations for every kind of nuclear bound-state and reaction problems, because the wave functions can be systematically improved to yield successively better description of the system under consideration and the asymptotic boundary conditions in all the channels can be satisfied in a very natural manner. In addition, it is important to note that the unified theory formulated here can also be used to investigate general properties of the nucleus and nuclear processes. For instance, we can examine in the frame of this theory the microscopic foundation of the nuclear optical model, the characteristics of compound-nucleus resonances (Breit-Wigner formulae), and so on. We shall discuss these topics in Chapter 8 and subsequent chapters.

Finally, we wish to mention that the general formulation of the unified nuclear theory presented here does not depend on the specific form of the nucleon-nucleon potential. Therefore, projection equations of the general form (4.19) should also be applicable in the description of many-particle reactions and bound states in other fields of physics, such as atomic physics, solid-state physics, and so on. The distinction in the application of the unified theory to different fields lies in the choice of the basis wave-function set into which the trial function of eq. (4.17a) is best expanded. As should be quite clear, the most appropriate choice for this basis set will depend very strongly on the special properties of the interaction potentials which govern the behaviour of the systems in these different fields of physics.

5. Bound-State Calculations

5.1. General Remarks

In this chapter, we apply the unified nuclear theory formulated in chapter 4 to study the properties of bound and quasibound states in specific nuclear systems. Our purpose is to show that with a phenomenological nucleon-nucleon potential, we can obtain a satisfactory description of not only the energy spectra in the low-excitation region but also the electromagnetic properties of these systems.

In section 5.2, we shall present two integration techniques which have been employed for the evaluation of matrix elements required for nuclear bound-state and reaction calculations. One of these techniques, the cluster-coordinate technique, has been extensively described in the literature; hence, we shall discuss it only very briefly here. The other technique, the generator-coordinate technique, has been utilized for practical calculations only quite recently and will therefore be described in more detail. Specific examples of bound-state calculations are given in sections 5.3, 5.4, and 5.5 for the nuclei ^6Li, ^{12}C, and ^7Be, respectively. The discussion of the ^6Li calculation will be a rather extensive one; in particular, we shall discuss the types of approximation which have been adopted to alleviate computational difficulties arising from the presence of the Jastrow anti-correlation factor in the trial wave function. Finally, in section 5.6, concluding remarks will be made.

5.2. Calculation of Matrix Elements

In the bound-state and reaction calculations, we are concerned with calculating matrix elements of the type

$$\langle \mathcal{A}\, \bar{\psi}_f | O | \mathcal{A}\, \bar{\psi}_i \rangle$$

where $\bar{\psi}_i$ and $\bar{\psi}_f$ are unantisymmetrized intial- and final-state wave functions, respectively, and O is an operator symmetrical in all the nucleon coordinates. As a first obvious simplification, we can write, using the Hermitian property of the antisymmetrization operator $\mathcal{A}$ and the relation given by eq. (4.20),

$$\langle \mathcal{A}\, \bar{\psi}_f | O | \mathcal{A}\, \bar{\psi}_i \rangle = A!\, \langle \mathcal{A}\, \bar{\psi}_f | O | \bar{\psi}_i \rangle = A!\, \langle \bar{\psi}_f | O | \mathcal{A}\, \bar{\psi}_i \rangle . \tag{5.1}$$

That is, for a symmetric operator O, it is necessary to carry out the antisymmetrization only on the initial-state wave function or on the final-state wave function, but not both. Alternatively, one can of course use antisymmetrized wave functions on both sides of the Dirac brackets, but take only one term from the symmetric operator O. For example, if O is represented by a sum of two-body operators,

$$O = \sum_{i<j=1}^{A} O_{ij}, \tag{5.2}$$

then one can also write

$$\langle \mathcal{A}\, \bar{\psi}_f | O | \mathcal{A}\, \bar{\psi}_i \rangle = \frac{A(A-1)}{2} \langle \mathcal{A}\, \bar{\psi}_f | O_{12} | \mathcal{A}\, \bar{\psi}_i \rangle . \tag{5.3}$$

Finally, we should mention that there is yet a third way which has occasionally been used [KR 69]. In this way, one calculates a matrix element of the form $\langle \bar{\psi}_f | O | \mathcal{A}\, \bar{\psi}_i \rangle$ by applying part of the permutations which $\mathcal{A}$ comprises to the initial-state wave function $\bar{\psi}_i$ and the other part to the final-state wave function $\bar{\psi}_f$.

Examples of matrix elements which occur in bound-state and reaction calculations are:

(i) Expectation value of the potential energy:

$$O = V = \sum_{i<j} V_{ij}; \quad \bar{\psi}_i = \bar{\psi}_f. \tag{5.4}$$

(ii) Expectation value of the kinetic energy in the center-of-mass system:

$$O = T = \sum_i \frac{1}{2M} p_i^2 - T_C = \sum_i \frac{1}{2M} p_i^2 - \frac{1}{2AM}\left(\sum_i p_i\right)^2$$
$$= \frac{1}{2AM} \sum_{i<j} (p_i - p_j)^2; \quad \bar{\psi}_i = \bar{\psi}_f. \tag{5.5}$$

We should mention that the expression of T as a sum of two-particle operators is often useful for explicit calculations.

(iii) Square of the normalization constant:

$$O = 1; \quad \bar{\psi}_i = \bar{\psi}_f. \tag{5.6}$$

(iv) Proton matter-density at the position $\bar{\mathbf{r}}'$:

$$O = \sum_i \delta(\bar{\mathbf{r}}_i - \bar{\mathbf{r}}') \frac{1}{2}(1 - \tau_{i3}); \quad \bar{\psi}_i = \bar{\psi}_f. \tag{5.7}$$

In eq. (5.7), the operator $\frac{1}{2}(1 - \tau_{i3})$ has an eigenvalue + 1 for a proton state and 0 for a neutron state. Also, it should be noted that $\bar{\mathbf{r}}_i$ and $\bar{\mathbf{r}}'$ are measured from the center of mass of the nucleus.

(v) Electric quadrupole transition matrix elements:

$$O = Q_{2\mu} = \sum_i \bar{r}_i^2 Y_{2\mu}^*(\bar{\theta}_i, \bar{\varphi}_i) \frac{1}{2}(1 - \tau_{i3}); \quad \bar{\psi}_i \neq \bar{\psi}_f. \tag{5.8}$$

In all these examples, the operators involved are translationally invariant and, hence, the total center-of-mass motion is irrelevant.

To illustrate the evaluation of the matrix elements, we shall use the nucleus ^{8}Be as an example. As was mentioned in section 5.1, two integration techniques have been used to evaluate these matrix elements and these will be discussed separately in subsections 5.2a and 5.2b.

5.2a. Evaluation of Matrix Elements by the Cluster-Coordinate Technique-Example of ^{8}Be

The evaluation of matrix elements by the cluster-coordinate technique has been thoroughly discussed by other authors (see, for example, ref. [CH 73]); hence, only a very brief description of this technique will be given here. It suffices to say that, in this technique, one introduces as independent spatial variables a set of internal and relative coordinates of the clusters. By choosing functions of Gaussian dependence for both the spatial part of the wave function and the spatial part of the nucleon-nucleon potential, multi-dimensional integrals involved can then be evaluated analytically. Because of the necessity to use a totally antisymmetrized wave function, the calculation of all the matrix elements is in general rather lengthy and tedious, but the procedure is straightforward and easy to understand.

The ^{8}Be wave function is assumed as

$$\begin{aligned}\psi(\tilde{r}_1,\ldots,\tilde{r}_8) &= A\,\bar{\psi}(\tilde{r}_1,\ldots,\tilde{r}_8)\\ &= A\,\{\phi_0(\bar{\mathbf{r}}_1,\ldots,\bar{\mathbf{r}}_4)\,\phi_0(\bar{\mathbf{r}}_5,\ldots,\bar{\mathbf{r}}_8)\,\chi(\mathbf{R})\,\xi(s,t)\,Z(\mathbf{R}_C)\},\end{aligned} \tag{5.9}$$

where

$$\phi_0(\bar{\mathbf{r}}_1,\ldots,\bar{\mathbf{r}}_4) = \exp\left[-\frac{a}{2}(\bar{\mathbf{r}}_1^2+\bar{\mathbf{r}}_2^2+\bar{\mathbf{r}}_3^2+\bar{\mathbf{r}}_4^2)\right], \tag{5.9a}$$

$$\chi(\mathbf{R}) = g_n(R)\,Y_{lm}(\theta,\varphi), \tag{5.9b}$$

$$\xi(s,t) = \alpha_1\nu_1\alpha_2\pi_2\beta_3\nu_3\beta_4\pi_4\alpha_5\nu_5\alpha_6\pi_6\beta_7\nu_7\beta_8\pi_8, \tag{5.9c}$$

$$Z(\mathbf{R}_C) = \exp(-4\,a\,\mathbf{R}_C^2), \tag{5.9d}$$

with the cluster coordinates $\bar{\mathbf{r}}_1, \bar{\mathbf{r}}_2, \bar{\mathbf{r}}_3, \bar{\mathbf{r}}_5, \bar{\mathbf{r}}_6, \bar{\mathbf{r}}_7$, $\mathbf{R}$ and $\mathbf{R}_C$ defined by eq. (3.8). The coordinates $\bar{\mathbf{r}}_4$ and $\bar{\mathbf{r}}_8$ are not independent coordinates; they are related to the other $\bar{\mathbf{r}}_i$'s as indicated also in eq. (3.8). Furthermore, it should be noted that in eq. (5.9b), the relative orbital angular-momentum quantum number l can only take on even integral values, because α-particles are bosons.

To perform a definite calculation, let us take for O a potential-energy operator consisting of a sum of two-nucleon interaction potentials; that is,

$$O = V = \sum_{i<j} V_{ij}, \tag{5.10}$$

where

$$V_{ij} = -V_0\,\exp[-\kappa(\mathbf{r}_i-\mathbf{r}_j)^2]\,[w(1+P_{ij}^r)], \tag{5.11}$$

with P_{ij}^r being a Majorana space-exchange operator which exchanges the spatial coordinates of nucleons i and j.† Proceeding now with the calculation of the expectation value of the potential-energy operator, we take the matrix element in the form $\langle\bar{\psi}|V|A\,\bar{\psi}\rangle$. Because of the presence of the antisymmetrization operator, the number of terms in the matrix element is quite large. But we shall now show that by using the symmetry property of the wave function, the number of terms that must be separately evaluated can be considerably reduced.

Since, in our example, the operator O contains no spin or isospin exchange operators, it cannot change the spin-isospin function of the unantisymmetrized final-state wave function on the left side of the Dirac bracket. Therefore, the only terms in the antisymmetrized initial-state wave function which contribute to the expectation value of O are the terms which have the same spin and isospin configuration as the unantisymmetrized wave function in the final state; if they do not have the same configuration as the final state, then they are orthogonal to this state and their contribution to the matrix element of O vanishes.††

---

† For the case of antisymmetrized wave functions, it is sometimes useful to write the space-exchange operator as $P_{ij}^r = -P_{ij}^\sigma P_{ij}^\tau$, where P_{ij}^σ and P_{ij}^τ are the spin- and isospin-exchange operators, respectively.

†† If the operator O does contain spin (or isospin) exchange operators, then in the matrix element only those terms in the antisymmetrized initial state contribute which have the same spin and isospin configuration as $P_{ij}^\sigma\bar{\psi}_f$ (or $P_{ij}^\tau\bar{\psi}_f$).

In our example where $\bar{\psi}_i = \bar{\psi}_f = \bar{\psi}$, inspection of the spin-isospin part of the wave function shows that only a limited number of permutations have this spin-isospin configuration unaltered. Of the terms which contribute, there is first the term in which no particles are exchanged, called the direct or no-exchange term, and the term where the two clusters are totally exchanged, which we call the four-particle exchange term. But because our ^{8}Be wave function is symmetric in the coordinates of the two α clusters, we observe that these two terms yield equal contribution to the matrix element. Going on, there are four possible non-vanishing one-particle exchange terms, $\{1 \leftrightarrow 5\}, \{2 \leftrightarrow 6\}$, etc. The contributions from these terms are equal to each other and equal to those from three-particle exchange terms, of which there are also four. This is again due to the symmetry character of our ^{8}Be wave function. These exchange terms have a negative sign, because an odd number of pair exchanges are performed. Finally, for the two-particle exchange terms the sign of the permutation is positive. Examples of non-vanishing two-particle exchange terms are $\{1 \leftrightarrow 5; 2 \leftrightarrow 6\}, \{2 \leftrightarrow 6; 4 \leftrightarrow 8\}$, and so on. Altogether, there are six of these exchange terms which, due to the symmetry property of our ^{8}Be wave function, yield again equal contribution to the matrix element.

The result for the expectation value of V thus becomes

$$\langle V\rangle = \frac{1}{N^2}\, 8! \int \phi^*(\mathbf{r}_1, \mathbf{r}_2, \mathbf{r}_3, \mathbf{r}_4; \mathbf{r}_5, \mathbf{r}_6, \mathbf{r}_7, \mathbf{r}_8) \Big[\sum_{i<j} V_{ij}\Big] \times [2\phi(\mathbf{r}_1, \mathbf{r}_2, \mathbf{r}_3, \mathbf{r}_4; \mathbf{r}_5, \mathbf{r}_6, \mathbf{r}_7, \mathbf{r}_8) - 8\phi(\mathbf{r}_5, \mathbf{r}_2, \mathbf{r}_3, \mathbf{r}_4; \mathbf{r}_1, \mathbf{r}_6, \mathbf{r}_7, \mathbf{r}_8) + 6\phi(\mathbf{r}_5, \mathbf{r}_6, \mathbf{r}_3, \mathbf{r}_4; \mathbf{r}_1, \mathbf{r}_2, \mathbf{r}_7, \mathbf{r}_8)]\, d\tau, \tag{5.12}$$

where N^2 is a normalization factor and ϕ denotes the spatial part of $\bar{\psi}$. The volume element $d\tau$ is given by

$$d\tau = d\mathbf{r}_1\, d\mathbf{r}_2 \ldots d\mathbf{r}_8 = 2^{12}\, d\bar{\mathbf{r}}_1\, d\bar{\mathbf{r}}_2\, d\bar{\mathbf{r}}_3\, d\bar{\mathbf{r}}_5\, d\bar{\mathbf{r}}_6\, d\bar{\mathbf{r}}_7\, d\mathbf{R}\, d\mathbf{R}_C, \tag{5.12a}$$

with the factor 2^{12} being the Jacobian of transformation.

For the evaluation of $\langle V\rangle$, we introduce parameter coordinates $\mathbf{R}'$ and $\mathbf{R}''$ for the relative coordinate $\mathbf{R}$ in eq. (5.9). That is, we write the spatial wave functions ϕ and ϕ^* as

$$\phi = \int \phi_0(\bar{\mathbf{r}}_1, \ldots, \bar{\mathbf{r}}_4)\phi_0(\bar{\mathbf{r}}_5, \ldots, \bar{\mathbf{r}}_8)\, \delta(\mathbf{R} - \mathbf{R}')\, \chi(\mathbf{R}')\, Z(\mathbf{R}_C)\, d\mathbf{R}', \tag{5.13a}$$

and

$$\phi^* = \int \phi_0^*(\bar{\mathbf{r}}_1, \ldots, \bar{\mathbf{r}}_4)\phi_0^*(\bar{\mathbf{r}}_5, \ldots, \bar{\mathbf{r}}_8)\, \delta(\mathbf{R} - \mathbf{R}'')\, \chi^*(\mathbf{R}'')\, Z^*(\mathbf{R}_C)\, d\mathbf{R}''. \tag{5.13b}$$

It should be mentioned that while the introduction of the integral form for the spatial wave function does not avoid any of the labor required in performing the integration, it does make the calculation somewhat more transparent. In addition, as will be seen especially from our later considerations on nuclear rotational states, the introduction of suitably chosen parameter coordinates can also lead to alternative expressions for the matrix elements, giving considerable insight into the structure of these elements.

We substitute now eqs. (5.13a) and (5.13b) into eq. (5.12) and carry out the integration over all nucleon spatial coordinates, but not the parameter coordinates $\mathbf{R}'$ and $\mathbf{R}''$. Then, after lengthy but straightforward integrations, we obtain [PE 60]

$$\begin{aligned}\langle V\rangle = 2(8!)(-w V_0)\frac{1}{N^2}(24 P_1 + 12 P_2 - 48 P_3 - 12 P_4 + 12 P_5 \\ - 96 P_6 + 24 P_7 - 12 P_8 + 96 P_9),\end{aligned} \tag{5.14}$$

where

$$P_1 = \left(\frac{\pi^6}{a^6}\right)^{3/2}\left(\frac{\pi}{\kappa + \frac{a}{2}}\right)^{3/2}\int |g_n(R')|^2\,|Y_{lm}(\theta',\varphi')|^2\, d\mathbf{R}',$$

$$P_2 = 2^3\left(\frac{\pi^6}{3a^6}\right)^{3/2}\left(\frac{\pi}{\kappa + \frac{2}{3}a}\right)^{3/2}\int |g_n(R')|^2\, e^{-\frac{2a\kappa}{2a+3\kappa}R'^2}\,|Y_{lm}(\theta',\varphi')|^2\, d\mathbf{R}',$$

$$P_3 = 2^6\left(\frac{\pi^5}{6a^5}\right)^{3/2}\left(\frac{\pi}{\kappa + \frac{1}{2}a}\right)^{3/2}\int g_n^*(R'')\,g_n(R')$$

$$\times\, e^{-\frac{a}{3}(5\mathbf{R}''^2 - 8\mathbf{R}''\cdot\mathbf{R}' + 5\mathbf{R}'^2)}\, Y_{lm}^*(\theta'',\varphi'')\,Y_{lm}(\theta',\varphi')\, d\mathbf{R}''\, d\mathbf{R}',$$

$$P_4 = 2^3\left(\frac{\pi^5}{a^5}\right)^{3/2}\left(\frac{\pi}{\kappa + \frac{3}{4}a}\right)^{3/2}\int g_n^*(R'')\,g_n(R')$$

$$\times\, e^{-a\left(\frac{5a+8\kappa}{3a+4\kappa}\mathbf{R}''^2 - \frac{8a+8\kappa}{3a+4\kappa}\mathbf{R}''\cdot\mathbf{R}' + \frac{5a+8\kappa}{3a+4\kappa}\mathbf{R}'^2\right)}\, Y_{lm}^*(\theta'',\varphi'')\,Y_{lm}(\theta',\varphi')\, d\mathbf{R}''\, d\mathbf{R}'$$

$$P_5 = 2^6\left(\frac{\pi^6}{3a^6}\right)^{3/2}\int g_n^*(R'')\,g_n(R')$$

$$\times\, e^{-(\frac{5}{3}a+4\kappa)\mathbf{R}''^2 + (\frac{8}{3}a+8\kappa)\mathbf{R}''\cdot\mathbf{R}' - (\frac{5}{3}a+4\kappa)\mathbf{R}'^2}\, Y_{lm}^*(\theta'',\varphi'')\,Y_{lm}(\theta',\varphi')\, d\mathbf{R}''\, d\mathbf{R}'$$

$$P_6 = 2^3\left(\frac{\pi^5}{a^5}\right)^{3/2}\left(\frac{\pi}{\kappa + \frac{3}{4}a}\right)^{3/2}\int g_n^*(R'')\,g_n(R')$$

$$\times\,\frac{1}{2}\Big[e^{-a\left(\frac{5a+8\kappa}{3a+4\kappa}\mathbf{R}''^2 - \frac{8a+16\kappa}{3a+4\kappa}\mathbf{R}''\cdot\mathbf{R}' + \frac{5a+12\kappa}{3a+4\kappa}\mathbf{R}'^2\right)}$$

$$+\, e^{-a\left(\frac{5a+12\kappa}{3a+4\kappa}\mathbf{R}''^2 - \frac{8a+16\kappa}{3a+4\kappa}\mathbf{R}''\cdot\mathbf{R}' + \frac{5a+8\kappa}{3a+4\kappa}\mathbf{R}'^2\right)}\Big]\, Y_{lm}^*(\theta'',\varphi'')\,Y_{lm}(\theta',\varphi')\, d\mathbf{R}''$$

$$P_7 = 2^3\left(\frac{\pi^5}{2a^5}\right)^{3/2}\left(\frac{\pi}{\kappa + \frac{1}{2}a}\right)^{3/2}\int g_n^*(R'')\,g_n(R')$$

$$\times\, e^{-a(\mathbf{R}''^2 + \mathbf{R}'^2)}\, Y_{lm}^*(\theta'',\varphi'')\,Y_{lm}(\theta',\varphi')\, d\mathbf{R}''\, d\mathbf{R}',$$

$$P_8 = 2^3 \left(\frac{\pi^5}{a^5}\right)^{3/2} \left(\frac{\pi}{\kappa + a}\right)^{3/2} \int g_n^* (R'') \, g_n (R')$$
$$\times \, e^{-a \left(\frac{a + 2\kappa}{a + \kappa} \mathbf{R}''^2 - \frac{2\kappa}{a + \kappa} \mathbf{R}'' \cdot \mathbf{R}' + \frac{a + 2\kappa}{a + \kappa} \mathbf{R}'^2\right)} Y_{lm}^* (\theta'', \varphi'') \, Y_{lm} (\theta', \varphi') \, d\mathbf{R}'' \, d\mathbf{R}',$$

$$P_9 = 2^6 \left(\frac{\pi^5}{6a^5}\right)^{3/2} \left(\frac{\pi}{\kappa + \frac{2}{3} a}\right)^{3/2} \int g_n^* (R'') \, g_n (R')$$
$$\times \, \frac{1}{2} \left[e^{-a \left(\frac{2a + 5\kappa}{2a + 3\kappa} \mathbf{R}''^2 + \mathbf{R}'^2\right)} + e^{-a \left(\mathbf{R}''^2 + \frac{2a + 5\kappa}{2a + 3\kappa} \mathbf{R}'^2\right)} \right] Y_{lm}^* (\theta'', \varphi'') \, Y_{lm} (\theta', \varphi') \, d\mathbf{R}'' \, d\mathbf{R}' \tag{5.15}$$

and N^2 is a normalization factor, given by

$$N^2 = 2 (8!) (A_0 - 4A_1 + 3A_2), \tag{5.16}$$

with

$$A_0 = \left(\frac{\pi^6}{a^6}\right)^{3/2} \left(\frac{2\pi}{a}\right)^{3/2} \int |g_n (R')|^2 \, |Y_{lm} (\theta', \varphi')|^2 \, d\mathbf{R}',$$

$$A_1 = 2^6 \left(\frac{\pi^5}{6a^5}\right)^{3/2} \left(\frac{2\pi}{a}\right)^{3/2} \int g_n^* (R'') \, g_n (R')$$
$$\times \, e^{-\frac{a}{3} (5\mathbf{R}''^2 - 8\mathbf{R}'' \cdot \mathbf{R}' + 5\mathbf{R}'^2)} Y_{lm}^* (\theta'', \varphi'') \, Y_{lm} (\theta', \varphi') \, d\mathbf{R}'' \, d\mathbf{R}',$$

$$A_2 = 2^3 \left(\frac{\pi^5}{2a^5}\right)^{3/2} \left(\frac{2\pi}{a}\right)^{3/2} \int g_n^* (R'') \, g_n (R')$$
$$\times \, e^{-a (\mathbf{R}''^2 + \mathbf{R}'^2)} Y_{lm}^* (\theta'', \varphi'') \, Y_{lm} (\theta', \varphi') \, d\mathbf{R}'' \, d\mathbf{R}'. \tag{5.17}$$

If we now assume

$$g_n (R) = R^n \, e^{-bR^2}, \tag{5.18}$$

then the integrals in eqs. (5.15) and (5.17) can be easily evaluated by using the following formulae:

$$\int R'^{2n} \, e^{-p\mathbf{R}'^2} \, |Y_{lm} (\theta', \varphi')|^2 \, d\mathbf{R}' = \frac{\sqrt{\pi}}{2^{n+2}} \, p^{-\frac{2n+3}{2}} \, (2n+1)!!,$$

$$\int R''^n R'^n \, e^{-p\mathbf{R}''^2 + q\mathbf{R}'' \cdot \mathbf{R}' - s\mathbf{R}'^2} \, Y_{lm}^* (\theta'', \varphi'') \, Y_{lm} (\theta', \varphi') \, d\mathbf{R}'' \, d\mathbf{R}'$$
$$= \frac{\pi^2 \, [(n + l + 1)!!]^2}{2^{n + l + 2} \, (2l + 1)!!} \, q^l \, (ps)^{-\frac{n + l + 3}{2}} \left(1 - \frac{q^2}{4ps}\right)^{-\frac{2n + 3}{2}}$$
$$\times \, {}_2F_1\left(\frac{-n + l}{2}, \frac{-n + l}{2}; l + \frac{3}{2}; \frac{q^2}{4ps}\right),$$

$$\int R''^n R'^n e^{-pR''^2 - sR'^2} Y^*_{lm}(\theta'',\varphi'') Y_{lm}(\theta',\varphi')\, d\mathbf{R}''\, d\mathbf{R}'$$

$$= (ps)^{-\frac{n+3}{2}} \frac{\pi^2}{2^{n+2}} [(n+1)!!]^2\, \delta_{l0}, \tag{5.19}$$

with ${}_2F_1$ being a hypergeometric function. In deriving eq. (5.19), we have used the expansion [MA 54]

$$e^{-\lambda \mathbf{R}''\cdot\mathbf{R}'} = 4\pi \sum_{l=0}^{\infty} \sum_{m=-l}^{l} (-1)^l \left(\frac{\pi}{2\lambda R'' R'}\right)^{\frac{1}{2}} I_{l+\frac{1}{2}}(\lambda R'' R')\, Y^*_{lm}(\theta',\varphi')\, Y_{lm}(\theta'',\varphi'') \tag{5.20}$$

where $I_{l+\frac{1}{2}}(\lambda R'' R')$ are modified Bessel functions which are related to Bessel functions of half integral order with imaginary argument $i\lambda R'' R'$.

The expectation value of the kinetic-energy operator can be calculated in a similar manner. First, one defines a quantity

$$N'^2 = 8!\,\langle \bar{\psi}' | A\, \bar{\psi} \rangle, \tag{5.21}$$

which is similar to the normalization factor N^2, but with $\bar{\psi}'$ containing internal width parameter a' and relative-motion width parameter b' instead of a and b as in $\bar{\psi}$. Then, one can obtain the kinetic-energy expectation value mainly by differentiating N'^2 with respect to a' and b' and performing some other relatively simple operations. As in the potential-energy calculation, the procedure is straightforward but rather tedious, and we refer to refs. [PE 60, PE 60a] for details.

5.2b. Evaluation of Matrix Elements by the Generator-Coordinate Technique – Example of ^{8}Be

The evaluation of matrix elements by the cluster-coordinate technique is usually a rather tedious procedure. The main reason is that, with the internal and relative coordinates of the clusters chosen as independent spatial variables, the Gaussian functions will have exponents which contain cross terms $\bar{\mathbf{r}}_i \cdot \bar{\mathbf{r}}_j$. For example, in the α-cluster case, the exponent

$$\begin{aligned} -\frac{a}{2}(\bar{r}_1^2 + \bar{r}_2^2 + \bar{r}_3^2 + \bar{r}_4^2) &= -\frac{a}{2}[\bar{r}_1^2 + \bar{r}_2^2 + \bar{r}_3^2 + (-\bar{\mathbf{r}}_1 - \bar{\mathbf{r}}_2 - \bar{\mathbf{r}}_3)^2] \\ &= -\frac{a}{2}(2\bar{r}_1^2 + 2\bar{r}_2^2 + 2\bar{r}_3^2 + 2\bar{\mathbf{r}}_1\cdot\bar{\mathbf{r}}_2 + 2\bar{\mathbf{r}}_1\cdot\bar{\mathbf{r}}_3 + 2\bar{\mathbf{r}}_2\cdot\bar{\mathbf{r}}_3) \end{aligned} \tag{5.22}$$

contains such cross terms, because the coordinate $\bar{\mathbf{r}}_4$ is not independent of the coordinates $\bar{\mathbf{r}}_1$, $\bar{\mathbf{r}}_2$, and $\bar{\mathbf{r}}_3$. Therefore, to compute the many-dimensional integrals which involve integrands of Gaussian functions multiplied by polynomials of spatial coordinates, one must first diagonalize the quadratic terms in the exponents by applying linear coordinate transformations. For systems containing a relatively small number of nucleons ($A \lesssim 8$), this is not too difficult. However, when one deals with systems which contain a rather large number of nucleons, this diagonalization procedure can become very tedious due to the antisymmetrization of the wave function, because for every permutation of nucleons in different clusters, one has to introduce a different coordinate transformation.

We shall therefore describe in this subsection another technique, the generator-coordinate technique [SU 72], which is especially useful when the number of nucleons contained in the system is rather large. In this technique, the essential idea is to express the wave function as a linear superposition of antisymmetrized products of single-particle wave functions or Slater determinants. Then, by employing well-developed techniques of dealing with product functions, one can usually carry out the analytical calculation of the matrix elements in a much simpler way than with the cluster-coordinate technique.

Let us start with a two-cluster function of the form

$$\psi = A\,\{\phi(1)\,\phi(2)\,\chi(\mathbf{R}_1 - \mathbf{R}_2)\,\xi(s, t)\}. \tag{5.23}$$

The cluster internal functions $\phi(1)$ and $\phi(2)$ are assumed to be products of functions, with each function describing the motion of a nucleon relative to the center of mass of the corresponding cluster. That is, we choose the internal functions $\phi(k)$ to have the form

$$\phi(k) = \prod_j \varphi_j(\mathbf{r}_j - \mathbf{R}_k), \tag{5.24}$$

with

$$\mathbf{R}_k = \frac{1}{n_k} \sum_j \mathbf{r}_j. \tag{5.25}$$

In eqs. (5.24) and (5.25), the index j goes from 1 to n_1 for $k = 1$ and $n_1 + 1$ to $n_1 + n_2$ for $k = 2$. The function φ_i is chosen such that the functions $\phi(1)$ and $\phi(2)$ describe properly the spatial behaviour of the clusters; it can be generated from an oscillator well, a Woods-Saxon well, or any other well which is appropriate.

Because of the presence of the cluster center-of-mass coordinates $\mathbf{R}_1$ and $\mathbf{R}_2$, the wave function ψ is not in the form of an antisymmetrized product of single-particle wave functions. However, by introducing an integral representation for ψ, we can show that the integrand can be represented by a sum of such products or, in other words, the integrand will contain no cross terms of the type $\mathbf{r}_i \cdot \mathbf{r}_j$. To show this, we rewrite ψ in the following form:

$$\begin{aligned}\psi = A \int &\{\phi(1, \mathbf{R}_1')\,\phi(2, \mathbf{R}_2')\,\delta(\mathbf{R}_1 - \mathbf{R}_1')\,\delta(\mathbf{R}_2 - \mathbf{R}_2')\,\xi(s, t)\}\\ &\times \chi(\mathbf{R}_1' - \mathbf{R}_2')\,d\mathbf{R}_1'\,d\mathbf{R}_2',\end{aligned} \tag{5.26}$$

where $\mathbf{R}_1'$ and $\mathbf{R}_2'$ are parameter coordinates on which, we repeat, the antisymmetrization operator A does not act. These parameter coordinates are included in the arguments of the internal wave functions to call attention to the fact that in $\phi(1, \mathbf{R}_1')$ and $\phi(2, \mathbf{R}_2')$, the parameter coordinates $\mathbf{R}_1'$ and $\mathbf{R}_2'$, instead of the cluster coordinates $\mathbf{R}_1$ and $\mathbf{R}_2$, are used (see eq. (5.24)); due to the presence of Dirac delta functions in eq. (5.26), this substitution is certainly allowed.

Let us consider now the term $\phi(k, \mathbf{R}_k') \, \delta(\mathbf{R}_k - \mathbf{R}_k')$ in eq. (5.26). If we represent $\delta(\mathbf{R}_k - \mathbf{R}_k')$ as

$$\delta(\mathbf{R}_k - \mathbf{R}_k') = \left(\frac{1}{2\pi}\right)^3 \int e^{i\mathbf{S}_k' \cdot (\mathbf{R}_k - \mathbf{R}_k')} \, d\mathbf{S}_k', \tag{5.27}$$

then

$$\begin{aligned} \phi(k, \mathbf{R}_k') \, \delta(\mathbf{R}_k - \mathbf{R}_k') &= \left(\frac{1}{2\pi}\right)^3 \int \phi(k, \mathbf{R}_k') \, e^{i\mathbf{S}_k' \cdot (\mathbf{R}_k - \mathbf{R}_k')} \, d\mathbf{S}_k' \\ &= \left(\frac{1}{2\pi}\right)^3 \int \left[\prod_j \varphi_j(\mathbf{r}_j - \mathbf{R}_k') \, e^{i\frac{1}{n_k}\mathbf{S}_k' \cdot \mathbf{r}_j}\right] e^{-i\mathbf{S}_k' \cdot \mathbf{R}_k'} \, d\mathbf{S}_k' \end{aligned} \tag{5.28}$$

represents an integral over a product of single-particle wave functions with respect to the nucleon coordinates $\mathbf{r}_j$. Using eq. (5.28), we obtain, therefore, for ψ the following expression:

$$\begin{aligned} \psi = \left(\frac{1}{2\pi}\right)^6 \int \mathcal{A} \Bigg\{ & \left[\prod_{j=1}^{n_1} \varphi_j(\mathbf{r}_j - \mathbf{R}_1') \, e^{i\frac{1}{n_1}\mathbf{S}_1' \cdot \mathbf{r}_j}\right] \\ & \times \left[\prod_{j=n_1+1}^{n_1+n_2} \varphi_j(\mathbf{r}_j - \mathbf{R}_2') \, e^{i\frac{1}{n_2}\mathbf{S}_2' \cdot \mathbf{r}_j}\right] \xi(s, t) \Bigg\} \\ & \times \chi(\mathbf{R}_1' - \mathbf{R}_2') \, e^{-i\mathbf{S}_1' \cdot \mathbf{R}_1' - i\mathbf{S}_2' \cdot \mathbf{R}_2'} \, d\mathbf{S}_1' \, d\mathbf{S}_2' \, d\mathbf{R}_1' \, d\mathbf{R}_2'. \end{aligned} \tag{5.29}$$

This shows that by introducing parameter coordinates $\mathbf{R}_1'$ and $\mathbf{R}_2'$ and generator coordinates $\mathbf{S}_1'$ and $\mathbf{S}_2'$, the wave function ψ can be represented as an integral over antisymmetrized products of single-particle wave functions or Slater determinants.

Even though our discussion above was based on a two-cluster wave function, it should be clear that it can be easily generalized to a wave function which describes the behaviour of a system composed of a larger number of clusters.

If the wave functions ψ_i and ψ_f are both represented in the form discussed above, then any matrix element $\langle \psi_f | O | \psi_i \rangle$ can be calculated by using well-known techniques involving Slater determinants. Consider an operator O which is composed of a sum of one- and two-particle operators, then the matrix element $\langle \psi_f | O | \psi_i \rangle$ becomes an integral over the generator coordinates $\mathbf{S}_1', \mathbf{S}_2', \ldots, \mathbf{S}_1'', \mathbf{S}_2'', \ldots$ and the parameter coordinates $\mathbf{R}_1', \mathbf{R}_2', \ldots, \mathbf{R}_1'', \mathbf{R}_2'', \ldots$, with its integrand consisting of a sum of products of one- and two-particle matrix elements of the form

$$\langle \hat{\varphi}_i | \hat{\varphi}_j \rangle, \quad \langle \hat{\varphi}_i | O_p | \hat{\varphi}_j \rangle, \quad \langle \hat{\varphi}_i \, \hat{\varphi}_j | O_{pq} | \hat{\varphi}_l \, \hat{\varphi}_m \rangle,$$

where the $\hat{\varphi}_i$'s are one-particle functions depending on the generator and parameter coordinates. The orthogonality relations between the $\hat{\varphi}_i$'s in the same cluster will usually reduce very drastically the number of terms which need to be evaluated.

We shall now consider in more detail the case where the cluster internal functions $\phi(k)$ in eq. (5.23) are oscillator cluster functions. This is an especially important case, since, as has been mentioned previously, the integrations can be carried out analytically. We should emphasize, however, that this assumption about the form of the cluster internal functions does not really impose any severe restriction on the flexibility required in treating the problem properly, because any cluster internal function can be well approximated by a linear superposition of oscillator cluster functions with different width parameters.

In this special case where oscillator cluster functions are adopted as cluster internal functions, we have then

$$\phi(k, \mathbf{R}_k') = \left[\prod_j h_j(\mathbf{r}_j - \mathbf{R}_k')\right] e^{-\frac{a_k}{2} \sum_j (\mathbf{r}_j - \mathbf{R}_k')^2}, \tag{5.30}$$

where the functions h_j are polynomials in single-particle coordinates $\mathbf{r}_j$.† With eq. (5.30), the integrand of the integral in eq. (5.28) becomes

$$\left[\prod_j \varphi_j\, e^{i\frac{1}{n_k}\mathbf{S}_k'\cdot\mathbf{r}_j}\right] e^{-i\mathbf{S}_k'\cdot\mathbf{R}_k'}$$

$$= \prod_j \left[h_j\, e^{i\frac{1}{n_k}\mathbf{S}_k'\cdot\mathbf{r}_j - i\frac{1}{n_k}\mathbf{S}_k'\cdot\mathbf{R}_k' - \frac{1}{2}a_k(\mathbf{r}_j - \mathbf{R}_k')^2}\right]$$

$$= \left\{\prod_j h_j\, e^{-\frac{1}{2}a_k\left[\mathbf{r}_j - (\mathbf{R}_k' + \frac{i}{n_k a_k}\mathbf{S}_k')\right]^2}\right\} e^{-\frac{1}{2}\frac{1}{n_k a_k}\mathbf{S}_k'^2}$$

$$= \left\{\prod_j h_j\, e^{-\frac{1}{2}a_k(\mathbf{r}_j - i\mathbf{Q}_k')^2}\right\} e^{\frac{1}{2}n_k a_k(\mathbf{R}_k' - i\mathbf{Q}_k')^2}, \tag{5.31}$$

where, for convenience, we have defined††

$$\mathbf{Q}_k' = \frac{1}{n_k a_k}\mathbf{S}_k' - i\mathbf{R}_k'. \tag{5.32}$$

By substituting eq. (5.31) into eq. (5.29), we obtain then the following expression for a two-cluster wave function ψ:

$$\psi = \left(\frac{n_1 a_1}{2\pi}\right)^3 \left(\frac{n_2 a_2}{2\pi}\right)^3 \int A\left\{\left[\prod_{j=1}^{n_1} h_j(\mathbf{r}_j - \mathbf{R}_1')\, e^{-\frac{1}{2}a_1(\mathbf{r}_j - i\mathbf{Q}_1')^2}\right]\right.$$

$$\left.\times \left[\prod_{j=n_1+1}^{n_1+n_2} h_j(\mathbf{r}_j - \mathbf{R}_2')\, e^{-\frac{1}{2}a_2(\mathbf{r}_j - i\mathbf{Q}_2')^2}\right]\xi(s, t)\right\} \tag{5.33}$$

$$\times\, \chi(\mathbf{R}_1' - \mathbf{R}_2')\, e^{\frac{1}{2}n_1 a_1(\mathbf{R}_1' - i\mathbf{Q}_1')^2 + \frac{1}{2}n_2 a_2(\mathbf{R}_2' - i\mathbf{Q}_2')^2}\, d\mathbf{Q}_1'\, d\mathbf{Q}_2'\, d\mathbf{R}_1'\, d\mathbf{R}_2'.$$

---

† It should be noted that the factor inside the square brackets can be replaced by $\prod_j h_j(\mathbf{r}_j)$, if the nucleons are in the lowest possible harmonic-oscillator shells. This is permissible, because the function $\phi(k, \mathbf{R}_k')$ is a part of a totally antisymmetrized wave function (for a further discussion, see section 7.3e).

†† The coordinates $\mathbf{Q}_k'$ will also be referred to as generator coordinates.

To demonstrate explicitly how the generator-coordinate technique works, we shall calculate the normalization factor N^2 in the ^{8}Be case. In this example where

$$\begin{aligned} n_1 &= n_2 = 4, \\ a_1 &= a_2 = a, \\ h_j &= 1, \end{aligned} \tag{5.34}$$

the wave function ψ of eq. (5.33) takes on the form

$$\begin{aligned} \psi = \left(\frac{2a}{\pi}\right)^6 \int A\left\{\prod_{j=1}^{4} e^{-\frac{1}{2}a(\mathbf{r}_j - i\mathbf{Q}_1')^2} \prod_{j=5}^{8} e^{-\frac{1}{2}a(\mathbf{r}_j - i\mathbf{Q}_2')^2} \xi(s,t)q\right\} \\ \times \chi(\mathbf{R}_1' - \mathbf{R}_2')\, e^{2a[(\mathbf{R}_1' - i\mathbf{Q}_1')^2 + (\mathbf{R}_2' - i\mathbf{Q}_2')^2]} \\ \times Z\left(\frac{\mathbf{R}_1' + \mathbf{R}_2'}{2}\right) d\mathbf{Q}_1'\, d\mathbf{Q}_2'\, d\mathbf{R}_1'\, d\mathbf{R}_2', \end{aligned} \tag{5.35}$$

where we have also included the center-of-mass wave function of eq. (5.9d). Because, in this particular case, the width parameters of the two clusters are the same, the wave function ψ can be simplified by defining

$$\begin{aligned} \mathbf{R}' &= \mathbf{R}_1' - \mathbf{R}_2', \quad \mathbf{R}_C' = \tfrac{1}{2}(\mathbf{R}_1' + \mathbf{R}_2'), \\ \mathbf{Q}' &= \mathbf{Q}_1' - \mathbf{Q}_2', \quad \mathbf{Q}_C' = \tfrac{1}{2}(\mathbf{Q}_1' + \mathbf{Q}_2'). \end{aligned} \tag{5.36}$$

Using these new coordinates, we obtain for ψ the following simplified form:

$$\begin{aligned} \psi = \left(\frac{2a}{\pi}\right)^6 \int A\left\{\prod_{j=1}^{4} e^{-\frac{1}{2}a(\mathbf{r}_j - i\mathbf{Q}_C' - \frac{1}{2}i\mathbf{Q}')^2} \prod_{j=5}^{8} e^{-\frac{1}{2}a(\mathbf{r}_j - i\mathbf{Q}_C' + \frac{1}{2}i\mathbf{Q}')^2} \xi(s,t)\right\} \\ \times \chi(\mathbf{R}')\, e^{a(\mathbf{R}' - i\mathbf{Q}')^2 - 4a\mathbf{Q}_C'^2}\, e^{-i8a\mathbf{Q}_C' \cdot \mathbf{R}_C'}\, d\mathbf{Q}'\, d\mathbf{R}'\, d\mathbf{Q}_C'\, d\mathbf{R}_C'. \end{aligned} \tag{5.37}$$

Carrying out first the integration over $\mathbf{R}_C'$ and then the integration over $\mathbf{Q}_C'$ yields

$$\begin{aligned} \psi = \left(\frac{a}{\pi}\right)^3 \int A\left\{\prod_{j=1}^{4} e^{-\frac{1}{2}a(\mathbf{r}_j - \frac{1}{2}i\mathbf{Q}')^2} \prod_{j=5}^{8} e^{-\frac{1}{2}a(\mathbf{r}_j + \frac{1}{2}i\mathbf{Q}')^2} \xi(s,t)\right\} \\ \times \chi(\mathbf{R}')\, e^{a(\mathbf{R}' - i\mathbf{Q}')^2}\, d\mathbf{Q}'\, d\mathbf{R}'. \end{aligned} \tag{5.38}$$

To obtain the normalization factor, we integrate over nucleon spatial coordinates and sum over nucleon spin and isospin coordinates; the result is

$$N^2 = 2(8!)(D_0 - 4D_1 + 3D_2), \tag{5.39}$$

where

$$\begin{aligned} D_0 = \left(\frac{a}{\pi}\right)^6 \int [W(\tfrac{1}{2}i\mathbf{Q}'', -\tfrac{1}{2}i\mathbf{Q}')]^4\, [W(-\tfrac{1}{2}i\mathbf{Q}'', \tfrac{1}{2}i\mathbf{Q}')]^4 \\ \times G(\mathbf{R}'', \mathbf{R}', \mathbf{Q}'', \mathbf{Q}')\, d\mathbf{R}''\, d\mathbf{R}'\, d\mathbf{Q}''\, d\mathbf{Q}', \end{aligned} \tag{5.39a}$$

$$D_1 = \left(\frac{a}{\pi}\right)^6 \int W(\tfrac{1}{2}\,iQ'', \tfrac{1}{2}\,iQ')\, W(-\tfrac{1}{2}\,iQ'', -\tfrac{1}{2}\,iQ')$$
$$\times [W(\tfrac{1}{2}\,iQ'', -\tfrac{1}{2}\,iQ')]^3\, [W(-\tfrac{1}{2}\,iQ'', \tfrac{1}{2}\,iQ')]^3 \tag{5.39b}$$
$$\times\, G(\mathbf{R}'', \mathbf{R}', \mathbf{Q}'', \mathbf{Q}')\, d\mathbf{R}''\, d\mathbf{R}'\, d\mathbf{Q}''\, d\mathbf{Q}',$$

$$D_2 = \left(\frac{a}{\pi}\right)^6 \int [W(\tfrac{1}{2}\,iQ'', \tfrac{1}{2}\,iQ')]^2\, [W(-\tfrac{1}{2}\,iQ'', -\tfrac{1}{2}\,iQ')]^2$$
$$\times [W(\tfrac{1}{2}\,iQ'', -\tfrac{1}{2}\,iQ')]^2\, [W(-\tfrac{1}{2}\,iQ'', \tfrac{1}{2}\,iQ')]^2 \tag{5.39c}$$
$$\times\, G(\mathbf{R}'', \mathbf{R}', \mathbf{Q}'', \mathbf{Q}')\, d\mathbf{R}'' d\mathbf{R}'\, d\mathbf{Q}''\, d\mathbf{Q}',$$

with

$$W(i\mathbf{K}, i\mathbf{L}) = \int e^{-\frac{a}{2}[(\mathbf{r}+i\mathbf{K})^2 + (\mathbf{r}+i\mathbf{L})^2]}\, d\mathbf{r} = \left(\frac{\pi}{a}\right)^{3/2} e^{\frac{a}{4}(\mathbf{K}-\mathbf{L})^2} \tag{5.40}$$

and

$$G(\mathbf{R}'', \mathbf{R}', \mathbf{Q}'', \mathbf{Q}') = e^{a[(\mathbf{R}''+i\mathbf{Q}'')^2 + (\mathbf{R}'-i\mathbf{Q}')^2]}\, \chi^*(\mathbf{R}'')\, \chi(\mathbf{R}'). \tag{5.41}$$

Next, we perform the integration over $\mathbf{Q}''$ and $\mathbf{Q}'$. For this integration, we shall illustrate with the one-particle exchange term D_1 only. It proceeds as follows:

$$D_1 = \left(\frac{a}{\pi}\right)^6 \left(\frac{\pi}{a}\right)^{12} \int e^{2\frac{a}{4}[\frac{1}{2}(\mathbf{Q}''-\mathbf{Q}')]^2}\, e^{6\,\frac{a}{4}[\frac{1}{2}(\mathbf{Q}''+\mathbf{Q}')]^2}$$
$$\times\, e^{a[\mathbf{R}''^2 + \mathbf{R}'^2 - \mathbf{Q}''^2 - \mathbf{Q}'^2 + 2i(\mathbf{R}''\cdot\mathbf{Q}'' - \mathbf{R}'\cdot\mathbf{Q}')]}\, \chi^*(\mathbf{R}'')\, \chi(\mathbf{R}')\, d\mathbf{R}''\, d\mathbf{R}'\, d\mathbf{Q}''\, d\mathbf{Q}'$$
$$= \left(\frac{\pi}{a}\right)^6 \int e^{-a[\frac{1}{2}\mathbf{Q}''^2 + \frac{1}{2}\mathbf{Q}'^2 - \frac{1}{2}\mathbf{Q}''\cdot\mathbf{Q}' - 2i(\mathbf{R}''\cdot\mathbf{Q}'' - \mathbf{R}'\cdot\mathbf{Q}')]} \tag{5.42}$$
$$\times\, e^{a(\mathbf{R}''^2 + \mathbf{R}'^2)}\, \chi^*(\mathbf{R}'')\, \chi(\mathbf{R}')\, d\mathbf{R}''\, d\mathbf{R}'\, d\mathbf{Q}''\, d\mathbf{Q}'$$
$$= \frac{8^2}{3^{3/2}} \left(\frac{\pi}{a}\right)^9 \int e^{-\frac{a}{3}(5\mathbf{R}''^2 - 8\mathbf{R}''\cdot\mathbf{R}' + 5\mathbf{R}'^2)}\, \chi^*(\mathbf{R}'')\, \chi(\mathbf{R}')\, d\mathbf{R}''\, d\mathbf{R}',$$

which is the same as A_1 of eq. (5.17). Similarly, we can show that D_0 and D_2 are equal to A_0 and A_2, respectively, as it must be.

In a system with larger clusters which contain more than just 1s nucleons or in a system where the clusters have different width parameters, the calculation with the generator-coordinate technique will be fairly complicated, although still quite tractable in many instances.

In the generator-coordinate technique, one has to eventually carry out the integration over the generator coordinates $\mathbf{S}_1', \mathbf{S}_2' \ldots, \mathbf{S}_1'', \mathbf{S}_2'', \ldots$ and the parameter coordinates $\mathbf{R}_1', \mathbf{R}_2', \ldots \mathbf{R}_1'', \mathbf{R}_2'' \ldots$. This means that the computational effort required for the diagonalization of the quadratic terms in the cluster internal coordinates in the cluster-coordinate technique is essentially replaced by that required for the diagonalization of the quadratic terms in the generator and parameter coordinates. Therefore, in comparison with the cluster-coordinate technique, the generator-coordinate technique of computing the matrix elements is particularly useful when the number of clusters is relatively small and the number of nucleons of which the clusters are composed is relatively large.

Finally, we should mention that the generator-coordinate technique discussed here is closely related to the method of Brink and others [BR 66, BR 68, DE 72, GI 73, HO 70, HO 72, TA 72, YU 72, ZA 71]. Therefore, the interested reader is also referred to these references for further details.

5.3. Ground and Low Excited States of ^{6}Li

5.3a. Introduction

In our first example, we study the properties of the ground and low excited states of ^{6}Li. As has been mentioned earlier, ^{6}Li is a particularly interesting case, because clustering effects are very important and, hence, one will be able to see in an especially clear way how systematic improvements in the trial wave function can refine the description of the behaviour of this nucleus.

Analyses of experimental data [BE 69a, BU 72, LI 71, SI 70, SU 67] showed that ^{6}Li has a larger charge radius (2.54 fm) than even the heavier nuclei ^{7}Li (2.39 fm) and ^{12}C (2.46 fm). Its charge distribution has a relatively long tail which cannot be described, for instance, by using single-particle wave functions of the lowest configuration in an oscillator potential. This is in contrast to the fact that for many other p-shell nuclei, the experimental results can be explained fairly well using wave functions of the configuration $(1s)^4 (1p)^{A-4}$ in such a potential well [EL 61].

The relatively large radius of ^{6}Li seems to be a typical clustering effect. For the ground state of this nucleus, we have seen from our earlier considerations that the predominant cluster structure is the $d + \alpha$ structure, with other structures such as $t + {}^3$He playing only a minor role.† In the $d + \alpha$ structure, the two clusters are bound together by only 1.47 MeV, which means that they are on the average rather far apart and behave more or less like free particles. For the deuteron cluster, we can thus expect that it will have a relatively long tail, just as a free deuteron does. Qualitatively, we can therefore understand the large diffuseness and charge radius of ^{6}Li as arising from the long tail of the deuteron cluster and the fact that the clusters are rather widely separated.

Another experimental observation which indicates strong clustering in ^{6}Li is the large electric quadrupole transition probability from the ground state ($J^\pi = 1^+, T = 0$) to the first excited state†† ($J^\pi = 3^+, T = 0$). The experimental value for the reduced transition probability B(E 2) obtained by *Eigenbrod* [EI 69] from electron inelastic scattering experiment is 25.6 fm^4, which is quite large compared to the value of about 3 fm^4 for single-particle transitions [BO 67a].

---

† We shall discuss later the question why the $t + {}^3$He cluster structure, which coincides with the $d + \alpha$ cluster structure in the oscillator model (see section 3.5c), plays a minor role in the ground and low excited states of ^{6}Li.

†† The first excited state of ^{6}Li is a quasibound state which lies below the Coulomb barrier. If the Coulomb interaction between protons were turned off, this state would become bound.

To explain these experimental results and the energy spectrum of ^{6}Li, one must use rather flexible generalized cluster wave functions.† In this section, we shall describe three ^{6}Li calculations which have been performed for this purpose. In the first calculation (section 5.3b), a rather complicated nucleon-nucleon potential containing a hard core was employed. The computational difficulty arising from the presence of a Jastrow anti-correlation factor was overcome by the use of a Monte-Carlo integration technique. The result obtained was fairly reasonable, but the amount of computing time needed was quite substantial even on a modern fast computer and it is apparent that a similar calculation will not be feasible for a system containing a larger number of nucleons. Therefore, in sections 5.3c and 5.3d, we describe two other calculations which were designed to circumvent the difficulty associated with an accurate treatment of the Jastrow factor. As will be seen, these two calculations do simplify the computation to a large extent, but still yield results which are quite satisfactory.

5.3b. Calculation With a Nucleon-Nucleon Potential Containing a Hard Core – Accurate Treatment of the Jastrow Factor

In this calculation [HE 66, SC 63a], a variational procedure was used to compute an upper bound for the ground-state energy of ^{6}Li. As is well known, this upper bound is given by the expectation value of the Hamiltonian

$$\langle H\rangle = \frac{\langle \psi | H | \psi\rangle}{\langle \psi | \psi\rangle}. \tag{5.43}$$

By choosing a trial wave function ψ which contains a fairly large number of linear and nonlinear variational parameters, one can in general expect to obtain a good value for the upper bound by minimizing $\langle H\rangle$ with respect to these parameters.††

The nucleon-nucleon potential used is purely central and has the form

$$\begin{aligned} V_{ij} &= \infty \quad , & (r < r_C) \\ &= \left(\frac{1+P^{\sigma}_{ij}}{2} V_t + \frac{1-P^{\sigma}_{ij}}{2} V_s\right)\left(\frac{1+P^{r}_{ij}}{2}\right) + \frac{e^2}{r_{ij}} \frac{1-\tau_{i3}}{2}\frac{1-\tau_{j3}}{2}, & (r \geqslant r_C) \end{aligned} \tag{5.44}$$

---

† It is in just such cases that the flexibility of the cluster description to incorporate one's physical knowledge into the formulation shows its usefulness. With an arbitrary complete set of wave functions, e. g., oscillator shell-model wave functions, the same collective effect can be achieved in principle by mixing in the proper amount of many higher excited states. However, because of the large number of variational amplitudes present, it would virtually be impossible to visualize the collective motion described in such a basis.

†† For excited states having the same quantum-number set as the ground state, one must vary the nonlinear parameters subject to the condition that the resultant wave function be orthogonal to the wave functions of all lower states. This is in contrast to the case where only linear variational parameters are used. In the latter case, such a precautionary procedure is not necessary, because the orthogonality condition between the resultant wave functions is automatically satisfied (see chapter 2). This is the main reason why, in most calculations, one prefers to use only linear variational parameters and linear variational functions.

where P_{ij}^{σ} and P_{ij}^{r} are the spin- and space-exchange operators, respectively. The quantities V_t and V_s are the triplet and singlet interactions in the even orbital angular-momentum states, given by

$$\begin{aligned} V_t(r_{ij}) &= -V_{0t} \exp[-\kappa_t(r_{ij}-r_C)], \\ V_s(r_{ij}) &= -V_{0s} \exp[-\kappa_s(r_{ij}-r_C)]. \end{aligned} \tag{5.45}$$

The parameters $r_C, V_{0t}, \kappa_t, V_{0s}, \kappa_s$ are adjusted to yield correct values for the two-nucleon effective-range parameters; they are found to be [SC 63]

$$\begin{aligned} r_C &= 0.35 \text{ fm}, & & \\ V_{0t} &= 434.0 \text{ MeV}, & V_{0s} &= 216.0 \text{ MeV}, \\ \kappa_t &= 2.4 \text{ fm}^{-1}, & \kappa_s &= 1.97 \text{ fm}^{-1}. \end{aligned} \tag{5.46}$$

In the odd orbital angular-momentum states, the interaction is taken as zero except for a hard core of radius r_C; this is reasonable, since it is known that the experimental n + p differential scattering cross section is nearly symmetrical with respect to 90°.

The potential of eq. (5.44) does not contain any spin-orbit or tensor component. However, it does yield reasonable values for the rms radii and the binding energies of the triton and the alpha particle [TA 65]. Therefore, it could be considered as an effective interaction which is suitable for calculations in light nuclear systems.

In order to obtain reliable results from the variational calculation, a rather flexible trial function was assumed. It has the form

$$\psi = A\left\{\phi_\alpha(\mathbf{r}_1,\mathbf{r}_2,\mathbf{r}_3,\mathbf{r}_4)\,\phi_d(\mathbf{r}_5,\mathbf{r}_6)\,\chi(\mathbf{R})\,\xi(s,t)\prod_{\substack{n=1,2,3,4\\ m=5,6}} f_2(r_{nm})\right\}, \tag{5.47}$$

where $\xi(s,t)$ is a spin-isospin function chosen to give S = 1 and T = 0. The last factor inside the braces is a cut-off factor which prevents nucleons in the two clusters from approaching each other too closely.

The α-cluster spatial function is taken as [SC 63]

$$\phi_\alpha = \left[\prod_{i<k=1}^{4} f_1(r_{ik})\right]\left[e^{-\frac{a_1}{8}\sum_{i<k=1}^{4} r_{ik}^2} + a_2\, e^{-\frac{a_3}{8}\sum_{i<k=1}^{4} r_{ik}^2}\right]. \tag{5.48}$$

The deuteron spatial function ϕ_d is equal to zero for $r < r_C$. Outside, it is generated by the equation

$$-\frac{\hbar^2}{M}\frac{d^2}{dr^2}(r\,\phi_d) + [V_t(r) - b_2](r\,\phi_d) = 0 \tag{5.49}$$

between $r = r_C$ and $r = b_1$. For $r > b_1$, the function

$$\phi_d(r) = \frac{B_1}{r}\left(e^{-b_3 r} - B_2\, e^{-B_3 r}\right) \tag{5.50}$$

is used. In eq. (5.50), the constants B_1, B_2, and B_3 are determined by the condition that ϕ_d and its first and second derivatives are continuous at $r = b_1$. With this trial wave function for the deuteron, the tail can be varied without any variation in the interior region of the deuteron cluster. This seems to be a desirable feature, since the interaction of the clusters might conceivably alter the tail of the deuteron-cluster function without having much effect in the interior region where the neutron-proton interaction dominates.

The long-range part of the relative-motion function is chosen as

$$\chi(\mathbf{R}) = R^2 \left(e^{-c_1 R^2} + c_2 e^{-c_3 R^2}\right) Y_{00}(\theta, \varphi), \tag{5.51}$$

where $\mathbf{R} = \mathbf{R}_d - \mathbf{R}_\alpha$ is the vector separation distance of the two clusters. The inclusion of a second Gaussian function is necessary, since the binding between the clusters is rather weak and, therefore, a single Gaussian function is not likely to describe the behaviour of the relative motion properly.

The cut-off function f_i with $i = 1$ or 2 is equal to zero for $r < r_C$. For $r > r_C$, it is generated by the equation

$$-\frac{\hbar^2}{M}\frac{d^2}{dr^2}(r f_i) + \left[\frac{1}{2}\gamma_i (V_t + V_s) - e_i\right](r f_i) = 0 \tag{5.52}$$

up to r_{0i} where f_i has its first maximum, normalized to unity. For values of r greater than r_{0i}, the function f_i is taken to be equal to one.

The expectation value of the six-nucleon Hamiltonian

$$H = \sum_{i=1}^{6} \frac{1}{2M} p_i^2 + \sum_{i<j=1}^{6} V_{ij} - T_C \tag{5.53}$$

was calculated by a Monte-Carlo technique [SC 62]. As has been mentioned, the use of this technique was the main reason why it was possible to adopt such a complicated wave function in the calculation.

The optimum values of the parameters occurring in various parts of the trial function are as follows [HE 66]:

alpha-cluster function: $a_1 = 1.17\ \mathrm{fm}^{-2}$, $a_2 = 0.25$, $a_3 = 0.52\ \mathrm{fm}^{-2}$,
$\gamma_1 = 0.68$, $e_1 = 0$,

deuteron-cluster function: $b_1 = 1.35$ fm, $b_2 = -2.22$ MeV, $b_3 = 0.34\ \mathrm{fm}^{-1}$,

relative-motion function: $c_1 = 0.18\ \mathrm{fm}^{-2}$, $c_2 = 0.50$, $c_3 = 0.065\ \mathrm{fm}^{-2}$,
$\gamma_2 = 1.0$, $e_2 = 0$. (5.54)

The parameters for the α-cluster wave function are the same as those for a free α particle [SC 63]. This indicates that, as expected, the α cluster has a small compressibility and, hence, the specific distortion effect mentioned in section 4.2a is not important. In the deuteron-cluster wave function, B_3 is much larger than b_3 and B_2 is rather small; hence, the asymptotic behaviour of this function is determined by b_3. As given in eq. (5.54), the value of b_3 variationally determined is larger than the corresponding value for a free deuteron of 0.23 fm^{-1}. Therefore, the deuteron cluster has a somewhat shorter tail as compared with a free deuteron. This is a consequence of the interaction between the nucleons in the two clusters, which restricts strongly the volume in which these nucleons can move. In other words, this calculation shows that because the deuteron cluster is easily deformable, the specific distortion effect is quite important for this particular cluster.

With the optimum wave function, the expectation value $\langle H \rangle$ is -29.9 ± 0.4 MeV, of which 0.9 MeV represents the interaction energy between the clusters.† Although this interaction energy is positive, we should point out that the minimum of $\langle H \rangle$ is quite well determined because of the Coulomb barrier which is about 1 MeV high. The rms radius of the charge distribution is calculated as 2.65 ± 0.15 fm, with the error coming mostly from the statistical uncertainty in the determination of the optimum values of the variational parameters. In obtaining this value, a finite charge distribution of the proton with a Gaussian shape and a rms radius of 0.72 fm has been included.

The experimental value for the interaction energy (i.e., the negative of the separation energy) is -1.47 MeV [AJ 66]. This shows that the agreement between calculation and experiment is only fair. To further improve the variational result, we feel that it is necessary to introduce more flexibility into the long-range part of the relative-motion function.†† Clearly, a systematic way to do this is by adding more and more Gaussian functions into eq. (5.51). However, the amount of computing time required in the Monte-Carlo calculation was already so large that a procedure of this type, requiring the variation of more nonlinear parameters, has to be considered as not very practical.

In addition to the interesting result about the specific distortion of the deuteron and the α clusters, this calculation also shows that the wave function of eqs. (5.47)–(5.52) reflects the saturation character of the nuclear forces quite satisfactorily. In this respect, we should emphasize that no parameters of this wave function were fixed phenomenologically; the internal width parameters along with the other parameters have all been varied completely freely to find the minimum of the energy expectation value $\langle H \rangle$. Thus, the saturation is the result of a consistent theoretical treatment beginning with the six-nucleon Schrödinger equation including hard-core forces and with dynamical effects, such as the radius-change effect and the anharmonicity effect discussed in section 3.7, taken into proper account.

There is another useful finding which should be mentioned here. In calculating expectation values and transition matrix elements, most of the difficulties arise from the

---

† With the wave function ϕ_α of eq. (5.48) and the parameters given in eq. (5.54), the expectation value of the α-particle Hamiltonian is -28.6 ± 0.2 MeV. This value is somewhat different from that reported in ref. [SC 63]; the new value is obtained with a much larger number of estimates in the Monte-Carlo calculation.

†† It is interesting to note that a modified calculation with $\chi(\mathbf{R})$ written as

$$\chi(\mathbf{R}) = R^2 (e^{-d_1 R} + d_2 e^{-d_3 R})\, Y_{00}(\theta, \varphi)$$

resulted in a decrease of the interaction energy by about 0.3 MeV (unpublished result of *R. C. Herndon* and *Y. C. Tang*).

Jastrow anti-correlation part of the wave function. The matrix elements of many one-particle operators, however, are insensitive to short-range correlations and, therefore, the Jastrow factor can be disregarded in calculating the values of these matrix elements. As an example, the charge distribution of ^{6}Li has been calculated with and without this factor and the result did indicate that the Jastrow factor has only minor influence [SC 63a]. Similarly, it is expected that such a simplification can also be adopted to calculate the matrix elements of other long-ranged operators, such as electromagnetic multipole transition operators and so on.

The main disadvantage of calculating with a Monte-Carlo technique is the large amount of computing time which one needs in order to obtain results with reasonably small statistical uncertainties. As is apparent, this computing time will increase rapidly with an increasing number of nucleons, with the consequence that, for larger nuclei, such calculations can no longer be carried out. Therefore, in the next two subsections, we shall discuss other calculations in which simplifications are made such that the energy expectation value $\langle H \rangle$ can be computed without the use of a time-consuming statistical technique.

5.3c. Calculation With a Nucleon-Nucleon Potential Containing a Soft Core – Approximate Treatment of the Jastrow Factor

In a variational calculation where the nucleon-nucleon potential contains a strong repulsive core, it is necessary to employ a trial wave function of the form

$$\psi = \mathcal{A}\{\phi_J \, \bar{\psi}_L\}, \tag{5.55}$$

where ϕ_J is a Jastrow anti-correlation factor (see section 3.7c), and $\bar{\psi}_L$ may be a product of single-particle wave functions or an unantisymmetrized cluster-type wave function describing long-range correlations between the nucleons. To simplify the calculation, one usually chooses the Jastrow factor to be a symmetric function in nucleon coordinates, i.e.,

$$\phi_J = \prod_{i<j=1}^{A} f(|\mathbf{r}_i - \mathbf{r}_j|). \tag{5.56}$$

With this particular choice, the Jastrow factor commutes with the antisymmetrization operator $\mathcal{A}$ and the wave function ψ may be written as

$$\psi = \phi_J \, \mathcal{A} \, \bar{\psi}_L = \phi_J \, \psi_L . \tag{5.57}$$

To perform bound-state and reaction calculations, one makes use again of the projection equation (2.3), which means that one will need to evaluate matrix element of the form

$$\langle \psi | H - E | \psi \rangle = \langle \psi | T + V - E | \psi \rangle = \langle \psi_L \, \phi_J | T + V - E | \phi_J \, \psi_L \rangle, \tag{5.58}$$

where

$$T = -\frac{\hbar^2}{2M} \sum_i \nabla_i^2 - T_C = -\frac{\hbar^2}{2AM} \sum_{i<j} (\nabla_i - \nabla_j)^2 \tag{5.59}$$

and

$$V = \sum_{i<j} V_{ij} = \sum_{i<j} (V_{ij}^N + V_{ij}^C), \tag{5.60}$$

with V_{ij}^N and V_{ij}^C being the nuclear and the Coulomb part of the nucleon-nucleon potential, respectively. Without the use of a Monte-Carlo technique, the evaluation of this matrix element is obviously a very complicated procedure. This is so, because even with a simple form of $f(r_{ij})$ given by

$$f(r_{ij}) = 1 + q(r_{ij}) = 1 + c e^{-d r_{ij}^2}, \tag{5.61}$$

the factor ϕ_J is a sum of very many terms (2^{15} terms for $A = 6$). Therefore, if one wishes to perform an analytic computation of $\langle \psi | H - E | \psi \rangle$, simplifying assumptions must be made. In this subsection, we shall discuss the assumptions adopted by *Hackenbroich* and his collaborators [HA 67, HU 68] in their bound-state and reaction calculations for nuclear systems with $A \leqslant 12$.

The assumptions made by *Hackenbroich et al.* were motivated by the fact that the average internucleon distance is appreciably larger than the range of the repulsive core and hence, the probability of finding three or more nucleons within the healing distance is rather small. With this in mind, they have used the approximation [HU 68]

$$\langle \psi_L \phi_J | H - E | \phi_J \psi_L \rangle \approx \langle \psi_L | H_{eff} - E | \psi_L \rangle, \tag{5.62}$$

where

$$H_{eff} = T_{eff} + V_{eff}, \tag{5.63}$$

with

$$T_{eff} = -\frac{\hbar^2}{2M} \left[\frac{1}{A} \sum_{i<j} (\nabla_i - \nabla_j)^2 + \frac{1}{2} \sum_{i<j} q(r_{ij}) (\nabla_i - \nabla_j)^2 q(r_{ij}) \right] \tag{5.64}$$

and

$$V_{eff} = \sum_{i<j} [f(r_{ij}) V_{ij}^N f(r_{ij}) + V_{ij}^C]. \tag{5.65}$$

To obtain the expression for T_{eff}, they have also used a procedure which was specifically designed to make T_{eff} into a positive-definite operator (for details, see ref. [HU 68]). In addition, these authors have made the further simplification that in calculating matrix elements of one-particle operators, such as matter-density operator and electromagnetic transition operators, the Jastrow factor can be entirely omitted.

The method of *Hackenbroich et al.* bears certain similarities to the Brueckner method in the treatment of the nuclear-matter problem [BR 58] and the method of cluster development in the lowest order [JA 55, IW 57]. However, as these authors themselves realized, their approximations are rather crude and lack clear mathematical foundation in the sense that there is no definite prescription as to how systematic improvements could be made. In addition, it has been found that this particular way of treating approximately the Jastrow factor leads to an underestimate of the nuclear binding energy. For example, with a nucleon-nucleon potential to be given by eqs. (5.66)–(5.68) below, the α-particle binding energy calculated with this method is 23.71 MeV [HU 68], which is appreciably smaller than the value of 27.74 MeV obtained in a calculation [AF 68] where the Jastrow factor has been taken into account properly. On the other hand, because clusters in light nuclei tend to be relatively far apart, it is reasonable to believe that, particularly for energy levels having the same cluster structure, the calculated excitation energies are likely to be quite good [HU 70, ST 71].

We shall now describe the ^{6}Li calculation of *Stöwe et al.* [ST 71] to show that the approximation method described above does lead to nearly correct values for excitation energies and a reasonable long-range nucleon correlation behaviour. In this calculation, the nucleon-nucleon potential has the form

$$V_{ij} = \left(\frac{1 + P_{ij}^{\sigma}}{2} V_t + \frac{1 - P_{ij}^{\sigma}}{2} V_s \right) \left(\frac{1 + P_{ij}^{r}}{2} \right) + \frac{e^2}{r_{ij}} \frac{1 - \tau_{i3}}{2} \frac{1 - \tau_{j3}}{2}, \qquad (5.66)$$

where

$$\begin{aligned} V_t(r_{ij}) &= \sum_{i=1}^{3} V_{ti} \exp(-\kappa_{ti} r_{ij}^2), \\ V_s(r_{ij}) &= \sum_{i=1}^{3} V_{si} \exp(-\kappa_{si} r_{ij}^2), \end{aligned} \qquad (5.67)$$

with

$$\begin{aligned} V_{t1} &= 600.0 \text{ MeV}, & \kappa_{t1} &= 5.5 \text{ fm}^{-2}, \\ V_{t2} &= -70.0 \text{ MeV}, & \kappa_{t2} &= 0.50 \text{ fm}^{-2}, \\ V_{t3} &= -27.6 \text{ MeV}, & \kappa_{t3} &= 0.38 \text{ fm}^{-2}, \\ V_{s1} &= 880.0 \text{ MeV}, & \kappa_{s1} &= 5.4 \text{ fm}^{-2}, \\ V_{s2} &= -70.0 \text{ MeV}, & \kappa_{s2} &= 0.64 \text{ fm}^{-2}, \\ V_{s3} &= -21.0 \text{ MeV}, & \kappa_{s3} &= 0.48 \text{ fm}^{-2}. \end{aligned} \qquad (5.68)$$

This potential was obtained by *Eikemeier* and *Hackenbroich* [EI 66] and fits reasonably well the empirical even-l phases up to an energy of about 300 MeV in the laboratory system.†

The function f (r) in the Jastrow factor is chosen to have the form of eq. (5.61) with

$$\begin{aligned} c &= -0.6, \\ d &= 3.0 \text{ fm}^{-2}. \end{aligned} \qquad (5.69)$$

† It is interesting to note that $f(r_{ij})\, V_{ij}^N\, f(r_{ij})$ with $f(r_{ij})$ given by eqs. (5.61) and (5.69) is rather similar to the nucleon-nucleon potential of *Volkov* [VO 65], which has been used in projected *Hartree-Fock* studies of light nuclei [BO 67].

For the function ψ_L, a superposition of 36 cluster wave functions is used, i.e.,

$$\psi_L = \sum_{i=1}^{36} A_{li}\,\hat{\varphi}_{li}, \tag{5.70}$$

where

$$\hat{\varphi}_{li} = \mathcal{A}\,\{\phi_\alpha(\beta_i)\,\phi_d(\gamma_i)\,\chi_l(\delta_i, k_i)\,\xi(s,t)\}, \tag{5.71}$$

with $\xi(s, t)$ being a spin-isospin function chosen to yield $T = 0$ and $S = 1$ after antisymmetrization. The internal spatial functions ϕ_α and ϕ_d and the relative-motion function χ_l are chosen as follows:

$$\phi_\alpha(\beta) = \exp\left[-\beta \sum_{j<k=1}^{4} (\mathbf{r}_j - \mathbf{r}_k)^2\right], \tag{5.72}$$

$$\phi_d(\gamma) = \exp\left[-\gamma(\mathbf{r}_5 - \mathbf{r}_6)^2\right], \tag{5.73}$$

$$\chi_l(\delta, k) = R^{l+2k}\, e^{-\delta R^2}\, Y_{ll}(\theta, \varphi). \tag{5.74}$$

From these equations, one sees that ψ_L is represented by a sum of $d + \alpha$ cluster functions with different internal width parameters and different radial dependences of their relative-motion functions, characterized by the parameters β_i, γ_i, δ_i, and k_i. These parameters are chosen in an appropriate manner according to one's physical intuition about the behaviour of the system and, once chosen, are kept fixed in the variational calculation.†

Because of the fact that only linear variational parameters A_{li} are contained in the function ψ_L, one obtains by solving the projection equation (2.3) a set of orthogonal wave functions, with the number in this set equal to the number of variational parameters. These wave functions can be used for the description of the lowest ^{6}Li states which possess the same set of quantum numbers. It should be pointed out, however, that as a consequence of using only a restricted number of variational parameters, the description becomes worse for those states with larger excitation energies. The reason for this is that the wave function describing the excited state is not orthogonal to the true eigenfunctions of all lower states, but only to functions that approximate these eigenfunctions.

Calculations were made for both $l = 0$ and $l = 2$. Because of the approximate treatment of the Jastrow factor, the value of the ground-state energy did not turn out to be too satisfactory, as expected. On the other hand, the calculated excitation energy of the lowest $l = 2$ state is 3.56 MeV, which compares quite favourably with the value of 3.8 MeV obtained by averaging the experimental excitation energies of the $^3D_3$, $^3D_2$, and $^3D_1$ levels in ^{6}Li, weighted according to the expectation value of $\mathbf{l}\cdot\mathbf{s}$ [AJ 66]. The rms charge radius of ^{6}Li was calculated to be 2.68 fm, which is somewhat large but still reasonably close to the experimentally determined value of 2.54 ± 0.05 fm [SU 67].

---

† For the choice of these parameters, see ref. [ST 71]. It should be noted that because of the large number of linear variational parameters A_{li} employed, the choice of β_i, γ_i, δ_i, and k_i is not very critical.

To show that the optimum trial wave function yields a reasonable long-range nucleon correlation behaviour, *Stöwe et al.* have also calculated elastic and inelastic form factors for electron scattering. As will be seen below, a comparison of these calculated form factors with experimentally determined values does indicate that cluster wave functions of eq. (5.70) with optimum values for the variational parameters give a fairly satisfactory description of the behaviour of ^{6}Li in its ground and low excited states.

Since the ground state of ^{6}Li has a total angular momentum equal to 1, the electron elastic-scattering cross section has contributions from longitudinal C0, C2, and transverse M1 interactions. However, because the static electric quadrupole moment of ^{6}Li is small and the experiment was performed at forward angles, the contributions of C2 and M1 terms are unimportant and the differential cross section is determined mainly by the C0 term. Therefore, in the plane-wave Born approximation, the cross section is given by

$$\frac{d\sigma}{d\Omega} = \left(\frac{d\sigma}{d\Omega}\right)_M |F_e|^2, \tag{5.75}$$

where $(d\sigma/d\Omega)_M$ is the differential cross section for Mott scattering and F_e is the elastic form factor from longitudinal C0 interaction.

The elastic form factor F_e can be written as

$$F_e = F_e^B F_p, \tag{5.76}$$

where F_e^B is the bare form factor, given by

$$F_e^B = \frac{1}{Z}\left\langle \psi_L^g \left| \sum_{i=1}^{A} j_0(q\bar{r}_i) \frac{1-\tau_{i3}}{2} \right| \psi_L^g \right\rangle, \tag{5.77}$$

with ψ_L^g being the normalized ground-state wave function of eq. (5.70) and $\bar{r}_i$ being the distance of the ith nucleon from the center of mass of ^{6}Li. The factor F_p in eq. (5.76) is the proton charge form factor, for which the expression given by *Janssens et al.* [JA 66] has been used.

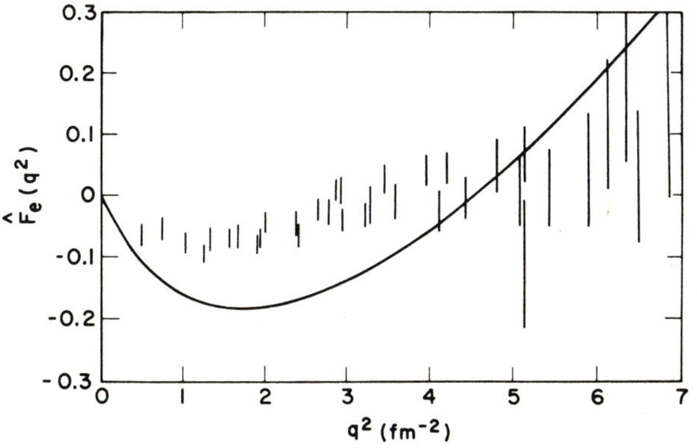

Fig. 6
$\hat{F}_e$ as a function of q^2.
(Adapted from ref. [ST 71].)

In fig. 6, the quantity

$$\hat{F}_e(q^2) = \ln\,[F_e(q^2)/e^{-d_1 q^2}] \qquad (5.78)$$

with $d_1 = 0.8815\ \text{fm}^2$ is shown as a function of the square of the momentum transfer q^2. The experimental data shown are those of *Suelzle et al.* [SU 67]. Here one sees that the agreement is fairly satisfactory. For small values of q^2, the calculated curve lies below the experimental points, which is an indication of the fact that the calculated value for the rms charge radius is larger than the value experimentally determined.

For electron inelastic scattering, whereby the nucleus ^{6}Li is excited from the ground state with $J_i = 1$ to the first excited state with $J_f = 3$, the cross section has its main contribution from the longitudinal C2 interaction, with the transverse E2 and higher multipole interactions playing a minor role [LI 71]. In the case of C2 transition, the absolute square of the bare inelastic form factor is given by [SC 54, WI 63]

$$|F_i^B|^2 = \frac{1}{Z^2}\,\frac{4\pi}{2J_i+1}\sum_{\mu}\sum_{M_i}\sum_{M_f}|\langle\,\psi_L^e\,|\sum_{i=1}^{A} j_2(q\bar{r}_i)\,Y_{2\mu}(\bar{\theta}_i,\bar{\varphi}_i)\,\frac{1-\tau_{i3}}{2}\,|\,\psi_L^g\,\rangle|^2, \qquad (5.79)$$

where ψ_L^e denotes the normalized first excited state wave function of eq. (5.70), and M_i and M_f denote the magnetic quantum numbers of the initial and final states, respectively.† Upon correcting for the finite size of the proton as indicated by eq. (5.76), one obtains then the inelastic form factor F_i which can be compared with experimental results.

The quantity

$$\hat{F}_i(q^2) = \ln\left[\left(\frac{225\,Z^2}{4\pi q^4}|F_i|^2\right)^{1/2} \Big/\ e^{-d_2 q^2}\right] \qquad (5.80)$$

with $d_2 = 1\ \text{fm}^2$ is plotted in fig. 7, where the experimental data of *Neuhausen* [NE 69] are also shown. From this figure it is seen that the agreement between calculation and

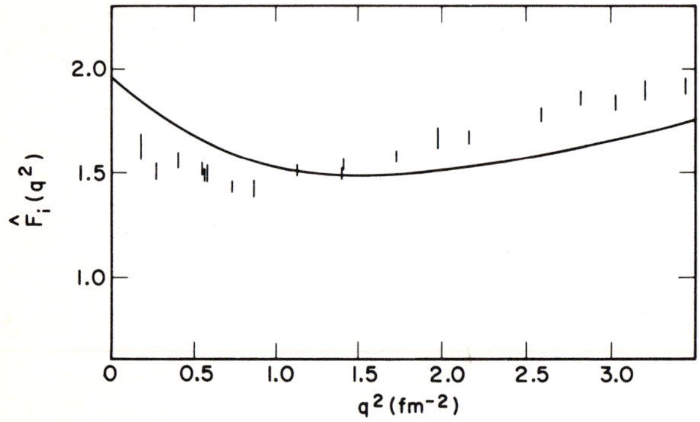

Fig. 7
$\hat{F}_i$ as a function of q^2.
(Adapted from ref. [ST 71].)

† From the bare inelastic form factor one can derive the reduced transition probability according to the relation

$$B(E2; J_i \to J_f) = \frac{225\,Z^2}{4\pi}\lim_{q\to 0}\frac{|F_i^B|^2}{q^4}\,.$$

experiment is again fairly satisfactory. The calculated value of B(E2) is 50.2 fm^4, which compares reasonably well with the value of 25.6 fm^4 obtained experimentally by *Eigenbrod* [EI 69]. In this respect, it is interesting to note that the shell-model value of B(E2) is only 3.36 fm^4 [BO 67a], which indicates indirectly that the wave function of eq. (5.70) does describe the behaviour of a system containing strong cluster correlations.

5.3d. Calculation With a Nucleon-Nucleon Potential Containing no Repulsive Core

In this subsection, we discuss another method [WI 66] which has been devised to avoid the computational difficulty associated with the repulsive core in the nucleon-nucleon potential. In this method, one employs a relatively simple smooth-varying nucleon-nucleon potential without repulsive core in the calculation. To prevent the nucleus from acquiring an unreasonably small size, the internal width parameters of the clusters are fixed, whenever necessary, such that the rms radii of the free clusters take on experimentally determined values. As is quite evident, this method should be particularly suited for calculations in lighter nuclear systems, because here the Pauli principle is rather effective in keeping the clusters relatively far apart and the clusters involved are more or less spin and isospin saturated.

The procedure of employing a non-saturating potential and fixing the internal width parameters of the clusters is rather similar in philosophy to what has been commonly done in shell-model calculations. There it has been found that if one uses the experimentally determined value for the radius of the nucleus as a restraining condition, then the total binding energy calculated with a non-saturating nucleon-nucleon potential will usually turn out to be fairly satisfactory [WI 54]. The reason for this is that the average distance between neighbouring nucleons is appreciably larger than twice the core radius and, consequently, the repulsive core has only a minor influence upon the total binding energy of a normal nucleus [BE 71].

The nucleon-nucleon potential commonly employed has the form [RE 69]

$$V_{ij} = \left(\frac{1 + P_{ij}^{\sigma}}{2} V_t + \frac{1 - P_{ij}^{\sigma}}{2} V_s\right)\left(\frac{u}{2} + \frac{2 - u}{2} P_{ij}^{r}\right) + \frac{e^2}{r_{ij}} \frac{1 - \tau_{i3}}{2} \frac{1 - \tau_{j3}}{2}, \tag{5.81}$$

where V_t and V_s are the triplet and singlet potentials in even orbital angular-momentum states, given by

$$\begin{aligned} V_t(r_{ij}) &= -V_{0t} \exp(-\kappa_t r_{ij}^2), \\ V_s(r_{ij}) &= -V_{0s} \exp(-\kappa_s r_{ij}^2). \end{aligned} \tag{5.82}$$

The parameters $V_{0t}, \kappa_t, V_{0s}, \kappa_s$ are adjusted to yield correct values for the nucleon-nucleon effective-range parameters; they are found to be

$$\begin{aligned} V_{0t} &= 66.92 \text{ MeV}, & V_{0s} &= 29.05 \text{ MeV}, \\ \kappa_t &= 0.415 \text{ fm}^{-2}, & \kappa_s &= 0.292 \text{ fm}^{-2}. \end{aligned} \tag{5.83}$$

The quantity u in eq. (5.81) which governs the amount of space-exchange mixture in the nucleon-nucleon potential is treated as an adjustable parameter. It is introduced for the purpose of compensating for possible deficiencies in the trial wave function, such as a lack of consideration of specific distortion effects and so on. Its value will be determined by requiring that a good over-all fit to the experimental data be obtained. It should be emphasized that the value of u so determined must be relatively close to 1, which corresponds to a Serber exchange mixture. In this sense, therefore, u can be considered as a monitoring parameter; if its value turns out to be considerably larger than 1, then one gets a rather clear indication that the trial wave function does not describe the behaviour of the system properly and an improved function with more flexibility must be used.

With the nucleon-nucleon potential of eqs. (5.81)–(5.83), the energy expectation values $\langle H \rangle$ and the rms radii R_{rms} for the deuteron, triton, and α clusters have been computed using a trial wave function which has a spatial part of the form

$$\phi = \sum_{i=1}^{N_C} B_i \exp\left(-\frac{1}{2} a_i \sum_{j=1}^{A} \bar{\mathbf{r}}_j^2\right), \tag{5.84}$$

where $\bar{\mathbf{r}}_j$ denotes as usual the cluster internal coordinate defined in a way as indicated in eqs. (3.8) and (3.9). The number of terms N_C in the sum is taken as three for the deuteron and triton cases and two for the α-cluster case. These numbers are sufficient, since it has been found that a larger value of N_C does not yield any significant improvement in the value of $\langle H \rangle$.

The results for $\langle H \rangle$ and R_{rms} are given in table 1, together with the optimum values of the variational parameters.† From this table it is seen that for the deuteron and triton clusters, the calculated values of both $\langle H \rangle$ and R_{rms} agree quite satisfactorily with the values experimentally determined [KO 74]. Therefore, with this simple nucleon-nucleon potential, even specific distortion effects of these clusters can be considered. For the α

Table 1. Values of $\langle H \rangle$ and R_{rms} for the deuteron, triton, and α clusters

Cluster	B_1	B_2	B_3	a_1 (fm^{-2})	a_2 (fm^{-2})	a_3 (fm^{-2})	$\langle H \rangle$ (MeV)	R_{rms} (fm)
d	1.0	3.631	5.746	0.0728	0.366	1.470	- 2.20	1.90
t	1.0	9.110	16.603	0.172	0.510	1.320	- 7.94	1.53
α	1.0	6.353		0.518	1.220		- 33.65	1.22

† The choice of a sum of symmetric Gaussian functions for the triton and the α cluster is somewhat restrictive. For example, a very good value of $\langle H \rangle$ for the triton cluster, obtained by a Monte-Carlo calculation using a trial wave function of the form given in Ref. [TA 65], is - 8.45 MeV. The rms radius obtained in this latter calculation is 1.58 fm, which is also somewhat different from that given in table 1.

cluster, the calculated value of R_{rms} is appreciably smaller than the experimental value of 1.48 fm [KO 74]. This means that it is not proper to use for the α cluster the wave function of eq. (5.84) with the parameters of table 1; rather, one has to take a phenomenological view by adjusting the parameters in the wave function such that the value of 1.48 fm for the rms radius is obtained. Of course, as a result of this adjustment, the specific distortion effect of the α cluster can no longer be properly considered. Fortunately, however, this is not too serious, because the ^{6}Li calculation described in section 5.3b shows that at least in light nuclear systems, the α-cluster specific distortion effect is not likely to have any profound influence on the behaviour of the system.

As an example of this type of calculation, we discuss now the ^{6}Li study of *Thompson* and *Tang* [TH 73]. In this calculation, the wave function is given by eqs. (4.13)–(4.15); it has a rather flexible form, containing both a linear variational function F(**R**) and a number of linear variational amplitudes A_{li}.

The spatial part of the α-cluster wave function is assumed as

$$\phi(\alpha) = \exp\left(-\frac{1}{2} a_\alpha \sum_{j=1}^{4} \bar{r}_j^2\right), \tag{5.85}$$

with the width parameter a_α chosen as 0.514 fm^{-2} which yields the desired value of 1.48 fm for the rms radius. For the spatial part of the deuteron-cluster wave function, the form of eq. (5.84), with the parameters B_i and a_i listed in table 1, is used.

The functions $\tilde{\phi}_i$ and g_{li} in the distortion function of eq. (4.15) are assumed to have the forms

$$\tilde{\phi}_i(\tilde{d}) = \exp\left(-\frac{1}{2} \tilde{\alpha}_i \sum_{j=5}^{6} \bar{r}_j^2\right) \alpha_5 \nu_5 \alpha_6 \pi_6 \tag{5.86}$$

and

$$g_{li}(R) = R^{n+1} \exp\left(-\frac{2}{3} \tilde{\beta}_i R^2\right). \tag{5.87}$$

The value of n in eq. (5.87) is chosen in the following way:

$$\begin{aligned} n &= 2, && \text{for } l = 0, \\ n &= 3, && \text{for } l = 1, \\ n &= l. && \text{for } l \geqslant 2. \end{aligned} \tag{5.88}$$

The rationale for choosing these particular values of n is contained in the discussion of section 3.4 and will not be further elaborated here. The choice of the nonlinear parameters $\tilde{\alpha}_i$ and $\tilde{\beta}_i$ in eqs. (5.86) and (5.87) is not too critical, as long as a large number of distortion functions is employed. A detailed study shows that the result will be sufficiently accurate if nine distortion functions (i.e., N = 9 in eq. (4.14)) are used, with $\tilde{\alpha}_i$ taking on the values 0.07, 0.25, and 1.20 fm^{-2}, and $\tilde{\beta}_i$ taking on the values 0.27, 0.45, and 0.63 fm^{-2}.

A careful study has also been performed to make certain that the freedom in the distortion functions does not create an unrealistic distortion of the α-cluster. This study is necessary, since the wave function of eq. (5.85) does not minimize the expectation value of the α-cluster Hamiltonian. The result shows that by using the same function $\phi(\alpha)$ in both ψ_0 of eq. (4.13) and $\hat{\varphi}_{li}$ of eq. (4.15), the addition of these distortion functions does indeed correct for the deuteron specific distortion effect, rather than causes a collapse of the α cluster.

The substitution of the ^{6}Li trial wave function into the projection equation (2.3) yields a set of coupled equations given by eqs. (4.16a) and (4.16b) in section 4.2b. If one now makes the expansion

$$F(\mathbf{R}) = \sum_l \frac{1}{R} f_l(R) P_l(\cos\theta) \tag{5.89}$$

and carries out some algebraic manipulation, then one obtains a rather complicated integrodifferential equation for $f_l(R)$. The separation energy E_S of the deuteron and the α clusters in the ground state of ^{6}Li can then be determined by solving this equation subject to the asymptotic boundary condition for a bound state. For the $l = 2$ excited state, a somewhat more complicated procedure has to be adopted, because this state is in fact only quasibound. The procedure which has been employed is to solve this equation using scattering boundary condition and then analyze the resultant phase shifts with an R-matrix formula [TH 73a] to obtain the resonance energy for this state.†

To obtain the experimental value of 1.47 MeV for E_S, it is found that the value of u in the nucleon-nucleon potential of eq. (5.81) needs to be equal to 0.925. This is gratifying, since in a similar calculation of $\alpha + \alpha$ scattering where the specific distortion effect is not very important, the value of u required is 0.92 [BR 71]. This indicates, therefore, that with the specific distortion effect taken into proper account, the same nucleon-nucleon potential can be used to yield a consistent and satisfactory treatment of both the $d + \alpha$ and the $\alpha + \alpha$ systems.

The excitation energy of the $l = 2$ state is calculated as 4.9 MeV. This is larger than the value of 3.8 MeV experimentally determined (see section 5.3c). However, considering the complexity of this problem, we feel that the agreement is still fairly satisfactory.

Another interesting finding is that, if the ground state of ^{6}Li is considered with the function ψ_0 left off from eq. (4.14), the calculated value of E_S is worse by about 2.2 MeV. This indicates that the behaviour of the relative motion is very important in determining the binding energy of the clusters in this particular case. Therefore, we can now understand why the calculation of section 5.3b yields a relatively poor result. In that calculation, the long-range part of the relative-motion function consists of a sum of only two Gaussian functions, which is probably not flexible enough to give a proper description of the relative motion between the two clusters when they are not too close together.

---

† The discussion of the quasibound $l = 2$ state should properly belong to Chapter 7 where scattering and reaction calculations are described. We include it here, simply because this state forms together with the ground state a rotational band.

5.3e. Summary

In this section, we have described three ^{6}Li calculations to illustrate the methods which have been employed in microscopic studies of light nuclear systems. The first calculation (subsection 5.3b) is basically the more desirable one, because it uses a nucleon-nucleon potential with a hard core and handles the Jastrow anti-correlation factor in a proper way. On the other hand, even with the variation of a large number of nonlinear parameters, the result obtained is only fair. In addition, it suffers from the fact that the amount of computing time required is quite large and, consequently, similar calculations for heavier systems are likely to be impractical.

The methods described in subsections 5.3c and 5.3d are designed to require much less computational effort, but they do contain features which are somewhat undesirable. In the method of subsection 5.3c, the way of approximately handling the Jastrow factor is rather arbitrary and lacks a clear mathematical basis. The method of subsection 5.3d employs a nucleon-nucleon potential containing no repulsive core and, consequently, has the defect of being necessary to fix the internal wave functions of certain clusters in order to prevent a collapse of the nucleus under consideration. However, these methods do yield wave functions which describe satisfactorily the long-range correlation behaviour of the nucleons; therefore, they have been used successfully to study a large number of nuclear bound-state and reaction problems where the short-range correlation structures are relatively unimportant.

Before we leave this section, we wish to discuss why the $t + {}^3$He cluster structure is not included in the trial wave functions of all these ^{6}Li calculations. The reason for this is as follows. Because of the Pauli principle, a large overlap between a $d + \alpha$ cluster structure and a $t + {}^3$He cluster structure exists in the region where the clusters penetrate each other strongly. As has been discussed in section 3.5, this overlap is in fact even complete in the limit of the oscillator shell model. Therefore, the addition of a $t + {}^3$He cluster function will not significantly affect the behaviour of the wave function in the region of strong cluster interaction. Now, if we go to the region in the configuration space where the nucleons do not all interact strongly with each other, then the probability of finding a large degree of $t + {}^3$He clustering must be rather small, because it is energetically unfavourable to break up a strongly bound α cluster [TA 62]. This means, therefore, that the amplitude of such a cluster configuration in the surface region of ^{6}Li is quite weak and, consequently, the $t + {}^3$He cluster configuration may be safely omitted from a variational calculation. In addition, it should be noted that the electromagnetic properties of the two ^{6}Li states discussed in subsection 5.3c are described by observables of long-range character which essentially probe only the surface part of ^{6}Li where the amplitude of the $t + {}^3$He configuration is small compared with that of the $d + \alpha$ configuration. Thus, a study of these electromagnetic properties will not reveal the small amount of $t + {}^3$He clustering which may actually be present in this surface region. On the other hand, if one is concerned with a process which is especially sensitive to the presence of $t + {}^3$He cluster structure in ^{6}Li, then it will be very useful to include also this structure in the trial wave function. An example which serves to illustrate this type of consideration will be given in a later chapter.

5.4. Low-Energy T = 0 States of ^{12}C

As second example we discuss the calculation of *Hutzelmeyer* and *Hackenbroich* [HU 68, HU 70] on the low-energy T = 0 states of ^{12}C, using the approximation method of subsection 5.3c. Because of the relatively large breakup energy of an α cluster, one can assume in a good approximation that these states have a 3 α cluster structure or, equivalently, an $\alpha + {}^8Be$ cluster structure, with no internal excitations for the α clusters. This description should be approximately correct up to an excitation energy of about 15 MeV where the first T = 1 state of ^{12}C appears [AJ 68].

The trial wave function is chosen to have the form of eq. (5.57), with the Jastrow factor given by eqs. (5.56), (5.61), and (5.69). For the function ψ_L, the following ansatz is used:

$$\psi_L = \sum_{i=1}^{N} A_{li}\, \hat{\varphi}_{li}, \tag{5.90}$$

where

$$\hat{\varphi}_{li} = \mathcal{A}\Big\{ \tilde{\phi}_{\alpha 1}(\beta_{i1})\, \tilde{\phi}_{\alpha 2}(\beta_{i2})\, \tilde{\phi}_{\alpha 3}(\beta_{i3}) \times \Big[\sum_{m_i, \hat{m}_i} C(l, l_i, \hat{l}_i; m, m_i, \hat{m}_i)\, \chi_{l_i m_i}(\mathbf{R})\, \hat{\chi}_{\hat{l}_i \hat{m}_i}(\hat{\mathbf{R}}) \Big] \Big\}, \tag{5.91}$$

with

$$\begin{aligned} \mathbf{R} &= \mathbf{R}_{\alpha 1} - \mathbf{R}_{\alpha 2}, \\ \hat{\mathbf{R}} &= \tfrac{1}{2}(\mathbf{R}_{\alpha 1} + \mathbf{R}_{\alpha 2}) - \mathbf{R}_{\alpha 3}, \end{aligned} \tag{5.92}$$

and $C(l, l_i, \hat{l}_i; m, m_i, \hat{m}_i)$ being a Clebsch-Gordan coefficient. In eq. (5.91), the internal function $\tilde{\phi}_{\alpha 1}(\beta_{i1})$ for the first α cluster is chosen to have a spatial part given by

$$\phi_{\alpha 1}(\beta_{i1}) = \exp\Big[-\beta_{i1} \sum_{p<q=1}^{4} (\mathbf{r}_p - \mathbf{r}_q)^2 \Big], \tag{5.93}$$

and the internal functions $\tilde{\phi}_{\alpha 2}(\beta_{i2})$ and $\tilde{\phi}_{\alpha 3}(\beta_{i3})$ for the other two α clusters are chosen in a similar manner. The relative-motion functions $\chi_{l_i m_i}$ and $\hat{\chi}_{\hat{l}_i \hat{m}_i}$ are assumed to have the forms

$$\chi_{l_i m_i}(\mathbf{R}) = R^{l_i + 2k_i}\, e^{-\delta_i R^2}\, Y_{l_i m_i}(\theta, \varphi) \tag{5.94}$$

and

$$\hat{\chi}_{\hat{l}_i \hat{m}_i}(\hat{\mathbf{R}}) = \hat{R}^{\hat{l}_i + 2\hat{k}_i}\, e^{-\hat{\delta}_i \hat{R}^2}\, Y_{\hat{l}_i \hat{m}_i}(\hat{\theta}, \hat{\varphi}). \tag{5.95}$$

The total angular momentum $J = l$ of a given ^{12}C state is obtained by coupling the orbital angular momentum l_i of the relative motion of the first two α clusters which form a ^8Be cluster to the orbital angular momentum $\hat{l}_i$ of the relative motion of the third α cluster with respect to the center of mass of the ^8Be core.

Because of computational problems, it was found necessary to use a relatively small number of $\hat{\varphi}_{li}$ in eq. (5.90).† As a consequence of this, the choice of the fixed nonlinear

† For example, the value of N used for the ground and second excited states of ^{12}C is equal to only 8.

parameters β_{ij} (j = 1, 2, 3), l_i, k_i, δ_i, $\hat{l}_i$, $\hat{k}_i$, and $\hat{\delta}_i$ becomes quite critical. Fortunately, there exists a collective boson-oscillator model [HU 69, KR 66] which yields a reasonable qualitative description of the low-lying T = 0 levels of ^{12}C and which can be used as a guide for the choice of these parameters. In recognition of this, *Hutzelmeyer* and *Hackenbroich* [HU 68, HU 70] have made extensive use of the features of this model and have consequently adopted the simplifying conditions

$$\beta_{i1} = \beta_{i2} = \beta_{i3} = \beta_i \qquad (5.96)$$

and

$$\hat{\delta}_i = \frac{4}{3}\,\delta_i. \qquad (5.97)$$

In addition, they have been able to choose values for the fixed nonlinear parameters which do seem to yield a reasonable T = 0 energy spectrum for ^{12}C even with a rather small number of $\hat{\varphi}_{li}$ in the trial function ψ_L.

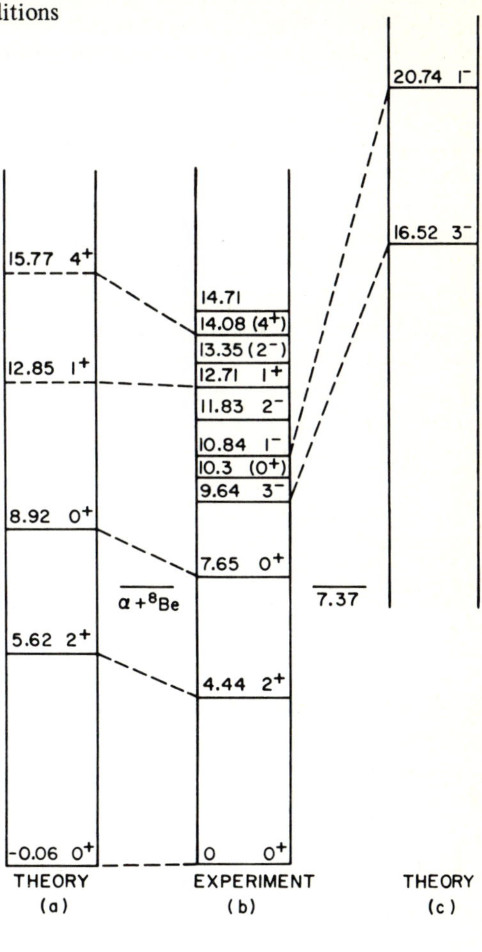

Fig. 8
T = 0 energy spectrum of ^{12}C: (a) calculated positive-parity states (states denoted by dashed lines are obtained with less accuracy in the calculation), (b) experimental levels, (c) calculated negative-parity states. The calculated energy spectrum is taken from ref. [HU 70], while the experimental energy spectrum is taken from ref. [AJ 68].

We discuss now some of the essential results of this calculation performed with the nucleon-nucleon potential of eqs. (5.66)–(5.68). In fig. 8, the calculated T = 0 energy spectrum of ^{12}C is compared with the energy spectrum experimentally determined [AJ 68]. To facilitate the comparison of excitation energies, the energy origin of the calculated spectrum is shifted such that the calculated and experimental $\alpha + {}^8Be$ threshold energies occur at the same vertical position. This adjustment is necessary, because with the approximate treatment of the Jastrow factor as discussed in subsection 5.3c, the binding energy of an α particle is only 23.71 MeV, which is considerably smaller than the experimental value of 28.3 MeV.

From fig. 8 it is seen that the calculated excitation energies of the positive-parity levels compare quite favourably with experimental values. In particular, the calculation yields an excited 0^+ state with a rather low excitation energy of 8.98 MeV. As is well known, a proper account of this level has not been achieved in many shell-model-type calculations [CO 65]. In order to see why the present calculation is successful in this respect, a close examination of the relative magnitudes of the variational amplitudes A_{li} has been made. This examination does yield an indication that the radius-change effect† and the anharmonicity effect discussed in section 3.7 are responsible in bringing this level down to its low-excited position.

The calculated separation energy of an α cluster in the ground state of ^{12}C is 7.43 MeV which agrees almost perfectly with the experimental value of 7.37 MeV. We feel, however, that this agreement is somewhat fortuitous for the following reason. The trial wave function employed in this calculation is not likely to be flexible enough to describe in a detailed manner the behaviour of ^{12}C in its gound state. For example, the condition represented by eq. (5.96), where the width parameters of the three α clusters are chosen as equal to one another, is definitely too restrictive. In a configuration where two α clusters penetrate each other strongly and the third cluster is relatively far away, it is certainly conceivable that the radius of the third α cluster should be smaller than the radii of the other two. Therefore, if one improves the trial wave function to take this effect into consideration, the separation energy will become larger. On the other hand, it should be noted that this calculation is performed with a purely central nucleon-nucleon potential which fits reasonably well the two-nucleon scattering data. As can be expected, an improved calculation using a more realistic potential containing tensor components will surely result in a value for the separation energy which is smaller than that quoted here.

In spite of our reservation about the reliability of the calculated separation energy, we do feel that the excitation energies obtained are fairly accurate. Especially for low-lying levels, it is reasonable to expect that the deficiencies mentioned above should affect the total energies of these levels in a more or less similar way.††

The negative-parity states do not come out as well as the positive-parity states. It can be seen from fig. 8 that their excitation energies are too large by 7 MeV or more. The reason for this is that in the trial function of eqs. (5.90) and (5.91), configurations which represent

---

† It should be mentioned that there is no clear distinction between the radius-change effect discussed in section 3.7 and the specific distortion effect discussed in section 4.2. Customarily, the former term has been used in bound-state calculations, while the latter term has been used in scattering and reaction calculations. In this monograph, we have more or less adhered to this usage.

†† There is some quantitative justification for this statement. The ^6Li calculations of ref. [TH 69, TH 73] show that the excitation energy of the $l = 2$ state depends less sensitively on the exchange mixture of the nucleon-nucleon potential (see footnote 18 of ref. [TH 69]) and the quality of the trial wave function (cf. subsection 5.3d with table 5 of ref. [TH 73]) than the separation energy in the $l = 0$ ground state does.

the breakup of an α cluster have not been included and, consequently, states with higher excitation energies are less well described.

Even though the calculated excitation energies of the negative-parity levels are too large, the energy ordering of the 1^- and 3^- levels does agree with experiment. This is an interesting finding, since the penetrating-orbit argument (see section 3.6) predicts that the 1^- level should have the lower energy. The penetrating-orbit prediction would have been correct if the Pauli principle were not important, as a calculation using an unantisymmetrized wave function has shown. Therefore, this calculation demonstrates convincingly the importance of using antisymmetrized wave functions in calculations on light nuclear systems and any qualitative argument based on unantisymmetrized wave functions must be used with great caution.

The prediction of the low-energy spectrum of ^{12}C in the framework of the shell model [CO 65] is quite different from that presented here. Therefore, it would be worthwhile to perform a calculation where the trial function is a linear superposition of 3 α cluster wave functions and oscillator shell-model wave functions. In this way, one could make it clear that the 3 α-cluster description of the low-energy ^{12}C states is in fact adequate for most purposes. In such a calculation, a technical problem will come up, since the trial function contains terms depending on different single-particle and collective coordinates. But this can be handled in a straightforward fashion by using the generator-coordinate technique discussed in subsection 5.2b, because this technique allows us to perform multi-dimensional integrations always with single-particle coordinates.

5.5. Low-Lying Levels of ^{7}Be

To illustrate further that the method of subsection 5.3d yields satisfactory results, we now describe a calculation on the low-lying levels in ^{7}Be [FU 75]. In this calculation, the nucleon-nucleon potential used is a superposition of a central potential and a spin-orbit potential; that is,

$$V_{ij} = \left[\left(\frac{1+P_{ij}^{\sigma}}{2}V_t + \frac{1-P_{ij}^{\sigma}}{2}V_s\right)\left(\frac{u}{2}+\frac{2-u}{2}P_{ij}^{r}\right) + \frac{e^2}{r_{ij}}\,\frac{1-\tau_{i3}}{2}\,\frac{1-\tau_{j3}}{2}\right]$$
$$+\left[-(V_\lambda + V_{\lambda\tau}\,\boldsymbol{\tau}_i\cdot\boldsymbol{\tau}_j)\,e^{-\lambda r_{ij}^2}\,(\mathbf{r}_i-\mathbf{r}_j)\times(\mathbf{p}_i-\mathbf{p}_j)\cdot(\boldsymbol{\sigma}_i+\boldsymbol{\sigma}_j)\,\frac{1}{2\hbar}\right], \tag{5.98}$$

where the central part is the same as the potential of eq. (5.81). The spin-orbit potential in eq. (5.98) is an effective non-central potential for calculations involving p-shell nuclei; it is obtained by fitting the p + α scattering data in the energy region below the reaction threshold. With

$$V_\lambda = -50\text{ MeV},\qquad V_{\lambda\tau} = 270\text{ MeV},\qquad \lambda = 2.0\text{ fm}^{-2}, \tag{5.99}$$

it was found [FU 75] that the experimental p + α differential cross-section, polarization, and spin-rotation-parameter data can be fitted to a high degree of accuracy. We should mention, of course, that a more desirable calculation will be to employ a nucleon-nucleon potential containing a strong tensor component, obtained by fitting the two-nucleon scattering data. However, such a calculation will be quite difficult if performed in a proper manner, because one must use complicated cluster internal wave functions in order to take full advantage of the attractive nature of the tensor interaction.

The low-lying levels of ^{7}Be are assumed to have an $\alpha + {}^3$He cluster structure (see subsection 3.6a). The wave functions describing these levels have therefore the form

$$\psi_{JlS}^{M_J} = A\left\{ \phi(\alpha)\,\phi(h)\left[\frac{1}{R}\,f_{Jl}(R)\,Y_{JlS}^{M_J}\right]\right\}, \tag{5.100}$$

where $Y_{JlS}^{M_J}$ is a spin-isospin-angle function, appropriate for $T = \frac{1}{2}$, $S = \frac{1}{2}$, and required values of orbital angular momentum l and total angular momentum J with z-component M_J. The α-cluster spatial wave function is chosen as given by eq. (5.85) with $a_\alpha = 0.514\ \mathrm{fm}^{-2}$. For the ^{3}He spatial wave function, one should properly use the function of eq. (5.84) with $N_C = 3$ and the parameters given in table 1. However, for the sake of reducing computing time, a simplified form of

$$\phi(h) = \exp\left(-\frac{1}{2}\,a_h \sum_{j=5}^{7} \bar{r}_j^2\right) \tag{5.101}$$

consisting of only one Gaussian function has been used in this calculation. The parameter a_h is then adjusted to yield a correct rms radius [KO 74] in accordance with data from electron scattering experiments. The value of a_h so determined is equal to $0.367\ \mathrm{fm}^{-2}$.

The expectation value of the ^{3}He Hamiltonian obtained with $\phi(h)$ of eq. (5.101) is -3.96 MeV, which is considerably larger than the value obtained with a more flexible many-Gaussian wave function. Therefore, the specific distortion effect cannot be properly considered with the choice of this simple internal wave function and one must adjust the exchange-mixture parameter u in order to obtain a correct value for the separation energy of the two clusters. Since it is expected that this effect will create an additional intercluster attractive interaction, one anticipates that the value of u so determined will be larger than the value of 0.925 in the d + α case, where the specific distortion effect has been explicitly considered.

By substituting the wave function $\psi_{JlS}^{M_J}$ of eq. (5.100) into the projection equation (2.3), one obtains in each (J, l) state an integrodifferential equation of the form

$$\left\{\frac{\hbar^2}{2\mu}\left[\frac{d^2}{dR^2} - \frac{l(l+1)}{R^2}\right] + E_R - V_N(R) - V_C(R) - \eta_{Jl}\,V_{SO}(R)\right\} f_{Jl}(R)$$
$$= \int_0^\infty \left[k_l^N(R,R') + k_l^C(R,R') + \eta_{Jl}\,k_l^{SO}(R,R')\right] f_{Jl}(R')\,dR', \tag{5.102}$$

where μ is the reduced mass, E_R is the relative energy of the two clusters in the center-of-mass system, and the quantities η_{Jl} are given by

$$\eta_{l+\frac{1}{2},l} = l, \qquad \eta_{l-\frac{1}{2},l} = -(l+1). \tag{5.103}$$

The functions V_N, V_C, and V_{SO} are the direct nuclear-central potential, the direct Coulomb potential, and the direct spin-orbit potential, respectively, while the functions k_l^N, k_l^C, and k_l^{SO} represent the kernels arising from the exchange character in the nucleon-nucleon potential and the antisymmetrization procedure.

Equation (5.102) is solved with appropriate boundary conditions to yield the separation energies in the $\frac{3}{2}^-$ ground state and the $\frac{1}{2}^-$ first excited state and the resonance energies of the $\frac{7}{2}^-$ and $\frac{5}{2}^-$ states. The value of u is adjusted such that the average separation energy in the $\frac{3}{2}^-$ and $\frac{1}{2}^-$ states is correctly obtained. This value turns out to be 0.984, which is indeed close to 1 and larger than the value of 0.925 in the $d + \alpha$ case.

The result is shown in fig. 9, where a comparison with experiment is also made. In this figure, the energy origin of the calculated spectrum is again shifted such that the calculated and experimental $^3He + {}^4He$ threshold energies occur at the same vertical position. Here one sees that the agreement between calculation and experiment is quite satisfactory. The relatively minor discrepancy in the level positions is probably due to the fact that, in this calculation, an effective nucleon-nucleon spin-orbit potential is used and the trial wave function does not include configurations where the ^3He cluster is excited.

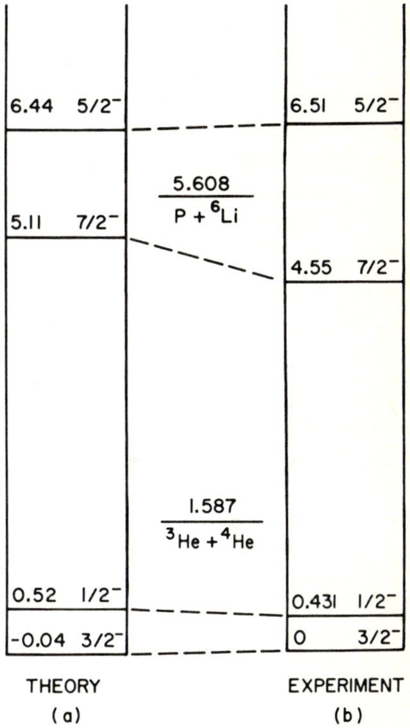

Fig. 9
Low-lying levels of ^7Be. (Adapted from ref. [FU 75].)

5.6. Concluding Remarks

The examples discussed in this chapter show that even for low-lying levels in light nuclei, the calculation can become quite involved if accurate results are desired. This has two main reasons. First, the trial wave function is required to contain an anti-correlation factor, because the nucleon-nucleon potential which describes the experimental data up to about 300 MeV in the laboratory system possesses a repulsive core. Due to the complicated nature of this factor, the calculation becomes very lengthy if this factor is to be treated in a proper manner. Second, the exchange terms arising from the process of antisymmetrizing the wave function are quite numerous even for a nuclear system where the number of nucleons involved is relatively small.

On the other hand, the relatively short-range character of the attractive part of the nucleon-nucleon potential, together with the Pauli principle, does simplify to a significant extent many considerations of nuclear properties. It has the effect that the correlation behaviour of the nucleons becomes not very sensitive to the specific form of this potential. For these correlations, it is mainly important that nucleons are fermions and that the nucleon-nucleon potential is on the average attractive enough to allow the formation of nucleon clusters with particular spin and isospin configurations in compliance with the Pauli principle.

The relative insensitivity of the nucleon correlation behaviour to the specific form of the nucleon-nucleon attractive potential means that all measurable quantities which depend mainly on this correlation are also rather insensitive to this specific form. Thus, one expects that these quantities can be described quite well even if one uses in the calculation potentials which are relatively simple but contain the main characteristics of the nucleon-nucleon interaction. Examples of such quantities are the nuclear charge density, the electromagnetic transition probability, and so on.

In many cases, it is even possible to analyze experimental data without performing any microscopic calculation. What one can do is to assume particular generalized cluster wave functions for various nuclear states and introduce into these wave functions a small number of parameters through which the internal and relative motions of the clusters can be varied. These parameters are then determined in such a way that experimental data on certain observables, such as electromagnetic transition probabilities, are described properly. Then one uses the resultant wave function to calculate other measurable quantities, such as electron-scattering elastic and inelastic form factors. Especially in light nuclei where the cluster structures of nuclear levels can be reliably guessed using energetical and other arguments, this semi-phenomenological approach usually works quite well and has been adopted for analyses of experimental results in particular by Neudatchin and collaborators [NE 69a].

As has been discussed extensively in section 5.3, the difficulty associated with the anti-correlation factor can be circumvented by using the methods of subsections 5.3c and 5.3d. In regard to the difficulty arising from the requirement of using totally antisymmetrized wave functions, we should mention that there are cases where one can obtain fairly reliable results by carrying out the antisymmetrization process in an approximate manner or even neglecting it entirely. To understand the situation under which this type of approximation becomes valid, it is necessary to study the relationship between the effect of antisymmetrization and the overlapping of clusters in coordinate or momentum space. Since the use of unantisymmetrized or partially antisymmetrized wave functions will obviously simplify practical calculations to a very large extent, we devote the next chapter exclusively to a discussion of this relationship.

6. Further Comments About the Pauli Principle

6.1. General Remarks

In section 3.5 we discussed, mainly in the frame of the oscillator model, the influence of the Pauli principle on nuclear wave functions. We now resume this discussion, but without the restriction of the oscillator assumption. We shall be especially interested in how the effect of the Pauli principle changes with increasing mutual penetration of the clusters. To accomplish this purpose, two detailed examples [HA 66] will be considered. In the first example, it will be seen how the Pauli principle usually causes clusters to differ more and more from the corresponding composite particles as the clusters increasingly penetrate each other. In the second example, however, we shall show that a cluster still remains energetically favoured even when embedded in a large nucleus where it is surrounded and penetrated by many other nucleons.†

6.2. Cluster Overlapping and Pauli Principle

As has been mentioned many times, one must be very careful in interpreting an antisymmetrized cluster wave function. In general, the meaning of such a wave function is quite different from that of its corresponding unantisymmetrized part. In this section, we shall examine this problem in great detail by comparing quantitatively the expectation values and transition matrix elements of observables calculated with both antisymmetrized and unantisymmetrized wave functions.

Consider, for simplicity, a two-cluster wave function of the form

$$\psi = A\,\bar{\psi} = A\,\{\tilde{\phi}_0(\tilde{\mathrm{A}})\,\tilde{\phi}_0(\tilde{\mathrm{B}})\,\chi(\mathbf{R}_\mathrm{A}-\mathbf{R}_\mathrm{B})\}\,. \tag{6.1}$$

By defining

$$\hat{\psi} = [A_\mathrm{A}\,\tilde{\phi}_0(\tilde{\mathrm{A}})]\,[A_\mathrm{B}\,\tilde{\phi}_0(\tilde{\mathrm{B}})]\,\chi(\mathbf{R}_\mathrm{A}-\mathbf{R}_\mathrm{B}), \tag{6.2}$$

where A_A and A_B are, respectively, antisymmetrization operators with respect to nucleon coordinates in clusters A and B, one can write eq. (6.1) as

$$\psi = A'\,\hat{\psi} = \hat{\psi} + \sum_{\mathrm{P}'} (-1)^{\mathrm{p}'}\,\mathrm{P}'\,\hat{\psi}\,. \tag{6.3}$$

In eq. (6.3), the operator P' denotes just those permutations interchanging nucleons in different clusters but not within the individual clusters themselves, and the sum extends over a sufficient number of these permutations to effect a complete antisymmetrization

† This energetical favouring of cluster formation in a large nucleus is very important for the explanation of asymmetric fission (see chapter 17).

of the cluster function.†, †† The meaning of the function $\hat{\psi}$ is quite easy to understand; it describes two composite particles, with wave functions $A_A \widetilde{\phi}_0(\widetilde{A})$ and $A_B \widetilde{\phi}_0(\widetilde{B})$, in relative motion with respect to each other. On the other hand, because of the presence of the antisymmetrization operator A', the meaning of the function ψ is not as clear. Our purpose here is in fact to discern the situations in which the functions ψ and $\hat{\psi}$ could be interpreted in approximately the same manner.

To proceed, we define a quantity

$$\eta = \frac{\left|\left\langle \sum_{P'} (-1)^{p'} P' \hat{\psi} \,\middle|\, \hat{\psi} \right\rangle\right|^2}{\left\langle \sum_{P'} (-1)^{p'} P' \hat{\psi} \,\middle|\, \sum_{P'} (-1)^{p'} P' \hat{\psi} \right\rangle \left\langle \hat{\psi} \,\middle|\, \hat{\psi} \right\rangle} \tag{6.4}$$

which will be referred to as the degree of overlapping between the clusters. From the way η is defined, it is clear that when η is much smaller than unity, the antisymmetrization needs only be carried out within the individual clusters in the calculation of most matrix elements.††† Only in this case does it make sense to consider the clusters as ordinary particles. Therefore, the degree of overlapping η is a quantitative measure of the influence of the Pauli principle upon the behaviour of the cluster function ψ.

Let us now make some general arguments as to how the quantity η depends upon the mutual penetration of the two clusters. Suppose first that the two clusters do not penetrate each other; that is, the most probable separation distance of the two cluster centroids, as determined from the function $\hat{\psi}$, is much larger than the sum of the radii of the two clusters. In this case, $P'\hat{\psi}$ as a function of the permuted nucleon spatial coordinates will be small in a region where $\hat{\psi}$ as a function of the nucleon coordinates is large, and vice versa. Consequently, η will be much less than unity. Suppose, on the other hand, that the

† For instance, in a d + t cluster state of ^{5}He, we have

$$\hat{\psi} = [A_t \widetilde{\phi}_{0t}(\widetilde{r}_1, \widetilde{r}_2, \widetilde{r}_3)]\,[A_d \widetilde{\phi}_{0d}(\widetilde{r}_4, \widetilde{r}_5)]\,\chi(\mathbf{R}_d - \mathbf{R}_t),$$

where A_t and A_d are given by

$$A_t = P_I - P_{12} - P_{13} - P_{23} + P_{13}P_{12} + P_{23}P_{12},$$

$$A_d = P_I - P_{45},$$

with P_I being the identity permutation and P_{ij} being a pair-exchange operator. Also, the operator A' in eq. (6.3) is represented by

$$A' = P_I + (-P_{14} - P_{15} - P_{24} - P_{25} - P_{34} - P_{35} + P_{14}P_{25} + P_{14}P_{35} + P_{24}P_{35})$$

in this particular case.

†† In the special case where A and B are identical clusters, the function $\hat{\psi}$ is re-defined to include also a term which corresponds to the interchange of all the nucleons in these two clusters and the summation in eq. (6.3) should then be modified to exclude a permutation term of this type. In the following discussion, we shall for simplicity always refer to $\hat{\psi}$ as an unantisymmetrized wave function.

††† We mention once more that in the case involving identical clusters, the term corresponding to the interchange of the clusters must also be considered.

most probable separation distance of the clusters is extremely small compared to the sum of the radii of the clusters. Then, by the uncertainty principle, the mean kinetic energy of the two clusters about each other becomes very large. In this semi-classical limit situation, we again expect the influence of the operator A' to be small. To see that this is in fact the case, we consider the cluster function in the momentum representation. Here we see that the internal momentum spread within a cluster is much smaller than the momentum spread of the relative motion.† Hence, $P' \hat{\psi}$ as a function of the permuted nucleon momenta is small when $\hat{\psi}$ as a function of the nucleon momenta is large, and vice versa. Therefore, the degree of overlapping η will be small despite the considerable spatial penetration of the two clusters.†† Between these two extremes, η will become large and we expect the behaviour of the antisymmetrized cluster function ψ to deviate considerably from the ordinary particle picture represented by the function $\hat{\psi}$.

To substantiate these general arguments, we consider the $\alpha + \alpha$ problem as a specific example. In this case, the cluster wave function (see eq. (5.9)) is given by

$$\psi(\bar{r}_1, \ldots, \bar{r}_8) = \hat{N}_{lm}\, A\left\{\exp\left(-\frac{a}{2}\sum_{i=1}^{4} \bar{r}_i^2\right) \exp\left(-\frac{a}{2}\sum_{i=5}^{8} \bar{r}_i^2\right) \times g_n(R)\, Y_{lm}(\theta, \varphi)\, \xi(s, t)\right\}, \tag{6.5}$$

where $g_n(R)$ is chosen to be $R^n \exp(-bR^2)$ as in eq. (5.18), and $\hat{N}_{lm}$ is a normalization constant. The ratio of the relative-motion width parameter b to the internal width parameter a is a measure of the spatial penetration of the clusters. For $b = a$ and $n = 4$, the wave function of eq. (6.5) becomes identical to the ^{8}Be oscillator cluster function, and the clusters penetrate each other strongly, i.e., η should be of the order of 1. For $b/a \ll 1$, the clusters are relatively far apart and η will acquire a value much smaller than unity.

We shall now compare the proton densities $\rho_P(r)$ calculated for various values of b/a, using the antisymmetrized wave function ψ of eq. (6.5) and the corresponding unantisymmetrized wave function $\hat{\psi}$. The values of n and l are chosen as equal to 4 and 0, respectively. The results are shown in fig. 10. Here we see that for relatively small penetration of the clusters with $b/a = 1/4$, the antisymmetrization A' does not influence very much the proton distribution. On the other hand, for larger penetration of the clusters with $b/a = 1$ and 4, the proton distribution is altered more noticeably by the antisymmetrization. As is shown in this figure, the change is always in the direction of reducing the proton density for small r values and increasing it at larger r values. This is understandable because, when the clusters begin to overlap, the Pauli principle forces nucleons of the same spin and isospin to repel each other. For b/a much larger than 5, the proton distributions again become essentially the same for both the antisymmetrized and the unantisymmetrized wave functions. The behaviour of the proton density is thus just what we had expected from our general arguments given above.

---

† This is obvious, because the width parameters in the momentum representation are proportional to the reciprocals of the respective width parameters in the position representation.

†† This second extreme case will not be realized in nature, because the large relative oscillation would destroy the internal cluster structure or cause the nucleus to decay into a system of free particles.

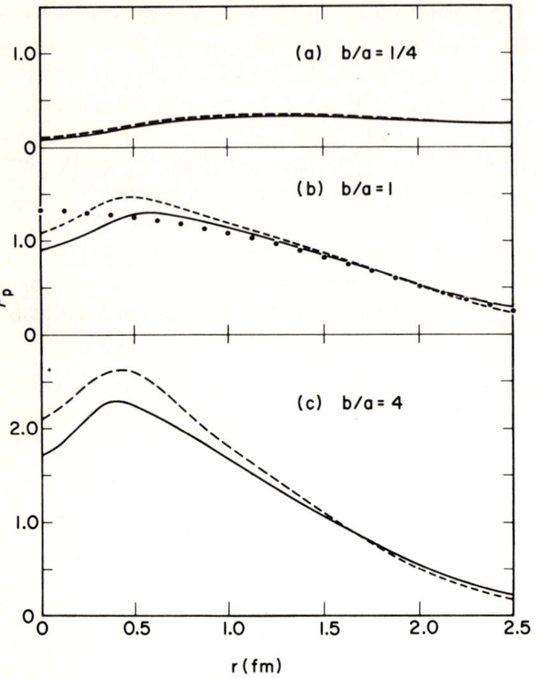

Fig. 10
Proton distributions of ^{8}Be with antisymmetrized (solid curves) and unantisymmetrized (dashed curves) wave functions: (a) b/a = 1/4, (b) b/a = 1, (c) b/a = 4. Solid dots in (b) represent proton densities obtained with $g_n(R)$ given by eq. (6.6).

Even though the proton distribution is significantly changed by the antisymmetrization procedure for b/a = 1 and 4, the change, especially in the oscillator limit case, does not seem to be as large as one might expect at the first moment. The reason for this is that the relative-motion function $g_n(R)$ is chosen to behave as R^4 for small intercluster distances and, consequently, the two α clusters cannot approach each other too closely even when the operator $\mathcal{A}'$ is not taken into consideration. If we choose for $g_n(R)$ the form

$$g_n(R) = \left(1 - \frac{8}{3} bR^2 + \frac{16}{15} b^2 R^4\right) \exp(-bR^2), \tag{6.6}$$

instead of $R^4 \exp(-bR^2)$, then the difference between the proton densities calculated with and without antisymmetrization becomes much larger (see fig. 10b), because now the wave function $\hat{\psi}$ does allow an appreciable probability for the two α clusters to approach each other. It should be noted that, in the oscillator limit case, the normalized antisymmetrized wave functions ψ obtained with $g_n(R) = R^4 \exp(-bR^2)$ and $g_n(R)$ given by eq. (6.6) are entirely identical to each other. Therefore, the results obtained here indicate that if, for computational reasons, one is compelled to use the simplified wave function $\hat{\psi}$ in a calculation of certain matrix elements, then the choice of $g_n(R)$ as $R^4 \exp(-bR^2)$ is to be preferred.

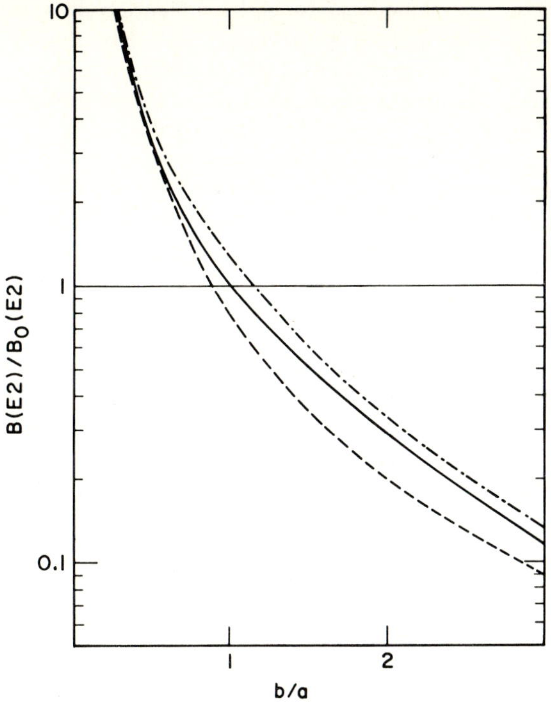

Fig. 11
Dependence of reduced E2 transition probability upon intercluster separation distance. The solid and dashed curves are obtained with antisymmetrized and unantisymmetrized wave functions, respectively. The dot-dashed curve is obtained by a procedure explained in the text.

Next, we compute the reduced transition probability B(E 2), given by (see eq. (5.8))

$$B(E2) = \frac{1}{2J_i + 1} \sum_{\mu} \sum_{M_i} \sum_{M_f} \left| \left\langle \psi_f \right| \sum_{i=1}^{A} \bar{r}_i^2 \, Y_{2\mu}^* (\bar{\theta}_i, \bar{\varphi}_i) \frac{1 - \tau_{i3}}{2} \left| \psi_i \right\rangle \right|^2 , \qquad (6.7)$$

from the 2^+ first excited state to the 0^+ ground state as a function of the parameter b/a. Here again, we shall compare the results obtained using both antisymmetrized and unantisymmetrized wave functions in order to see the influence of the antisymmetrization operator $\mathcal{A}'$. In fig. 11, the ratio $B(E\,2)/B_0(E\,2)$ is plotted versus b/a, where $B_0(E\,2)$ is the value of B(E 2) in the oscillator limit case with b/a = 1. To perform the calculation, we have made the assumption that the radial parts of the relative-motion functions in the initial and final states are the same.

The solid and dashed curves in fig. 11 represent the results calculated using antisymmetrized and unantisymmetrized wave functions, respectively. For b/a appreciably smaller than 1, it is seen that the two curves nearly coincide, as is expected. When b and a become similar in magnitude, we do observe that these curves deviate from each other, indicating that the antisymmetrization operator $\mathcal{A}'$ has a noticeable influence. This influence is, however, not too large, which is a consequence of the fact that the E2 transition operator contains a factor $\bar{r}_i^2$ and is, therefore, a long-range operator. Finally, as b/a takes on values larger than about 5 (not shown in the figure), the effect of antisymmetrization becomes again quite small, just as in the case of the proton density.

To show more clearly that the antisymmetrization operator A' has a relatively minor influence on the matrix elements of long-range operators even when a and b have similar magnitudes, we have calculated the B(E2) value by computing the transition matrix element with unantisymmetrized wave functions, but the initial- and final-state normalization factors with antisymmetrized wave functions. The result is shown by the dot-dashed curve in fig. 11. Here one sees that the deviation from the solid curve is quite small for all values of b/a. This means that the antisymmetrization has a much larger influence on the normalization factors than on the E2 transition matrix element. The reason for this is rather obvious; the normalization operator is not as long-ranged as the E2 transition operator and, therefore, the normalization factors depend more sensitively on the wave functions describing the interior of the nucleus where the antisymmetrization effect is particularly important.

From the results of these calculations, we may conclude that in the surface region of the nucleus, where the separation distances of clusters are relatively large, the clusters can behave approximately like the corresponding free particles. This is not so in the interior of the nucleus, where the degree of overlapping between the clusters is quite appreciable. In other words, the correlations among the nucleons are expected to be particularly strong in the surface region; as a consequence of this, the nuclear surface may have a rather granular structure.

The calculations with the $\alpha + \alpha$ cluster wave functions discussed above have borne out our original arguments that, excepting when the spatial penetration of the clusters is slight or when their relative momentum is great, the degree of cluster overlapping η is large and the behaviour of the antisymmetrized cluster function can be quite different from that of the unantisymmetrized function. We emphasize that these arguments are general and not restricted to this specific example.

Therefore, in speaking about clusters, one has to remember that nucleons are indistinguishable. No individual nucleon can be said to belong to a particular cluster, whereas this cluster may well always contain the same number of nucleons. One may imagine a continuous exchange of nucleons going on between the clusters (from the time required for a nucleon to traverse the nucleus we infer the exchange time to be around 10^{-23} sec). This nucleon exchange decreases very rapidly with decreasing spatial penetration of the clusters and vanishes when the clusters are completely separated. The adequate way to express this is by utilizing the antisymmetrization operator A.

We apply now the above considerations to a specific example which deals with the process of γ transition from the 2^+ first excited state to the 0^+ ground state in ^{20}Ne. From this example, we shall see that, because of the structure of these states, the antisymmetrization operator A' can be omitted in good approximation and, consequently, the calculation can be simplified to a very large extent.

Due to the stable closed-shell structure of the α and the ^{16}O clusters, it is reasonable to presume that the ground state and the low excited states of ^{20}Ne can be described by an $\alpha + {}^{16}$O cluster structure without appreciable internal excitations of the clusters. Then, because of the Pauli exclusion principle and the considerations given above, the relative-motion functions for these states should be appropriately chosen as proportional to R^8 for small intercluster distances. Therefore, even for low-excited states, the two clusters are relatively well separated and the separation distance is fairly well defined. As a consequence of this, one can approximately describe these ^{20}Ne states as positive-parity rotational levels of a rather rigid rotator and omit the antisymmetrization operator A' in the calculation of the rotational energies and the electromagnetic transition probabilities.

The neglect of the operator A' in the wave function allows us to derive a simple relation between the excitation energy of the 2^+ state and the electric quadrupole transition probability from this 2^+ state to the 0^+ ground state. As is well known, the E 2 transition probability is given by

$$T(E2) = \frac{12\pi}{225}\frac{e^2}{\hbar}\, k^5\, B(E2), \tag{6.8}$$

where B (E 2) is given by eq. (6.7). Because of the spherically symmetric structure of the clusters involved, eq. (6.7) may be simplified and one obtains

$$\begin{aligned} B(E2) &= \left(\frac{8}{5}\right)^2 |\langle \psi_f(J_f = 0, M_f = 0)\,|R^2\, Y^*_{22}(\theta,\varphi)|\, \psi_i(J_i = 2, M_i = 2)\rangle|^2 \\ &= \frac{64}{25}\frac{1}{4\pi}\langle R^2\rangle^2, \end{aligned} \tag{6.9}$$

where $\langle R^2\rangle$ is the mean square separation distance between the α and ^{16}O clusters which form the rigid rotator.

In the frame of the rigid-rotator model, the excitation energy of the 2^+ state is related to the moment of inertia J and may be written as

$$\Delta E = 6\,\frac{\hbar^2}{2J}, \tag{6.10}$$

where

$$\frac{1}{2J} = \frac{1}{2\mu}\left\langle\frac{1}{R^2}\right\rangle = \frac{5}{32\,M}\left\langle\frac{1}{R^2}\right\rangle, \tag{6.11}$$

with M being the mass of a nucleon. By combining eqs. (6.8)–(6.11) and by setting

$$\left\langle\frac{1}{R^2}\right\rangle = 1/\langle R^2\rangle \tag{6.12}$$

which follows from the use of the rigid-rotator model, one obtains

$$T(E2) = \frac{3}{100}\frac{(\Delta E)^3}{c^5\,\hbar^2}\left(\frac{e}{M}\right)^2. \tag{6.13}$$

To check how well this relation is satisfied, we substitute into eq. (6.13) the experimental value of 1.63 MeV for ΔE and compute T (E 2). The resultant calculated value for T (E 2) is 1.6×10^{12} sec^{-1}, which compares very well with the experimental value of between 0.9×10^{12} and 2.3×10^{12} sec^{-1} [AJ 66].

In later chapters, we shall discuss further examples where the influence of the Pauli principle can be neglected in good approximation, with the consequence that the calculations can be greatly simplified.

6.3. Energetical Favouring of a Cluster Inside a Large Nucleus

Although cluster overlapping destroys appreciably the simple picture of intranuclear structures as composite particles moving about one another, one important property remains which fundamentally influences all nuclear-structure and reaction problems. If a cluster with a tightly bound closed-shell or near closed-shell configuration is surrounded by many other nucleons, then, even if the degree of overlapping is large, this cluster correlation usually remains energetically favoured over any kind of state in which this cluster is broken up. In this section, we wish to justify this assertion in some detail.

To understand this continued energetical preference of cluster correlations, we consider a simple one-dimensional model where the essential features creating the effect can be seen especially clearly. This model consists of a Fermi sea of nucleons into which we introduce a cluster of a finite number of nucleons. The motion of the nucleons in the Fermi sea we describe by single-particle wave functions, i. e., eigenfunctions in an infinite potential well of width L. The internal motion of the nucleons in the cluster we describe also by single-particle eigenfunctions, but in an infinite potential well of width l much smaller than L. We have not used a more realistic oscillator potential to build up the cluster, because we wish to keep the calculation as simple as possible.

The wave function of the total system is given by

$$\psi = A' \hat{\psi} = A' \{\hat{\phi}_0 \hat{\phi}_1\}, \tag{6.14}$$

where $\hat{\phi}_0$ and $\hat{\phi}_1$ are antisymmetrized functions describing the motion of the nucleons in the Fermi sea and in the cluster, respectively. The normalized wave function $\hat{\phi}_0$ has the form

$$\hat{\phi}_0 = \frac{1}{\sqrt{A_0!}} \left(\frac{2}{L}\right)^{A_0/2} A_0 \left\{ \prod_{n=1}^{A_0/4} \prod_{i=1}^{4} \sin(K_n x_{ni})\, \xi_i(s_{ni}, t_{ni}) \right\}, \quad \left(-\frac{L}{2} \leqslant x_{ni} \leqslant \frac{L}{2}\right) \tag{6.15}$$

with

$$K_n = \frac{2\pi n}{L}. \qquad (n = 1, 2, 3, \ldots) \tag{6.16}$$

In eq. (6.15), the quantity A_0 denotes the number of nucleons in the Fermi sea and the ξ_i are spin-isospin functions. In writing the wave function $\hat{\phi}_0$, it is assumed that the four different spin-isospin states, corresponding to every spatial state characterized by the quantum number n, are all filled up. For the normalized wave function $\hat{\phi}_1$ one writes, analogous to eq. (6.15),

$$\hat{\phi}_1 = \frac{1}{\sqrt{A_1!}} \left(\frac{2}{l}\right)^{A_1/2} A_1 \left\{ \prod_{m=1}^{A_1/4} \prod_{i=1}^{4} \sin(k_m x_{mi})\, \xi_i(s_{mi}, t_{mi}) \right\}, \quad \left(-\frac{l}{2} \leqslant x_{mi} \leqslant \frac{l}{2}\right) \tag{6.17}$$

with

$$k_m = \frac{2\pi m}{l}, \qquad (m = 1, 2, 3, \ldots) \tag{6.18}$$

and A_1 being the number of nucleons in the cluster.

For simplicity, we have assumed that all single-particle wave functions in $\hat{\phi}_0$ and $\hat{\phi}_1$ vanish at the origin. That means we use only single-particle wave functions with odd parity. This restriction does not influence the generality of our considerations, because the following discussion can be applied in exactly the same way with even-parity states. Also, in eqs. (6.15) and (6.17), we have assumed that in the ground state of the Fermi sea and of the cluster, the lowest single-particle levels of the potential wells are filled up, in a way allowed by the Pauli principle.

One can easily recognize that if a nucleon in the cluster is described by a single-particle wave function which has a wave number k_m smaller than or comparable to the wave number $K_F = 2\pi A_0/4L$ of the Fermi limit, then its behaviour will be profoundly influenced by the presence of the Fermi sea through the antisymmetrization operator $\mathcal{A}'$. On the other hand, if it is described by a wave function with k_m appreciably larger than K_F, then the operator $\mathcal{A}'$ will have only very little influence.

To see this in more detail, we expand the single-particle wave functions of the cluster in terms of single particle wave functions of the Fermi sea. That is, we write

$$\varphi_1(k_m; x) = \sqrt{\frac{2}{l}} \sin k_m x = \sum_{n=1}^{\infty} a_{mn} \sqrt{\frac{2}{L}} \sin K_n x. \tag{6.19}$$

The expansion coefficient a_{mn} is given by

$$a_{mn} = \int_{-l/2}^{l/2} \frac{2}{\sqrt{lL}} \sin(k_m x) \sin(K_n x)\, dx = (-1)^m \frac{4}{\sqrt{lL}} \frac{k_m \sin \frac{K_n l}{2}}{K_n^2 - k_m^2}, \tag{6.20}$$

which approaches $\sqrt{l/L}$ as K_n becomes close to k_m. Now it can be easily seen that the behaviour of the total system can also be described by the wave function

$$\psi_C = \mathcal{A}' \hat{\psi}_C = \mathcal{A}' \{\hat{\phi}_0 \hat{\phi}_{1C}\}, \tag{6.21}$$

where

$$\hat{\phi}_{1C} = \hat{N}_1 \frac{1}{\sqrt{A_1!}} \mathcal{A}_1 \left\{ \prod_{m=1}^{A_1/4} \prod_{i=1}^{4} \varphi_{1C}(k_m; x_{mi})\, \xi_i(s_{mi}, t_{mi}) \right\}, \tag{6.22}$$

with

$$\begin{aligned} \varphi_{1C}(k_m; x) &= \sum_{n=K_F L/2\pi}^{\infty} a_{mn} \sqrt{\frac{2}{L}} \sin K_n x, \quad \left(-\frac{L}{2} \leqslant x \leqslant \frac{L}{2}\right) \\ &= 0, \qquad \left(|x| > \frac{L}{2}\right) \end{aligned} \tag{6.23}$$

and $\hat{N}_1$ being a normalization constant. Because of the presence of the antisymmetrization operator $\mathcal{A}'$, one can easily see that the wave function ψ of eq. (6.14) and the wave function ψ_C of eq. (6.21) differ by only a constant factor and are, therefore, equivalent to each other.

We shall examine now the difference in behaviour of the single-particle wave functions φ_1 and φ_{1C} in the two limit cases where $k_m \gg K_F$ and $k_m < K_F$. In the first case, we assume further that the wave-number difference $\Delta k = 2\pi/l$ of two successive levels in the cluster is much smaller than $(k_m - K_F)$, i.e., many levels of the cluster lie above the Fermi sea.

In the first case, a_{mn} has its maximum value of about $\sqrt{l/L}$ when K_n is approximately equal to k_m. The wave-number width of the considered level is of the order of Δk, because already for

$$K_n \approx k_m \pm \frac{\pi}{l}, \tag{6.24}$$

a_{mn}^2 decreases by a factor of about 2.5. Therefore, for such a value of k_m, the two wave functions φ_1 and φ_{1C} are practically the same and the presence of the Fermi sea has virtually no influence upon the behaviour of this single-particle wave function of the cluster.

In the second case where $k_m < K_F$, the situation is quite different. Here, because the coefficient a_{mn} has an appreciable magnitude for $n < K_F L/2\pi$, the single-particle wave functions φ_1 and φ_{1C} have rather different behaviour. To show this more clearly, let us further assume that $(K_F - k_m)$ is much larger than Δk. This means that we consider a cluster nucleon which is deeply embedded into the Fermi sea. For the altered single-particle wave function of this cluster level, we obtain then

$$\begin{aligned} \varphi_{1C}(k_m; x) &\approx \frac{L}{2\pi} \int_{K_F}^{\infty} (-1)^m \frac{4}{\sqrt{lL}} \frac{k_m \sin\frac{Kl}{2}}{K^2 - k_m^2} \sqrt{\frac{2}{L}} \sin Kx \, dK \\ &\propto \left[\frac{\sin K_F\left(\frac{l}{2} + x\right)}{\frac{l}{2} + x} - \frac{\sin K_F\left(\frac{l}{2} - x\right)}{\frac{l}{2} - x} \right] \end{aligned} \tag{6.25}$$

in the range of x from $-L/2$ to $L/2$. From this equation, one sees that φ_{1C} is much more smeared out than the function φ_1 which is confined within the region with $-\frac{l}{2} \leqslant x \leqslant \frac{l}{2}$.

The effect of the antisymmetrization operator A' on the total wave function of the system will now be discussed. The most important thing to note is that, even though the wave functions ψ and ψ_C are equivalent, the antisymmetrization operator A' has a smaller influence in ψ_C than in ψ. In other words, the difference in behaviour between ψ_C and its unantisymmetrized part $\hat{\psi}_C$ is appreciably smaller than that between ψ and its unantisymmetrized part $\hat{\psi}$. The reason for this in that, in constructing the single-particle wave functions φ_{1C}, we have already used the fact that the operator A' is present in the total wave function to eliminate terms in eq. (6.19) with $n < K_F L/2\pi$ (compare eq. (6.19) with eq. (6.23)). In fact, in the discussion of the proton density in section 6.2, we have encountered a similar situation. There it was demonstrated quantitatively that the operator A' has a smaller influence when the radial part of the relative-motion functions is chosen as $R^4 \exp(-bR^2)$ instead of the function $g_n(R)$ of eq. (6.6). Therefore, as a first approximation, we can consider the functions ψ_C and $\hat{\psi}_C$ as having rather similar behaviour. Accepting this approximation, we can then understand quite clearly the difference in behaviour between the functions ψ and $\hat{\psi}$. Thus, while $\hat{\psi}$ describes a system consisting of a Fermi sea and a small cluster of extension l, the function ψ describes instead a system with a much more extended cluster embedded in a Fermi sea. To state it in another way, the antisymmetrization operator A' has the effect of reducing very much the spatial correlations among the cluster nucleons as described by the un-antisymmetrized wave function $\hat{\psi}$.

We use these results now for the discussion of the energetical behaviour of the system consisting of the Fermi sea and the cluster. The normalized wave function of this system can be written as

$$\psi_{Cn} = \left(\frac{A_0!\,A_1!}{A!}\right)^{1/2} \psi_C, \tag{6.26}$$

where ψ_C is given by eq. (6.21) and $A = A_0 + A_1$. It should be noted that, in contrast to the functions φ_1 of eq. (6.19), the single-particle wave functions φ_{1C} in the cluster function $\hat{\phi}_{1C}$ are not mutually orthogonal to one another. However, they are now all orthogonal to the single-particle wave functions of the Fermi sea. In addition, if the highest cluster levels have $k_m \gg K_F$, then the single-particle functions for these levels are also approximately orthogonal to all other φ_{1C}.

We consider first the potential energy of the system. If we assume two-particle forces V_{ik} as interaction forces between the nucleons, then the potential-energy expectation value is given by

$$\overline{V} = \left\langle \psi_{Cn} \left| \frac{1}{2} \sum_{i=1}^{A} \sum_{k \neq i}^{A} V_{ik} \right| \psi_{Cn} \right\rangle. \tag{6.27}$$

By using eqs. (6.26) and (6.21), we can write $\overline{V}$ as a sum of three terms, i.e.,

$$\begin{aligned} \overline{V} = {} & \left\langle \hat{\phi}_0 \left| \frac{1}{2} \sum_{i}^{A_0} \sum_{k \neq i}^{A_0} V_{ik} \right| \hat{\phi}_0 \right\rangle + \left\langle \hat{\phi}_{1C} \left| \frac{1}{2} \sum_{i}^{A_1} \sum_{k \neq i}^{A_1} V_{ik} \right| \hat{\phi}_{1C} \right\rangle \\ & + \left\langle A' \hat{\phi}_0 \hat{\phi}_{1C} \left| \sum_{i}^{A_0} \sum_{k}^{A_1} V_{ik} \right| \hat{\phi}_0 \hat{\phi}_{1C} \right\rangle, \end{aligned} \tag{6.28}$$

where we have introduced the notation

$$\sum_{i}^{A_0} \equiv \sum_{i=1}^{A_0} \quad \text{and} \quad \sum_{k}^{A_1} \equiv \sum_{k=A_0+1}^{A}.$$

In eq. (6.28), the first term $(\overline{V}_0)$ and the second term $(\overline{V}_1)$ represent the potential energies of the Fermi sea and of the cluster, respectively, while the third term $(\overline{V}_{01})$ represents the interaction potential energy of the cluster with the Fermi sea. In deriving this equation, we have used the fact that the single-particle wave functions contained in

$\hat{\phi}_0$ are orthogonal to those contained in $\hat{\phi}_{1C}$. Therefore, any exchange term between the Fermi sea and the cluster makes no contribution to the first and the second term of eq. (6.28).

For further calculations, we assume a simple form for our two-nucleon potential V_{ik}. To simulate the short-range character of nuclear forces, we choose a δ-function potential

$$V_{ik} = -U\,\delta(x_i - x_k), \tag{6.29}$$

which has often been used in nuclear physics to simplify nuclear-structure calculations.

By introducing the nucleon density

$$\rho_0(x) = \left\langle \hat{\phi}_0 \middle| \sum_i^{A_0} \delta(x_i - x) \middle| \hat{\phi}_0 \right\rangle \tag{6.30}$$

of the Fermi sea and the nucleon density

$$\rho_1(x) = \left\langle \hat{\phi}_{1C} \middle| \sum_k^{A_1} \delta(x_k - x) \middle| \hat{\phi}_{1C} \right\rangle \tag{6.31}$$

of the cluster, one obtains

$$\overline{V}_{01} = -\frac{3}{4}\,U \int \rho_0(x)\,\rho_1(x)\,dx \approx -\frac{3}{4}\,U \int \rho_1(x)\,\frac{A_0}{L}\,dx = -\frac{3}{4}\,U\,\frac{A_0\,A_1}{L}. \tag{6.32}$$

The factor $\frac{3}{4}$ in eq. (6.32) appears, due to cancellations from the exchange terms in the antisymmetrized function $A'\{\hat{\phi}_0\,\hat{\phi}_{1C}\}$; it takes into account the fact that nucleons with the same spin and isospin configuration cannot appear at the same spatial location. We see that $\overline{V}_{01}$ does not depend on the specific form of the cluster wave function $\hat{\phi}_{1C}$, as long as the number of nucleons A_0 in the Fermi sea is sufficiently large.

For the evaluation of $\overline{V}_1$, we introduce a pair-correlation function $w_1(x_r, x_s)$,† defined as

$$w_1(x_r, x_s) = \left\langle \hat{\phi}_{1C} \middle| \frac{1}{2} \sum_i^{A_1} \sum_{k \neq i}^{A_1} \delta(x_i - x_r)\,\delta(x_k - x_s) \middle| \hat{\phi}_{1C} \right\rangle. \tag{6.33}$$

Using this function, we obtain

$$\overline{V}_1 = \int V_{rs}\,w_1(x_r, x_s)\,dx_r\,dx_s = -U \int_{-L/2}^{L/2} w_1(x_r, x_r)\,dx_r. \tag{6.34}$$

† The quantity $w_1(x_r, x_s)\,dx_r\,dx_s$ denotes the probability of finding a cluster nucleon between x_r and $x_r + dx_r$ and at the same time another cluster nucleon between x_s and $x_s + dx_s$. This probability is usually not physically measurable, because when the cluster is greatly overlapped by surrounding nucleons, it is not possible to say definitely which nucleons are participating in the cluster correlation at a given moment. Nevertheless, $w_1(x_r, x_s)\,dx_r\,dx_s$ is well defined and enters into the calculation of the cluster potential energy.

Equation (6.34) leads to quite significant results. The potential energy of the cluster depends on the integral of the pair-correlation function $w_1(x_r, x_r)$ for the cluster nucleons. This integral has a rather small value if $w_1(x_r, x_r)$ is constant over the region $-L/2 \leqslant x_r \leqslant L/2$. It has a much larger value if $w_1(x_r, x_r)$ is sharply peaked with a large maximum value somewhere in this region. But this means that the density of cluster nucleons must also be sharply peaked with a large maximum value. Hence, the stronger the cluster correlation (here determined by the cluster well-parameter l), the greater is the contribution of the cluster to the potential energy.†

Opposing this tendency of the potential energy to cluster the nucleons strongly together are the kinetic energy, the Pauli principle, and the repulsive core in the nucleon-nucleon potential (we have not included repulsive-core forces explicitly in the calculation; clearly, they do not affect the over-all nature of eq. (6.34). But as they are one of the principal sources of saturation, we discuss their role here). These three effects limit, often drastically, how tightly the cluster nucleons may be packed together. As we decrease the cluster well-width l, the kinetic energy of the cluster nucleons increases due to the uncertainty principle.†† The Pauli principle forbids nucleons in the same spin and isospin state to come close to one another. No nucleons may come so close that they penetrate deeply into the repulsive-core regions. Since both the Pauli principle and the repulsive-core forces serve to limit the available space in which the nucleons may move, they show up here mainly as an increased contribution to the kinetic energy. The nucleons will pack together as closely as possible until the rise in kinetic energy offsets any further gain in potential energy from the short-range attractive part of the nuclear forces. This important conclusion shows that clusters remain energetically favoured even when surrounded and appreciably overlapped by many other nucleons.

We consider now more quantitatively the energetical behaviour of the cluster in the presence of the Fermi sea. In order to investigate this behaviour, we shall make the assumption that the highest occupied levels of the cluster (which determine, e. g., if the cluster has a closed-shell structure or not) lie well above the occupied levels of the Fermi sea. We emphasize that this assumption is not necessary for our previous conclusion about the energetical preference for the formation of clusters; equations (6.32) and (6.34) are derived without it.

With this assumption, we can write the cluster function $\hat{\phi}_{1C}$ as

$$\hat{\phi}_{1C} = \left(\frac{A_2!\, A_3!}{A_1!}\right)^{1/2} \mathcal{A}'' \{\hat{\phi}_{2C}\, \hat{\phi}_{3C}\}, \tag{6.35}$$

---

† It is obvious that, for this result, the short-range character of the nuclear forces is essential.

†† The expectation value of the kinetic-energy operator is given by

$$\bar{T} = \left\langle \hat{\phi}_0 \middle| \sum_i^{A_0} T_i \middle| \hat{\phi}_0 \right\rangle + \left\langle \hat{\phi}_{1C} \middle| \sum_i^{A_1} T_i \middle| \hat{\phi}_{1C} \right\rangle .$$

Note that because the kinetic-energy operator is a single-particle operator, any exchange term between the Fermi sea and the cluster has no contribution. Therefore, we can discuss here the qualitative behaviour of the kinetic energy by using only the uncertainty principle.

with

$$A_2 + A_3 = A_1 . \tag{6.36}$$

The function $\hat{\phi}_{2C}$ is an antisymmetrized product of single-particle wave functions $\varphi_{1C}\xi_i$ of all occupied cluster levels except the highest ones, with A_2 being the number of these lower cluster levels. The function $\hat{\phi}_{3C}$ is an antisymmetrized product of single-particle wave functions of the remaining $(A_1 - A_2)$ occupied higher cluster levels. Both $\hat{\phi}_{2C}$ and $\hat{\phi}_{3C}$ are normalized to one. The symbol A'' denotes the antisymmetrization operator which interchanges nucleons occupying the lower levels with those occupying the higher levels. To keep the notation simple, it will always be understood that, before antisymmetrization, the first A_0 nucleons are in the Fermi sea, the nucleons from $A_0 + 1$ to $A_0 + A_2$ constitute the group of A_2 lower-energy cluster nucleons, and the nucleons $A_0 + A_2 + 1$ to A constitute the group of A_3 higher-energy cluster nucleons.

Based on our discussion above, the normalized wave function $\hat{\phi}_{3C}$ can be written in good approximation as

$$\hat{\phi}_{3C} = \frac{1}{\sqrt{A_3!}} \left(\frac{2}{l}\right)^{A_3/2} A_3 \left\{ \prod_{m}^{A_3/4} \prod_{i=1}^{4} \sin(k_m x_{mi})\, \xi_i(s_{mi}, t_{mi}) \right\} \qquad \left(-\frac{l}{2} \leqslant x_{mi} \leqslant \frac{l}{2}\right) \tag{6.37}$$

The single-particle wave functions in eq. (6.37) are orthogonal to each other.† Further, they are approximately orthogonal to the single-particle wave functions of the nucleons in the Fermi sea and, in a lesser extent, to the single-particle wave functions contained in $\hat{\phi}_{2C}$.

Using eq. (6.35) and the orthogonality property of the single-particle wave functions just mentioned, we can write the kinetic energy of the cluster as

$$\overline{T}_1 = \left\langle \hat{\phi}_{1C} \middle| \sum_i^{A_1} T_i \middle| \hat{\phi}_{1C} \right\rangle = \left\langle \hat{\phi}_{2C} \middle| \sum_i^{A_2} T_i \middle| \hat{\phi}_{2C} \right\rangle + \left\langle \hat{\phi}_{3C} \middle| \sum_i^{A_3} T_i \middle| \hat{\phi}_{3C} \right\rangle . \tag{6.38}$$

For the second term in eq. (6.38), we obtain by using eq. (6.37)

$$\left\langle \hat{\phi}_{3C} \middle| \sum_i^{A_3} T_i \middle| \hat{\phi}_{3C} \right\rangle = 4 \frac{\hbar^2}{2M} \sum_m^{A_3/4} k_m^2 , \tag{6.39}$$

where the index m goes over the quantum numbers characterizing the spatial behaviour of the $A_3/4$ highest occupied levels of the cluster with 4-fold degeneracy.

We consider now the cluster potential energy. Here again, we employ the ansatz for $\hat{\phi}_{1C}$ as given by eq. (6.35). With this we obtain, analogous to eqs. (6.28), (6.32), and (6.34),

$$\overline{V}_1 = -U \int w_2(x_r, x_r)\, dx_r - \frac{3}{8} U \int [\rho_3(x)]^2\, dx - \frac{3}{4} U \int \rho_2(x)\, \rho_3(x)\, dx , \tag{6.40}$$

† We mention once more that the single-particle wave functions involved in $\hat{\phi}_{2C}$ are not orthogonal to each other.

where the nucleon density functions $\rho_2(x)$ and $\rho_3(x)$ and the pair-correlation function $w_2(x_r, x_s)$ are defined in a similar way as in eqs. (6.30), (6.31), and (6.33). In eq. (6.40), the first term and the second term describe the interaction among the lower-energy (which we shall call "inner") cluster nucleons and among the higher-energy ("outer") cluster nucleons, respectively. The third term represents the interaction between the inner and outer nucleons.

A measure for the correlation strength of a cluster configuration is the breakup energy which one needs to remove a nucleon from one of the highest occupied cluster levels to an unoccupied state of the Fermi sea. For this breakup energy, the contribution of the kinetic energy can be described in good approximation by the variation of the last term in eq. (6.38) only. This kinetic-energy variation is approximately the same as in the case where one considers the breaking up of the cluster without a surrounding Fermi sea, except that, in this latter case, no restriction due to the Pauli principle exists as to which state one can remove the cluster nucleon to. We shall see later that this restriction does not play an important role in the determination of those quantities in which we are interested.

Next, we consider the potential-energy contribution to the breakup energy. We see from eq. (6.40) that the terms which will change are those representing the interaction among just the outer cluster nucleons themselves and the interaction between the inner cluster nucleons and the outer cluster nucleons. Analogous to the kinetic-energy change, the change in the interaction energy of the outer cluster nucleons is in good approximation the same as in the case where the Fermi sea is not present. As for the change in the interaction energy between the inner and the outer nucleons, it is determined by the variation of the third term in eq. (6.40), i.e.,

$$\Delta\overline{V}_{23} = -\frac{3}{4}\,U\int \rho_2(x)\,\Delta\rho_3(x)\,dx. \qquad (6.41)$$

Because of the Pauli principle, $\rho_2(x)$ has a relatively slow variation over the whole spatial region between $-L/2$ and $L/2$. Together with the fact that

$$\int \Delta\rho_3(x)\,dx = 0, \qquad (6.42)$$

this has the consequence that $\Delta\overline{V}_{23}$ has a rather small value. Finally, it should be noted that the interaction energy $\overline{V}_{01}$ of the cluster with the Fermi sea is independent of the cluster nucleon-density variation $\Delta\rho_1(x)$ (see eq. (6.32)) and, therefore, also of $\Delta\rho_3(x)$.

Our considerations show that the energy necessary to remove one or several nucleons from the highest shells of a cluster is mainly determined by the change in the kinetic energy of eq. (6.39) and by the change in the interaction potential energy among the higher-energy or outer cluster nucleons. These changes depend essentially only on the properties of the highest single-particle levels of the cluster. As has been discussed, the properties of these levels are approximately the same as those of the corresponding free particle. Therefore, the breakup energy of the cluster will not be sensitively dependent on whether the cluster is surrounded by other nucleons or not. So we see in this calculable, though special, case of a many-nucleon system, the trend found in light nuclei concerning the energetical behaviour of clusters remains valid.

Our discussion is based on the assumption that the highest occupied levels of the cluster lie well above the occupied levels of the Fermi sea. This may seem arbitrary at first, but is actually fairly realistic. In a practical application to a system consisting of a small cluster and a large cluster, one will normally find that the two clusters oscillate against each other with a rather large number of oscillation quanta. Therefore, in our simplified example, it is appropriate to simulate this situation by assuming the cluster nucleons to have rather large average kinetic energy and, consequently, an appreciable probability of filling single-particle levels with relatively high excitation energies.

We have now to ask if the arguments for the energetical behaviour of clusters, discussed above, are restricted to our special example or do they hold in general. The essential point is that the attractive δ-forces want to pack the nucleons as tightly as possible. While we do not have δ-forces in reality, we do have average attractive nuclear forces of finite but rather short range. Hence, eq. (6.29) does reflect the principal trend of the real forces. A second point is that we have not included the repulsive-core part of the nuclear forces explicitly in our considerations. However, suppose we start with a Fermi sea of uncorrelated nucleons representing nuclear matter. Then, as is well known, the average density of nuclear matter is such that there is only a small probability for nucleons to come so close together as to penetrate into the repulsive-core region [DE 58, GO 58]. Hence, aside from determining the equilibrium density, repulsive-core effects are not very significant for the wave function of the Fermi sea and especially not for its longer-range correlations. On the other hand, it can be easily shown [WI 66] that, for short-range forces, the potential energy of the Fermi sea depends on the integral of the square of the nucleon density. Because of this, we can anticipate immediately that the potential energy will be decreased by introducing cluster correlations, thereby increasing the density in some parts of the Fermi sea and decreasing it in other parts. Eventually, of course, the density in the cluster can be made sufficiently great that repulsive-core effects will become significant, as well as the rise in kinetic energy and Pauli-principle effects. But before we have reached this point, we have already shown the energetical favouring for the existence of clusters which influence especially the longer-range correlation behaviour. Hence, the repulsive-core part of the nuclear forces does not prevent the formation of clusters; rather, it serves primarily only to help limit the final strength of cluster correlations. For this reason, our use of simple attractive δ-forces is justified.

Now let us consider the generality of our model, a cluster in a Fermi sea. The first point is that the results obtained above (see, e. g., eq. (6.34)) are not dependent on the form of the well used to build up the cluster nor on the fact that the model is not three-dimensional. Although the calculation is considerably more complicated, essentially the same results are obtained if the cluster is constructed with an oscillator well. Nor does the Fermi-sea approximation introduce any essential restriction. If we were to narrow the width of the Fermi sea and use a large but realistic number of nucleons, the same over-all behaviour would recur. In fact, if the larger and smaller clusters are described in the frame of the oscillator cluster model, the general results are unchanged, as an explicit but lengthy calculation shows. In particular, the potential-energy contribution of the cluster remains intimately related to the density of nucleons in the cluster.

Finally, we discuss what general inferences may be drawn regarding the most energetically favoured cluster configuration and the related cluster breakup energy. We begin with a plausible suggestion which we then refine. After antisymmetrization with the Fermi sea, the original cluster density is reduced considerably, since the cluster nucleons (particularly the "inner" nucleons) become smeared out over the Fermi sea. It is plausible that this reduced density will be maximally peaked if the original cluster before antisymmetization with the Fermi sea was as densely packed as possible. From this we can draw an immediate conclusion. If our plausibility argument is correct, then closed-shell clusters and near closed-shell clusters in the Fermi sea will usually be energetically favoured over nonclosed-shell clusters because, even after antisymmetrization, closed-shell and near closed-shell clusters will be more densely packed than nonclosed-shell clusters.

Now let us refine the above argument. If we expand an arbitrary antisymmetrized cluster function in terms of, for example, oscillator shell-model functions, the inner (low-energy) nucleons will occupy completely filled oscillator shells. Nothing more can be done to correlate these inner nucleons further without breaking up the shells. On the other hand, the outer nucleons occupy unfilled shells and so can be placed in a superposition of states which packs them tightly. Now we refer back to our consideration of the breakup energy of a cluster in the Fermi sea. Under the assumption made there (that they lie well above the Fermi sea), the single-particle levels of the outer nucleons are negligibly affected by antisymmetrization with the Fermi sea. Hence, they are nearly identical to those of a free cluster. Consequently, the outer nucleons of the cluster in the Fermi sea will be most tightly packed if the cluster is a magic-number cluster. As has just been pointed out, since the outer nucleons are the only ones which can be significantly packed anyway, then closed-shell clusters are more energetically favoured than nonclosed-shell clusters. Suppose we now relax our assumption so that the outer-nucleon levels need not lie far from the top of the Fermi sea. It still remains that the outer nucleons are the ones least affected by antisymmetrization and, therefore, it remains plausible that in a closed-shell cluster, the nucleons are packed more closely than in a nonclosed-shell cluster even when surrounded by other nucleons.

A further refinement yields some idea about the breakup energy of the cluster. In a shell-model expansion of the antisymmetrized cluster function, it is well known that the attractive nuclear forces act most effectively between nucleons in the same shell. Then the energy required to remove the last nucleons from the cluster will be greatest if the last shell is filled. However, if the last shell is filled or almost filled, the outer nucleons provide a significant part of the energy binding the last nucleons to the cluster. Since the outer nucleons are the ones least smeared out by the antisymmetrization, we expect for clusters, and especially for closed-shell and near closed-shell clusters, that the breakup energy to remove the last nucleons will not deviate too greatly from that for the corresponding free cluster. In particular, it is expected that the difference in breakup energies for neighbouring clusters will not depend too much on whether the clusters are surrounded by other nucleons or not.

Hence, we conclude generally that closed-shell and near closed-shell clusters usually remain energetically preferred even when highly overlapped by many surrounding nucleons, and that the breakup energies of clusters and particularly the breakup-energy differences in neighbouring clusters are similar to those when the clusters are free particles.

We shall now briefly apply our results to some specific examples. In the case of light nuclei, one can explicitly show, as we have often discussed already, that closed-shell or nearly closed-shell clusters, such as triton clusters, α clusters, ^{16}O clusters, etc., in their ground states remain energetically favoured compared with other clusters, even if they are surrounded by other nucleons. Deuteron clusters are usually even more tightly bound when they are embedded in other nucleons than when they are free particles (see subsection 3.6b). As for heavy clusters, such explicit calculations are much more difficult, and we have to rely at the moment on the results of our general considerations.

As an example of heavy-cluster behaviour, let us qualitatively examine the effects of cluster overlapping on a heavy closed-shell cluster within a larger nucleus. For simplicity, we consider a cluster having 50 neutrons. As far as cluster overlapping is concerned, we can neglect the presence of the protons. Let us assume that our heavy compound nucleus has a total number of 132 neutrons, so that the other cluster, having 82 neutrons, can also be a closed-shell cluster. These particular magic-number clusters are of interest in connection with asymmetric fission. As usual, to indicate what remains of a cluster correlation after antisymmetrization, we shall consider our example simultaneously in several representations: in the oscillator shell model, in the oscillator cluster model, and with generalized cluster functions.

In the oscillator shell model, the energetically lowest state allowed by the Pauli principle for 132 neutrons has the first six oscillator shells filled and the seventh partially occupied by the last 20 neutrons (see appendix B). This state has 540 oscillator quanta of energy. If this oscillator shell-model function is now considered in conjunction with the nuclear Hamiltonian, then the noncentral parts of the nuclear forces will strongly favour the outside nucleons to have $\boldsymbol{l}$ parallel with $\mathbf{s}$ as much as possible. Hence, 14 of the 20 outside neutrons will occupy the 14 possible single-particle levels of the $1\,i_{13/2}$ subshell in which $\boldsymbol{l}$ and $\mathbf{s}$ are parallel. These 14 neutrons along with the 112 neutrons filling the first six oscillator shells comprise the nuclear closed-shell configuration of 126 neutrons [GO 55]. The remaining 6 neutrons occupy other subshells of the 7th oscillator shell.

In the oscillator cluster model, we construct the internal functions of the clusters with single-particle functions of an oscillator well.† Therefore, due to the Pauli principle, a closed-shell cluster of 50 neutrons must have 130 oscillator quanta of internal energy. If we take into account the nuclear noncentral forces, the ten outside neutrons in the unfilled 5th oscillator shell will occupy the 10 levels of the $1\,g_{9/2}$ subshell in which $\boldsymbol{l}$ and $\mathbf{s}$ are alligned, thereby completing the nuclear 50-neutron closed shell. Similarly, a closed-shell cluster of 82 neutrons will have 270 oscillator quanta of internal energy, and the twelve neutrons in the unfilled 6th oscillator shell will occupy the 12 levels of the $1\,h_{11/2}$ subshell. The lowest relative oscillation of the two clusters must have at least 140 oscillator quanta of energy; for we have seen in the oscillator-shell model that a wave function for 132 neutrons of less than 540 oscillator quanta of excitation energy vanishes under antisymmetrization.

† We use an oscillator well again to construct the internal function, instead of a square well as in the cluster plus Fermi-sea example, in order to facilitate comparison with the oscillator shell model.

In reality, this relative oscillation will have more than 140 oscillator quanta of excitation energy. The reason for this is as follows. The formation of large closed-shell clusters with 50 and 82 neutrons will demand such strong correlations among the nucleons that, in the oscillator shell-model representation, more single-particle states are needed for such a description than are available in the first seven oscillator shells. Because of the Pauli principle, this means that single-particle states of the 8th and higher oscillator shells are also required to describe these very strong correlations. As a result of this, particularly the single-particle states of the 6th oscillator shell and of the $1\,i_{13/2}$ subshell will not be completely filled, and a large increase in the kinetic energy of the relative oscillation is expected.

The main point of the above discussion is that closed-shell clusters with 50 and 82 neutrons can only be present if the resultant compound nucleus is excited. As a consequence, one expects that this excited compound-nucleus configuration will usually have a stronger internal deformation than the ground-state configuration. It should be noted, however, that because the antisymmetrization reduces the differences between different structures, the excitation energy and the internal deformation for this excited configuration will be smaller than what one would expect for a similar cluster structure if one considers the nucleons as distinguishable.† This means that one can obtain such a compound-nucleus configuration containing these two closed-shell clusters by only moderately breaking up the higher oscillator shells and the $1\,i_{13/2}$ subshell.

If we now leave the oscillator assumption and use generalized cluster functions to describe even larger deformations of the compound nucleus, then with increasing deformation the energetical preference of the 50- and 82-neutron magic-number clusters as substructures will further increase. To see this, consider an expansion of the generalized cluster function describing the deformed nucleus into oscillator shell-model functions. As the deformation becomes larger, one will find that terms in which the lowest single-particle levels are not completely filled and higher single-particle levels are occupied will appear with larger amplitude. All the nucleons in unfilled shells can now be correlated so as to increase the nucleon density within the limits of the Pauli principle and, by the general form of eq. (6.34), the potential energy will decrease (without an increase in the kinetic energy). This shows, therefore, that with increasing deformation the nucleon correlations within the clusters will approach more and more the correlations of free particles corresponding to the considered clusters.

Our discussion in this example shows that the average kinetic energy of relative motion between unexcited or not too highly excited clusters can be rather large. Therefore, the assumption that the highest single-particle levels of the clusters are not influenced appreciably by the Pauli principle is quite valid for such nuclear configurations. As has already been mentioned several times, one consequence of this is that the breakup-energy difference in neighbouring clusters is almost independent of whether the clusters are surrounded by other nucleons or not. We emphasize that this conclusion is quite useful and will play an important role in our future considerations on nuclear fission in chapter 17.

† As has been discussed, the antisymmetrization smears out the nucleon density of mutually penetrating clusters, thereby reducing the internal deformation of the compound-nucleus configuration.

7. Scattering and Reaction Calculations

7.1. General Remarks

In this chapter, we apply the unified theory formulated in chapter 4 to study nuclear scattering and reaction problems. As was mentioned there, the procedure involved in the application of this theory is quite straightforward. One assumes a trial function which contains a sufficiently large number of linear variational functions and linear variational amplitudes required for a proper description of the behaviour of the system under consideration. By using the projection equation (2.3) and the integration techniques described in section 5.2, a set of coupled equations satisfied by these linear functions and amplitudes is derived. These equations are then solved, subject to appropriate asymptotic boundary conditions, to yield the S-matrix and, subsequently, the differential scattering and reaction cross sections and other quantities of interest.

A brief discussion on how to derive the set of coupled equations is given in section 7.2. In this section, we shall first (subsections 7.2a and 7.2b) illustrate the procedure involved by considering the $n + \alpha$ scattering problem of subsection 4.2a in more detail. Then (subsection 7.2c), we shall discuss a slightly different way of solving scattering and reaction problems based on using Hulthén-Kohn-type variational functions [WU 62]. The results obtained by applying the unified theory to a number of specific problems will be described in section 7.3. As will be seen, these particular problems are chosen because they serve to illustrate the importance of including specific distortion effects, reaction channels, etc. in the calculation. Finally, in section 7.4, concluding remarks will be made.

7.2. Derivation of Coupled Equations

7.2a. Single-Channel Problem

For illustration, the $n + \alpha$ scattering problem in the single-channel approximation will be used [TH 71]. In this approximation, the wave function is written as

$$\psi = A\{\phi_0(1234)\, F(\mathbf{r}_5 - \mathbf{R}_\alpha)\, \xi(s, t)\}, \tag{7.1}$$

where $\mathbf{R}_\alpha$ is the position vector of the center-of-mass of the α cluster. The function $\phi_0(1234)$ describes the spatial behaviour of the α cluster and is assumed as

$$\phi_0(1234) = \exp\left[-\frac{a}{2}\sum_{i=1}^{4}(\mathbf{r}_i - \mathbf{R}_\alpha)^2\right]. \tag{7.2}$$

The function $\xi(s, t)$ is an appropriate spin-isospin function, given by

$$\xi(s, t) = [\alpha_1 \nu_1 \alpha_2 \pi_2 \beta_3 \nu_3 \beta_4 \pi_4]\, \alpha_5 \nu_5. \tag{7.3}$$

In the above equation, the function inside the square brackets describes the spin and isospin configuration of the α cluster before antisymmetrization. As has been explained previously, it is an eigenfunction of the operators $S_{\alpha z}$ and $T_{\alpha 3}$, but not of the operators $\mathbf{S}_\alpha^2$ and $\mathbf{T}_\alpha^2$. However, one can easily see that, because of the symmetrical structure of ϕ_0 (1234), it does become an eigenfunction with $S_\alpha = 0$ and $T_\alpha = 0$ after antisymmetrization. Therefore, with the choice of ξ as given by eq. (7.3), the function ψ is a proper eigenfunction of $\mathbf{S}^2$ and $\mathbf{T}^2$ with quantum numbers $S = \frac{1}{2}$ and $T = \frac{1}{2}$.

The Hamiltonian of the system is given by eq. (4.6), i.e.,

$$H = \sum_{i=1}^{5} \frac{1}{2M} p_i^2 + \sum_{i<j=1}^{5} V_{ij} - T_C. \tag{7.4}$$

For the nucleon-nucleon potential, we use

$$V_{ij} = -V_0 \exp(-\kappa r_{ij}^2)(w - m P_{ij}^\sigma P_{ij}^\tau + b P_{ij}^\sigma - h P_{ij}^\tau) + \frac{e^2}{r_{ij}} \frac{1-\tau_{i3}}{2} \frac{1-\tau_{j3}}{2}, \tag{7.5}$$

with P_{ij}^σ and P_{ij}^τ being the usual spin and isospin exchange operators, respectively.

For the evaluation of multi-dimensional integrals appearing in this problem, one can use either the cluster-coordinate technique described in subsection 5.2a or the generator-coordinate technique described in subsection 5.2b. With the cluster-coordinate technique, one defines three independent α-cluster internal coordinates $\bar{\mathbf{r}}_i$ (i = 1, 2, 3), the relative coordinate $\mathbf{R}$, and the center-of-mass coordinate $\mathbf{R}_C$ in the following way:

$$\begin{aligned} \bar{\mathbf{r}}_i &= \mathbf{r}_i - \mathbf{R}_\alpha, \\ \mathbf{R} &= \mathbf{r}_5 - \mathbf{R}_\alpha, \\ \mathbf{R}_C &= \tfrac{1}{5}(\mathbf{r}_5 + 4\mathbf{R}_\alpha). \end{aligned} \tag{7.6}$$

By using these coordinates, the kinetic-energy part of H can then be written as

$$\sum_{i=1}^{5} \frac{1}{2M} p_i^2 - T_C = T_\alpha - \frac{\hbar^2}{2\mu} \nabla_{\mathbf{R}}^2, \tag{7.7}$$

with T_α being the α-cluster internal kinetic-energy operator. From eq. (7.7), one sees that the total center-of-mass coordinate $\mathbf{R}_C$ does not appear. Therefore, because of the particular structure of the wave function ψ, the degree of freedom associated with the center-of-mass motion can be handled easily and will not be further considered.

The scattering function $F(\mathbf{R})$ in eq. (7.1) is determined from the projection equation (2.3). By writing the variation of ψ as

$$\delta\psi = \int \delta F(\mathbf{R}') \mathcal{A}\{\phi_0(1234)\,\delta(\mathbf{R}-\mathbf{R}')\,\xi(s,t)\}\, d\mathbf{R}', \tag{7.8}$$

one obtains, by using eq. (2.3),

$$\langle \phi_0(1234)\,\delta(\mathbf{R}-\mathbf{R}')\,\xi(s,t) | H - E | \mathcal{A}\{\phi_0(1234)\,F(\mathbf{R})\,\xi(s,t)\}\rangle = 0. \tag{7.9}$$

The summation over spin and isospin coordinates in eq. (7.9) can be performed in a manner similar to that described in subsection 5.2a for the ^{8}Be problem. In our present case, one can easily see that in the antisymmetrized wave function on the ket side, only two terms yield nonvanishing contributions. These are the direct or no-exchange term and the one-particle exchange term where nucleon 1 is interchanged with nucleon 5. Therefore, after this summation, eq. (7.9) is reduced to

$$\begin{aligned}&\int \phi_0\,(1234)\,\delta\,(\mathbf{R}-\mathbf{R}')\,(H_0-\mathrm{E})\,\phi(1234;5)\,\mathrm{d}\bar{\mathbf{r}}_1\,\mathrm{d}\bar{\mathbf{r}}_2\,\mathrm{d}\bar{\mathbf{r}}_3\,\mathrm{d}\mathbf{R}\\ &=\int \phi_0\,(1234)\,\delta\,(\mathbf{R}-\mathbf{R}')\,(H_1-\mathrm{E})\,\phi(5234;1)\,\mathrm{d}\bar{\mathbf{r}}_1\,\mathrm{d}\bar{\mathbf{r}}_2\,\mathrm{d}\bar{\mathbf{r}}_3\,\mathrm{d}\mathbf{R},\end{aligned} \tag{7.10}$$

where we have introduced the notation

$$\phi(1234;5)=\phi_0\,(1234)\,\mathrm{F}\,(\mathbf{R}), \tag{7.11}$$

and

$$\begin{aligned}\phi(5234;1)=\mathrm{P}^{\mathrm{r}}_{15}\,\phi(1234;5)=\int \phi_0\,(1234)\exp\left[-\frac{2}{5}\,\mathrm{a}\,(\mathbf{R}^2-\mathbf{R}''^2)\right]\mathrm{F}\,(\mathbf{R}'')\\ \times\,\delta\left(\mathbf{R}''+\frac{1}{4}\mathbf{R}-\frac{5}{4}\bar{\mathbf{r}}_1\right)\,\mathrm{d}\mathbf{R}'',\end{aligned} \tag{7.12}$$

with P^{r} being the space-exchange operator. Also, in eq. (7.10), the operators H_0 and H_1 are defined as

$$H_0=\mathrm{T}_\alpha-\frac{\hbar^2}{2\mu}\,\nabla^2_{\mathbf{R}}+\frac{\mathrm{e}^2}{\mathrm{r}_{24}}-\mathrm{V}_0\,(6\,\mathrm{w}+6\,\mathrm{m})\,\mathrm{v}_{12}-\mathrm{V}_0\,(4\,\mathrm{w}-\mathrm{m}+2\,\mathrm{b}-2\,\mathrm{h})\,\mathrm{v}_{15}, \tag{7.13}$$

and

$$\begin{aligned}H_1=\mathrm{T}_\alpha-\frac{\hbar^2}{2\mu}\,\nabla^2_{\mathbf{R}}+\frac{\mathrm{e}^2}{\mathrm{r}_{24}}+\mathrm{V}_0\,(-\,\mathrm{w}+4\,\mathrm{m}-2\,\mathrm{b}+2\,\mathrm{h})\,\mathrm{v}_{15}\\ +\,\mathrm{V}_0\,(-\,3\,\mathrm{w}-3\,\mathrm{m})\,(\mathrm{v}_{12}+\mathrm{v}_{25})+\mathrm{V}_0\,(-\,3\,\mathrm{w}-3\,\mathrm{m})\,\mathrm{v}_{23},\end{aligned} \tag{7.14}$$

with v_{ij} being the form factor of the nucleon-nucleon potential, given by

$$\mathrm{v}_{ij}=\exp(-\,\kappa\,\mathrm{r}^2_{ij}). \tag{7.15}$$

It should be noted that, in eq. (7.10), the left side gives rise to the usual direct terms, while the right side contains the nonlocal exchange terms.

Equation (7.10) can be further simplified by carrying out the integration on the left side and using the equation

$$\mathrm{E}=\mathrm{E}_\alpha+\mathrm{E}_\mathrm{R}, \tag{7.16}$$

where E_R is the total kinetic energy of the neutron and the α particle at large separation in the center-of-mass system, and E_α is the internal energy of the α particle, given by

$$\mathrm{E}_\alpha=\frac{1}{\mathrm{N}^2}\int \phi_0\,(1234)\left[\mathrm{T}_\alpha-\mathrm{V}_0\,(6\,\mathrm{w}+6\,\mathrm{m})\,\mathrm{v}_{12}+\frac{\mathrm{e}^2}{\mathrm{r}_{24}}\right]\phi_0\,(1234)\,\mathrm{d}\bar{\mathbf{r}}_1\,\mathrm{d}\bar{\mathbf{r}}_2\,\mathrm{d}\bar{\mathbf{r}}_3, \tag{7.17}$$

with

$$\mathrm{N}=\left(\frac{\pi^3}{4\,\mathrm{a}^3}\right)^{3/4}. \tag{7.18}$$

The result is

$$\left[-\frac{\hbar^2}{2\mu}\nabla^2_{\mathbf{R}'} - E_R + V_D(\mathbf{R}')\right]F(\mathbf{R}') + K_1 + K_2 + K_3 = 0, \tag{7.19}$$

with

$$V_D = -V_0(4w - m + 2b - 2h)\frac{1}{N^2}\int \phi_0(1234)\,\delta(\mathbf{R}-\mathbf{R}') \times v_{15}\,\phi_0(1234)\,d\bar{\mathbf{r}}_1\,d\bar{\mathbf{r}}_2\,d\bar{\mathbf{r}}_3\,d\mathbf{R}, \tag{7.20}$$

$$K_1 = \frac{1}{N^2}\int \phi_0(1234)\,\delta(\mathbf{R}-\mathbf{R}')\left(\frac{\hbar^2}{2\mu}\nabla^2_{\mathbf{R}} - T_\alpha\right)\phi(5234;1)\,d\bar{\mathbf{r}}_1\,d\bar{\mathbf{r}}_2\,d\bar{\mathbf{r}}_3\,d\mathbf{R}, \tag{7.21}$$

$$K_2 = \frac{1}{N^2}\int \phi_0(1234)\,\delta(\mathbf{R}-\mathbf{R}')\Big[-V_0(-w + 4m - 2b + 2h)\,v_{15} - V_0(-3w - 3m)(v_{12} + v_{25}) - V_0(-3w - 3m)\,v_{23} - \frac{e^2}{r_{24}}\Big]\phi(5234;1)\,d\bar{\mathbf{r}}_1\,d\bar{\mathbf{r}}_2\,d\bar{\mathbf{r}}_3\,d\mathbf{R}, \tag{7.22}$$

and

$$K_3 = E\frac{1}{N^2}\int \phi_0(1234)\,\delta(\mathbf{R}-\mathbf{R}')\,\phi(5234;1)\,d\bar{\mathbf{r}}_1\,d\bar{\mathbf{r}}_2\,d\bar{\mathbf{r}}_3\,d\mathbf{R}. \tag{7.23}$$

The quantity V_D is the familiar direct potential between the α particle and the neutron; as is evident from eq. (7.17), it is obtained by folding the direct part of the nucleon-nucleon potential into the α-particle matter-density distribution [GR 68]. The quantities K_1, K_2, and K_3 are nonlocal terms, arising from the use of a totally antisymmetric wave function in our calculation.

To complete the formulation, we explicitly carry out the spatial integrations in eqs. (7.20)–(7.23). This yields

$$V_D = -V_0(4w - m + 2b - 2h)\left(\frac{4a}{4a + 3\kappa}\right)^{3/2} \exp\left(-\frac{4a\kappa}{4a + 3\kappa}\,\mathbf{R}'^2\right), \tag{7.24}$$

and

$$K_i = \int K_i(\mathbf{R}', \mathbf{R}'')\,F(\mathbf{R}'')\,d\mathbf{R}'', \tag{7.25}$$

where

$$K_1(\mathbf{R}', \mathbf{R}'') = -\frac{\hbar^2}{2M}\left(\frac{4}{5}\right)^3\left(\frac{4a}{3\pi}\right)^{3/2}\left[\frac{47}{5}a - \frac{1216}{1125}a^2(\mathbf{R}'^2 + \mathbf{R}''^2) - \frac{1568}{1125}a^2\,\mathbf{R}'\cdot\mathbf{R}''\right] \times \exp\left[-\frac{34}{75}a(\mathbf{R}'^2 + \mathbf{R}''^2) - \frac{32}{75}a\,\mathbf{R}'\cdot\mathbf{R}''\right], \tag{7.26}$$

$$K_2(\mathbf{R}',\mathbf{R}'') = -V_0\left(\frac{4}{5}\right)^3\left(\frac{4a}{3\pi}\right)^{3/2}$$
$$\times\Bigg\{(-w+4m-2b+2h)\exp\left[-\frac{34a+48\kappa}{75}(\mathbf{R}'^2+\mathbf{R}''^2)-\frac{32a-96\kappa}{75}\mathbf{R}'\cdot\mathbf{R}''\right]$$
$$+(-3w-3m)\left(\frac{3a}{3a+2\kappa}\right)^{3/2}$$
$$\times\left[\exp\left(-\frac{34a^2+28a\kappa}{75a+50\kappa}\mathbf{R}'^2-\frac{34a^2+108a\kappa}{75a+50\kappa}\mathbf{R}''^2-\frac{32a^2+64a\kappa}{75a+50\kappa}\mathbf{R}'\cdot\mathbf{R}''\right)\right.$$
$$\left.+\exp\left(-\frac{34a^2+108a\kappa}{75a+50\kappa}\mathbf{R}'^2-\frac{34a^2+28a\kappa}{75a+50\kappa}\mathbf{R}''^2-\frac{32a^2+64a\kappa}{75a+50\kappa}\mathbf{R}'\cdot\mathbf{R}''\right)\right]$$
$$+(-3w-3m)\left(\frac{a}{a+2\kappa}\right)^{3/2}\exp\left[-\frac{34}{75}a(\mathbf{R}'^2+\mathbf{R}''^2)-\frac{32}{75}a\,\mathbf{R}'\cdot\mathbf{R}''\right]\Bigg\}$$
$$-e^2\left(\frac{4}{5}\right)^3\left(\frac{4a}{3\pi}\right)^{3/2}\left(\frac{2a}{\pi}\right)^{1/2}\exp\left[-\frac{34}{75}a(\mathbf{R}'^2+\mathbf{R}''^2)-\frac{32}{75}a\,\mathbf{R}'\cdot\mathbf{R}''\right], \qquad (7.27)$$

and

$$K_3(\mathbf{R}',\mathbf{R}'') = E\left(\frac{4}{5}\right)^3\left(\frac{4a}{3\pi}\right)^{3/2}\exp\left[-\frac{34}{75}a(\mathbf{R}'^2+\mathbf{R}''^2)-\frac{32}{75}a\,\mathbf{R}'\cdot\mathbf{R}''\right]. \qquad (7.28)$$

One sees from eqs. (7.26)–(7.28) that, because of antisymmetrization, the effective interaction between the neutron and the α particle is not only nonlocal but also explicitly energy-dependent.

From eq. (7.19) and eqs. (7.24)–(7.28), it is seen that the total integrodifferential operator which acts on the relative-motion function $F(\mathbf{R}')$ is not only translationally invariant but also rotationally invariant. This of course must be the case, because the original five-nucleon Hamiltonian of eq. (7.4) and the cluster internal function ϕ_0 (1234) possess such properties. Therefore, to solve eq. (7.19), it is useful to make the following partial-wave expansions for $F(\mathbf{R}')$ and $K_i(\mathbf{R}',\mathbf{R}'')$:

$$F(\mathbf{R}') = \sum_{l=0}^{\infty}\frac{1}{R'}f_l(R')P_l(\cos\theta'), \qquad (7.29)$$

and

$$K_i(\mathbf{R}',\mathbf{R}'') = \sum_{l=0}^{\infty}\frac{2l+1}{4\pi}\frac{1}{R'R''}k_{il}(R',R'')P_l(\cos\gamma), \qquad (7.30)$$

where γ denotes the angle between $\mathbf{R}'$ and $\mathbf{R}''$. Now, by substituting eqs. (7.29) and (7.30) into eq. (7.19) and making use of the orthonormality relations for spherical harmonics, we obtain

$$\left\{-\frac{\hbar^2}{2\mu}\left[\frac{d^2}{dR'^2}-\frac{l(l+1)}{R'^2}\right]-E_R+V_D(R')\right\}f_l(R')+\int_0^{\infty}k_l(R',R'')f_l(R'')\,dR''=0, \qquad (7.31)$$

with

$$k_l(R', R'') = \sum_{i=1}^{3} k_{il}(R', R''). \tag{7.32}$$

Thus, we see that, by making a partial-wave expansion, eq. (7.19) is decomposed into a set of ordinary integrodifferential equations for the radial functions $f_l(R')$.† As is well known, this is always convenient for numerical computations, because these equations depend only on a single variable R′.

As has been mentioned, another way to perform multi-dimensional integrals involved in eq. (7.9) is by using the generator-coordinate technique described in subsection 5.2b. Since this technique will be very useful in solving scattering and reaction problems involving medium- and heavy-weight nuclei, we shall indicate briefly how the calculation in the n + α case proceeds when this particular technique is used.

With the generator-coordinate technique, one writes in eq. (7.9)

$$\begin{aligned}\phi_0(1234)\,\delta(\mathbf{R}-\mathbf{R}')\,\xi(s,t) &= \exp\left[-\frac{a}{2}\sum_{i=1}^{4}(\mathbf{r}_i-\mathbf{R}_\alpha)^2\right]\delta(\mathbf{R}-\mathbf{R}')\,\xi(s,t)\\ &= \int \exp\left[-\frac{a}{2}\sum_{i=1}^{4}(\mathbf{r}_i-\mathbf{R}'_\alpha)^2\right]\xi(s,t)\,\delta(\mathbf{R}_\alpha-\mathbf{R}'_\alpha)\,\delta(\mathbf{R}-\mathbf{R}')\,d\mathbf{R}'_\alpha\end{aligned} \tag{7.33}$$

and

$$\begin{aligned}&A\{\phi_0(1234)\,F(\mathbf{R})\,\xi(s,t)\}\\ &= A\int\left\{\exp\left[-\frac{a}{2}\sum_{i=1}^{4}(\mathbf{r}_i-\mathbf{R}''_\alpha)^2\right]\times\xi(s,t)\,\delta(\mathbf{R}_\alpha-\mathbf{R}''_\alpha)\,\delta(\mathbf{R}-\mathbf{R}'')\right\}F(\mathbf{R}'')\,d\mathbf{R}''_\alpha\,d\mathbf{R}''.\end{aligned} \tag{7.34}$$

If one introduces now for the δ-functions integral representations of the form given by eq. (5.27), then the right sides of eqs. (7.33) and (7.34) can be represented as integrals over products of single-particle wave functions, because, as was explained in subsection 5.2b, cross terms of the type $\mathbf{r}_i \cdot \mathbf{r}_j$ no longer appear.

The derivation of the integrodifferential equation (7.19) is achieved by carrying out the integrations over the nucleon coordinates $\mathbf{r}_i$, the generator coordinates $\mathbf{S}'_\alpha$, $\mathbf{S}''_\alpha$, $\mathbf{S}'$, and $\mathbf{S}''$ which are contained in the integral representations of the δ-functions, and the parameter coordinates $\mathbf{R}'_\alpha$ and $\mathbf{R}''_\alpha$, but not the parameter coordinates $\mathbf{R}'$ and $\mathbf{R}''$. We should mention that, in this derivation, it is useful to multiply the wave function ψ by a term which localizes the center-of-mass of the n + α system in order to prevent the appearance of a trivial divergent integral. The reason for this is that, without such a term, the wave function ψ represents a state in which the amplitude of the center-of-mass motion is constant over the whole space.

† An alternative procedure is to make at the beginning a partial wave expansion for F(**R**) in the wave function ψ of eq. (7.1). Substitution of this wave function and a corresponding expression for δψ into the projection equation (2.3) then leads directly to eq. (7.31). (See also subsection 7.2b).

In a more complicated system, the wave function will contain several relative-motion functions as linear variational functions and a number of bound-state-type distortion functions with linear variational amplitudes. For such a system, the derivation of the coupled equations will require a great deal of laborious effort; however, it should be clear that the procedure involved in this derivation will remain the same as in this simple n + α single-channel problem.

The general procedure also does not change, if the nucleon-nucleon potential contains noncentral and repulsive-core components. The derivation of the coupled equations will of course be more complicated, because it is now necessary to include in the wave function a Jastrow anti-correlation factor, as has been mentioned many times in previous chapters.

7.2b. Coupled-Channel Problem

As an example of a coupled-channel calculation, we discuss the coupling of the n + α channel with the d + t channel. For illustration, we shall consider only the case with $J^\pi = \frac{3}{2}^+$, with the n + α configuration having $l = 2$ and $S = \frac{1}{2}$ and the d + t configuration having $l = 0$ and $S = \frac{3}{2}$.

The wave function of the total system is written as

$$\begin{aligned}\psi = A\Big\{ &\phi_0(\alpha) \frac{f_1(\hat{R}_1)}{\hat{R}_1} \left[-\sqrt{\tfrac{1}{5}}\, Y_{21}(\hat{\theta}_1, \hat{\varphi}_1)\, \alpha_5 + \sqrt{\tfrac{4}{5}}\, Y_{22}(\hat{\theta}_1, \hat{\varphi}_1)\, \beta_5 \right] \\ &\times (\alpha_1 \nu_1 \alpha_2 \pi_2 \beta_3 \nu_3 \beta_4 \pi_4)\, \nu_5 \Big\} \\ +A\Big\{ &\phi_0(t)\, \phi_0(d) \frac{f_2(\hat{R}_2)}{\hat{R}_2}\, Y_{00}(\hat{\theta}_2, \hat{\varphi}_2) (\alpha_1 \nu_1 \alpha_2 \pi_2 \beta_3 \nu_3)(\alpha_4 \nu_4 \alpha_5 \pi_5) \Big\},\end{aligned} \tag{7.35}$$

where $(\hat{R}_1, \hat{\theta}_1, \hat{\varphi}_1) = \hat{\mathbf{R}}_1$ and $(\hat{R}_2, \hat{\theta}_2, \hat{\varphi}_2) = \hat{\mathbf{R}}_2$ denote the channel radii as defined by Eq. (4.9). The expression inside the square brackets in eq. (7.35) is a spin-angle function which describes the coupling of $l = 2$ and $S = \frac{1}{2}$ into $J = \frac{3}{2}$ and $M_J = \frac{3}{2}$. The radial functions $f_1(\hat{R}_1)$ and $f_2(\hat{R}_2)$ are linear variational functions and will be obtained by solving the projection equation (2.3).

By substituting eq. (7.35) into eq. (2.3) we obtain, after a lengthy procedure similar to that described in subsection 7.2a, the following coupled integrodifferential equations:

$$\begin{aligned}&\left[-\frac{\hbar^2}{2\mu_1} \left(\frac{d^2}{d\hat{R}_1'^2} - \frac{6}{\hat{R}_1'^2} \right) + V_{D1}(\hat{R}_1') \right] f_1(\hat{R}_1') \\ &+ \int_0^\infty k_{11}(\hat{R}_1', \hat{R}_1'')\, f_1(\hat{R}_1'')\, d\hat{R}_1'' + \int_0^\infty k_{12}(\hat{R}_1', \hat{R}_2'')\, f_2(\hat{R}_2'')\, d\hat{R}_2'' = E_{R1}\, f_1(\hat{R}_1'), \\ &\left[-\frac{\hbar^2}{2\mu_2} \frac{d^2}{d\hat{R}_2'^2} + V_{D2}(\hat{R}_2') \right] f_2(\hat{R}_2') \\ &+ \int_0^\infty k_{21}(\hat{R}_2', \hat{R}_1'')\, f_1(\hat{R}_1'')\, d\hat{R}_1'' + \int_0^\infty k_{22}(\hat{R}_2', \hat{R}_2'')\, f_2(\hat{R}_2'')\, d\hat{R}_2'' = E_{R2}\, f_2(\hat{R}_2'),\end{aligned} \tag{7.36}$$

where E_{R1} and E_{R2} are defined by eq. (4.11). Because of the hermiticity of the Hamiltonian operator, the direct potentials V_{D1} and V_{D2} are real, and the kernel functions satisfy the relations

$$
\begin{aligned}
k_{11}(\hat{R}_1', \hat{R}_1'') &= k_{11}^*(\hat{R}_1'', \hat{R}_1'),\\
k_{22}(\hat{R}_2', \hat{R}_2'') &= k_{22}^*(\hat{R}_2'', \hat{R}_2'),\\
k_{12}(\hat{R}_1', \hat{R}_2'') &= k_{21}^*(\hat{R}_2'', \hat{R}_1').
\end{aligned} \tag{7.37}
$$

We should mention here that in order to obtain nonvanishing coupling kernels k_{12} and k_{21} in this particular system, one must include two-nucleon tensor forces in the Hamiltonian of the system.

To carry out the multi-dimensional integrations leading to coupled integrodifferential equations of the type given by eq. (7.36), one can again use either the cluster-coordinate technique or the generator-coordinate technique. At the present time, there exist a number of reaction calculations employing the cluster-coordinate technique (see, e.g., ref. [CH 73a]), but none employing the generator-coordinate technique. We are of the opinion, however, that the generator-coordinate technique is likely to be more convenient for such calculations. The reason for this is that, in the cluster-coordinate technique, it is necessary to use different sets of cluster coordinates in different channels and then introduce linear transformations to relate these coordinates. In the generator-coordinate technique, such a complication is avoided, because only nucleon coordinates, in addition to generator and parameter coordinates, are used. Therefore, we feel that, especially for problems involving clusters which contain a relatively large number of nucleons, the generator-coordinate technique would require less labour and should be considered for adoption in future reaction calculations.

We shall now show explicitly how the law of current conservation, or the unitarity of the S-matrix, follows from coupled-channel equations of the form given by eq. (4.10). To show this, we multiply the first of eq. (4.10) by $F_1^*(\hat{\mathbf{R}}_1')$, the complex-conjugate equation by $F_1(\hat{\mathbf{R}}_1')$, subtract the resultant equations from each other, and integrate over $\hat{\mathbf{R}}_1'$. The result is

$$
\begin{aligned}
&-\frac{\hbar^2}{2\mu_1}\int [F_1^*(\hat{\mathbf{R}}_1')\,\nabla_1'^2\,F_1(\hat{\mathbf{R}}_1') - F_1(\hat{\mathbf{R}}_1')\,\nabla_1'^2\,F_1^*(\hat{\mathbf{R}}_1')]\,d\hat{\mathbf{R}}_1'\\
&= \int [F_1(\hat{\mathbf{R}}_1')\,K_{12}^*(\hat{\mathbf{R}}_1', \hat{\mathbf{R}}_2'')\,F_2^*(\hat{\mathbf{R}}_2'') - F_1^*(\hat{\mathbf{R}}_1')\,K_{12}(\hat{\mathbf{R}}_1', \hat{\mathbf{R}}_2'')\,F_2(\hat{\mathbf{R}}_2'')]\,d\hat{\mathbf{R}}_1'\,d\hat{\mathbf{R}}_2'',
\end{aligned} \tag{7.38}
$$

where we have used the fact that V_{D1} is real and $K_{11}(\hat{\mathbf{R}}_1', \hat{\mathbf{R}}_1'')$ is a hermitian kernel function. By carrying out the same procedure for the second of eq. (4.10), we obtain similarly

$$
\begin{aligned}
&-\frac{\hbar^2}{2\mu_2}\int [F_2^*(\hat{\mathbf{R}}_2')\,\nabla_2'^2\,F_2(\hat{\mathbf{R}}_2') - F_2(\hat{\mathbf{R}}_2')\,\nabla_2'^2\,F_2^*(\hat{\mathbf{R}}_2')]\,d\hat{\mathbf{R}}_2'\\
&= \int [F_2(\hat{\mathbf{R}}_2')\,K_{21}^*(\hat{\mathbf{R}}_2', \hat{\mathbf{R}}_1'')\,F_1^*(\hat{\mathbf{R}}_1'') - F_2^*(\hat{\mathbf{R}}_2')\,K_{21}(\hat{\mathbf{R}}_2', \hat{\mathbf{R}}_1'')\,F_1(\hat{\mathbf{R}}_1'')]\,d\hat{\mathbf{R}}_2'\,d\hat{\mathbf{R}}_1''.
\end{aligned} \tag{7.39}
$$

Adding eqs. (7.38) and (7.39) and defining

$$\mathbf{j}_i = \frac{\hbar}{2\mu_i i}\,[F_i^*(\hat{\mathbf{R}}_i')\,\nabla_i F_i(\hat{\mathbf{R}}_i') - F_i(\hat{\mathbf{R}}_i')\,\nabla_i F_i^*(\hat{\mathbf{R}}_i')], \tag{7.40}$$

we get

$$\oint \mathbf{j}_1 \cdot \mathbf{dS}_1 + \oint \mathbf{j}_2 \cdot \mathbf{dS}_2 = 0, \tag{7.41}$$

where $\mathbf{dS}_1$ and $\mathbf{dS}_2$ are differential surface elements. In deriving eq. (7.41), we have made use of the divergence theorem and the fact that the coupling kernels K_{12} and K_{21} are finite-ranged and adjoint to each other (see eq. (4.12)).

It is easy to see that, for an arbitrary number of open channels, one obtains

$$\sum_i \oint \mathbf{j}_i \cdot \mathbf{dS}_i = 0. \tag{7.42}$$

Furthermore, it should be noted that in calculations where bound-state-type distortion functions are included, eq. (7.42) also does not change, because distortion functions are finite-normalizable wave functions.

One further point should be mentioned here. In deriving the coupled-channel equations of the form given by eq. (4.10) which contains coupling kernels adjoint to each other, it is necessary that the relative normalization of the different channel wave functions be carried out properly. Only when this is done can one interpret the relative-motion functions as relative probability amplitudes, with the consequence that the current conservation law, as represented by eq. (7.42), becomes valid.

To see this, we consider a two-cluster wave function of the form

$$\psi = \hat{N}\mathcal{A}'\{[\mathcal{A}_1\tilde{\phi}(\tilde{1})]\,[\mathcal{A}_2\tilde{\phi}(\tilde{2})]\,\chi(\mathbf{R}_1 - \mathbf{R}_2)\}, \tag{7.43}$$

where $\mathcal{A}_1\tilde{\phi}(\tilde{1})$ and $\mathcal{A}_2\tilde{\phi}(\tilde{2})$ are internal functions of the two clusters and are both normalized to 1. The function $\chi(\mathbf{R}_1 - \mathbf{R}_2)$ describes a relative-motion wave packet, which is assumed to be nonzero only when the two clusters do not penetrate each other. In addition, we normalize $\chi(\mathbf{R})$ in such a way that $\int|\chi(\mathbf{R})|^2\,d\mathbf{R} = 1$; with such a normalization, we can interpret $\chi(\mathbf{R})$ as the probability amplitude of finding the two clusters at a vector distance $\mathbf{R}$ with respect to each other. Now, because of the non-overlapping of the clusters, terms in the wave function produced by permutations of nucleons between the clusters are all orthogonal to one another. Thus, one obtains

$$\langle\psi|\psi\rangle = \hat{N}^2\,\frac{A!}{A_1!\,A_2!}\int|\chi(\mathbf{R})|^2\,d\mathbf{R} = \hat{N}^2\,\frac{A!}{A_1!\,A_2!}, \tag{7.44a}$$

where A_1 and $A_2 = (A-A_1)$ are the number of nucleons in the two clusters and the factor $A!/(A_1!\,A_2!)$ is the number of permutations produced by the antisymmetrization operator $\mathcal{A}'$. If we choose $\langle\psi|\psi\rangle = 1$, then the normalization factor $\hat{N}$ is given by

$$\hat{N} = \left(\frac{A_1!\,A_2!}{A!}\right)^{1/2}. \tag{7.44b}$$

From this one sees that the normalization factor depends on the distribution of the A nucleons among the two clusters. Consequently, the relative normalization of the channel wave functions in a coupled-channel formulation must be chosen with this fact taken properly into consideration (see also ref. [CH 73a]).

7.2c. Reaction Calculations Using Hulthén-Kohn-Type Variational Functions

In our formulation of nuclear-reaction problems as discussed in chapter 4, the functions which describe the relative motion of clusters in open channels are treated as continuous variational amplitudes. The result of this is that the reaction problem is represented by a set of coupled integrodifferential and integral equations (see eq. (4.19)). As has been mentioned many times, this formulation of the reaction problem has the advantage that the solution of these coupled equations leads always to a unitary S-matrix, even if we restrict the number of cluster configurations in the ansatz for the trial wave function. Therefore, this particular formulation with linear variational functions is especially suited for the investigation of general consequences of the unified theory presented here, as for instance, the examination of the microscopic foundation of the various phenomenological nuclear-reaction theories, the derivation of the Breit-Wigner formula, and so on.

For explicit calculation of S-matrix elements, it is sometimes convenient to employ a slight variation of the above approach by adopting, instead of linear variational functions, only discrete linear variational amplitudes [HA 69, WI 69, ZA 71a, BA 73a]. To explain how this can be done, let us consider the case where a nucleus A is elastically scattered by another nucleus B, such as $\alpha + {}^{16}O$ scattering. For simplicity, we shall neglect the Coulomb interaction and reaction effects, and consider only scattering in $l = 0$ states. The wave function of this system is then written as

$$\psi = \psi_{in} + \psi_{out}, \tag{7.45}$$

where ψ_{in} is chosen to describe mainly the behaviour in the interior of the compound nucleus and ψ_{out} is chosen to describe mainly the behaviour of the system where the two clusters are fairly well separated. For ψ_{in}, we use

$$\psi_{in} = \sum_{m=1}^{N_i} a_m \hat{\varphi}_m, \tag{7.46}$$

with $\hat{\varphi}_m$ being totally antisymmetrized normalizable wave functions. The functions $\hat{\varphi}_m$ can be chosen to be in any convenient representation; for heavier nuclei, the most convenient representation is usually the translationally invariant shell-model representation, including Hill-Wheeler type functions (see 14.7–14.9). To improve the flexibility of ψ_{in}, it is sometimes useful to include also generalized cluster wave-function terms which can describe the phenomenon of cluster formation in the surface region of the compound nucleus. An important point to note is that, because the Pauli principle reduces considerably the difference between different structures, the number of terms, N_i, in the function ψ_{in} can frequently be chosen to be rather small; as a consequence of this, the computational effort required can be greatly reduced. For the choice of ψ_{out}, we make use of the fact that, when the clusters are well separated, specific distortion effects can be neglected.

Thus, as a good approximation, we can simply use

$$\psi_{out} = \mathcal{A}\left\{ \tilde{\phi}(\tilde{A})\,\tilde{\phi}(\tilde{B})\,g(R)\left[\frac{\sin kR}{R} + b_0\,\frac{\cos kR}{R}\right]\right\}, \tag{7.47}$$

where $R = |\mathbf{R}_A - \mathbf{R}_B|$ and k denotes the wave number of the relative motion between the nuclei A and B in the asymptotic region. The quantity b_0 is a variational parameter; its value determines the scattering phase shift δ_0 through the relation

$$\delta_0 = \tan^{-1} b_0 . \tag{7.48}$$

Finally, the function g(R) is a cutoff function which can be chosen, for example, to have the form

$$g(R) = 1 - \exp\left[-\left(\frac{R}{R_0}\right)^n\right], \tag{7.49}$$

with R_0 having a magnitude of the order of the compound-nucleus radius and n being an integer appreciably larger than 1. This cutoff function is introduced such that the function ψ_{out} becomes vanishingly small in the interior of the compound nucleus.

For the determination of the variational amplitudes a_m ($m = 1$ to N_i) and b_0 in the Hulthén-Kohn-type trial function of eqs. (7.45)–(7.47), one again uses the projection equation (2.3). By substituting ψ and the variation of ψ into this projection equation, one obtains the following set of inhomogeneous linear algebraic equations:

$$\sum_{m=1}^{N_i} a_m \langle \varphi_n | H - E | \hat{\varphi}_m \rangle + b_0 \left\langle \varphi_n | H - E | \mathcal{A}\left\{ \tilde{\phi}(\tilde{A})\,\tilde{\phi}(\tilde{B})\,g(R)\,\frac{\cos kR}{R}\right\}\right\rangle$$

$$= -\left\langle \varphi_n | H - E | \mathcal{A}\left\{ \tilde{\phi}(\tilde{A})\,\tilde{\phi}(\tilde{B})\,g(R)\,\frac{\sin kR}{R}\right\}\right\rangle, \qquad (n = 1 \text{ to } N_i)$$

$$\sum_{m=1}^{N_i} a_m \left\langle \tilde{\phi}(\tilde{A})\,\tilde{\phi}(\tilde{B})\,g(R)\,\frac{\cos kR}{R}\, | H - E | \hat{\varphi}_m \right\rangle$$

$$+ b_0 \left\langle \tilde{\phi}(\tilde{A})\,\tilde{\phi}(\tilde{B})\,g(R)\,\frac{\cos kR}{R}\, | H - E | \mathcal{A}\left\{ \tilde{\phi}(\tilde{A})\,\tilde{\phi}(\tilde{B})\,g(R)\,\frac{\cos kR}{R}\right\}\right\rangle$$

$$= -\left\langle \tilde{\phi}(\tilde{A})\,\tilde{\phi}(\tilde{B})\,g(R)\,\frac{\cos kR}{R}\, | H - E | \mathcal{A}\left\{ \tilde{\phi}(\tilde{A})\,\tilde{\phi}(\tilde{B})\,g(R)\,\frac{\sin kR}{R}\right\}\right\rangle, \tag{7.50}$$

where φ_m is the unantisymmetrized part of $\hat{\varphi}_m$. The solution of these equations then yields b_0 and, subsequently, the phase shift δ_0.

If one chooses the cutoff radius R_0 to be somewhat larger than the radius of the compound nucleus, then in calculating the coupling terms

$$\left\langle \varphi_n | H - E | \mathcal{A}\left\{ \tilde{\phi}(\tilde{A})\,\tilde{\phi}(\tilde{B})\,g(R)\,\frac{\cos kR}{R}\right\}\right\rangle$$

and

$$\left\langle \varphi_n | H - E | \mathcal{A}\left\{ \tilde{\phi}(\tilde{A})\,\tilde{\phi}(\tilde{B})\,g(R)\,\frac{\sin kR}{R}\right\}\right\rangle,$$

one can neglect antisymmetrization terms which correspond to the interchange of nucleons in different clusters or, equivalently, the interchange of nucleons in the interior and the surface region of the compound nucleus.† This shows therefore that, by choosing ψ in this way, the calculation in many cases can be simplified to a considerable extent. Indeed, it is quite obvious that this simplification in the antisymmetrization procedure can become very useful when the number of nucleons involved is large. In heavy-ion reactions, for instance, it is almost certain that, without such a simplification, any microscopic study will very likely be impracticable.

If one has to take into consideration reaction channels in addition to the elastic-scattering channel, then one must add into ψ_{out} terms which describe the appearance of outgoing waves in these reaction channels. These terms will also contain cutoff functions and linear variational amplitudes as multiplicative factors. By using again the projection equation (2.3), one can then solve for these variational amplitudes and obtain consequently the S-matrix elements which determine the elastic and reaction cross sections.

Since, in the method discussed here, the total wave function contains an inhomogeneous term, the solutions of the projection equation (2.3) do not automatically fulfill the law of current conservation. In other words, calculations based on this method will not yield automatically a unitary S-matrix. However, there are ways, such as the Kato-correction [WU 62], which show how one can correct this deficiency afterwards. On the other hand, one can also use the deviation of the S-matrix from unitarity as one possibility to test the quality of the solutions.††

To obtain satisfactory results, it is clear that one must choose N_i in eq. (7.46) large enough such that the resultant wave function with optimum variational amplitudes yields at least a good description of the behaviour of the system in the surface region of the compound nucleus. Especially in this surface region where both ψ_{in} and ψ_{out} have appreciable magnitudes, the Pauli principle plays an important role, because, as has been emphasized many times before, it serves to reduce the difference between different structures, as, for example, between ψ_{in} and ψ_{out}.

7.3. Quantitative Results

In this section, we discuss the results obtained in a number of scattering and reaction calculations. In these calculations, the simplifying methods of subsections 5.3c and 5.3d are used. As in the bound-state case, the adoption of these methods is necessary, because an accurate treatment of the Jastrow anti-correlation factor is very nearly an impossible task.

---

† If the function φ_n is a shell-model wave function for example, these nucleons are, respectively, those which occupy the inner and outer shells. Also, it is useful to mention here that, if R_0 is chosen to be substantially larger than the compound-nucleus radius, then even the short-ranged nuclear interaction between nucleons in different clusters can be neglected. We should point out, however, that such a choice of R_0 is, in general, not desirable, because one then has to adopt a much more flexible form for the function ψ_{in} by using a large number of $\hat{\varphi}_m$.

†† In a linear-expansion method, spurious resonances may occur [SC 61]. These resonances are however well understood [NU 69] and, hence, pose no problem.

The calculations to be discussed in subsections 7.3a–f are performed with the method of subsection 5.3d, while those to be discussed in subsections 7.3g–i are performed with the method of 5.3c. A large number of examples are discussed here, because, as has been mentioned, each one of these examples serves to illustrate some interesting points, such as the importance of specific distortion effects, the influence of noncentral forces, the effects of reaction channels on elastic-scattering cross sections, and so on.

7.3a. $^3He + \alpha$ Elastic Scattering

The formulation of the $^3He + \alpha$ single-channel problem [FU 75] is given in section 5.5.† In the present case of elastic scattering where the relative energy E_R is greater than zero, eq. (5.102) is solved subject to the boundary condition that in the asymptotic region,

$$f_{Jl}(R) \sim \sin\left(kR - \frac{1}{2}l\pi - \eta \ln 2kR + \sigma_l + \delta_{Jl}\right), \tag{7.51}$$

where σ_l and δ_{Jl} are Coulomb and nuclear phase shifts, respectively, and $\eta = z_1 z_2 e^2/\hbar v$, with v being the relative velocity of the two nuclei at infinity. Using these phase shifts, one then obtains the spin-independent scattering amplitude $g(\theta)$ and the spin-dependent scattering amplitude $h(\theta)$ by the following equations [GI 64]:

$$g(\theta) = f_C + \frac{1}{k}\sum_l [(l+1)\, e^{i\delta_l^+} \sin\delta_l^+ + l\, e^{i\delta_l^-} \sin\delta_l^-]\, e^{2i\sigma_l} P_l(\cos\theta), \tag{7.52}$$

$$h(\theta) = \frac{i}{k}\sum_l [e^{i\delta_l^+} \sin\delta_l^+ - e^{i\delta_l^-} \sin\delta_l^-]\, e^{2i\sigma_l} \sin\theta \frac{dP_l(\cos\theta)}{d(\cos\theta)}, \tag{7.53}$$

where f_C is the pure Coulomb scattering amplitude, and δ_l^+ and δ_l^- denote nuclear phase shifts for $J = l + \frac{1}{2}$ and $J = l - \frac{1}{2}$, respectively. In terms of these scattering amplitudes, the differential cross section $\sigma(\theta)$, the polarization $P(\theta)$, and the spin-rotation parameter $\beta(\theta)$ are given by the relations

$$\sigma(\theta) = |g|^2 + |h|^2,$$

$$P(\theta) = \frac{2\,\mathrm{Re}(g^* h)}{|g|^2 + |h|^2},$$

$$\beta(\theta) = \tan^{-1}\left[\frac{2\,\mathrm{Im}(gh^*)}{|g|^2 - |h|^2}\right]. \tag{7.54}$$

Calculated phase shifts for l up to 5 are shown by solid and dashed curves in fig. 12. As a comparison, the real parts of the empirical phase shifts obtained by *Boykin et al.* [BO 72] at energies below 3 MeV and by *Hardy et al.* [HA 72] at energies above 3 MeV are also shown. Here it is seen that, in general, the agreement between calculated and empirical

† $^3He + \alpha$ calculations with a purely central nucleon-nucleon potential have been reported in refs. [BR 68a, KO 74, TA 63].

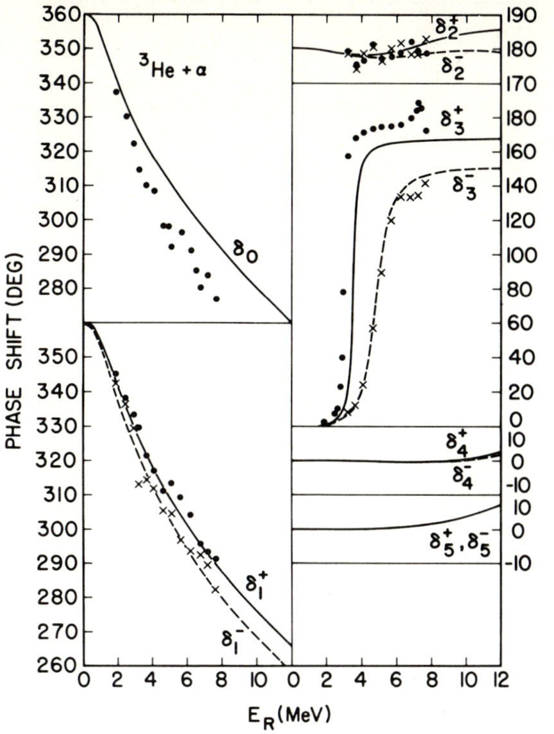

Fig. 12
Comparison of calculated phase shifts for ^{3}He + α scattering with the real parts of the empirical phase shifts given in refs. [B0 72, HA 72]. The solid dots represent empirical values for $J = l + 1/2$, and the crosses represent those for $J = l - 1/2$.

phase shifts is quite satisfactory. The calculated phase shifts in $l = 0$ states are somewhat too large, which is probably due to the fact that, in this calculation, a simple one-Gaussian internal wave function for the ^{3}He cluster and a nucleon-nucleon potential without a repulsive core have been used.

From fig. 12, one also sees that the calculated resonance energy for the $^2F_{7/2}$ level is about 0.5 MeV too large (see also fig. 9). At present, we are not certain of the source for this discrepancy, but it may be because, in this calculation, only the ^{3}He + α cluster configuration is considered. Experimentally, there is a $\frac{7}{2}^-$ level at 9.3 MeV excitation [SP 67] which has both a large reduced width for breakup into a proton and ^{6}Li in its first excited state (^{6}Li*) and a significant reduced width for α-particle emission. It is possible, therefore, that the $\frac{7}{2}^-$ level found experimentally at 4.6 MeV may contain a significant amount of p + ^{6}Li* structure, and an improved calculation with the p + ^{6}Li* channel taken explicitly into consideration might lead to a better agreement with experiment. For the $\frac{5}{2}^-$ level, the resonance energy calculated here does agree quite well with the value determined experimentally. In this case, a more complicated calculation involving the p + ^{6}Li or p + ^{6}Li* channel is not likely to have a large influence on this resonance energy, because the nearby $\frac{5}{2}^-$ level at 7.2 MeV has an α-particle reduced width which is less than 1 % of that for the $^2F_{5/2}$ level and, hence, the coupling between these two levels should be rather weak.

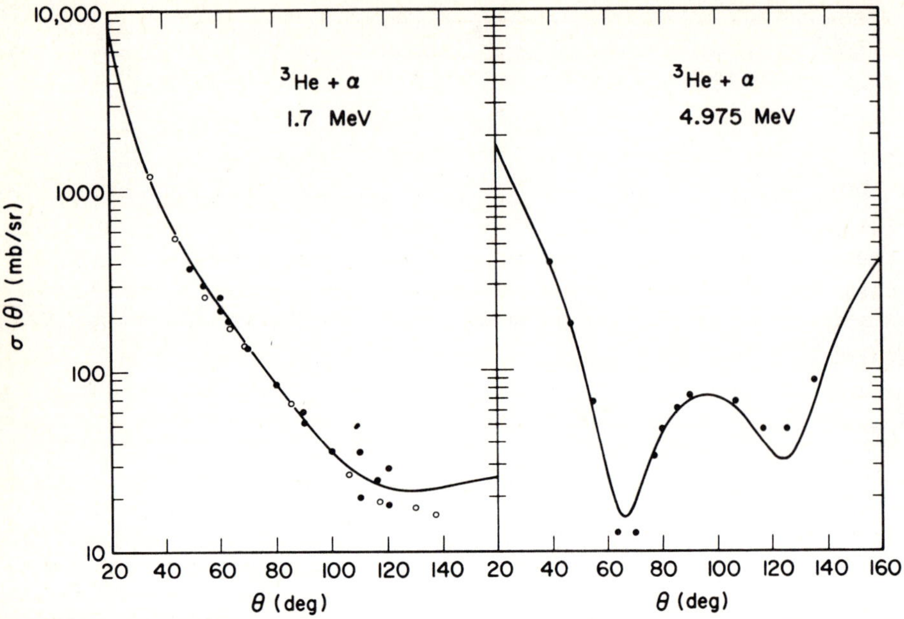

Fig. 13. Comparison of calculated differential cross sections for ^{3}He + α scattering with experimental data at 1.7 and 4.975 MeV. The data at 1.7 MeV are those of refs. [CH 71, MI 58], while the data at 4.975 MeV are those of ref. [SP 67a].

In fig. 13, a comparison of calculated differential cross sections with experimental data [CH 71, MI 58, SP 67a] at 1.7 and 4.975 MeV, where the reaction cross section is either zero or small, is shown. Here, one sees that the agreement between calculation and experiment is quite good, which is of course merely a reflection of the fact that the real parts of the empirical phase shifts are rather well reproduced.

The calculation also shows that, in the region where the two clusters penetrate each other strongly, the scattering wave function in each (J, l) state has the following properties [BR 68a, OK 66, TA 69]: (i) except for an energy-dependent amplitude, it varies only slightly as the energy E_R increases from 0 to about 20 MeV, and (ii) it becomes quite similar to the usual oscillator shell-model wave function in its lowest configuration. These properties of the wave function can be used to simplify certain calculations, as the example in the next subsection will show.

7.3b. $l = 0$ Phase Shift in $\alpha + \alpha$ Scattering

In this subsection, we show explicitly that for the calculation of S-matrix elements in certain cases, the method of subsection 7.2c (hereafter referred to as the method of linear amplitudes) requires much less computational effort than the method discussed in subsections 7.2a and 7.2b (hereafter referred to as the resonating-group method) and is yet capable of yielding rather satisfactory results. To show this, we shall consider the case of

$\alpha + \alpha$ scattering as an example, and compare the $l = 0$ phase shifts calculated using both of these methods.

In the resonating-group method, the wave function in $l = 0$ states is written as

$$\psi_{RG} = A\left\{\phi_0(\alpha_1)\,\phi_0(\alpha_2)\,\frac{f_0(R)}{R}\,P_0(\cos\theta)\,\xi(s,t)\,Z(\mathbf{R}_C)\right\}, \tag{7.55}$$

where the functions $\phi_0(\alpha)$, $\xi(s, t)$, and $Z(\mathbf{R}_C)$ are given by eqs. (5.9a), (5.9c), and (5.9d), respectively. The width parameter a of the α clusters is chosen as $0.543\ \text{fm}^{-2}$, which yields a value of 1.44 fm for the rms radius of the nucleon distribution in the α particle. The function $Z(\mathbf{R}_C)$ is a center-of-mass wave function which is commonly left off for convenience, but is included here in order to facilitate a comparison with the wave function in the method of linear amplitudes.

By using the procedure described in subsection 7.2a, one obtains the following integrodifferential equation which the linear variational function f_0 satisfies:

$$\left[\frac{\hbar^2}{2\mu}\frac{d^2}{dR^2} + E_R - V_N(R) - V_C(R)\right] f_0(R) = \int_0^\infty k_0(R, R')\, f_0(R')\, dR', \tag{7.56}$$

where $V_N(R)$, $V_C(R)$, and $k_0(R, R')$ are the direct nuclear potential, the direct Coulomb potential, and the $l = 0$ partial-wave kernel function, respectively. The nucleon-nucleon potential used is given by eq. (7.5) with the exchange constants w, m, b, and h satisfying the relations

$$\begin{aligned} w + m + b + h &= 1, \\ w + m - b - h &= x, \end{aligned} \tag{7.57}$$

where x is the ratio of the s-wave singlet to triplet interaction. By choosing

$$\begin{aligned} V_0 &= 72.98\ \text{MeV}, \\ \kappa &= 0.46\ \text{fm}^{-2}, \\ x &= 0.63, \end{aligned} \tag{7.58}$$

this potential yields a reasonable fit to the nucleon-nucleon effective-range parameters [TH 67]. Further, in view of the fact that the experimental two-nucleon scattering data favour a near-Serber exchange mixture, it is convenient to express the potential V_{ij} as a linear combination of a Serber potential V_S and a Rosenfeld potential V_R, i.e.,

$$V_{ij} = y\,V_S + (1 - y)\,V_R, \tag{7.59}$$

where V_S and V_R are given by Eq. (7.5) together with the condition

$$w = m, \quad b = h, \tag{7.60}$$

for the Serber potential and the condition

$$m = 2b, \quad h = 2w, \tag{7.61}$$

for the Rosenfeld potential. In this calculation, the value of y is chosen as equal to 0.94, such that the resultant $l = 0$ phase shift exhibits a resonance behaviour similar to that determined empirically.

To solve the integrodifferential equation (7.56), one divides the region of integration into two parts, separated at a point $R = R_M$ such that $k_0(R, R')$ has vanishingly small values for $R > R_M$. In the region $R \leqslant R_M$, the kernel is tabulated in the form of a $N_e \times N_e$ matrix at intervals ϵ in R and R′. The integrodifferential equation is then converted into a set of N_e simultaneous algebraic equations and solved using a method which has been described in detail by *Robertson* [RO 56]. In the region $R > R_M$, the kernel is set as zero and eq. (7.56) becomes an ordinary differential equation which can be solved by using a method of *Fox* and *Goodwin* [FO 49]. The resultant function $f_0(R)$ is then finally matched to Coulomb functions at a distance which is large enough to fulfill the requirement of an asymptotic-expansion method for Coulomb functions given by *Fröberg* [FR 55].

In single-channel calculations involving light nuclei, the smallest number of N_e needed for numerical accuracy is about 30. This means, therefore, that the computational effort required for such calculations using the resonating-group method is essentially that for solving a set of about 30 simultaneous algebraic equations. In the $\alpha + \alpha$ case, the result so obtained for the $l = 0$ phase shift δ_0 is shown by the solid curve in fig. 14. Here one sees that, even with a relatively simple trial wave function given by eq. (7.55), all the features possessed by the empirical phase shift are rather well reproduced.

In the method of linear amplitudes, the wave function is, as discussed in subsection 7.2c, written in the form

$$\psi_{LA} = \psi_{in} + \psi_{out}. \tag{7.62}$$

The function ψ_{in} is taken to be the ^{8}Be shell-model wave function of configuration $(1s)^4\,(1p)^4$ in an oscillator potential well. The width parameter of this oscillator well is chosen as equal to the width parameter of the α clusters, in order to make a comparison with the wave function ψ_{RG} of eq. (7.55) as simple as possible. The function ψ_{out} has the form of eq. (7.47), except that the sine and cosine functions are replaced by the $l = 0$ regular and irregular Coulomb functions (F_0 and G_0), respectively.

With the choice of ψ_{in} and ψ_{out} discussed above, the total wave function ψ_{LA} is, therefore, given by

$$\psi_{LA} = a_1\,\hat{\phi}(^8\mathrm{Be}) + A\left\{\phi_0(\alpha_1)\,\phi_0(\alpha_2)\,g(R)\left[\frac{F_0(R)}{R} + b_0\frac{G_0(R)}{R}\right]\right.$$
$$\left.\times\, P_0(\cos\theta)\,\xi(s,t)\,Z(\mathbf{R}_C)\right\}, \tag{7.63}$$

where a_1 and b_0 are linear variational amplitudes. The cutoff function $g(R)$ is chosen as

$$g(R) = \left[1 + \exp\left(\frac{R_0 - R}{C}\right)\right]^{-1}, \tag{7.64}$$

with†

$$R_0 = 3.0\ \mathrm{fm}, \quad C = 0.3\ \mathrm{fm}. \tag{7.65}$$

† Because R_0 is chosen to be relatively small, it is necessary to antisymmetrize between nucleons in the two α clusters when coupling terms of the type given in eq. (7.50) are computed.

Also, it should be noted that a multiplicative function $Z(\mathbf{R}_C)$ is included in ψ_{out}; this is necessary, because the ^{8}Be function $\hat{\phi}$ (^{8}Be) contains such a center-of-mass wave function.

The linear amplitudes a_1 and b_0 are obtained by solving inhomogeneous algebraic equations of the type given by eq. (7.50). Because, in this particular case, the number of these equations is only two, the computational effort required is considerably less than that required in the resonating-group method.

The phase shift δ_0 obtained by the method of linear amplitudes [FE 70] is shown as a function of energy by the dashed curve in fig. 14. From this figure, one sees that, even with the simple ansatz for ψ_{in} as given in eq. (7.63), an appreciable difference between the results of these two calculations does not occur until E_R becomes larger than about 15 MeV. The reason for this is rather simple. As was mentioned in subsection 7.3a, an investigation in the ^{3}He + α case revealed that the wave function in the compound-nucleus region is quite similar to the oscillator shell-model wave function in its lowest configuration and, in addition, varies only slightly with E_R in the low-energy region. Therefore, the choice of an oscillator shell-model function for ψ_{in} is in fact a rather appropriate one, as long as E_R is appreciably smaller than the depth of the effective interaction potential between the clusters, which is about equal to 100 MeV in the $\alpha + \alpha$ case [BR 71, TH 69a].

A further point is worth mentioning. The energy expectation value of ψ_{in} plus the internal binding energies of the two α particles is equal to 12 MeV, which is much larger than the value of 0.094 MeV [AJ 66] for the resonance energy of the $l = 0$ ground state in ^{8}Be. However, by the addition of the function ψ_{out}, the phase-shift behaviour as shown in fig. 14 does agree fairly well with that determined from an analysis of the experimental data. This shows, therefore, that the quality of the wave function in the surface region of the compound nucleus is very important, and a satisfactory account of the experimental result could be obtained only when a proper description of the behaviour of the system in this surface region is made.

In conclusion, this example shows that, with the method of linear amplitudes, reasonably accurate results can be obtained with only a small number of variational

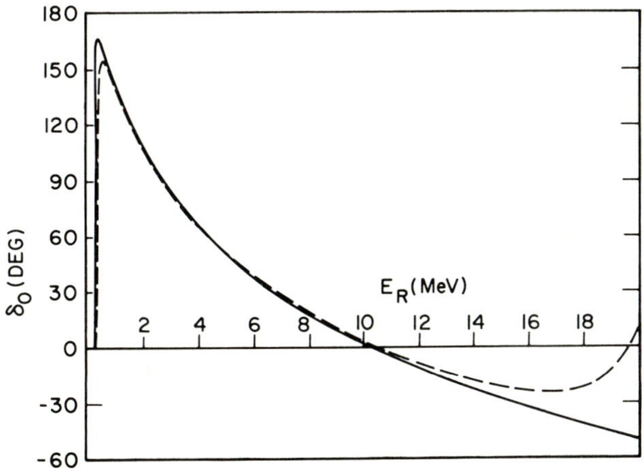

Fig. 14
$\alpha + \alpha$ $l = 0$ phase shifts as a function of energy. The solid curve represents the result obtained with the resonating-group method, while the dashed curve represents the result obtained with the method of linear amplitudes. (Adapted from ref. [FE 70].)

amplitudes. On the other hand, at least in this particular example, the $l = 0$ phase shifts obtained in the higher-energy region ($E_R \gtrsim 15$ MeV) with two variational amplitudes are considerably worse than those obtained with the resonating-group method.† Therefore, our present opinion is that the choice between these two methods is by no means a clear-cut one and will depend on many factors, such as the complexity of the problem, the availability of computer time, and so on.

7.3c. Specific Distortion Effects in d + α Scattering

In subsection 4.2a, it was mentioned that, in studying the properties of a nuclear system, the specific distortion effect needs in general to be considered in order to yield quantitatively satisfactory results. Here we shall examine this effect in detail by making a careful investigation of the problem of d + α scattering [JA 69, TH 73]. We choose to study this particular problem mainly for two reasons. First, the deuteron has a small binding energy and, hence, the deuteron cluster is expected to be rather strongly distorted when it overlaps appreciably with the α cluster. Second, as has been explained in subsection 5.3d, the specific distortion effect of the deuteron cluster can be properly considered by using the simple nucleon-nucleon potential of eqs. (5.81)–(5.83), which contains no repulsive core. In addition, it is noted that the high rigidity of the α particle renders its specific distortion quite small compared to that of the deuteron [TH 74]; therefore, the calculation can be considerably simplified by neglecting the α-cluster specific distortion effect as a good approximation.

The formulation of the d + α problem with specific distortion effect taken into account is given in subsections 4.2b and 5.3d. It suffices to say that what one does here is to solve the integrodifferential equation for $f_l(R)$ subject to the scattering asymptotic boundary condition. In this way, one obtains the phase shifts δ_l in various orbital angular-momentum states. In figs. 15 and 16, we show the effect of including the nine distortion functions mentioned in subsection 5.3d on even-l and odd-l phase shifts for energies in the range of 0–28 MeV. In these figures, the solid curves represent the results of the calculation with distortion functions, while the dashed curves show the no-distortion results. The data points represent the real parts of the empirical phase shifts determined by *McIntyre* and *Haeberli* [MC 67], *Schmelzbach et al.* [SC 72], and *Darriulat et al.* [DA 67]. The $l = 4$, 5, and 6 phase shifts are rather small in the energy range considered here and specific distortion effects have not been included in their computation; hence, only dashed curves are shown for these phase shifts in these two figures.

The salient features contained in figs. 15 and 16 are as follows:

(i) In the d + α system, specific distortion effects are important in the even-l cases, but much less so in the odd-l cases. This is connected with the fact that, in the oscillator model, two nucleons in the 1p shell can only form $l = 0$ and $l = 2$ d + α states.

(ii) Phase shifts calculated in the no-distortion approximation are consistently smaller than those calculated with distortion functions. In particular, it is noted that, in the $l = 0$ no-

† The behaviour of the dashed curve for $E_R > 15$ MeV may signify the existence of a spurious resonance at about 22 MeV (see subsection 7.2c).

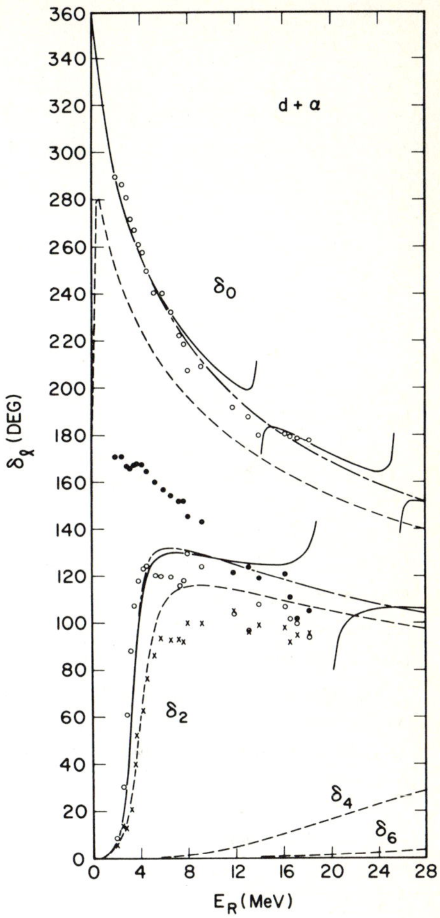

Fig. 15
d + α even-*l* phase shifts in the energy range 0–28 MeV. The solid curves represent the results of a calculation with distortion functions, while the dashed curves show the no-distortion results. The dot-dashed curves represent the result of a calculation with reaction effects taken into account as described in the text. Experimental data points are those of refs. [DA67, SC 72].

distortion case, the phase-shift behaviour indicates the nonexistence of a bound d + α system, but only the appearance of a resonance structure at about 0.34 MeV.
(iii) In the $l = 0$ case, the phase shifts calculated with distortion functions agree quite well with the empirical phases over the entire energy range considered. For the other phase shifts, the lack of a non-central component in our nucleon-nucleon potential and, consequently, the lack of splitting of the phases does not really allow a detailed comparison with the empirical phases to be made; however, one can still see from figs. 15 and 16 that, in general, the agreement is rather satisfactory.

From figs. 15 and 16, it is also noted that the addition of bound-state-type distortion functions into the calculation results in the appearance of resonance structures (plotted as breaks in the curves for convenience) in the energy region above 14 MeV. It should be emphasized, however, that the appearance of these resonances is expected in a calculation of this type and one must not consider their presence as indicating the existence of true resonance states in the compound system. In fact, it has been pointed out in particular by *Schmid* [SC 72a] and by *Schwager* [SC 73] that this type of resonance structure merely reflects the omission of open reaction channels in the calculation. Indeed, these spurious resonances can be easily eliminated by even a rather crude consideration of reaction effects. This can be done, for example, by phenomenologically adding an imaginary part to the direct nuclear potential and allowing the characteristic levels, introduced into the theory through the use of distortion functions, to decay into the neglected channels. In fig. 15, the dot-dashed curves represent the result of such a calculation [TH 74a] in which

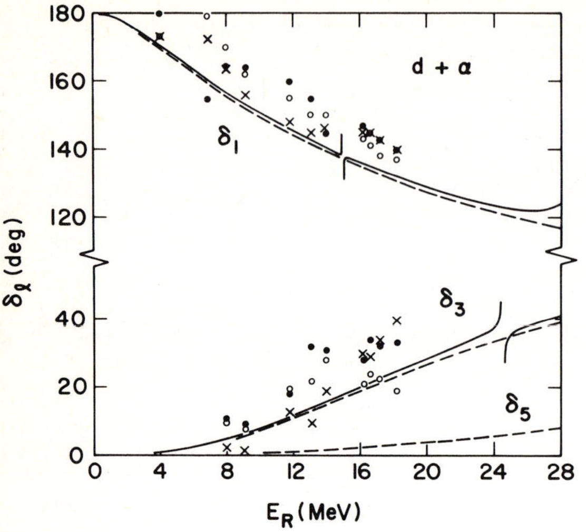

Fig. 16
d + α odd-l phase shifts in the energy range 0–28 MeV. The solid curves represent the results of a calculation with distortion functions, while the dashed curves show the no-distortion results. Experimental data points are those of refs. [DA 67, MC 67].

the experimental reaction cross-section data [AL 57, OH 64] are also taken into consideration. From this figure, one sees that the addition of absorption effects into the calculation does eliminate spurious resonances in both $l = 0$ and $l = 2$ partial waves. Phase shifts in $l = 1$ and $l = 3$ partial waves obtained from this calculation are not shown, but here also spurious resonances are entirely eliminated.

In conclusion, we feel that this calculation has demonstrated the importance of the specific distortion effect and, consequently, a proper consideration of this effect should be included in all calculations where a deuteron cluster, or even a ^{3}H or ^{3}He cluster, is involved.

7.3d. Effect of Reaction Channels on ^{3}He + ^{3}He Scattering Cross Sections

As an example of a coupled-channel calculation, we discuss the case of ^{3}He + ^{3}He elastic scattering with the α + 2p channel taken into account [SU 68]. In this example, the main interest is to see how the inclusion of a reaction channel influences the ^{3}He + ^{3}He differential scattering cross section; hence, the result will be compared with that of a single-channel calculation [TH 67] where, as in the ^{3}He + α case discussed in subsection 7.3a, only the exchange distortion of the ^{3}He clusters is taken into account through the use of a totally antisymmetrized wave function.†

In the low-energy region where the ^{3}He + ^{3}He relative energy in the center-of-mass system is less than about 10 MeV, the α + 2p channel which will have the largest influence is the one with the α cluster unexcited and the diproton in an $^1S_0$ state. Qualitatively, one can understand this in the following way: at low energies, the two ^{3}He clusters can come close to each other in an $l = 0$ state only when their spins are anti-alligned. This means that the compound nucleus ^{6}Be is essentially formed in a state with total orbital and spin

† It should be noted that, by the inclusion of a reaction channel, the specific distortion of the ^{3}He clusters is also partially taken into consideration.

angular momenta both equal to zero. Consequently, in a calculation where a purely central nucleon-nucleon potential is employed, the diproton should be in a singlet state with even orbital angular-momentum quantum number. Furthermore, because the protons in an ^{3}He cluster are known to be mainly in a relative s state, it is certainly reasonable to presume that the two protons in the α + 2p channel should have zero relative orbital angular momentum. Therefore, it is anticipated that the α + 2p channel which influences most significantly the low-energy ^{3}He + ^{3}He scattering must be the one where the diproton has an $^1S_0$ configuration.†

With the above-mentioned channels taken into consideration, the wave function is written as

$$\psi^s = A\{\phi(h_1)\,\phi(h_2)\,F_1^s(\mathbf{R}_{h_1} - \mathbf{R}_{h_2})\,\xi_1^s(s,t)\} + A\{\phi(\alpha)\,\phi(2p)\,F_2^s(\mathbf{R}_\alpha - \mathbf{R}_{2p})\,\xi_2^s(s,t)\} \qquad (7.66)$$

in the singlet (S = 0) state, and

$$\psi^t = A\{\phi(h_1)\,\phi(h_2)\,F_1^t(\mathbf{R}_{h_1} - \mathbf{R}_{h_2})\,\xi_1^t(s,t)\} \qquad (7.67)$$

in the triplet (S = 1) state. In eqs. (7.66) and (7.67), the channel indices 1 and 2 denote the ^{3}He + ^{3}He and the α + 2p channels, respectively. The functions $\phi(h)$ and $\phi(\alpha)$ are the internal spatial wave functions of the ^{3}He cluster and the α cluster, respectively; they have the forms given by eqs. (5.101) and (5.85), with $a_h = 0.36\ fm^{-2}$ and $a_\alpha = 0.543\ fm^{-2}$. The function $\phi(2p)$ describes the spatial structure of the diproton and is chosen also to have a Gaussian form, i.e.,

$$\phi(2p) = \exp\left[-\frac{1}{2}a_{2p}\sum_{i=5}^{6}(\mathbf{r}_i - \mathbf{R}_{2p})^2\right]. \qquad (7.68)$$

Because the diproton does not exist as a bound free particle, the choice of the width parameter a_{2p} poses a problem.†† In this calculation, it is simply taken to be the same as the width parameter of the ^{3}He nucleus. This is certainly a rather arbitrary choice, but the calculation does show that the influence of the α + 2p channel on the ^{3}He + ^{3}He differential scattering cross section is rather insensitive to the diproton radius.

The linear variational functions F_1^s, F_2^s, and F_1^t in eqs. (7.66) and (7.67) are determined by following the procedures outlined in subsections 7.2a and 7.2b. Once these functions are determined, one can then calculate the S-matrix and, subsequently, the ^{3}He + ^{3}He differential scattering cross section. By using the nucleon-nucleon potential of eqs. (7.5), (7.57), and (7.58) with the exchange-mixture parameter y of eq. (7.59) chosen as 1.2, the result obtained at an energy of 10 MeV is shown in fig. 17. In this figure, the solid

† The experimental data of *Bacher* [BA 67] show that the ^{3}He + ^{3}He reaction cross section begins to increase rapidly at a center-of-mass energy of about 8 MeV. This indicates that, at energies greater than about 8 MeV, other channels, such as the ^{3}He + d + p channel, should also be taken into consideration, if a detailed agreement with experiment for the ^{3}He + ^{3}He differential scattering cross section is desired.

†† A proper way is to consider the α + 2p channel as a three-particle channel.

and dashed curves represent, respectively, the results of the coupled-channel calculation described here and the single-channel calculation performed previously [TH 67]. A comparison between these curves indicates that the inclusion of the $\alpha + 2p$ channel increases the differential cross section in the forward angular region by almost a factor of 2, but has almost no effect on the diffraction peak at about 60°.

From fig. 17, one also sees that the agreement between calculation and experiment is only fair. The reason for this is as follows. In the compound nucleus ^{6}Be, there is a resonance level at about 25 MeV excitation [JE 70, TH 68, TH 73a], which has a structure of two ^{3}He clusters in $l = 3$ relative motion. This means that the ^{3}He + ^{3}He, $l = 3$ phase shift is quite rapidly increasing in the energy region around 13 MeV. Therefore, since the coupled-channel calculation described here is performed at a nearby energy of 10 MeV, it is likely that even relatively minor deficiencies in the theory will cause rather large discrepancies between calculated and experimental differential cross sections. For example, one of the deficiencies in those calculations leading to the solid and dashed curves in fig. 17 is the omission of exchange Coulomb interactions between the clusters. To see if this omission has any large effect, a calculation has been made which takes into account the exchange Coulomb interaction between the ^{3}He clusters in the $l = 3$ state [SU 68a]. The result is shown by the dot-dashed curve in fig. 17. Here it is seen that even this minor modification does result in a significant improvement in agreement between calculation and experiment in the angular region around the 60° diffraction maximum.

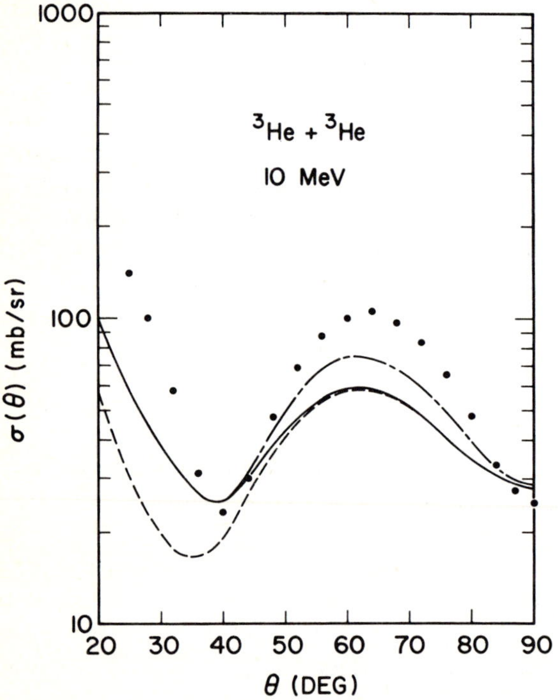

Fig. 17
^{3}He + ^{3}He differential scattering cross section at 10 MeV. The solid and dashed curves represent the results of the coupled-channel and the single-channel calculations, respectively. The dot-dashed curve represents the result of a coupled-channel calculation with the exchange Coulomb interaction in the $l = 3$ state taken into account. Experimental data are those of ref. [JE 70]. (Adapted from ref. [SU 68a].)

Even though the coupled-channel calculation described here contains many simplifications, the result does indicate that reaction channels should in general be included, if quantitatively reliable results for the differential elastic-scattering cross section are desired.

7.3e. $\alpha + {}^{16}O$ Scattering – Utilization of the Generator-Coordinate Technique

As next example, we consider the case of $\alpha + {}^{16}O$ elastic scattering [SU 72].† In this case, the number of nucleons is so large that to perform the multi-dimensional integrations involved becomes a major problem. One of the purposes of this calculation is therefore to show that, by using the generator-coordinate technique described in subsection 5.2b, a microscopic calculation on such a rather complicated system can still be carried out and, consequently, it can be realistically expected that one will be able to study in a similar way even more complicated systems, such as $\alpha + {}^{40}Ca$ scattering and so on.

The wave function for the $\alpha + {}^{16}O$ system is assumed as

$$\psi = A\{\tilde{\phi}(\tilde{\alpha})\,\tilde{\phi}(^{16}\tilde{O})\,F(\mathbf{R}_\alpha - \mathbf{R}_o)\}, \tag{7.69}$$

where $\tilde{\phi}(\tilde{\alpha})$ is the α-cluster internal wave function with its spatial part given by eq. (5.85), and

$$\tilde{\phi}(^{16}\tilde{O}) = \left[\prod_{i=5}^{20} h_i(\mathbf{r}_i)\right] \exp\left[-\frac{a_0}{2}\sum_{i=5}^{20}(\mathbf{r}_i - \mathbf{R}_o)^2\right] \xi_0(s_5, \ldots, t_{20}), \tag{7.70}$$

with

$$\mathbf{R}_o = \frac{1}{16}\sum_{i=5}^{20}\mathbf{r}_i \tag{7.71}$$

and $\xi_0(s_5, \ldots, t_{20})$ being an appropriate spin-isospin function for $S_0 = 0$ and $T_0 = 0$. The functions $h_i(\mathbf{r}_i)$ are polynomials; they are equal to 1, $x_i + iy_i$, z_i, and $x_i - iy_i$ for i = 5–8, 9–12, 13–16, and 17–20, respectively. It should be noted that, even though the argument of the function h_i is the nucleon coordinate $\mathbf{r}_i$, the antisymmetrized ^{16}O wave function $A\tilde{\phi}(^{16}\tilde{O})$ depends only on the relative coordinates $(\mathbf{r}_i - \mathbf{r}_j)$ of the nucleons. To show this, one multiplies $A\tilde{\phi}(^{16}\tilde{O})$ by a center-of-mass function which is an 1s oscillator function having the same frequency as that of the nucleon motion in the ^{16}O cluster. Then, after some simple algebraic manipulations, one obtains the following relation:

$$[A\tilde{\phi}(^{16}O)]\exp(-8a_0\mathbf{R}_0^2) = A\left\{\left[\prod_{i=5}^{20} h_i(\mathbf{r}_i)\right]\exp\left(-\frac{a_0}{2}\sum_{i=5}^{20}\mathbf{r}_i^2\right)\xi_0(s_5, \ldots, t_{20})\right\}. \tag{7.72}$$

In the above equation, the right-hand side is a wave function which describes the lowest shell-model configuration $(1s)^4\,(1p)^{12}$ for 16 nucleons moving in an oscillator well of width parameter a_0. Therefore, as is well known, the total center-of-mass motion is described by the function $\exp(-8a_0\mathbf{R}_0^2)$ just introduced. This means that no dependence on the center-of-mass coordinate $\mathbf{R}_0$ can be contained in $A\tilde{\phi}(^{16}\tilde{O})$; consequently, $A\tilde{\phi}(^{16}\tilde{O})$ must be translationally invariant and can depend only on the relative spatial coordinates of the nucleons.

---

† A microscopic $\alpha + {}^{16}O$ calculation using the resonating-group method has also been carried out by *Matsuse et al.* [MA 75, KA 74a]. The result obtained by these authors is quite similar to that given in ref. [SU 72].

To make the above discussion even more explicit, let us consider the ground state of ^{5}He in the oscillator shell model. In this case, the wave function is given by

$$\psi(^5\mathrm{He}) = A\,\bar{\psi}(^5\mathrm{He}) = A\left\{(x_5 + iy_5)\exp\left(-\frac{a}{2}\sum_{i=1}^{5} r_i^2\right)\alpha_1\nu_1\alpha_2\pi_2\beta_3\nu_3\beta_4\pi_4\alpha_5\nu_5\right\}. \quad (7.73)$$

By using eq. (3.28) one can also write $\psi(^5\mathrm{He})$ as

$$\psi(^5\mathrm{He}) \sim A\{(1 - P_{15})\,\bar{\psi}\} = A\left\{[(x_5 - x_1) + i(y_5 - y_1)]\exp\left(-\frac{a}{2}\sum_{i=1}^{5} r_i^2\right)\right.$$
$$\left.\times\,\alpha_1\nu_1\alpha_2\pi_2\beta_3\nu_3\beta_4\pi_4\alpha_5\nu_5\right\}. \quad (7.74)$$

Thus, one sees that the polynomial part of $\bar{\psi}(^5\mathrm{He})$, and consequently $\psi(^5\mathrm{He})$, depends only on the relative spatial coordinates of the nucleons, as discussed in the above paragraph.

Now, let us go back to the discussion of $\alpha + {}^{16}$O scattering. For simplicity, the calculation was carried out with the assumption that the width parameters of the α cluster and the ^{16}O cluster take on the same value. This common width parameter a was chosen as equal to 0.32 fm^{-2}; with this value, the charge radius of ^{16}O is correctly given by the shell-model wave function of configuration $(1s)^4\,(1p)^{12}$ [HO 57]. It should be mentioned that a calculation without the assumption of equal width parameter for the clusters will be only slightly more complicated [SU 72, SU 76].

The nucleon-nucleon potential used in this calculation was that of eqs. (7.5), (7.57), and (7.58). The exchange-mixture parameter y of eq. (7.59) was then adjusted to yield a best over-all agreement with the experimental data in the low-energy region up to about 20 MeV. The resultant value of y turned out to be 0.72, which is appreciably smaller than the value of 1 for a Serber potential. The reason for this rather small value of y is probably that in this case where the number of nucleons involved is quite large, the use of a nucleon-nucleon potential without a repulsive core is not too appropriate.

In fig. 18, the calculated differential cross section (solid curve) is compared with the experimental data (dashed curve) [CO 59] at a center-of-mass energy of 14.7 MeV. Here one sees that, even with the simplifications made in this calculation, the agreement between theory and experiment is still quite satisfactory. In particular, one notes that the feature of the experimental differential cross section in the backward angular region is quite well reproduced, which is of course a consequence of the fact that the wave function used is totally antisymmetric [TH 71].

One important point which is mainly responsible for this relatively good agreement between theory and experiment and which will be discussed in more detail later should already be mentioned here. Due to the indistinguishability of the nucleons, the lifetime of a cluster surrounded by many nucleons is much longer than it would be if nucleons were distinguishable. If the latter were in fact the case, then the α cluster would be destroyed immediately on the ^{16}O surface and one would observe even in scattering processes of small clusters on relatively light nuclei only the phenomenon of diffraction scattering in the elastic-scattering cross section. This means that the reaction cross section would be so large that every α cluster which penetrates into the ^{16}O nucleus would be absorbed and the compound nucleus will lose its energy mainly by evaporation of nucleons or composite particles which are not the same as the bombarding particle.

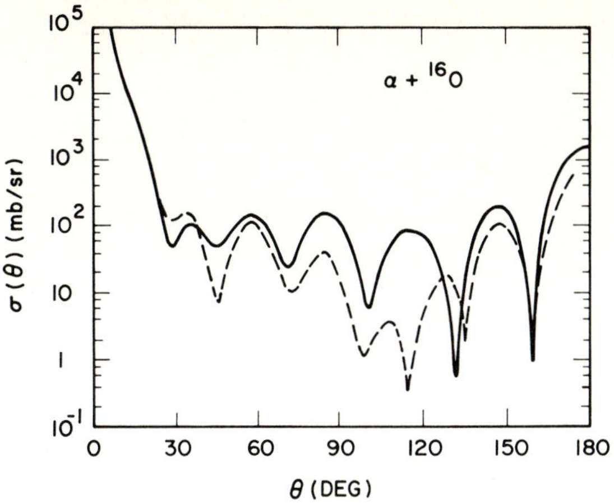

Fig. 18
Comparison of calculated $\alpha + {}^{16}O$ differential cross section at 14.7 MeV with experimental data. The dashed curve is a line drawn through the experimental data points of ref. [CO 59]. (Adapted from ref. [SU 72].)

7.3f. p + α Scattering Around $\frac{3}{2}^+$ Resonance Level in ^{5}Li

In this example, we consider the scattering of protons by α particles in the energy region around the $\frac{3}{2}^+$ resonance level in ^{5}Li [HE 69]. This is an interesting example, because it is known that this $\frac{3}{2}^+$ level at 16.65 MeV excitation does not have predominantly a cluster structure consisting of a proton plus an unexcited α cluster [KU 55, PE 60a]. In fact, a single-channel p + α scattering calculation [RE 70] which employs a nucleon-nucleon potential including a spin-orbit component shows that, besides the well-known $\frac{3}{2}^-$ ground state and the $\frac{1}{2}^-$ first excited state, there are no other relatively sharp resonance levels in ^{5}Li which have predominantly a p + α cluster structure. Therefore, it follows from energetical reasoning that the $\frac{3}{2}^+$ level in ^{5}Li must have a d + ^{3}He cluster structure, with the deuteron cluster in its triplet spin state.† Furthermore, it is expected that the relative orbital angular momentum between the deuteron and the ^{3}He clusters should be equal to zero, because then the two clusters can penetrate each other strongly to take maximum advantage of the attractive nuclear interaction.

Single-channel d + ^{3}He calculation [CH 73b] shows further that there is a very sharp $^4S_{3/2}$ resonance level occurring very near the d + ^{3}He threshold. Therefore, in refining the single-channel p + α calculation, it is most important to consider the d + ^{3}He channel with the relative orbital angular momentum between the clusters equal to 0 and the total spin angular momentum of the clusters equal to $\frac{3}{2}$.

For simplicity, only the $J^\pi = \frac{3}{2}^+$ state with the two channels mentioned above will be considered. In this simplified problem, the formulation is then given by that of subsection 7.2b, except for a trivial change in the atomic numbers of the clusters involved. As is mentioned there, a nucleon-nucleon potential with a tensor component is required in this calculation; this is necessary, because central and spin-orbit potentials cannot effect any

† In the d + ^{3}He cluster structure, the total spin angular-momentum quantum number S can be either $\frac{3}{2}$ or $\frac{1}{2}$. The $S = \frac{3}{2}$ state is expected to have a lower excitation energy, because the nucleon-nucleon potential is more attractive in triplet-spin state than in singlet-spin state.

coupling between the $p + \alpha$ channel having $l = 2$ and $S = \frac{1}{2}$ and the $d + {}^3He$ channel having $l = 0$ and $S = \frac{3}{2}$.

The nucleon-nucleon potential used has the form

$$V_{ij} = V_{ij}^c + V_{ij}^{so} + V_{ij}^T, \tag{7.75}$$

where V_{ij}^c, V_{ij}^{so}, and V_{ij}^T denote the central, spin-orbit, and tensor components, respectively. The central part V_{ij}^c is given by eqs. (7.5), (7.57), and (7.58), with the exchange-mixture parameter y of eq. (7.59) chosen as 0.94. For the spin-orbit potential, the one obtained by *Wurster* [WU 67] in an analysis of the ${}^7Be$ bound-state problem is used; it is given by the second part on the right side of eq. (5.98), with

$$V_\lambda = 30 \text{ MeV}, \quad V_{\lambda\tau} = 0, \quad \lambda = 1.063 \text{ fm}^{-2}. \tag{7.76}$$

The tensor potential used is a simplified version of that given in ref. [EI 71]; it has the form

$$V_{ij}^T = \frac{1}{2}(1 + P_{ij}^r)\,[3(\sigma_i \cdot \mathbf{r}_{ij})(\sigma_j \cdot \mathbf{r}_{ij}) - (\sigma_i \cdot \sigma_j)\, r_{ij}^2] \sum_{k=1}^{2} V_k^T \exp(-\kappa_k^T r_{ij}^2), \tag{7.77}$$

with

$$\begin{aligned} V_1^T &= -100.94 \text{ MeV} - \text{fm}^{-2}, & V_2^T &= -1.18 \text{ MeV} - \text{fm}^{-2}, \\ \kappa_1^T &= 1.047 \text{ fm}^{-2}, & \kappa_2^T &= 0.203 \text{ fm}^{-2}. \end{aligned} \tag{7.78}$$

The spatial parts of the cluster internal wave functions are assumed as usual to have simple Gaussian forms in order to facilitate the evaluation of multi-dimensional integrals. The functions $\phi_0(\alpha)$ and $\phi_0(h)$ are chosen as one-Gaussian functions of the forms given by eqs. (5.85) and (5.101), respectively, with $a_\alpha = 0.54 \text{ fm}^{-2}$ and $a_h = 0.69 \text{ fm}^{-2}$. The deuteron-cluster function $\phi_0(d)$ is chosen to have a more flexible form given by eq. (5.84) with $N_C = 3$; the parameters are

$$\begin{aligned} B_1 &= 1.395, & B_2 &= -2.77, & B_3 &= 1.926, \\ a_1 &= 0.32 \text{ fm}^{-2}, & a_2 &= 0.44 \text{ fm}^{-2}, & a_3 &= 0.60 \text{ fm}^{-2}. \end{aligned} \tag{7.79}$$

With these functions, the calculated $d + {}^3He$ threshold energy is 21.18 MeV, which is somewhat larger than the value of 18.35 MeV determined experimentally.

Equation (7.36) is solved subject to appropriate boundary conditions to yield the scattering functions $f_1(\hat{R}_1)$ and $f_2(\hat{R}_2)$ and, subsequently, the S-matrix in the $\frac{3}{2}^+$ state. Below the $d + {}^3He$ threshold, the S-matrix is a 1×1 matrix, given by

$$S = \exp(2i\,\delta_1), \tag{7.80}$$

where δ_1 is the $p + \alpha$ nuclear phase shift. Above the $d + {}^3He$ threshold, the S-matrix is a 2×2 matrix which is characterized by three real quantities. If one chooses to specify these three quantities as the $p + \alpha$ phase shift δ_1, the $d + {}^3He$ phase shift δ_2, and the reflection coefficient τ, then the S-matrix is written as

$$S = \begin{pmatrix} \tau \exp(2i\,\delta_1) & i(1-\tau^2)^{1/2} \exp[i(\delta_1 + \delta_2)] \\ i(1-\tau^2)^{1/2} \exp[i(\delta_1 + \delta_2)] & \tau \exp(2i\,\delta_2) \end{pmatrix}. \tag{7.81}$$

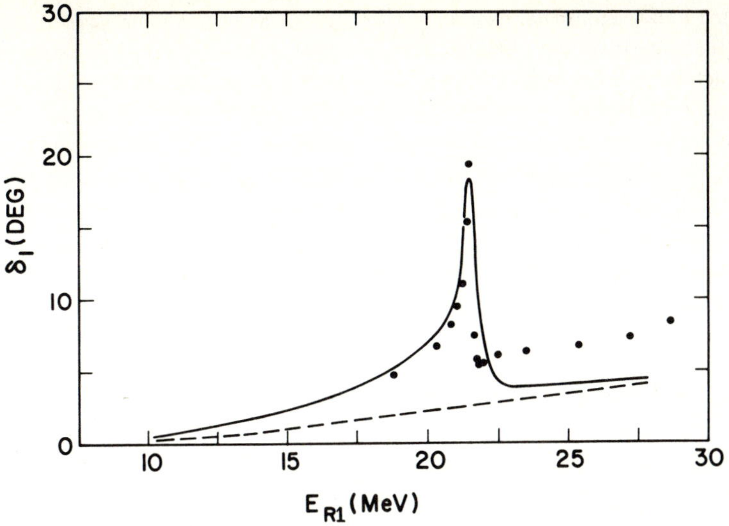

Fig. 19. Comparison of calculated and empirical values for the $p + \alpha$ phase shift δ_1. The solid and dashed curves represent the results of the coupled-channel and the single-channel calculations, respectively. Empirical data points are those of ref. [PL 72]; they are plotted with an energy shift in order to compensate for mismatch between calculated and experimental $d + {}^3He$ threshold energies. (Adapted from ref. [HE 69].)

A comparison between calculated and empirical values for δ_1 is shown in fig. 19. In this figure, the solid and dashed curves represent the results of the coupled-channel and the single-channel calculations, respectively. The empirical data points are those obtained by *Plattner et al.* [PL 72]; they are plotted with an energy shift of 2.83 MeV toward the right in order to compensate for the mismatch between calculated and experimental $d + {}^3He$ threshold energies. As is seen, the single-channel calculation (V_{ij}^T is set as zero) does not yield any resonance behaviour in δ_1, whereas the result of the coupled-channel calculation agrees fairly well with empirical values. On the higher-energy side of the resonance, the calculated values are too small, which is probably a consequence of the fact that $d + {}^3He$ channels with $l = 2$ and channel spins $\frac{1}{2}$ and $\frac{3}{2}$ are not considered in this calculation.

If one's sole purpose is to learn the influence of the $d + {}^3He$ channel on $p + \alpha$ elastic scattering in the energy region around the $\frac{3}{2}^+$ resonance rather than to obtain the $\alpha(p, d)\,{}^3He$ reaction cross section, then a simpler calculation can be made in which one represents the $d + {}^3He$ channel by an appropriate bound-state-type distortion function with a linear variational amplitude. Such a bound-state-type function could be obtained, for instance, by considering the $\frac{3}{2}^+$ level as a $d + {}^3He$ quasibound state and using a variational technique to solve for its wave function. This is a reasonable way to select the distortion function, because the $\frac{3}{2}^+$ level is deeply embedded under the Coulomb barrier and, therefore, behaves very much like a bound state.

Before we proceed to discuss further examples, we wish to make some remarks of general character. In a nuclear system, it occurs that, after some broad levels, there appear resonance levels which have rather small level widths. In fact, this has been observed in many nuclei in addition to the case of ^{5}Li discussed here. The explanation is always the same as in the ^{5}Li case, namely, that the sudden change in level width marks a change in cluster structure. In ^{5}Li, for instance, this is the change from the $p + \alpha$ cluster structure to the $d + {}^3$He cluster structure.

The fact that the $\frac{3}{2}^+$ level in ^{5}Li has a small width can be understood in the following qualitative way. The $d + {}^3$He effective interaction in the $l = 0$ state is stronger than even the $p + \alpha$ effective interaction in the $l = 1$ state [WI 66]. Therefore, since this resonance level occurs very close to the $d + {}^3$He threshold, the probability of decaying into a deuteron plus a ^{3}He particle is rather small. In addition, it also cannot decay readily into a proton plus an α particle; this is so, since this level is coupled to the $p + \alpha$ channel only through a relatively weak tensor force.

Another example is ^{8}Be where, after the very broad 4^+ level, many narrow levels in the energy region between 16 and 20 MeV follow. In these higher states, one of the α clusters is broken up into a triton cluster plus a proton or a ^{3}He cluster plus a neutron, with the consequence that the new clusters have a stronger mutual effective interaction than even the two unexcited α clusters.

It can even happen that the effective interaction between the clusters is strong enough to form a bound state which cannot decay into the corresponding free particles. Such a case occurs, for example, for the 16.63 MeV level in ^{8}Be. For this level, the energy width is determined essentially by the small decay probability into two α particles. Other example for this would be the $\frac{3}{2}^+$ level in ^{5}Li if protons had no Coulomb interaction. In this latter case, the $\frac{3}{2}^+$ level would become particle stable, if the coupling to the $p + \alpha$ channel is neglected.

7.3g. $\alpha + \alpha$ Scattering with Specific Distortion Effect and a Nucleon-Nucleon Potential Containing a Repulsive Core

As an example of a calculation using a nucleon-nucleon potential containing a repulsive core, we describe the case of $\alpha + \alpha$ scattering studied by *Niem et al.* [NI 71]. In this example, the nucleon-nucleon potential employed is that of eqs. (5.66)–(5.68) and the associated Jastrow factor is given by eqs. (5.56), (5.61), and (5.69). To make the calculation feasible, the method described in subsection 5.3c for an approximate treatment of the Jastrow factor is again adopted.

The specific distortion effect is taken into account by including in the calculation the $\alpha + \alpha^*$ channel, where α^* denotes the first excited $T = 0$, $J^\pi = 0^+$ level in ^{4}He [ME 68]. For simplicity, the system α^* is considered as a quasibound structure, in order to avoid the complication of having to treat the $\alpha + \alpha^*$ channel as a three-body breakup channel. This is a reasonable assumption, since the main purpose of this calculation is to study the specific distortion effect on $\alpha + \alpha$ elastic scattering, rather than to calculate the $\alpha(\alpha, pt)\alpha$ reaction cross section.

The long-range part ψ_L of the wave function given by eq. (5.57) is written as

$$\psi_L = A\{\phi(\alpha_1)\,\phi(\alpha_2)\,F_1(\mathbf{R})\,\xi(s,t)\} + A\{\phi(\alpha_1)\,\phi(\alpha_2^*)\,F_2(\mathbf{R})\,\xi(s,t)\}, \qquad (7.82)$$

where ξ is an appropriate spin-isospin function for S = 0 and T = 0, and **R** is given by eqs. (3.8) and (3.9). The functions $\phi(\alpha)$ and $\phi(\alpha^*)$ describe the spatial behaviour of α and α^*, respectively; they are assumed to have the forms

$$\begin{aligned} \phi(\alpha) &= \sum_{i=1}^{2} B_i \exp\left(-\frac{1}{2} a_i \sum_{j=1}^{4} \bar{\mathbf{r}}_j^2\right), \\ \phi(\alpha^*) &= \sum_{i=1}^{2} C_i \exp\left(-\frac{1}{2} a_i \sum_{j=1}^{4} \bar{\mathbf{r}}_j^2\right). \end{aligned} \qquad (7.83)$$

By using the Ritz variational principle, the parameters a_i, B_i, and C_i are determined. These parameters are

$$\begin{aligned} a_1 &= 0.168\ \text{fm}^{-2}, & a_2 &= 0.648\ \text{fm}^{-2}, \\ B_1 &= 4.0316 \times 10^{-4}, & B_2 &= 3.1854 \times 10^{-2}, \\ C_1 &= -1.8231 \times 10^{-3}, & C_2 &= 2.2379 \times 10^{-2}. \end{aligned} \qquad (7.84)$$

It should be noted that, because the nonlinear parameters in $\phi(\alpha)$ and $\phi(\alpha^*)$ are chosen to be the same, the orthogonality of these two functions follows easily from the Ritz variational procedure.

The relative-motion functions $F_1(\mathbf{R})$ and $F_2(\mathbf{R})$ satisfy coupled integrodifferential equations of the form given by eq. (4.10). In this particular calculation, these equations are solved by using a technique very similar to that discussed in subsection 7.2c. The results for the $\alpha + \alpha$ phase shifts in $l = 0$, 2, and 4 states are shown in fig. 20. In this figure, the solid and dashed curves represent, respectively, phase shifts calculated with and without the $\alpha + \alpha^*$ channel taken into consideration. The empirical data points are those of refs. [CH 73c, HE 56, NI 58, TO 63]. From this figure one sees that, because of the low compressibility of the α cluster, the specific distortion effect has only a moderate influence. Even in the $l = 0$ state where this effect is most important, the increase in phase shift over the single-channel result is less than 10°. However, as can be seen, the agreement between calculated and empirical values does become significantly improved when this effect is considered. In $l = 2$ and 4 states, specific distortion effects are even less important; hence, only results with $\alpha + \alpha^*$ channel included in the calculation are shown in fig. 20.

Instead of describing the specific distortion of the α clusters by adding $\alpha + \alpha^*$ structures to the wave function, one can also add, similar to the $d + \alpha$ case discussed in subsection 7.3c, $\alpha + \alpha$ bound structures where the α cluster has a rms radius either larger or smaller than that of a free α particle. The results of both calculations are in fact very similar to each other, as a recent calculation has shown [TH 74]. This is, of course, rather to be expected because, as has been mentioned many times previously, the Pauli principle has the effect that it reduces the difference between different structures to a large extent.

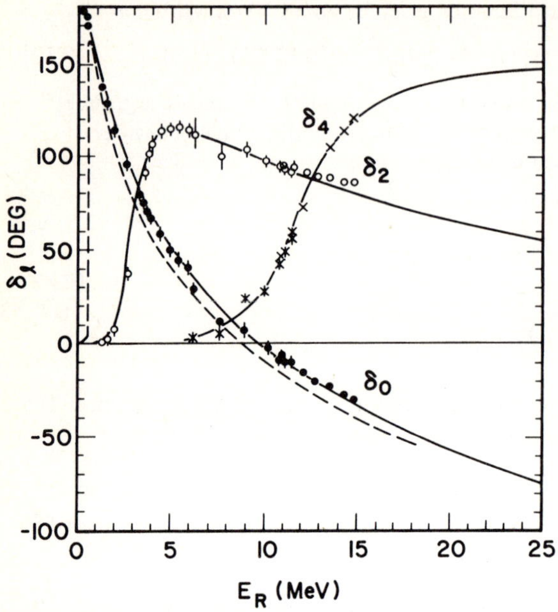

Fig. 20
Comparison of calculated and empirical $\alpha + \alpha$ phase shifts. The solid and dashed curves represent results obtained with and without the $\alpha + \alpha^*$ channel, respectively. Empirical data points are those of refs. [CH 73c, HE 56, NI 58, TO 63]. (Adapted from ref. [NI 71].)

7.3h. p + ^{3}He and n + t Scattering Calculations

To explain the experimental polarization as well as the differential scattering cross-section data, it is necessary to employ a noncentral nuclear potential in the calculation. Here we describe the p + ^{3}He [HE 72] and n + t [HA 71] scattering calculations performed with the nucleon-nucleon potential of eq. (7.75), with V_{ij}^c given by eqs. (5.66)–(5.68) and V_{ij}^{so} and V_{ij}^T being those obtained by *Eikemeier* and *Hackenbroich* [EI 71] from an analysis of the two-nucleon scattering data.

The formulation of these scattering problems is very similar to those described in subsections 7.3a–i; hence, only some essential points will be mentioned. With a noncentral nucleon-nucleon potential, the total angular momentum J and the parity π remain as good quantum numbers, but the orbital and the spin angular momenta are no longer constants of motion. Therefore, in a given (J, π) state, the wave function is in general given by a combination of states with different values of l and S. For example, in the $J^\pi = 1^-$ state, the wave function is a linear superposition of a $^1P_1$ function and a $^3P_1$ function, while in the $J^\pi = 1^+$ state, it is a linear superposition of a $^3S_1$ function and a $^3D_1$ function. Thus, because of the mixing of these orbital and spin angular-momentum quantum states, it is necessary to solve coupled equations even in a pure (J, π) state and the solution of these elastic-scattering problems is somewhat complicated.

The spatial parts of the triton and the ^{3}He internal wave functions are assumed to have the form

$$\phi = \sum_{i=1}^{2} B_i \exp\left(-\frac{1}{2} a_i \sum_{j=1}^{3} \bar{r}_j^2\right). \tag{7.85}$$

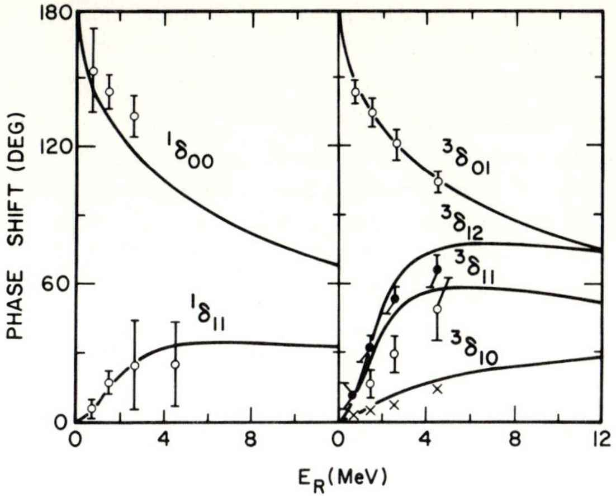

Fig. 21

Phase shifts $^{2s+1}\delta_{lJ}$ for n + t elastic scattering. Empirical data points are those of ref. [TO 66]. (Adapted from ref. [HA 71].)

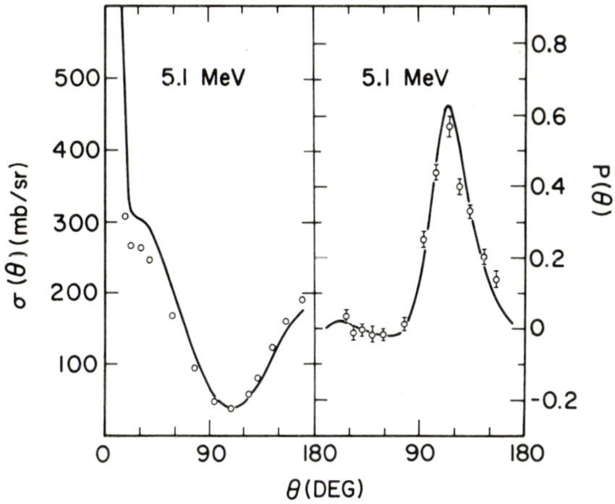

Fig. 22

Comparison of calculated and experimental differential cross sections and proton polarizations at an energy of 5.1 MeV in the p + ^{3}He case. Experimental data are those of ref. [MC 64, Mo 69]. (Adapted from ref. [HE 72].)

Table 2. Values of the parameters B_i and a_i in the triton and ^{3}He spatial wave functions.

cluster	B_1	B_2	a_1 (fm^{-2})	a_2 (fm^{-2})
t	0.2000	0.7138	0.24	0.66
^{3}He	0.2042	0.6938	0.24	0.66

The parameters B_i and a_i are determined by using the Ritz variational method, and the resultant values are listed in table 2. With these parameters, the calculated triton and ^{3}He binding energies are equal to 4.6 and 3.9 MeV, respectively.

In these calculations, the specific distortion of the triton or the ^{3}He cluster is also taken into consideration. As has been explained previously, this is achieved by the addition of distortion functions to improve the behaviour of the wave function in the region of configuration space where the clusters penetrate strongly into each other.

For relative energies up to 12 MeV, elements of the S-matrix are calculated for partial waves with $l \leqslant 3$. The result shows that, except for the spin-mixing parameter in the state with $J^\pi = 1^-$, all other spin- and tensor-mixing parameters are rather small. In fig. 21, a comparison in the n + t case is made between the calculated phase shifts ${}^{2S+1}\delta_{lJ}$ and the empirical values obtained by *Tombrello* [TO 66]. From this figure it is seen that the agreement is quite good. In fig. 22, calculated differential scattering cross-section and polarization results in the p + ^{3}He case are shown at an energy of 5.1 MeV. Here also, one notes that the agreement with the experimental data of refs. [MC 64, MO 69] is rather satisfactory.

7.3i. Coupled-Channel Study of t (p, n) ^{3}He Reaction

As last example, we describe the t (p, n) ^{3}He reaction calculation of *Hackenbroich* and collaborators [HA 72a, HE 73, HE 72a]. In this calculation, the p + t and n + ^{3}He channels are explicitly considered;† the d + d and many-particle breakup channels are neglected for simplicity. The nucleon-nucleon potential and the spatial parts of the cluster internal functions employed are the same as those used in the p + ^{3}He and n + t scattering calculations of the previous subsection.

Because of the coupling between the p + t and the n + ^{3}He channels through the nuclear and Coulomb interactions, the number of coupled equations in each (J, π) state doubles that in the p + ^{3}He or n + t case discussed in subsection 7.3h. In the $J^\pi = 0^+$ state, for example, the number of coupled equations is equal to two, as against one in the previously considered case of p + ^{3}He or n + t scattering. Therefore, in this particular state, the S-matrix is a 2 X 2 matrix specified by three parameters, namely, the p + t phase shift ${}^1\delta^p_{00}$, the n + ^{3}He phase shift ${}^1\delta^n_{00}$, and the reflection coefficient ${}^1\tau_{00}$.††

In this example, it is of course also possible to perform the calculation using eigenfunctions of the operator $\mathbf{T}^2$ with eigenvalues T = 0 and 1. These eigenfunctions are linear combinations of the p + t and n + ^{3}He wave functions and are coupled together by the Coulomb part of the nucleon-nucleon interaction. As is quite evident, the use of these functions should be particularly convenient in studying the resonance levels of the com-

† The wave functions for the p + t and n + ^{3}He systems are not eigenfunctions of the operator $\mathbf{T}^2$, especially in the asymptotic region. This is in contrast to the p + ^{3}He and n + t systems discussed in subsection 7.3h. For these latter systems, the wave functions are eigenfunctions of the operator $\mathbf{T}^2$.

†† The notation for the phase shift has the same meaning as that used in subsection 7.3h; namely, the upper-left superscript indicates the spin multiplicity and the subscripts indicate the values of l and J.

pound nucleus ^4He. Because of the long-range character of the Coulomb potential, the wave functions describing these resonance levels in the compound-nucleus region should be in good approximation eigenfunctions of the operator $\mathbf{T}^2$, and the coupling of the channel wave functions with different isobaric spins will take place mostly only in the nuclear surface region.†

To make certain that the calculation yields reliable results, distortion functions are again included for the purpose of improving the behaviour of the wave function in the region of strong interaction. These distortion functions can be chosen to have good isobaric spin quantum numbers, as follows from the discussion given in the above paragraph.

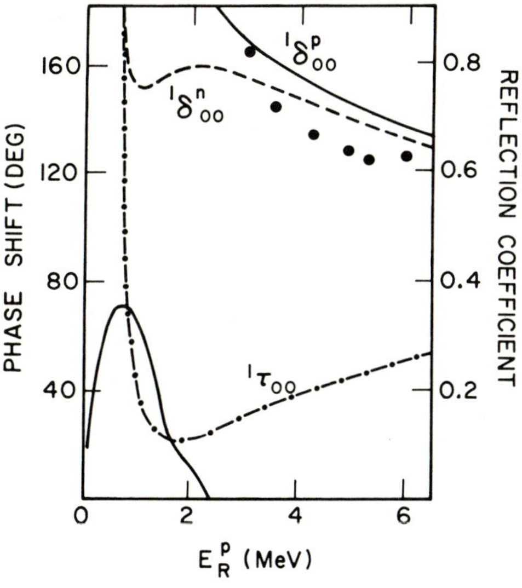

Fig. 23
Phase shifts and reflection coefficient in the $J^\pi = 0^+$ state. Empirical data points are those of ref. [KA 74]. (Adapted from ref. [HE 72a]

In fig. 23, the calculated phase shifts and reflection coefficient in the $J^\pi = 0^+$ state are shown as a function of E_R^p, the relative energy of the proton and the triton cluster in the center-of-mass system. The empirical phase-shift points are those of *Kankowsky* and *Fick* [KA 74]. Here one sees that, at low energies, the p + t phase-shift curve does exhibit the well-known resonance behaviour. At higher energies, the agreement between theory and experiment is not very close, but still reasonably satisfactory. The calculated reflection coefficient has a broad minimum around 2 MeV, which corresponds to the broad maximum in the t(p, n) ^3He reaction cross section found experimentally [SE 60].

A comparison between calculation and experiment for various differential cross-section and polarization results is shown in fig. 24. From this figure, it is seen that, except for relatively minor discrepancies, the agreement is indeed quite satisfactory.

---

† An appreciable mixing of nuclear states with different isobaric spins can take place in the compound-nucleus region, only if these states have similar excitation energies and all other quantum numbers being the same. Such a case occurs, for example, in ^8Be where some excited states are described by strong admixtures of eigenfunctions of the operator $\mathbf{T}^2$ (see chapter 10).

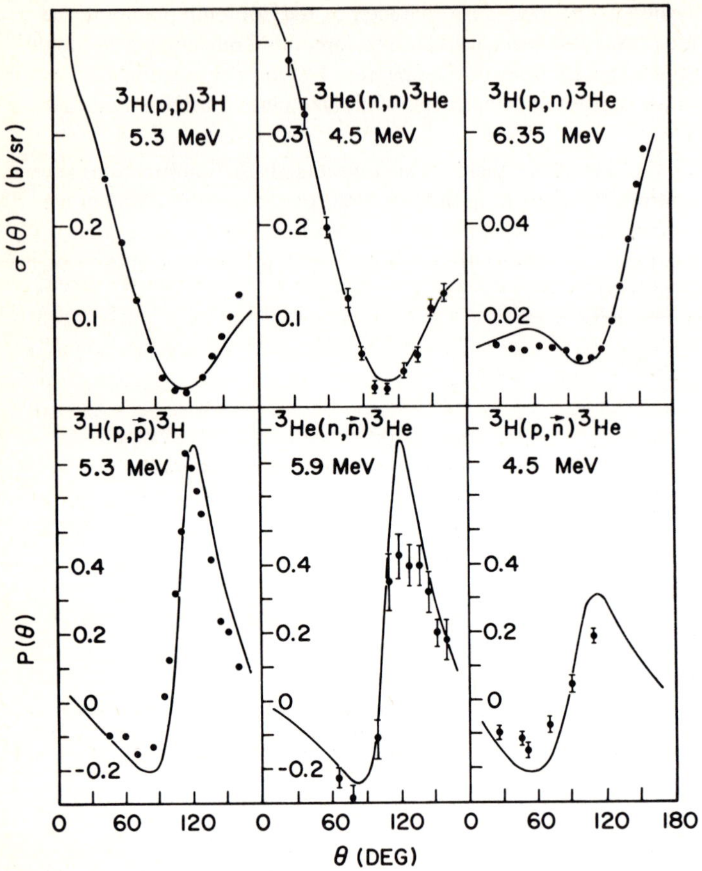

Fig. 24

Comparison between calculation and experiment for various differential cross-section and polarization results. (Adapted from ref. [HE 73].)

7.4. Concluding Remarks

The procedure used here to treat nuclear scattering and reaction problems is in principle a consistent and systematic approximation method to solve quantitatively the many-particle Schrödinger equation. As has been mentioned previously, this procedure consists in the adoption of a nucleon-nucleon potential which fits the scattering data up to about 300 MeV in the laboratory system and then solving the projection equation (2.3) in a truncated Hilbert space. In this way, one obtains an approximate solution which can be systematically improved by enlarging the truncated space through the addition of suitably chosen bound-structure and open-channel terms to the trial function describing the behaviour of the system under consideration.

In practice, it is often necessary to adopt some simplifications in order to make the calculation practicable. Thus, one either employs a nucleon-nucleon potential without repulsive core as described in subsection 5.3d or treats the Jastrow anti-correlation factor in an approximate fashion as described in subsection 5.3c. With the adoption of either one of these simplifications, it is inevitable that certain deficiencies will be introduced into the theory. However, the results from the large number of calculations described in section 7.3 do indicate that these deficiencies are likely minor and can be tolerated at this stage of development.

We should point out that the noncentral potentials used in the calculations of subsections 7.3f, h, and i are rather unconventional. These potentials contain a central part which alone yields the correct two-nucleon effective-range parameters and, consequently, the deuteron binding energy. Thus, the addition of a tensor component will undoubtedly lower the deuteron energy eigenvalue beyond the experimental value for the ground-state energy. To avoid such overbinding for the deuteron and other clusters, the internal functions of the clusters are therefore chosen to have simple forms for which the tensor potential has zero expectation value. In this way, one obtains fairly correct cluster internal energies and can, at the same time, study reaction processes for which these noncentral forces are mainly responsible, such as the transition from the $n + \alpha$ to the $d + t$ channel and so on. Quite clearly, this procedure of using a "hybrid" nucleon-nucleon potential and simplified cluster internal functions will inject some inaccuracies into the results of these calculations; however, a more consistent treatment of using a "correct" nucleon-nucleon potential and the resultant complicated cluster internal functions is not likely to be feasible from the computational viewpoint at the present moment.

Another minor deficiency contained in the calculations of section 7.3 is the lack of close agreement between calculated and experimental cluster binding energies. As a consequence, the calculated threshold energies do not match too well with experimental values and the calculated reaction cross sections can become rather inaccurate in energy regions where reaction channels have just opened. In fact, even in single-channel scattering calculations, such a defect in binding-energy values can have a rather significant influence on the phase shifts. This is so because, as has been pointed out previously, the kernel function $K(\mathbf{R}', \mathbf{R}'')$ depends explicitly on the total energy E of the system. For example, in the $\alpha + \alpha$ problem discussed in subsection 7.3g, the calculated α-particle binding energy is only 21.5 MeV, which is 6.8 MeV smaller than the value experimentally determined. In this case, an examination based on the energy dependence of the kernel function reveals that a mismatch in binding energy of this magnitude can create a difference of about 5° in the $l = 0$, 2, and 4 phase shifts at E_R of 15 MeV. Therefore, even in this system where the calculation seems to contain most of the necessary features, the agreement between theory and experiment, as exhibited in fig. 20, is somewhat fortuitous and further improvements must be made if more reliable results are to be obtained.

In spite of these deficiencies, it is our contention that the examples in section 7.3 serve to show that the procedure adopted here, based on the unified theory formulated in chapter 4, is practical and does lead to basically reliable results. Hence, it can be used not only to explain experimentally observed phenomena but also to study more fundamental questions as those to be discussed in the following chapters.

8. Introductory Considerations About the Derivation of General Nuclear Properties

8.1. General Remarks

Until now we have applied the unified nuclear theory formulated in chapter 4 mainly to numerical studies of bound-state and reaction problems. In the following chapters, we shall show how this theory can also be used for discussions of more general character, as, for instance, the derivation of Breit-Wigner resonance formulae, direct-reaction theories, and so on.

In this chapter, we shall describe the guiding principle for all these derivations. This principle is that one separates in a suitable way the total wave function ψ into different components ψ_k and its variation $\delta\psi$ into corresponding components $\delta\psi_k$; that is, one writes

$$\psi = \sum_{k=1}^{N} \psi_k \tag{8.1}$$

and

$$\delta\psi = \sum_{k=1}^{N} \delta\psi_k. \tag{8.2}$$

Because of arbitary variations of the linear variational functions and discrete variational amplitudes contained in ψ_k, every variation $\delta\psi_k$ defines a certain subspace of the total Hilbert space of the nuclear system. As will be seen in the following chapters, the great advantage of dividing the total Hilbert space into various subspaces is that this division can be chosen flexibly to provide a most convenient basis for discussing a certain particular problem in a nuclear system.

By substituting eqs. (8.1) and (8.2) into the projection equation (2.3), one obtains the following set of coupled equations:

$$\left\langle \delta\psi_r \middle| H-E \middle| \sum_{k=1}^{N} \psi_k \right\rangle = 0, \quad (r = 1, \ldots, N) \tag{8.3}$$

with the index r taking on integral values from 1 to N. These equations describe, of course, the coupling of wave functions in the subspaces of the Hilbert space with each other, through the action of the operator (H − E). Because the function ψ_k may contain an arbitrary number of variational functions and amplitudes, every equation in eq. (8.3) represents in general a coupled set of linear integrodifferential equations and linear integral equations of the form given by eq. (4.19).

In subsection 7.2c it was discussed that, for quantitative calculations, it is often convenient to transform the coupled set of integrodifferential and integral equations into a set of linear, algebraic, inhomogeneous equations. For the consideration of general properties of nuclear reactions, this is not useful because, in so doing, the S-matrix becomes no longer completely unitary. In such general considerations, it is more appropriate to work with the coupled set of integrodifferential and integral equations, obtained by letting the relative-motion functions in the open channels be varied in a completely arbitrary manner. As has been discussed in chapter 2 and subsection 7.2b, the solutions of these coupled equations always lead to a unitary S-matrix, even if one uses a restricted number of variational functions and amplitudes or, in other words, treats the nuclear system under consideration only approximately.

To see in more detail what the splitting of the total wave function ψ and its variation $\delta\psi$ into different terms ψ_K and $\delta\psi_K$ means, we consider again the ^{8}Be system as a specific example. In this case, if the purpose is to study the influence of ^{8}Be compound states on the $\alpha + \alpha$ scattering amplitude in the energy region where only the $\alpha + \alpha$ channel is open, then one splits ψ and $\delta\psi$ into two parts, i.e.,

$$\begin{aligned} \psi &= \psi_1 + \psi_2, \\ \delta\psi &= \delta\psi_1 + \delta\psi_2. \end{aligned} \tag{8.4}$$

The function ψ_1, which describes mainly the behaviour of the system in the $\alpha + \alpha$ elastic channel, is written in the form of eq. (4.17a) as

$$\psi_1 = A'\{\hat{\phi}_0(\tilde{\alpha}_1)\,\hat{\phi}_0(\tilde{\alpha}_2)\,F(\mathbf{R})\}, \tag{8.5}$$

with $\hat{\phi}_0(\tilde{\alpha}_1)$ being the antisymmetric ground-state eigenfunction of the α-particle Hamiltonian and $F(\mathbf{R})$ being an arbitrary variational function. The function ψ_2 is chosen to describe the behaviour in the compound-nucleus region of ^{8}Be; it is given by

$$\psi_2 = \sum_i a_i\,\hat{\varphi}_i, \tag{8.6}$$

with $\hat{\varphi}_i$ being bound-state-type functions and a_i being discrete linear variational amplitudes.† With these forms for ψ_1 and ψ_2, one sees that the variation $\delta\psi_1$ defines the subspace which contains all possible two-α cluster configurations with no internally excited α clusters, and the variation $\delta\psi_2$ defines the subspace which is spanned by the basis wave functions $\hat{\varphi}_i$. It should be noted again that the choice of $\hat{\varphi}_i$ is quite flexible; these functions need not be orthogonal to each other, but only be linearly independent.

In the equations

$$\begin{aligned} \langle\delta\psi_1|H-E|\psi_1+\psi_2\rangle &= 0, \\ \langle\delta\psi_2|H-E|\psi_1+\psi_2\rangle &= 0, \end{aligned} \tag{8.7}$$

which one obtains by substituting eq. (8.4) into eq. (2.3), the first equation represents an integrodifferential equation for the relative-motion function $F(\mathbf{R})$ and the discrete

† With a special ansatz for ψ_2 and approximate internal wave functions for the α clusters, such a splitting for ψ has already been used in the $\alpha + \alpha$ scattering calculation described in subsection 7.3g.

variational amplitudes a_i. This equation, in which F(**R**) and a_i are contained linearly, comes from an arbitrary variation of the relative-motion function F(**R**) in ψ_1. It describes the coupling of the bound-state-type functions $\hat{\varphi}_i$ with the $\alpha + \alpha$ channel wave function. The second equation of eq. (8.7) represents a set of linear integral equations for F(**R**) and a_i. This set of equations comes from the variation of a_i in ψ_2 and describes the mutual coupling of the functions $\hat{\varphi}_i$ and the coupling of the $\alpha + \alpha$ channel with these $\hat{\varphi}_i$.

Next, let us consider energy regions where, besides the $\alpha + \alpha$ channel, other channels are also open. In this case, if the purpose is to examine collectively the influence of the bound-state-type functions and these other open channels on the $\alpha + \alpha$ elastic-scattering cross section, then one again separates ψ into ψ_1 and ψ_2, but with ψ_2 including also wave functions for these other open channels. Substitution into the projection equation (2.3) then yields similarly a coupled set of equations represented by eq. (8.7), but the second equation will now include, besides integral equations coming from the variation of the amplitudes a_i, also integrodifferential equations which result from the variation of relative-motion functions contained in the open-channel functions of ψ_2.

On the other hand, if one wishes to investigate especially the transition probabilities from the elastic channel to these other open channels, then it is more appropriate to separate ψ into three parts, i. e.,

$$\psi = \psi_1 + \psi_2 + \psi_3, \tag{8.8}$$

where ψ_1 describes the elastic channel, ψ_2 is a set of bound-state-type functions, and ψ_3 describes the other open channels. In this case one obtains, instead of eq. (8.7), the following set of equations:

$$\begin{aligned} \langle \delta \psi_1 | H - E | \psi_1 + \psi_2 + \psi_3 \rangle &= 0, \\ \langle \delta \psi_2 | H - E | \psi_1 + \psi_2 + \psi_3 \rangle &= 0, \\ \langle \delta \psi_3 | H - E | \psi_1 + \psi_2 + \psi_3 \rangle &= 0. \end{aligned} \tag{8.9}$$

In this set, the first and third equations represent sets of linear integrodifferential equations which come from arbitrary variations of the relative-motion functions in the elastic channel and the other open channels. The second equation represents a set of integral equations which results from the arbitrary variation of the amplitudes a_i.

With the above discussion, it should be clear how one can generalize these considerations to write the wave function ψ and its variation $\delta\psi$ of any nuclear system – or, indeed, any many-particle system – in such a way that the set of coupled equations resulting from an application of the projection equation (2.3) is especially suited for the study of these nuclear properties which one is interested in at the moment.

8.2. Introduction of Effective Hamiltonians

In many investigations, it is useful to eliminate formally some of the open-channel and bound-structure terms such that they and their variations appear no longer in the set of coupled equations for the nuclear system. As is quite well known, this can be achieved by the use of resolvent operators which belong to the Hilbert subspaces for these terms.

If $\delta\psi_s$ represents an arbitrary variation in such a subspace (the s-subspace), then the resolvent operator G_s of this subspace is defined by the equation

$$\langle\delta\psi_s|H-E|G_s = \langle\delta\psi_s|. \qquad (8.10)$$

As is discussed in many textbooks (see, e.g., ref. [RO 67]), eq. (8.10) may be solved by using the eigenfunction-expansion technique. The eigenfunction will be denoted by ψ_s^α, with the index α including the energy eigenvalue E_s^α and all other quantum numbers λ_α necessary for a complete specification of this function. It satisfies the homogeneous equation

$$\langle\delta\psi_s|H-E_s^\alpha|\psi_s^\alpha\rangle = 0. \qquad (8.11)$$

The energy spectrum for E_s^α may have both a discrete and a continuous part. The functions ψ_s^α which correspond to discrete energy eigenvalues are the bound-state solutions of eq. (8.11). The other solutions which have continuous energy eigenvalues are the ones which describe scattering and reaction processes in the limited wave-function space defined by the variation $\delta\psi_s$.

If the energy E is in the discrete part of the spectrum, then the resolvent operator is given in terms of these eigenfunctions as

$$G_s = \mathop{\mathsf{S}}_{\alpha} \frac{|\psi_s^\alpha\rangle\langle\psi_s^\alpha|}{E_s^\alpha - E}, \qquad (8.12)$$

where the symbol S denotes summation over discrete quantum numbers and integration over continuous quantum numbers. The functions ψ_s^α must be course be orthonormalized by using either the Kronecker δ-function or the Dirac δ-function, depending upon whether the quantum number is discrete or continuous. In addition, these functions satisfy the closure relation

$$\mathop{\mathsf{S}}_{\alpha} |\psi_s^\alpha\rangle\langle\psi_s^\alpha| = 1_s, \qquad (8.13)$$

with 1_s being a unit operator in the s-subspace.

For an energy in the continuous part of the spectrum, the resolvent operator depends upon the boundary condition imposed in the asymptotic region. With an outgoing-wave boundary condition, the spectral representation of the resolvent operator has the familiar form

$$G_s = \mathop{\mathsf{S}}_{\alpha} \frac{|\psi_s^\alpha\rangle\langle\psi_s^\alpha|}{E_s^\alpha - E - i\epsilon}, \quad (\epsilon > 0) \qquad (8.14)$$

where ϵ is a positive infinitesimal number which approaches zero after the integration over E_s^α is made.†

† Equation (8.14) is also valid if the s-subspace contains not only wave functions describing two-particle decays but also wave functions describing three- and more-particle decays. Because of the complicated nature of the wave functions in the asymptotic regions of the three- and more-particle channels, the construction of a complete set of ψ_s^α will, in general, be quite difficult.

In the set of coupled equations given by eq. (8.3), one can formally eliminate the term ψ_N by writing it as

$$|\psi_N\rangle = -G_N\,|H-E|\,\psi'\rangle, \tag{8.15}$$

with

$$\psi' = \sum_{k=1}^{N-1} \psi_k. \tag{8.16}$$

That one can represent ψ_N in the form of eq. (8.15) may be seen easily. By using the equation which defines the resolvent operator G_N, one finds that, with ψ_N given in this way, the equation

$$\langle\delta\,\psi_N\,|H-E|\,\psi\rangle = \langle\delta\,\psi_N\,|H-E|\,\psi_N + \psi'\rangle = 0 \tag{8.17}$$

is indeed satisfied. Now, by introducing the effective Hamiltonian

$$H' = H - (H-E)\,G_N\,(H-E), \tag{8.18}$$

one can then write the (N − 1) remaining equations of eq. (8.3) as

$$\langle\delta\,\psi_r\,|H'-E|\,\psi'\rangle = 0, \quad (r = 1, \ldots, N-1) \tag{8.19}$$

where the index r now takes on only integral values from 1 to (N − 1). Thus one sees that, with the introduction of H', the term ψ_N and its variation $\delta\,\psi_N$ are formally eliminated from the set of coupled equations represented by eq. (8.3).

In a similar manner, one can eliminate a second term ψ_v of ψ by defining an effective Hamiltonian

$$H'' = H' - (H'-E)\,G'_v\,(H'-E), \tag{8.20}$$

with G'_v given by an equation similar to eq. (8.10), but with H replaced by H'. By this procedure one reduces the number of coupled equations which describe the behaviour of the nuclear system under consideration step by step. The price which one has to pay is of course that the resultant effective Hamiltonian becomes more and more complicated.

In the case where one eliminates only closed-channel or bound-state terms, the effective Hamiltonian remains as a hermitian operator. This is so, because the resolvent operator given by eq. (8.12) obviously has the property that G_s is equal to G_s^+. One the other hand, if one eliminates a term which contains open-channel functions, then the resolvent operator and, consequently, the effective Hamiltonian are no longer hermitian operators. To see this, one considers eq. (8.14). In this case, E_s^α is a continuous variable and one can write

$$\frac{1}{E_s^\alpha - E - i\epsilon} = P\frac{1}{E_s^\alpha - E} + i\pi\,\delta\,(E_s^\alpha - E), \tag{8.21}$$

where the symbol P stands for the principal value. If one substitutes eq. (8.21) into eq. (8.14), then one sees immediately that G_s splits into a hermitian and an antihermitian part. As for the consequences which follow from the introduction of a nonhermitian effective Hamiltonian, we shall discuss at appropriate places where the appearance of such an operator influences our physical considerations.

An application of the above procedure to eliminate terms in the total wave function of the system has been used very often in bound-state variational calculations. In such calculations, one uses in the trial function a restricted number of terms containing discrete linear variational amplitudes; in other words, one solves the problem by using trial functions which belong to a limited subspace of the total Hilbert space. As is shown by our considerations given above, this limitation in the choice of trial functions can be compensated by introducing an effective Hamiltonian which takes into account the omitted bound-state configurations.

In practice, one does not calculate the effective Hamiltonian in the manner described above, because to do so is equivalent to solving the Schrödinger equation exactly or, in other words, to carrying out a variational calculation in the complete Hilbert space. Rather, one adopts a semi-phenomenological approach by introducing effective two-nucleon forces or effective two-nucleon interaction matrix elements [KU 68] which are obtained by fitting essentially the energy spectrum in a neighbouring nucleus. These effective forces or matrix elements are then used to calcultae a level scheme of similar structure in the nucleus under consideration. In shell-model studies, for instance, such an effective-interaction method has been used for the calculation of level schemes in p-shell [CO 65] and other major-shell nuclei [DE 63, HA 70, MC 73]. That this method usually works quite well is connected with the fact that nuclear forces are rather short-ranged and, therefore, the effective nucleon-nucleon interaction for similar configurations varies only slightly among a group of neighbouring nuclei.

8.3. Elimination of Linear Dependencies

As already mentioned often, the different terms ψ_K which the wave function ψ of the nuclear system is composed of are usually chosen as not orthogonal to each other. Thus, it can happen that some of these chosen terms may turn out to be not even linearly independent. In such cases, computational difficulties will arise because the coupled equations which describe the behaviour of the nuclear system are then also not linearly independent of each other, with the consequence that the wave function ψ can no longer be numerically determined. As an example of this, let us consider again the case of $\alpha + \alpha$ scattering at energies near the ground-state resonance of ^{8}Be. To demonstrate the occurrence of linear dependencies, we shall express the total wave function as a sum of two terms ψ_1 and ψ_2. The function ψ_1 is chosen to describe mainly the open-channel part of the $\alpha + \alpha$ system; it is written as

$$\psi_1 = A'\{\hat{\phi}_0(\tilde{\alpha}_1)\,\hat{\phi}_0(\tilde{\alpha}_2)\;F(\mathbf{R})\} = \int A'\{\hat{\phi}_0(\tilde{\alpha}_1)\,\hat{\phi}_0(\tilde{\alpha}_2)\,\delta(\mathbf{R}-\mathbf{R}')\}\;F(\mathbf{R}')\,d\mathbf{R}', \quad (8.22)$$

with $F(\mathbf{R})$ being an arbitrary variational function. The function ψ_2 describes the ^{8}Be resonance in the compound-nucleus region and is given by

$$\psi_2 = a\,A'\{\hat{\phi}_0(\tilde{\alpha}_1)\,\hat{\phi}_0(\tilde{\alpha}_2)\,\chi(\mathbf{R})\}, \quad (8.23)$$

with a being a discrete variational amplitude and $\chi(\mathbf{R})$ being an appropriately chosen finite-normalizable relative-motion function. With these choices of ψ_1 and ψ_2, it is quite evident that there will be computational problems, because ψ_2 can clearly be expanded as a linear superposition of the basis wave functions $A'\{\hat{\phi}_0(\tilde{\alpha}_1)\,\hat{\phi}_0(\tilde{\alpha}_2)\,\delta(\mathbf{R}-\mathbf{R}')\}$ of ψ_1 and, consequently, there is no unique solution for the amplitude a.

The prevention of such linear dependencies is in principle very simple. Here we shall briefly discuss the only important case where one or several bound-structure terms ψ_s are linearly dependent on the basis functions of one or several open-channel wave functions. The example just described is clearly a case in this category. Now, let us denote the open-channel wave functions as ψ_r^0. Then what one does is to define new open-channel wave functions $\tilde{\psi}_r^0$ by projecting out the bound-structure terms ψ_s from ψ_r^0 in the following manner:

$$\tilde{\psi}_r^0 = (1-P)\,\psi_r^0, \tag{8.24}$$

where the projection operator P is defined as

$$P = \sum_s |\tilde{\psi}_s\rangle\langle\tilde{\psi}_s|, \tag{8.25}$$

with $\tilde{\psi}_s$ representing orthonormalized basis functions for the bound-structure terms ψ_s. Because of their special structure, the operator P and its complementary operator $Q = (1-P)$ have the following properties, as can be proven easily:

$$\begin{aligned} &P + Q = 1,\ \ P^2 = P,\ \ Q^2 = Q,\ \ P = P^+,\ \ Q = Q^+,\\ &PQ = QP = 0,\\ &P\psi_r^0 = \sum_s a_{rs}\,\tilde{\psi}_s,\\ &Q\psi_r^0 = \psi_r^0 - \sum_s a_{rs}\,\tilde{\psi}_s. \end{aligned} \tag{8.26}$$

In the above equations, the a_{rs} are the amplitudes of the bound-structure functions $\tilde{\psi}_s$ which one wants to eliminate from ψ_r^0.†

Similarly, one defines the variation $\delta\tilde{\psi}_r^0$ as

$$\delta\tilde{\psi}_r^0 = (1-P)\,\delta\psi_r^0. \tag{8.27}$$

With these definitions for $\tilde{\psi}_r^0$ and $\delta\tilde{\psi}_r^0$, the derivation of the general properties of a nuclear system can be carried out in almost exactly the same way as in the case where

† The adoption of the notation P and Q may create some confusion. In our case, these operators serve only to eliminate linear dependencies in the wave function. They are not introduced to project out wave functions with correct boundary conditions from a complete set of functions as, for instance, is done in the theory of *Feshbach* [FE 58, FE 62]. We wish to emphasize again that, in the unified theory presented here, we have always adopted wave functions with correct asymptotic boundary conditions from the very beginning.

no linear dependencies exist in the wave function ψ. Later, we shall demonstrate this explicitly in some cases where the use of this projection formalism is particularly appropriate in bringing out the interesting features of certain discussions.[†]

From a computational viewpoint, it is sometimes useful to perform the projection procedure even when linear dependencies do not strictly exist in the wave function. For example, if one or several bound-structure terms can be expressed, not exactly, but in a good approximation as a linear superposition of the basis functions of open-channel wave functions, then the use of the projection technique, as prescribed by eqs. (8.24) and (8.25), will improve greatly the accuracy in a numerical calculation.

8.4. Concluding Remarks

In this chapter, we have outlined the general procedure which we shall use to discuss various nuclear properties in the following chapters. This procedure consists in the splitting up of the wave function ψ and its variation $\delta\psi$ into constituent parts ψ_k and $\delta\psi_k$ in such a way that the resultant set of coupled equations becomes particularly suitable for a discussion of the problem under consideration.

To bring out as clearly as possible the important features in our discussion, we shall simplify the formulation in the following chapters to a maximum extent. Thus, unless otherwise stated, the derivations will be performed by assuming that the clusters have no internal angular momenta and no charge. For a generalization to the case involving charged clusters with nonzero internal angular momenta, the procedure is tedious but straightforward, and has been discussed in many textbooks (see, e.g., ref. [WU 62]). In addition, we shall for simplicity not write out the Jastrow anti-correlation factor explicitly but, as has been done frequently until now, simply assume its presence understood.

In the following chapters, we shall again discuss many numerical examples. In these examples, the internal angular momenta and charges of the clusters will be explicitly taken into account. We should emphasize, however, that these examples are presented mainly to illustrate the general nuclear properties under consideration, rather than to show any detailed agreement with experimental data. Therefore, simplifications will frequently be made for the purpose of stressing the basic ideas involved, even though the introduction of such simplifications may result in some compromise in fit to experimental results.

[†] A simple example demonstrating the use of the projection technique is given in appendix C.

9. Breit-Wigner Resonance Formulae

9.1. General Remarks

In this chapter, we shall use the unified theory formulated in chapter 4 and the general procedure described in chapter 8 to discuss the Breit-Wigner resonance formulae. We shall start our discussion in section 9.2 with the simplest case where there is an isolated resonance and only the elastic channel is open. This discussion will be a rather extensive one in order to explain the underlying ideas as clearly as possible. It is also quite general, in the sense that it can be applied not only to the case with incident nucleons but also to the case with arbitrary composite bombarding particles, in contrast to the discussion given in the book of *Mahaux* and *Weidenmüller* [MA 69]. Then, we shall proceed to discuss in sections 9.3, 9.4, and 9.5 more complicated situations where overlapping resonances exist and where inelastic and rearrangement channels are also open.

In addition, we shall make a brief discussion in section 9.6 of the energy dependence of the partial level width in the case where the resonance under consideration occurs near a threshold. This discussion is useful for the purpose of understanding the threshold behaviour of the resonance scattering or reaction amplitude and the applicability of the Breit-Wigner formula in this situation.

9.2. Single-Level Resonance Formula for Pure Elastic-Scattering

9.2a. Derivation of the Resonance Formula

In the case where the elastic channel is the only open channel, we separate the wave function ψ into two parts, i. e.,

$$\psi = \psi_D + \psi_C, \tag{9.1}$$

with

$$\psi_D = A' \{\hat{\phi}(\widetilde{A})\, \hat{\phi}(\widetilde{B})\, F(\mathbf{R}_A - \mathbf{R}_B)\} \tag{9.2}$$

and

$$\psi_C = \sum_i a_i \hat{\varphi}_i. \tag{9.3}$$

In the open-channel function ψ_D, the cluster internal functions $\hat{\phi}(\widetilde{A})$ and $\hat{\phi}(\widetilde{B})$ describe the exact structure of the bombarding particle and the target nucleus in the asymptotic region. The antisymmetrized functions $\hat{\varphi}_i$ in ψ_C are bound-state-type functions which, together with ψ_D, serve to describe the behaviour of the system in the region of the compound nucleus.

At a first glance, it seems that the adoption of the ansatz of eqs. (9.1)–(9.3) for ψ might preclude a true representation of the exact solution of our problem, because this ansatz already selects a particular subspace of the complete Hilbert space for our A-particle system. But this is not the case if one demands that the functions $\hat{\varphi}_i$ form a complete set of normalizable wave functions in a finite region of configuration space which is larger than that corresponding to the compound nucleus.† If this is done, then clearly one can describe in a correct way the behaviour of the system both inside the compound nucleus and in the surface region. Therefore, by using the ansatz for ψ given by eqs. (9.1)–(9.3), the exact wave function can be obtained in the whole configuration space, as long as no other but the elastic channel is open.

To continue our discussion, we substitute eq. (9.1) into the projection equation (2.3) and obtain the following set of coupled equations:

$$\langle \delta \psi_D | H - E | \psi_D \rangle + \langle \delta \psi_D | H - E | \psi_C \rangle = 0, \tag{9.4a}$$

$$\langle \delta \psi_C | H - E | \psi_D \rangle + \langle \delta \psi_C | H - E | \psi_C \rangle = 0. \tag{9.4b}$$

As was discussed in chapter 8, one can express ψ_C in terms of ψ_D by introducing the resolvent operator G_C for the bound-structure terms. The result is

$$| \psi_C \rangle = - G_C | H - E | \psi_D \rangle, \tag{9.5}$$

where

$$G_C = \sum_\alpha \frac{| \psi_C^\alpha \rangle \langle \psi_C^\alpha |}{E_C^\alpha - E}, \tag{9.6}$$

with the states ψ_C^α being orthonormalized solutions of the homogeneous equation

$$\langle \delta \psi_C | H - E_C^\alpha | \psi_C^\alpha \rangle = 0. \tag{9.7}$$

Upon substituting eq. (9.5) into eq. (9.4a), one obtains then the following equation for the open-channel part ψ_D of the wave function:

$$\langle \delta \psi_D | H - E | \psi_D \rangle = \sum_\alpha \frac{\langle \delta \psi_D | H - E | \psi_C^\alpha \rangle \langle \psi_C^\alpha | H - E | \psi_D \rangle}{E_C^\alpha - E}. \tag{9.8}$$

Equation (9.8) is quite interesting. It shows that, in addition to the interaction obtained in the no-distortion approximation, the two nuclei A and B also interact through a sum of nonlocal, energy-dependent, and separable potentials which arises from the introduction of bound-structure terms represented by the wave functions ψ_C^α. As will be seen below, these nonlocal potentials give rise to resonance structures in the wave function and, consequently, the scattering cross section.

---

† Except for functions of the form

$$A' \{ \hat{\phi}(\tilde{A}) \, \hat{\phi}(\tilde{B}) \, \chi(\mathbf{R}_A - \mathbf{R}_B) \}$$

which have an open-channel configuration with χ satisfying bound-state-type asymptotic boundary conditions. Because of the presence of ψ_D in ψ, these functions can be completely or partially excluded from the set of functions $\hat{\varphi}_i$.

Now let us consider an isolated resonance which occurs in an energy region around E_C^1. For this we write eq. (9.8) in the following form:

$$\langle \delta \psi_D | H'(E) - E | \psi_D \rangle = \frac{\langle \delta \psi_D | H - E | \psi_C^1 \rangle \langle \psi_C^1 | H - E | \psi_D \rangle}{E_C^1 - E}, \tag{9.9}$$

where $H'(E)$ is an energy-dependent effective Hamiltonian, given by

$$H'(E) = H - \sum_{\alpha \neq 1} \frac{|H - E| \psi_C^\alpha \rangle \langle \psi_C^\alpha | H - E|}{E_C^\alpha - E}. \tag{9.10}$$

Equation (9.9) may be solved by using the resolvent operator $G_D'^{+}(E)$ for $H'(E)$ in the subspace of the elastic channel with outgoing-wave boundary conditions. As was discussed in the previous chapter, this resolvent operator may be represented in terms of a complete set of orthonormalized solutions of the homogeneous equation

$$\langle \delta \psi_D | H'(E) - E_D^\beta | \psi_D^\beta (E) \rangle = 0, \tag{9.11}$$

where the eigenvalues E_D^β and the eigenfunctions ψ_D^β are both functions of E. With these solutions, the spectral representation of $G_D'^{+}(E)$ is then given by

$$G_D'^{+}(E) = \mathop{S}_{\beta} \frac{|\psi_D^\beta (E)\rangle \langle \psi_D^\beta (E)|}{E_D^\beta - E - i\epsilon}. \qquad (\epsilon > 0) \tag{9.12}$$

We need $G_D'^{+}(E)$ in an energy region where the spectrum of E_D^β is continuous, i.e., where the elastic channel is open. In this region, $G_D'^{+}(E)$ is a nonhermitian operator due to the presence of the factor $-i\epsilon$ in the denominator, which guarantees an outgoing-wave boundary condition for this operator.

With the resolvent operator $G_D'^{+}(E)$, we obtain from eq. (9.9)

$$|\psi_D\rangle = |\psi_D'^{+}(E)\rangle + \frac{G_D'^{+}(E) |H - E| \psi_C^1 \rangle \langle \psi_C^1 | H - E | \psi_D \rangle}{E_C^1 - E}, \tag{9.13}$$

where $\psi_D'^{+}(E)$ is a solution of the homogeneous equation (9.11) with $E_D^\beta = E$ and has an asymptotic form of an incoming plane wave plus an outgoing spherical wave. Equation (9.13) can be solved easily for ψ_D and the result is

$$|\psi_D\rangle = |\psi_D'^{+}(E)\rangle + G_D'^{+}(E) |H - E| \psi_C^1 \rangle \frac{\langle \psi_C^1 | H - E | \psi_D'^{+}(E)\rangle}{E_C^1 - E - \langle \psi_C^1 | (H - E)\, G_D'^{+}(E)\, (H - E) | \psi_C^1 \rangle}. \tag{9.14}$$

From eq. (9.14), the single-level Breit-Wigner formula can be derived. To show this, one performs partial-wave expansions of $\psi_D'^{+}(E)$ and $G_D'^{+}(E)$, and considers only the wave which has the same spin j and parity π as the resonance state under consideration. Further, by using the identity

$$\frac{1}{E_D^\beta - E - i\epsilon} = P \frac{1}{E_D^\beta - E} + i\pi\delta(E_D^\beta - E), \tag{9.15}$$

we can separate the nonhermitian operator $G'^{+}_{Dj}(E)$ into a hermitian and an antihermitian part, i. e.,†

$$\begin{aligned} G'^{+}_{Dj}(E) &= \underset{\beta}{S}\, P \frac{1}{E^{\beta}_{D} - E} |\psi^{\beta}_{Dj}(E)\rangle\langle\psi^{\beta}_{Dj}(E)| + i\pi |\psi'^{+}_{Dj}(E)\rangle\langle\psi'^{+}_{Dj}(E)| \\ &= G'^{P}_{Dj}(E) + i\pi \bar{G}'_{Dj}(E). \end{aligned} \tag{9.16}$$

Using the above equation, one finds then from the asymptotic behaviour of eq. (9.14) that the outgoing partial wave of total spin j is multiplied by the phase factor

$$\begin{aligned} A_j &= e^{2i\delta'_j(E)} + \frac{i\Gamma_j(E)\, e^{2i\delta'_j(E)}}{E^1_C - E - \Delta_j(E) - \frac{1}{2} i\Gamma_j(E)} \\ &= e^{2i\delta'_j(E)} \left[\frac{E^1_C - E - \Delta_j(E) + \frac{1}{2} i\Gamma_j(E)}{E^1_C - E - \Delta_j(E) - \frac{1}{2} i\Gamma_j(E)} \right], \end{aligned} \tag{9.17}$$

where $\Delta_j(E)$ and $\Gamma_j(E)$ are real quantities, given by

$$\Delta_j(E) = \langle\psi^1_C|(H - E)\, G'^{P}_{Dj}(E)\,(H - E)|\psi^1_C\rangle \tag{9.18}$$

and

$$\Gamma_j(E) = 2\pi |\langle\psi^1_C|H - E|\psi'^{+}_{Dj}(E)\rangle|^2, \tag{9.19}$$

and $e^{2i\delta'_j(E)}$ is the phase factor of the outgoing wave contained in the asymptotic form of $\psi'^{+}_{Dj}(E)$, i. e., the phase factor which near the considered resonance belongs to potential or background scattering. The relations represented by eqs. (9.17)–(9.19) are exact and are valid for every scattering energy provided that only the elastic-scattering channel is open; in other words, they are valid even at scattering energies far away from any resonance.

Equation (9.17) looks already like the Breit-Wigner resonance formula for pure elastic-scattering. The only but essential difference to this formula is that $\Delta_j(E)$ and $\Gamma_j(E)$ are energy-dependent. To eliminate this difference, it is necessary to make certain approximations. We have to use here, for the first time, the fact that we are considering an isolated, relatively sharp resonance. For such a resonance level, all other resonances lie energetically so far away that $\delta'_j(E)$ and $\Gamma_j(E)$ can be considered to be approximately energy-independent over the energy width of this level. Thus, we first define the resonance energy E_r as the energy at which the real part of the denominator in eq. (9.17) becomes zero, i. e., by the equation

$$E^1_C - E_r - \Delta_j(E_r) = 0. \tag{9.20}$$

† The function $\psi^{\beta}_{Dj}(E)$ appearing in eq. (9.16) is normalized according to the equation

$$\langle\psi^{\beta}_{Dj}|\psi^{\gamma}_{Dj}\rangle = \delta(E^{\beta}_{D} - E^{\gamma}_{D}).$$

Then we make the approximation that, in the neighbourhood of E_r,

$$\Delta_j(E) \approx \Delta_j(E_r) + (E - E_r)\left(\frac{d\Delta_j}{dE}\right)_{E=E_r}, \tag{9.21}$$

$$\Gamma_j(E) \approx \Gamma_j(E_r). \tag{9.22}$$

Substitution of eqs. (9.20)–(9.22) into eq. (9.17) then yields the following Breit-Wigner formula:

$$\begin{aligned} A_j &\approx e^{2i\delta_j'(E)}\left(1 + \frac{i\Gamma_{jr}}{E_r - E - i\frac{1}{2}\Gamma_{jr}}\right) \\ &= e^{2i\delta_j'(E)}\left(\frac{E_r - E + i\frac{1}{2}\Gamma_{jr}}{E_r - E - i\frac{1}{2}\Gamma_{jr}}\right), \end{aligned} \tag{9.23}$$

with

$$\Gamma_{jr} = \left[\frac{\Gamma_j(E)}{1 + \frac{d}{dE}\Delta_j(E)}\right]_{E=E_r}. \tag{9.24}$$

9.2b. Discussion of the Resonance Formula

The splitting of the total wave function ψ into an open-channel part ψ_D and a compound-nucleus part ψ_C is not unique. For instance, the choice of the finite volume in the configuration space in which ψ_C is defined is in principle arbitrary. Furthermore, one can add to ψ_C normalizable terms of open-channel configuration multiplied with linear variational parameters.† As can be seen easily, these freedoms will lead to nonunique values for the quantities E_C^1, $\Delta_j(E)$ and $\Gamma_j(E)$. But, especially for relatively sharp resonances, Γ_{jr} and E_r will be quite uniquely determined. This comes from the fact that the total phase shift as an observable magnitude is certainly well defined. Therefore, if the potential or background scattering phase shift is given, the resonance scattering phase shift, as determined from the phase factor inside the parentheses of eq. (9.23), and consequently E_r and Γ_{jr} are also given. This means, for example, that for Γ_{jr} the arbitrariness of $\Gamma_j(E)$ is compensated by the arbitrariness in the denominator of eq. (9.24). This can be shown explicitly, as has been done by Benöhr [BE 68].

Even though the arbitrariness in Γ_{jr} and E_r is much smaller than that in E_C^1, $\Gamma_j(E)$, and $\Delta_j(E)$, it does not disappear completely. This can be seen in the simplest way by using a projection formalism similar to that sketched in chapter 8 (see also appendix C). One projects out from the open-channel wave function ψ_D and its variation $\delta\psi_D$ the

† With such manipulations one can make $\Gamma_j(E)$ for an arbitrary energy E as small as one desires. But one needs for this always unrealistically large volumes in the configuration space which are much larger than the volume of the compound nucleus. If one chooses the volume in which ψ_C is defined to be of the order of the compound-nucleus volume, then $\Gamma_j(E)$ will have a magnitude similar to that of Γ_{jr}. In the following discussion, we shall therefore always choose such a volume in our definition of the compound-nucleus part ψ_C. For more details, see refs. [BE 68, BE 69].

bound structure which describes essentially the resonance state in question in the compound region. By this the potential or background scattering phase shift is changed. But, as already mentioned, the total phase shift as an observable magnitude is certainly fixed, therefore the resonance scattering phase shift will also be changed. On the other hand, it should be noted that the degree of arbitrariness associated with this type of freedom is rather small when the bound structure to be projected out is not well represented by a wave function having an open-channel configuration. This is so, because in this case the potential scattering corresponds essentially to the scattering from a hard sphere, with the consequence that the amplitude of the open-channel wave function is quite small in the compound-nucleus region.

9.2c. Existence of Sharp Resonances

We shall now discuss under what conditions sharp resonances appear. The main condition which must be fulfilled is that the function ψ_C^1 which essentially describes, near the resonance energy, the behaviour of the nuclear system in the compound region is weakly coupled to the open channels. This can be seen from eq. (9.19). There exist three kinds of wave function in the compound-nucleus region which are characteristically different from each other and which satisfy the above condition. The compound configuration of a realistic nuclear resonance is in general described by either one of these types of wave function or a mixture of them.

In the first case, the total wave function for the resonance state in the compound region consists almost completely of a superposition of very many different bound structures and the component which belongs to the subspace of the open-channel term is very small. This means that the wave function ψ_b, which describes the superposition of these many bound structures, satisfies in very good approximation the Schrödinger equation, i.e.,

$$\langle \delta \psi | H - E_b | \psi_b \rangle \approx 0, \tag{9.25}$$

where $\delta \psi$ is an arbitrary variation in the complete Hilbert space of the nuclear system and E_b is the expectation value of the Hamiltonian H with respect to ψ_b. Using eq. (9.25) and the fact that the overlap between ψ_b and the open-channel term is small, we then find from eq. (9.19) that $\Gamma_j(E)$ and, consequently, Γ_{jr} become very small.

Examples for such states are the resonance states excited by low-energy neutron scattering from medium-weight and heavy nuclei. In these states which have excitation energies of about 6 to 8 MeV (the separation energy of a nucleon), the amplitudes belonging to the subspaces of open-channel terms are very small because, at these rather high excitation energies of a heavy compound nucleus, the wave function is expected to consist mainly of a superposition of very many bound-structure or closed-channel configurations.

For the second kind of resonance wave functions in the coumpound region, the appearance of sharp resonance levels is associated with the existence of approximate transition selection rules. In this case, the resonance wave function and the open-channel wave functions are specified by partly different values for a set of nearly good quantum numbers.† As a consequence of this, the coupling between these functions is produced by only an interaction potential H_1 which can come, for instance, from the Coulomb forces, the tensor and the spin-orbit forces. Since this interaction potential H_1 is in general a small part of the total Hamiltonian H, one expects that the transition matrix element

$$\langle \psi_C^1 | H - E | \psi_{Dj}^{\prime +} \rangle \approx \langle \psi_C^1 | H_1 | \psi_{Dj}^{\prime +} \rangle \tag{9.26}$$

can become quite small.

An example for such a resonance is the 15.07 MeV $\frac{3}{2}^-$ level in ^{13}N which has an isobaric spin $T = \frac{3}{2}$ [AJ 70]. This level can be excited by p + ^{12}C elastic scattering. But the coupling is necessarily weak, because the p + ^{12}C channel has $T = \frac{1}{2}$ and hence the coupling is caused by only the Coulomb interaction. The width of this level (1.16 keV) is very small, even though it can also decay into the $\alpha + {}^9B$ channel. But this latter channel has again an isobaric spin of $T = \frac{1}{2}$, with the consequence that the Coulomb interaction is once more the only agent effective for the transition.

Another example is the ^5He, $\frac{3}{2}^+$ state at 16.7 MeV which can be reached from the $n + \alpha$ channel via the tensor force which produces an internal spin-flip from $S = \frac{1}{2}$ to $S = \frac{3}{2}$ and an orbital angular-momentum change of 2 units. Here again, the relative weakness of this interaction potential is responsible for the small width of this level.††

We come now to the third extreme case where the configuration in the compound region is completely of the open-channel form. An example for such a resonance is in very good approximation the ^8Be ground state which was already mentioned in chapter 8 in connection with the projection formalism sketched there. Because the treatment of such kind of resonances, which we shall call potential resonances, is a very good example for the application of this projection formalism, we shall discuss it in more detail.

Sharp potential resonances can only appear, if the compound region of the scattering system is shielded by a high Coulomb or centrifugal barrier from the outside region where only the long-ranged mutual Coulomb interaction is present. Under these circumstances, the wave function in the compound region can be well described by a normalizable bound-state-type function ψ_C^1 which falls off rapidly in the potential barrier. Because of this latter property of ψ_C^1, the coupling of the resonance state to the outside region is therefore weak and the width of the state becomes rather small.

In the case of a pure potential resonance, the function ψ_C^1 depends linearly on the basis wave functions which define the open-channel Hilbert subspace and therefore, for the derivation of the Breit-Wigner formula, the projection formalism discussed in chapter 8 has to be applied. This can be seen from eq. (9.14). In this equation, one can write the factor $\langle \psi_C^1 | H - E | \psi_D^{\prime +}(E) \rangle$ as $\langle \psi_C^1 | H'(E) \dot{-} E | \psi_D^{\prime +}(E) \rangle$, where $H'(E)$ is defined by

---

† They can differ, for instance, in the total spin and orbital angular momenta, or they can have different total isobaric spins.

†† Because this state is deeply embedded under the d + t Coulomb barrier, its decay probability into a deuteron and a triton is small.

eq. (9.10). This is possible, because $\langle \psi_C^1 | H - E | \psi_C^n \rangle$ is equal to zero for $n \neq 1$, as follows from the projection equation (9.7). Then, due to the fact that $\psi_D'^+$ obeys the equation

$$\langle \delta \psi_D | H'(E) - E | \psi_D'^+(E) \rangle = 0 \tag{9.27}$$

and ψ_C^1 belongs completely to the Hilbert subspace of the open-channel functions, the quantity $\langle \psi_C^1 | H'(E) - E | \psi_D'^+(E) \rangle$ must be equal to zero. Therefore, in this particular case, the term in eq. (9.14) which describes the coupling of the resonance state with the open-channel function vanishes and a Breit-Wigner resonance formula can no longer be derived.

To apply the projection formalism, one defines $|\tilde{\psi}_D\rangle$ and $\langle \delta \tilde{\psi}_D|$ as follows:

$$|\tilde{\psi}_D\rangle = (1 - P)|\psi_D\rangle = Q|\psi_D\rangle, \tag{9.28a}$$

$$\langle \delta \tilde{\psi}_D| = \langle \delta \psi_D|(1 - P) = \langle \delta \psi_D| Q, \tag{9.28b}$$

with

$$P = |\psi_C^1\rangle \langle \psi_C^1| \tag{9.29}$$

and

$$Q = 1 - P. \tag{9.30}$$

With the above definitions, one can then write

$$|\tilde{\psi}\rangle = |\tilde{\psi}_D\rangle + |\psi_C\rangle, \tag{9.31a}$$

$$\langle \delta \tilde{\psi}| = \langle \delta \tilde{\psi}_D| + \langle \delta \psi_C|. \tag{9.31b}$$

It should be noted that $\langle \delta \tilde{\psi}|$ of eq. (9.31b) spans the same Hilbert subspace as does $\langle \delta \psi|$; that is, the variation $\langle \delta \tilde{\psi}|$ is as general as the variation $\langle \delta \psi|$ before the introduction of the projection operator Q. Using eq. (9.31), one obtains from the projection equation (2.3)

$$\langle \delta \psi_D | Q(H - E) Q | \psi_D \rangle + \langle \delta \psi_D | Q(H - E) | \psi_C \rangle = 0, \tag{9.32a}$$

$$\langle \delta \psi_C | (H - E) Q | \psi_D \rangle + \langle \delta \psi_C | H - E | \psi_C \rangle = 0. \tag{9.32b}$$

Equation (9.32) can now be treated in the same way as eq. (9.4), except that, in the equations which are derived from eq. (9.4), the resolvent operator $G_D'^+(E)$ of the elastic channel is to be replaced by the resolvent operator $\tilde{G}_D'^+(E)$, defined by the equation

$$\langle \delta \tilde{\psi}_D | H'(E) - E | \tilde{G}_D'^+(E) = \langle \delta \tilde{\psi}_D |. \tag{9.33}$$

In analogy to eq. (9.12), the spectral representation of $\tilde{G}_D'^+(E)$ is given by

$$\tilde{G}_D'^+(E) = \mathop{S}_{\beta} \frac{|\tilde{\psi}_D^\beta(E)\rangle \langle \tilde{\psi}_D^\beta(E)|}{E_D^\beta - E - i\epsilon}, \tag{9.34}$$

but this time the functions $|\tilde{\psi}_D^\beta(E)\rangle$ are solutions of the homogeneous equation

$$\langle \delta \tilde{\psi}_D | H'(E) - E_D^\beta | \tilde{\psi}_D^\beta \rangle = 0, \tag{9.35}$$

which is only defined in a reduced subspace of the elastic-channel Hilbert subspace. The important point to note is that, as a consequence of projecting out the function ψ_C^1, the function $\tilde{\psi}_D'^+ = \tilde{\psi}_D^\beta$ ($E_D^\beta = E$) does not show any resonance behaviour around E_r, in contrast to the function $\psi_D'^+(E)$.†

Because the replacement of $G_D'^+(E)$ by $\tilde{G}_D'^+(E)$ is the only change required in the application of the projection formalism, one obtains again the Breit-Wigner formula in the form of eq. (9.23), with the shift function $\tilde{\Delta}_j(E)$ and the width function $\tilde{\Gamma}_j(E)$ given by expressions similar to eqs. (9.18) and (9.19). As has already been stated, one sees here that $\tilde{\Gamma}_j(E)$ becomes very small if the nuclear interaction region is shielded by a high potential barrier from the outside. This is so, since the main contribution to the factor $\langle \psi_C^1 | H - E | \tilde{\psi}_{Dj}'^+(E) \rangle$ comes from the surface region where ψ_C^1 has a small magnitude.

Realistic resonance wave functions are of either one of these three kinds or a mixture of them. The sharp resonances observed in medium heavy and heavy nuclei are usually of the first kind where the wave functions in the compound region consist mainly of a superposition of very many bound structures. The reason for this is that in the compound nuclei formed, the excitation energies are at least of the order of 6 to 8 MeV (separation energy of a nucleon) and these excitation energies can be quickly distributed among the many degrees of freedom available in a heavy system. In addition, it should be noted that these sharp resonances are mainly ones excited by low-energy neutron scattering, because then the excitation energy involved will not be too large. At higher neutron bombarding energies, many more channels will become open and the resonances will become much less sharp.

In light nuclei, the sharp resonances are either approximate potential resonances or resonances where the decay to the open channels is inhibited by approximate transition selection rules. In other words, they are in good approximation resonances of the second and third kinds discussed above. Completely pure potential resonances, i. e., resonances where the wave functions in the compound region can be described exclusively by open-channel terms, are not expected to exist in reality.

Some of the resonances which are in good approximation potential resonances can be observed in low-energy $\alpha + \alpha$, $^3H + \alpha$, and $^3He + \alpha$ scattering. These resonances have been discussed quantitatively in chapter 7. Low-energy resonances in $d + \alpha$ scattering can only be considered more or less crudely as potential resonances. In this latter case, the large degree of specific distortion of the deuteron cluster in the compound region necessitates the addition of structures other than the open-channel structure to the total wave function in order to obtain a good description of these resonances (see subsection 7.3c).

The projection formalism used here for the description of pure potential resonances can certainly be applied to all kinds of resonances. As is clear, one can always project out from the elastic open-channel function a term which describes as well as possible the

† Even if one projects out a function ψ_C^1 which gives only a fair description of the behaviour of the resonance state in the compound region, the potential scattering amplitude will usually not show any resonance behaviour in the neighbourhood of E_r.

behaviour of the resonance in the compound region.† By this procedure, the energy dependence of the potential-scattering phase shift becomes as small as possible. In other words, the projection formalism can be used to smooth out the energy dependence of the potential-scattering amplitude which is determined from the function $\widetilde{\psi}_D'^{+}(E)$. This makes it especially clear that, as has been mentioned before, the splitting of the total phase shift into a potential-scattering part and a resonance-scattering part is not completely unique.

By such a projection formalism, it is always possible to obtain a form for the integral $\langle \psi_C^1 | H - E | \widetilde{\psi}_{Dj}'^{+}(E) \rangle$ in $\widetilde{\Gamma}_j(E)$, such that the contribution to this integral comes mainly from the surface region. This is true not only in the single-channel case discussed here, but also in the many-channel case to be discussed in section 9.4.

All these different kinds of resonances which have been discussed have one feature in common: for sharp resonances to appear, it is necessary that the resonance wave functions in the interaction or compound region behave very much like bound-state wave functions.

At higher excitation energies of the compound nuclei, other channels besides the elastic channel will be open. If the resonances remain narrow, then our above considerations about the level width, the energy shift, and the behaviour of the resonance wave function in the compound region remain essentially valid. Only the shape of the resonance phase shift in the elastic channel as a function of the energy can depend drastically on the presence of these other open channels, a point which we shall discuss in detail in sections 9.4 and 9.5.

In this connection it should be mentioned that in the case where more than one channel is open, it often happens especially in light nuclei that the same resonance may be a potential resonance with respect to one channel but another kind of resonance with respect to another channel. An example for this is the 16.7 MeV resonance in ^5He. For the d + t channel it is approximately a potential resonance, but for the n + α channel it is a resonance whose decay into this channel is inhibited by approximate transition selection rules.

One additional point should be mentioned. For narrow resonances, the resonance wave functions in the compound region have an appreciable amplitude only for energies around E_r. This can be seen from eq. (9.5) where ψ_C is represented by a linear superposition of the bound structures ψ_C^n. The amplitude a_1 of ψ_C^1 is given by

$$a_1 = -\frac{\langle \psi_C^1 | H - E | \psi_D \rangle}{E_C^1 - E} . \tag{9.36}$$

† To illustrate how one can obtain a good resonance wave function ψ_C^1 in the compound region, we consider again the ground state of ^8Be. By omitting the mutual Coulomb interaction of the α clusters, one can turn this resonance state into a bound state. The resultant bound-state function is then expected to represent satisfactorily the resonance wave function in the compound region, because this particular resonance is deeply imbedded under the Coulomb barrier and hence the total wave function as a function of the α + α separation distance must fall off rapidly in this barrier region.

By using eq. (9.14) for ψ_D, one obtains after selecting the partial wave ψ_{Dj} which has the same spin and parity as ψ_C^1 the following expression:

$$a_1 = -\frac{\langle \psi_C^1 | H - E | \psi_{Dj}'^+ \rangle}{E_C^1 - E - \Delta_j(E) - \frac{1}{2}\Gamma_j(E)} . \tag{9.37}$$

Thus one realizes that for sharp resonances, if ψ_C^1 is constructed to describe satisfactorily the behaviour of the resonance state in the compound region, a_1 becomes appreciably different from zero only for E around E_r.

9.2d. Discussion of a Simple Resonance Model

In this subsection, we shall illustrate by a simple model the procedure discussed in subsection 9.2a and the validity of the linear approximation (eqs. (9.21) and (9.22)) required in the derivation of the Breit-Wigner formula of eq. (9.23).

In this model study [BE 69], we consider the $n + \alpha$ system in the $l = 0$, $S = \frac{1}{2}$ state. The wave function is taken to be given by eq. (4.8), with the cluster internal functions chosen as single-Gaussian functions having a common width parameter of 0.543 fm^{-2}. The nucleon-nucleon potential† employed is a Serber potential which has the form of eq. (7.5) with the parameters given by eqs. (7.57) and (7.58), except that the depth V_0 is adjusted such that the $\frac{1}{2}^+$ resonance state to be investigated occurs near but below the d + t threshold. The value of V_0 so chosen is equal to 120 MeV; with this value, the d + t threshold occurs at 39.29 MeV and a single-channel d + t calculation reveals the presence of an $l = 0$, $S = \frac{1}{2}$ bound state at an energy of 2.99 MeV below this threshold.

It should be mentioned that with the adoption of the wave function given by eq. (4.8), the behaviour of the $n + \alpha$ system cannot be described in an exact way even below the d + t threshold. To allow of a representation of the exact solution of this problem, terms involving excited deuteron and triton clusters must further be added to the wave function. Thus the approximation associated with the use of eq. (4.8) is that ψ_C is represented by a superposition of a complete set of functions only in the d + t channel subspace, with the amplitudes of these functions determined as functions of the scattering energy in the $n + \alpha$ channel.

The coupled-channel equation (4.10) is solved at energies below the d + t threshold. The resultant phase shift δ_0 in the $l = 0$ state is shown as a function of the relative energy E_{R1} by the solid curve in fig. 25. To see how well the Breit-Wigner formula of eq. (9.23) works, we have adjusted the parameters E_r and Γ_{jr} in eq. (9.23) to fit as well as possible the rapid-rising part of the calculated phase-shift curve. The result obtained with $E_r = E_\alpha + 38.24$ MeV and $\Gamma_{jr} = 0.28$ MeV is shown by the crosses in fig. 25. As is seen, the agreement between these two calculations is generally quite good, with the discrepancy occurring essentially only at energies near the d + t threshold. The main reason for this discrepancy is that, in the two-channel calculation, one takes into account at energies

† For simplicity, the Coulomb potential is neglected.

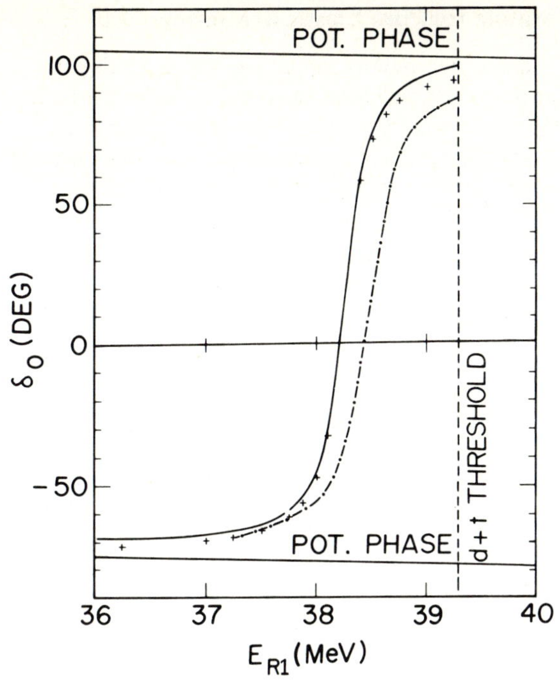

Fig. 25
Phase shift δ_0 as a function of the relative energy E_{R1}. The solid curve represents the result of a two-channel calculation using the wave function of eq. (4.8), while the dashed curve represents the result of a simpler calculation using the wave function of eq. (9.38). The crosses represent a fit to the two-channel result using the Breit-Wigner formula of eq. (9.23). (Adapted from ref. [BE 69].)

above the d + t threshold automatically inelastic processes resulting from the opening of the d + t channel. This means that, at these energies, the n + α phase shift will become complex as a consequence of transitions between these two channels. In the simple Breit-Wigner parametrization described in subsection 9.2a, these transitions can of course not be taken into consideration.

We have further performed a calculation in which the wave function is given by

$$\psi = A\{\phi(\alpha)\, F_l\, (\hat{\mathbf{R}}_1)\, \xi_1\, (s, t)\} + a\, \psi_B, \tag{9.38}$$

where ψ_B is the $l = 0$, $S = \frac{1}{2}$ bound state wave function obtained in a d + t single-channel calculation, and a is a variational amplitude. The purpose of this calculation is to see how good the approximation is when ψ_C is represented by just one reasonably chosen function. The result is shown by the dot-dashed curve in fig. 25, where one finds that the agreement with the two-channel result is rather satisfactory. This means that, as long as the resonance is not too near the threshold, one can indeed use this type of simplified calculation to replace a more tedious many-channel study. Resonance parameters E_r and Γ_{jr} are calculated by using eqs. (9.20) and (9.24) (see ref. [BE 68] for details). The resultant phase shifts obtained by substituting the values of these parameters into eq. (9.23) agree very well with the phase shifts calculated from eq. (9.17), indicating that the linear approximation represented by eqs. (9.21) and (9.22) is quite valid in this case.

9.3. Many-Level Resonance Formula for Pure Elastic-Scattering

In the case where there are several close-lying and overlapping resonance levels with the same quantum numbers, the approximations which we have used to derive the Breit-Wigner formula of eq. (9.23) from the exact formula of eq. (9.17) are no longer valid. Here the energy dependence of the potential-scattering phase shift becomes so strong that it can no longer be approximately neglected in the energy region around the resonance level.

To derive the phase factor A_j in this case, it is useful to include in the potential-scattering amplitude only contributions from those resonances which lie far away from the group of resonance levels under consideration. Thus we divide ψ_C in eq. (9.1) into two parts, i. e.,

$$\psi_C = \psi_C^R + \psi_C^B, \tag{9.39}$$

where

$$\psi_C^R = \sum_{l=1}^{n_0} a_l \, \psi_C^l, \tag{9.40a}$$

$$\psi_C^B = \sum_{l=n_0+1}^{\infty} a_l \, \psi_C^l, \tag{9.40b}$$

with ψ_C^l being again the orthonormalized solutions of eq. (9.7). With this division we see that ψ_C^R is a wave function in the subspace of the n_0 functions ψ_C^l which describe this group of close-lying resonance levels, and ψ_C^B is a wave function in the subspace of the remaining ψ_C^l. Therefore, it follows that these two functions satisfy the relations

$$\langle \psi_C^R | \psi_C^B \rangle = 0 \tag{9.41}$$

and

$$\langle \psi_C^R | H | \psi_C^B \rangle = 0. \tag{9.42}$$

Using these relations, we obtain then from eq. (9.4) the following set of equations:

$$\langle \delta \psi_D | H - E | \psi_D \rangle + \langle \delta \psi_D | H - E | \psi_C^R \rangle + \langle \delta \psi_D | H - E | \psi_C^B \rangle = 0, \tag{9.43a}$$

$$\langle \delta \psi_C^R | H - E | \psi_D \rangle + \langle \delta \psi_C^R | H - E | \psi_C^R \rangle = 0, \tag{9.43b}$$

$$\langle \delta \psi_C^B | H - E | \psi_D \rangle + \langle \delta \psi_C^B | H - E | \psi_C^B \rangle = 0. \tag{9.43c}$$

From eq. (9.43c) one finds that

$$\psi_C^B = - \sum_{l=n_0+1}^{\infty} \frac{\langle \psi_C^l | H - E | \psi_D \rangle}{E_C^l - E} \, \psi_C^l. \tag{9.44}$$

Substituting this expression for ψ_C^B into eq. (9.43a) yields

$$\langle \delta \psi_D | H''(E) - E | \psi_D \rangle = - \langle \delta \psi_D | H - E | \psi_C^R \rangle, \tag{9.45}$$

with

$$H''(E) = H - \sum_{l = n_0 + 1}^{\infty} \frac{|H - E|\psi_C^l\rangle\langle\psi_C^l|H - E|}{E_C^l - E}. \tag{9.46}$$

Because we are considering here again the case where the elastic channel is the only open channel, the effective Hamiltonian $H''(E)$ is a hermitian operator. As is quite evident, the motivation for the introduction of this effective Hamiltonian is that the potential or background scattering amplitude will now include only contributions from distant energy levels.

By introducing, in a manner analogous to that described previously, the resolvent operator $G_D''^+(E)$ for $H''(E)$ and selecting the partial wave which has the same quantum numbers as the resonance states under consideration, one obtains from eq. (9.45) the equation

$$|\psi_{Dj}(E)\rangle = |\psi_{Dj}''^+(E)\rangle - G_{Dj}''^+(E)|H - E|\psi_C^R\rangle. \tag{9.47}$$

Substituting this into eq. (9.43b), we finally arrive at the following relation for ψ_C^R:

$$\langle\delta\psi_C^R|H - (H-E)\,G_{Dj}''^+(E)\,(H-E) - E|\psi_C^R\rangle = -\langle\delta\psi_C^R|H - E|\psi_{Dj}''^+(E)\rangle. \tag{9.48}$$

To obtain a new wave-function basis which is especially suited for our consideration and in which ψ_C^R can be expanded, we separate, similar to eq. (9.16), $G_{Dj}''^+(E)$ into a hermitian and an antihermitian part, i. e.,

$$G_{Dj}''^+(E) = G_{Dj}''^P(E) + i\pi\,\bar{G}_{Dj}''(E), \tag{9.49}$$

and consider the expression

$$\langle\delta\psi_C^R|H - (H-E)\,G_{Dj}''^P\,(H-E) - E_\lambda(E)|\psi_C^{R\lambda}(E)\rangle = 0, \tag{9.50}$$

which defines n_0 new basis wave functions $\psi_C^{R\lambda}(E)$ with eigenvalues E_λ. These latter functions form an energy-dependent complete orthonormalized set in the subspace of ψ_C^R, and in terms of which we can write

$$\psi_C^R = \sum_{\lambda=1}^{n_0} b_\lambda(E)\,\psi_C^{R\lambda}(E). \tag{9.51}$$

Using eqs. (9.48)–(9.50), we find the following set of equations for the expansion coefficients b_λ:

$$(E_\lambda - E)\,b_\lambda - i\pi\gamma_\lambda \sum_{\mu=1}^{n_0} b_\mu\,\gamma_\mu^* = -\gamma_\lambda \tag{9.52}$$

with

$$\gamma_\lambda(E) = \langle\psi_C^{R\lambda}(E)|H - E|\psi_{Dj}''^+(E)\rangle. \tag{9.53}$$

Equation (9.52) can be easily solved; the result is

$$b_\lambda = \frac{\gamma_\lambda}{E_\lambda - E}\;\frac{1}{i\pi \sum\limits_{\mu=1}^{n_0} \dfrac{\gamma_\mu\gamma_\mu^*}{E_\mu - E} - 1}. \tag{9.54}$$

Using the above expression for b_λ, we obtain from the asymptotic behaviour of eq. (9.47) the phase factor

$$A_j = e^{2i\delta_j''(E)} \frac{1 + i\pi \sum\limits_{\lambda=1}^{n_0} \frac{\gamma_\lambda \gamma_\lambda^*}{E_\lambda - E}}{1 - i\pi \sum\limits_{\lambda=1}^{n_0} \frac{\gamma_\lambda \gamma_\lambda^*}{E_\lambda - E}}. \tag{9.55}$$

Now, because of the use of the effective Hamiltonian $H''(E)$, only the influence of distant resonance levels is included in the phase factor $e^{2i\delta_j''(E)}$ of the potential scattering. Also, one sees that eq. (9.17) for an isolated resonance level is contained in eq. (9.55) as a special case.†

Equation (9.55) can also be written in the form

$$A_j = e^{2i\delta_j''(E)} \frac{\prod\limits_{\nu=1}^{n_0} [\epsilon_\nu^*(E) - E]}{\prod\limits_{\nu=1}^{n_0} [\epsilon_\nu(E) - E]} = e^{2i\delta_j''(E)} \left[1 + \sum_{\nu=1}^{n_0} \frac{B_\nu}{\epsilon_\nu(E) - E}\right], \tag{9.56}$$

where the ϵ_ν's and the B_ν's are complex energy-dependent quantities. By performing a contour integration in the complex E plane, it can be easily shown that the relation

$$\sum_{\nu=1}^{n_0} B_\nu = -2i \sum_{\nu=1}^{n_0} \operatorname{Im} \epsilon_\nu \tag{9.57}$$

exists.†† Similar to the case of an isolated resonance, one can now define the resonance energy E_r^ν by the equation

$$\operatorname{Re} \epsilon_\nu (E_r^\nu) - E_r^\nu = 0 \tag{9.58}$$

and the level width Γ_r^ν as

$$\Gamma_r^\nu = -2 \left[\frac{\operatorname{Im} \epsilon_\nu(E)}{1 - \frac{d}{dE} \operatorname{Re} \epsilon_\nu(E)}\right]_{E = E_r^\nu}. \tag{9.59}$$

Also, it should be mentioned that, when the elastic channel is the only open channel, the general features in the case of overlapping resonances are the same as those in the case of a single isolated level. For instance, the amplitudes of the resonance structures in the compound region are again appreciably different from zero only for energies around E_r^ν, a feature which can be easily seen from eq. (9.54) for b_λ.

† It should be noted that the energy shift which comes from the coupling to the open channel (see eq. (9.50)) is already included in E_λ. Thus, in going to the one-level case, one has to observe that E_1 in eq. (9.55) is equal to $E_C^1 - \Delta_j(E)$ in eq. (9.17).

†† In contrast to the one isolated-level case, the B_ν's are in general not purely imaginary.

9.4. Single-Level Resonance Formula Including Inelastic and Rearrangement Processes

In the previous two sections, it has been assumed that the elastic channel is the only open channel. Now we shall discuss the case where, besides the elastic channel, inelastic and rearrangement channels are also open. In this section, we shall first derive the resonance formula for the simpler case of an isolated resonance; in the next section, the more general case of overlapping resonances will then be studied. To investigate the validity and the usefulness of this resonance formula, we shall again consider a specific example where, in particular, the general behaviour of a resonance occurring near a reaction threshold will be examined.

9.4a. Derivation of the Resonance Formula

As in eq. (9.1), the wave function ψ is divided into two parts, i. e.,

$$\psi = \psi_D + \psi_C, \tag{9.60}$$

where

$$\psi_D = \sum_{p=1}^{m} \psi_{Dp}, \tag{9.61}$$

with p being the channel index. To consider the effect of an isolated resonance which occurs in an energy region around E_C^1, we proceed in the same manner as described in subsection 9.2a and obtain the following set of coupled equations:

$$\langle \delta \psi_{Ds} | H' - E | \sum_{p=1}^{m} \psi_{Dp} \rangle = \frac{\langle \delta \psi_{Ds} | H - E | \psi_C^1 \rangle \langle \psi_C^1 | H - E | \sum_{p=1}^{m} \psi_{Dp} \rangle}{E_C^1 - E}, \quad (s = 1, \ldots, m) \tag{9.62}$$

with

$$H'(E) = H - \sum_{\alpha \neq 1} \frac{|H - E| \psi_C^\alpha \rangle \langle \psi_C^\alpha | H - E|}{E_C^\alpha - E}. \tag{9.63}$$

By introducing the resolvent operators for the various open channels, one can solve eq. (9.62) to obtain formal solutions for ψ_{Ds} which describe the resonance scattering and reaction processes. Here we shall discuss only the case where the direct coupling between the open channels can be neglected and the transitions occur, to a good approximation, only through the formation of the resonance state ψ_C^1. This approximation is certainly justified if the resonance occurs in an energy region where all or all but one open channels are shielded from the compound region by high potential barriers.

With the omission of direct-coupling terms $\langle \delta \psi_{Ds} | H' - E | \psi_{Dr} \rangle$ $(s \neq r)$ between the open channels, we obtain from eq. (9.62)

$$\langle \delta \psi_{Ds} | H' - E | \psi_{Ds} \rangle = - a \langle \delta \psi_{Ds} | H - E | \psi_C^1 \rangle, \tag{9.64}$$

where

$$a = - \frac{\left\langle \psi_C^1 | H - E | \sum_{p=1}^{m} \psi_{Dp} \right\rangle}{E_C^1 - E}. \tag{9.65}$$

If we denote the wave function for the incident channel as ψ_{Di}, then by solving eq. (9.64) we find

$$|\psi_n\rangle = \delta_{in} |\psi_n'^{+}\rangle + \frac{G_n'^{+} | H - E | \psi_C^1 \rangle \langle \psi_C^1 | H - E | \psi_i'^{+} \rangle}{E_C^1 - E - \sum_{p=1}^{m} \langle \psi_C^1 | (H - E) \, G_p'^{+} (H - E) | \psi_C^1 \rangle}, \tag{9.66}$$

where $G_n'^{+}$ is the resolvent operator with outgoing-wave boundary condition belonging to channel n.† Now, by decomposing $\psi_n'^{+}$ and $G_n'^{+}$ again into partial waves of given spin j and parity π, one obtains then the following expression for the asymptotic amplitude $A_{j,in}$ in the nth channel:

$$A_{j,\,in} = \left(\frac{k_i}{\mu_i}\right)^{1/2} \left(\frac{\mu_n}{k_n}\right)^{1/2} \left[\delta_{in} \, e^{2i\delta_{jn}'(E)} + \frac{i\, 2\pi \, \gamma_{ji}(E) \, \gamma_{jn}^{*}(E) \, e^{2i\delta_{jn}'(E)}}{E_C^1 - E - \Delta_j(E) - \frac{1}{2} i \sum_{p=1}^{m} \Gamma_{jp}(E)} \right], \tag{9.67}$$

where k_n and μ_n denote, respectively, the wave number and the reduced mass in channel n, and

$$\Delta_j(E) = \langle \psi_C^1 | (H - E) \left(\sum_{p=1}^{m} G_{jp}'^{P} \right) (H - E) | \psi_C^1 \rangle, \tag{9.68}$$

$$\gamma_{jn}(E) = \langle \psi_C^1 | H - E | \psi_{jn}'^{+}(E) \rangle, \tag{9.69}$$

$$\Gamma_{jn}(E) = 2\pi |\gamma_{jn}|^2. \tag{9.70}$$

In eq. (9.67), the factors $(k_i/\mu_i)^{1/2}$ and $(\mu_n/k_n)^{1/2}$ appear as a consequence of the normalization condition

$$\langle \psi_{jn}^{\beta} | \psi_{jn}^{\gamma} \rangle = \delta(E^{\beta} - E^{\gamma}) \tag{9.71}$$

for the function ψ_{jn}^{β}, which occurs in the spectral representation for $G_{jn}'^{+}$. Also, one can easily see that eq. (9.67) reduces to eq. (9.17) in the case where there is only one open channel.

---

† For simplicity in notation, we have dropped the index D in ψ_{Dn} and so on.

The amplitude $A_{j,\,in}$ can be written in a somewhat different form which will be useful for later considerations. By studying the asymptotic behaviour of the radial part for the relative-motion function contained in ψ'^{+}_{jn}, one sees from eq. (9.69) that γ_{jn} can be expressed as

$$\gamma_{jn} = -i\,\bar{\gamma}_{jn}\, e^{i\delta'_{jn}}, \tag{9.72}$$

with $\bar{\gamma}_{jn}$ being a real quantity. Substituting this expression into eq. (9.67), one obtains then

$$A_{j,\,in} = \left(\frac{v_i}{v_n}\right)^{1/2} e^{i(\delta'_{ji} + \delta'_{jn})} \left[\delta_{in} + \frac{i\,2\pi\,\bar{\gamma}_{ji}\,\bar{\gamma}_{jn}}{E_C^1 - E - \Delta_j(E) - \frac{1}{2}\,i\sum_{p=1}^{m}\Gamma_{jp}(E)}\right], \tag{9.73}$$

with

$$v_i = \hbar k_i/\mu_i \tag{9.74}$$

and a similar expression for v_n. In the neighbourhood of sharp resonances, linear approximations for $\Delta_j(E)$ and $\Gamma_{jn}(E)$ can again be made (see eqs. (9.21) and (9.22)) and the result obtained for $A_{j,\,in}$ after making these approximations is

$$A_{j,\,in} = \left(\frac{v_i}{v_n}\right)^{1/2} e^{i(\delta'_{ji} + \delta'_{jn})} \left(\delta_{in} + \frac{i\,2\pi\,\bar{\gamma}_{ji,\,r}\,\bar{\gamma}_{jn,\,r}}{E_r - E - \frac{1}{2}\,i\,\Gamma_{jr}}\right), \tag{9.75}$$

with

$$\bar{\gamma}_{jn,\,r} = \left\{\frac{\bar{\gamma}_{jn}(E)}{\left[1 + \frac{d}{dE}\Delta_j(E)\right]^{1/2}}\right\}_{E = E_r}, \tag{9.76}$$

$$\Gamma_{jr} = \sum_{p=1}^{m} \Gamma_{jp,\,r} = \sum_{p=1}^{m} 2\pi\,(\bar{\gamma}_{jp,\,r})^2, \tag{9.77}$$

and the resonance energy E_r defined by eq. (9.20).

From eq. (9.75), the total elastic cross section $\sigma_{j,\,ii}$ and the various reaction cross sections $\sigma_{j,\,in}$ can be obtained. They are given by

$$\sigma_{j,\,ii} = C\,\frac{1}{4\,k_i^2}\,|A_{j,\,ii} - 1|^2 = C\,\frac{1}{k_i^2}\left|\sin\delta'_{ji}\,e^{i\delta'_{ji}} + \frac{\frac{1}{2}\,\Gamma_{ji,\,r}\,e^{2i\delta'_{ji}}}{E_r - E - \frac{1}{2}\,i\,\Gamma_{jr}}\right|^2, \tag{9.78}$$

$$\sigma_{j,\,in} = C\,\frac{1}{4\,k_i^2}\,\frac{v_n}{v_i}\,|A_{j,\,in}|^2 = C\,\frac{1}{4\,k_i^2}\,\frac{\Gamma_{ji,\,r}\,\Gamma_{jn,\,r}}{(E_r - E)^2 + \frac{1}{4}\,(\Gamma_{jr})^2}, \tag{9.79}$$

where C is an energy-independent kinematical factor.† One sees that eqs. (9.78) and (9.79) have the usual forms for these cross sections if all reactions proceed through the formation of the resonance state. In addition, it can be shown easily from eqs. (9.75)–(9.79) that the flux of incoming particles is equal to the flux of outgoing particles and that the reciprocity theorem is fulfilled.

---

† In the simple case where the spins of the two interacting nuclei are assumed as zero, this factor is just $4\pi(2l + 1)$, with l being the relative orbital angular-momentum quantum number.

9.4b. Application of the Resonance Formula to a Specific Example Involving Two Open Channels

To illustrate the utility of the resonance formula derived in the above subsection, we consider once more the ^{5}He system in the excitation-energy region around the 16.7 MeV, $\frac{3}{2}^+$ state. For simplicity, we shall again, as in subsection 7.2b, take into account only the n + α channel with $l = 2$ and $S = \frac{1}{2}$ and the d + t channel with $l = 0$ and $S = \frac{3}{2}$. Here it will be seen that even though the calculated $\frac{3}{2}^+$ state is only moderately sharp, the omission of direct coupling between the open channels is still justified and the resonance formula yields results in good agreement with those obtained from a coupled-channel calculation.†

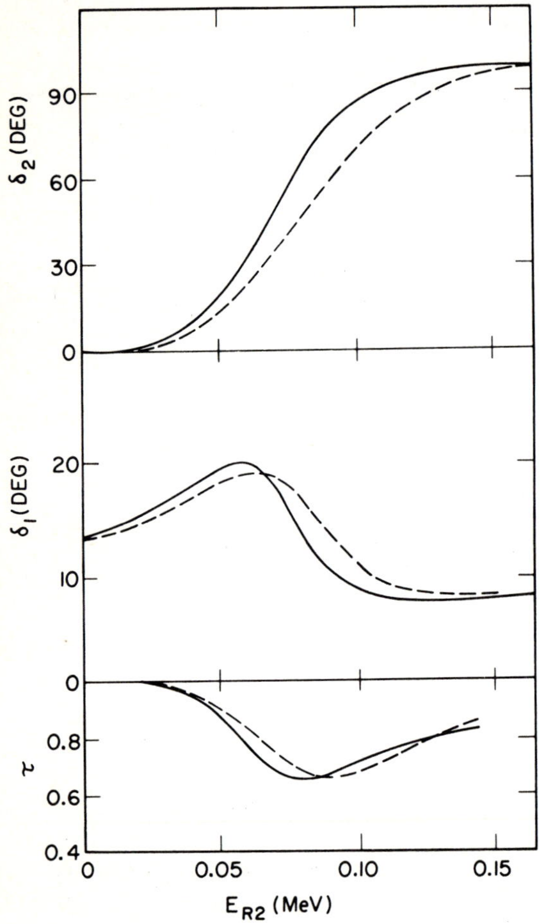

Fig. 26
d + t phase shift δ_2, n + α phase shift δ_1, and reflection coefficient τ as a function of the relative energy E_{R2} in the d + t channel. The solid curves represent the results of a two-channel calculation, while the dashed curves represent the results obtained by using the resonance formula. (Adapted from ref. [SC 72b].)

† For a detailed discussion of the coupled-channel calculation and the calculation using the resonance formula, see ref. [SC 72b].

Instead of the amplitudes $A_{j,\,in}$, we use now the S-matrix elements $S_{j,\,in}$ which are related to $A_{j,\,in}$ by the simple relation†

$$S_{in} = \left(\frac{v_n}{v_i}\right)^{1/2} A_{in}. \tag{9.80}$$

From eq. (9.73), one sees easily that the S-matrix is unitary and symmetric,†† as it certainly must be.

As was mentioned in subsection 7.3f, the S-matrix in the $\frac{3}{2}^+$ state is a 2×2 matrix of the form given by eq. (7.81). The three real quantities specifying this matrix is the n + α phase shift δ_1, the d + t phase shift δ_2, and the reflection coefficient τ. In fig. 26, we compare the values of these quantities shown as a function of the relative energy E_{R2} in the d + t channel, obtained from a complete two-channel calculation (solid curves) and by using the resonance formula of eqs. (9.73) and (9.80) (dashed curves). In the latter calculation, the function ψ_C^1 is obtained from a single-channel d + t calculation with the Coulomb interaction between the clusters turned off. Because of this special way in constructing ψ_C^1, it is necessary to perform in the d + t channel a projection procedure, as discussed in subsection 9.2c.

One sees from fig. 26 that the results of these two calculations agree reasonably well with each other. The reason for this is that in this case the resonance lies fairly close to the d + t threshold and, hence, the d + t channel is shielded rather strongly from the compound region by the Coulomb barrier. Consequently, the condition for the omission of direct coupling between the open channels is well satisfied.†††

We shall now discuss the general reason why the δ_1 and δ_2 phase-shift curves in fig. 26 have so different shapes. For this we consider in detail the mathematical structure of the S-matrix elements S_{11} and S_{22}. From eqs. (9.75) and (9.80) one obtains, after leaving away all indices unnecessary for our present discussion,

$$S_{11} = e^{2i\delta_1'(E)} \left[1 + \frac{i\Gamma_1}{E_r - E - \frac{1}{2}i\Gamma}\right], \tag{9.81a}$$

$$S_{22} = e^{2i\delta_2'(E)} \left[1 + \frac{i\Gamma_2}{E_r - E - \frac{1}{2}i\Gamma}\right], \tag{9.81b}$$

with

$$\Gamma = \Gamma_1 + \Gamma_2. \tag{9.82}$$

---

† From now on, we shall omit for brevity the spin index j whenever no confusion may arise.

†† This is connected with the fact that the Hamiltonian of our system is not only hermitian but also real.

††† When the resonance lies very close to the d + t threshold, some complication will arise – a point which we shall discuss later in detail.

At first, we consider the factor appearing within the square brackets of eq. (9.81a). This factor may be written as

$$\bar{S}_{11} = 1 + \frac{i\Gamma_1}{E_r - E - \frac{1}{2} i\Gamma} = \frac{\Gamma_2}{\Gamma} + \frac{\Gamma_1}{\Gamma} \frac{E - E_r - i\frac{\Gamma}{2}}{E - E_r + i\frac{\Gamma}{2}}. \tag{9.83}$$

Equation (9.83) shows that $\bar{S}_{11}$ as a function of E is described in the complex plane by a circle of radius Γ_1/Γ around Γ_2/Γ as its center. It touches the unit circle on the positive real axis for $E = \pm\infty$. As E goes from $-\infty$ to $+\infty$, the phase of the factor $(E - E_r - i\frac{\Gamma}{2})/(E - E_r + i\frac{\Gamma}{2})$ goes from 0 to 2π.

To study the behaviour of $\bar{S}_{11}$ for different combinations of Γ_1 and Γ_2 but subject to the condition of eq. (9.82), we introduce the parameter $x = \Gamma_1/\Gamma$ which has a range from 0 to 1. With this parameter, we can write

$$\bar{S}_{11} = \tau\, e^{2i\bar{\delta}_1} = (1 - x) + x \frac{E - E_r - i\frac{\Gamma}{2}}{E - E_r + i\frac{\Gamma}{2}} \tag{9.84}$$

From the above equation, one sees immediately that $\bar{S}_{11}$ as a function of E shows a principally different behaviour for x between $\frac{1}{2}$ and 1 and for x between 0 and $\frac{1}{2}$. In fig. 27, we show the energy dependence of the phase $\bar{\delta}_1$ for different values of x. For x between $\frac{1}{2}$ and 1, $\bar{\delta}_1$ increases with increasing energy $E\,(-\infty \leqslant E \leqslant +\infty)$ from 0 to π and is equal to $\pi/2$ for $E = E_r$. Also, it can be easily seen from eq. (9.84) that the curve for $\bar{\delta}_1$ is reflection-symmetrical with respect to the point with $E = E_r$ and $\bar{\delta}_1 = \pi/2$. In the case where x is between 0 and $\frac{1}{2}$, $\bar{\delta}_1$ goes from 0 at $E = -\infty$ to a maximum value smaller than $\pi/4$ and becomes equal to zero again for $E = E_r$. At E greater than E_r, $\bar{\delta}_1$ reaches a negative minimum value and approaches zero once more as E approaches $+\infty$.

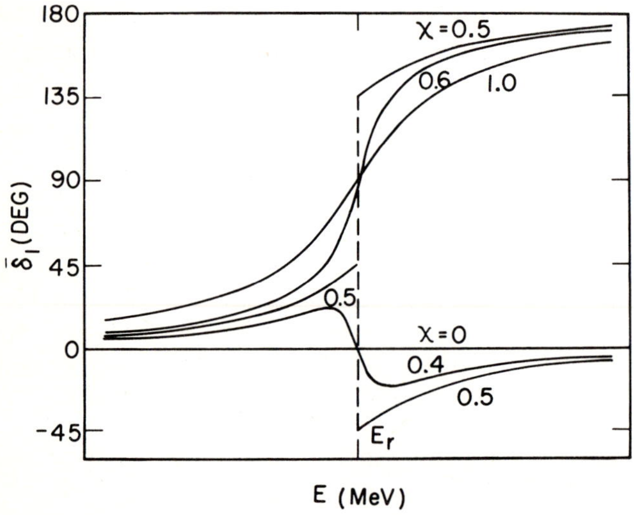

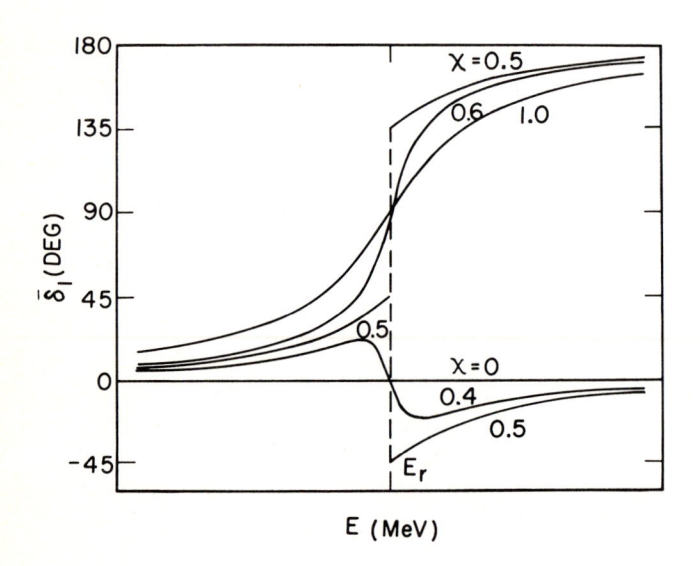

Fig. 27
Energy dependence of $\bar{\delta}_1$ for various values of x.

Also, the curve for $\bar{\delta}_1$ is reflection-symmetrical this time with respect to the point with $E = E_r$ and $\bar{\delta}_1 = 0$. For $x = \frac{1}{2}$, $\bar{\delta}_1$ has a discontinuity of $\pm \pi/2$ at $E = E_r$. However, it should be noted that, in spite of this discontinuity, the elastic-scattering cross section still has a continuous behaviour at this particular value of E.

Similarly, one can discuss the energy dependence of the phase $\bar{\delta}_2$ contained in $\bar{S}_{22}$. For this one has only to replace $(1 - x)$ in eq. (9.84) by x and vice versa.

The above discussion shows that, because of the relation $\Gamma_1/\Gamma + \Gamma_2/\Gamma = 1$, the energetical behaviour of the phase shift $\bar{\delta}_1$ is very different from that of the phase shift $\bar{\delta}_2$. If $\bar{\delta}_1$ shows a step-like behaviour, then $\bar{\delta}_2$ will show a dispersion-like behaviour, and vice versa. In more physical terms one can also express this in the following manner. The step-like behaviour appears always in the channel where the resonance state has a larger partial energy width, whereas the dispersion-like behaviour is associated with the other channel where the partial width has a smaller value. Because of the well-known connection between the energy width and the decay probability, one can also say that the step-like behaviour of the phase shift occurs in the channel into which the resonance state decays with the larger probability and the dispersion-like behaviour occurs in the channel where the decay probability of the resonance state is smaller.

If a resonance level happens to lie just above the threshold of one of the open channels, then the phase-shift behaviour in the various channels will depend very much on how far this level is from this threshold. The reason for this is that such a resonance will usually be shielded from the outside region by a relatively high Coulomb or centrifugal barrier† and, hence, the partial decay width will be a sensitive function of the relative energy of the decay products in this particular channel.

It is interesting to note that one may observe a reversal in the behaviour of the elastic-scattering phase shift in certain mirror nuclei where in the region of the resonance decay there exist at least two open channels in both nuclei and where the resonance lies just above the threshold of one of the channels. For such a mirror pair, it may happen that, in one of the nuclei, the resonance lies so close to the threshold that the partial decay width into this particular channel is very small and the phase shift shows a dispersion-like behaviour, while in the other nucleus, the resonance lies much farther from the threshold and consequently the partial decay width is much larger and the phase shift shows instead a step-like behaviour. An example which almost shows such a large change in the phase-shift behaviour occurs around the often-mentioned $\frac{3}{2}^+$ mirror states of ^{5}He and ^{5}Li at excitation energies around 17 MeV.†† In ^{5}Li where the energetical distance of the resonance from the d + ^{3}He threshold is 262 keV, one observes in the d + ^{3}He phase shift a well-formed step-like behaviour. On the other hand, in ^{5}He where the resonance lies only 68 keV above the d + t threshold, the situation becomes quite close to the case with $x = \frac{1}{2}$ (see fig. 27) for which the partial widths for decay into the $n + \alpha$ channel and the d + t

† In the case of $l = 0$ neutron emission, there will be no Coulomb or centrifugal barrier. But even here, because of the large reflection at the nuclear surface, the decay probability will decrease very much as the energy of the emitted neutron becomes smaller.

†† Very probably, one may also observe such a change in the n + ^{6}Li and p + ^{6}Li phase-shift behaviour in the energy region around the second $\frac{7}{2}^-$ mirror levels of ^{7}Li and ^{7}Be.

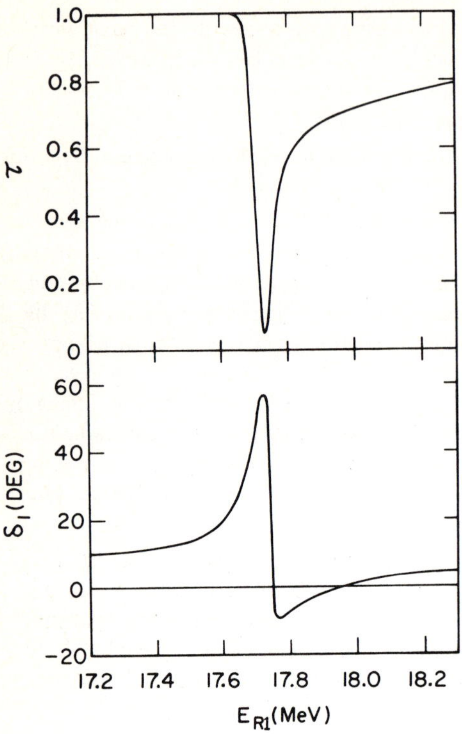

Fig. 28
Reflection coefficient τ and n + α phase shift δ_1 in the $J^\pi = 3/2^+$ state as a function of the relative energy E_{R1} in the n + α channel. (Adapted from ref. [HO 66].)

channel are approximately equal to each other. This can be seen especially clearly by examining the value of the reflection coefficient τ at $E = E_r$. When $x = \frac{1}{2}$, one finds from eq. (9.84) that τ becomes zero at this particular energy. In fig. 28, we show the empirically determined n + α phase shift δ_1 and the reflection coefficient τ in the energy region around the $\frac{3}{2}^+$ resonance [HO 66]. From this figure, it is indeed seen that τ is very small when E is equal to E_r.

So far, we have discussed the behaviour of the resonance-scattering phase shifts $\bar{\delta}_1$ and $\bar{\delta}_2$. To study the behaviour of the total phase shifts δ_1 and δ_2, given by

$$\delta_i = \bar{\delta}_i + \delta_i', \quad (i = 1, 2) \tag{9.85}$$

we have to take into account also the influence of the potential-scattering phase shifts δ_1' and δ_2'. This is a simple matter, however. Because potential-scattering phase shifts generally have a rather smooth energy dependence, one sees easily that, while the curves for δ_i and $\bar{\delta}_i$ (i = 1, 2) will have somewhat different shapes,† the qualitative features for δ_1 and δ_2 are similar to those discussed above for $\bar{\delta}_1$ and $\bar{\delta}_2$.

If a resonance is deeply imbedded under a potential barrier and decays mainly into a channel whose threshold lies just below this resonance, then the approximations which

† For example, the d + t phase shift δ_2 shown in fig. 26 rises to only about 100° in the resonance region, although $\bar{\delta}_2$ reaches nearly 180° as the energy becomes large.

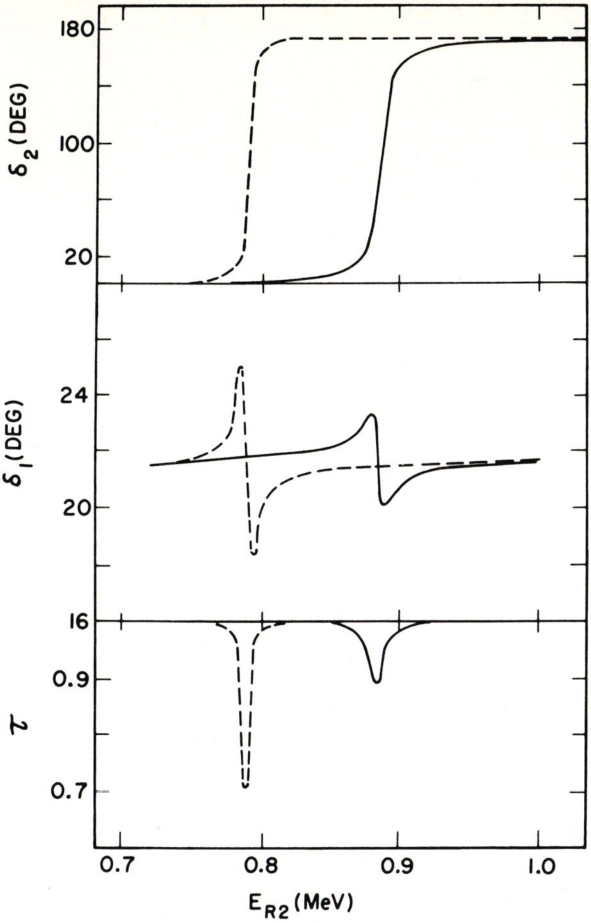

Fig. 29
d + t phase shift δ_2, n + α phase shift δ_1, and reflection coefficient τ as a function of the relative energy E_{R2} in the d + t channel. The solid curves represent the result of a two-channel calculation, while the dashed curves represent the result obtained by using the resonance formula. (Adapted from ref. [SC 72b].)

one makes to compute the phase shifts and the reflection coefficient must be carefully examined. To see this explicitly, the ^{5}He problem is again considered [SC 72b]. The strengths of the nuclear and Coulomb forces are arbitrarily adjusted such that a resonance state occurs very close to the d + t threshold and is deeply imbedded under the Coulomb barrier.† The results for δ_1, δ_2, and τ are shown in fig. 29, where the solid curves represent the result of a two-channel calculation and the dashed curves represent the result obtained by using the resonance formula of eq. (9.73), i.e., by omitting the direct coupling between the d + t and the n + α channels. From this figure it is seen that the

† The ratio $|\Delta E_1/\Delta E_2|$, where ΔE_1 is the energy of the resonance state measured from the top of the Coulomb barrier in the d + t channel and ΔE_2 is the energy of the resonance state measured from the d + t threshold, is increased by a factor of about 3 over the corresponding value in the example which yields the results shown in fig. 26.

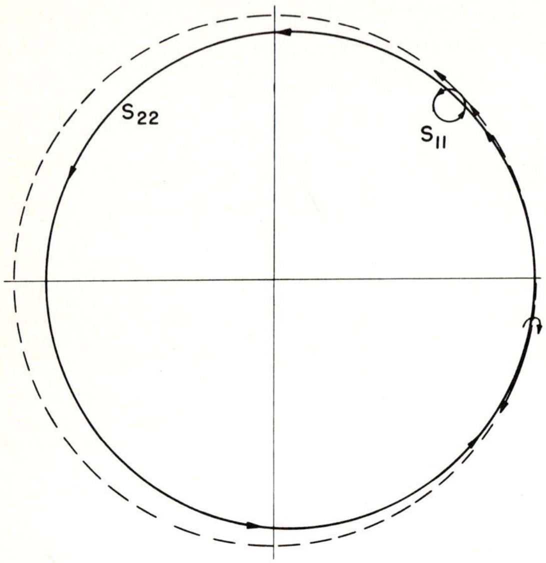

Fig. 30
Argand diagram for S_{11} and S_{22}.

resonance energies from these two calculations are quite similar, but the calculated widths do substantially differ. The reason for this discrepancy in the width is mainly that the calculated resonance energies are somewhat, although slightly, different. Due to the fact that, in this calculation, the energetical distance of the resonance from the d + t threshold is small compared with that from the top of the barrier, the phase shifts δ_1 and δ_2 and the reflection coefficient τ become very sensitive to this energetical distance. If one adjusts the interaction forces in the calculation where the direct-coupling term is omitted in such a way that one obtains the same resonance energy as in the two-channel calculation, then one finds that this discrepancy in the width mostly disappears.

The plot of S-matrix elements in the complex plane is called an Argand diagram. In fig. 30, the Argand diagram for S_{11} and S_{22} obtained with the two-channel calculation in the ^5He case just discussed is shown as an illustrative example.

In a completely analogous manner, one can also study the resonance behaviour of S-matrix elements in the case where more than two channels are open. Here then is an interesting possibility that, in the energy region around a relatively sharp resonance, the phase shifts in all the channels show a dispersion-like behaviour. This will happen when none of the partial widths has a value larger than half of the total width. As to whether such a case has actually been observed, we do not know at this moment.

9.5. Mutual Influence of Resonance Levels in Inelastic and Rearrangement Processes

In this section, we extend our considerations in the previous section to the case of overlapping resonances. For simplicity, only the case with two overlapping resonances and two open channels will be studied; the generalization to the case with more than two resonances and more than two open channels is straightforward and easy to perform. We shall start our discussion in subsection 9.5a by first deriving the general formula and then apply this formula to a specific example in subsection 9.5b.

9.5a. Derivation of a Two-Level Breit-Wigner Formula

Because the derivation of the resonance formula and the assumptions involved in the derivation are very similar to those given in previous sections, our discussion here will be relatively brief.

With the assumption of only two open channels, we write the wave function ψ as

$$\psi = \psi_D + \psi_C = \psi_{D1} + \psi_{D2} + \psi_C, \tag{9.86}$$

where ψ_C is written in the form of eq. (9.39), i.e.,

$$\psi_C = \psi_C^R + \psi_C^B, \tag{9.87}$$

with

$$\psi_C^R = a_1 \psi_C^1 + a_2 \psi_C^2 \tag{9.88}$$

and

$$\psi_C^B = \sum_{l=3}^{\infty} a_l \psi_C^l. \tag{9.89}$$

In eq. (9.88), the functions ψ_C^1 and ψ_C^2 describe the two resonance states which have the same quantum numbers; these two states lie energetically close to each other, but are well separated from all other resonance states which possess the same quantum-number set. By substituting eq. (9.86) into the projection equation (2.3), we obtain after some manipulation the following set of coupled equations:

$$\langle \delta \psi_D | H''(E) - E | \psi_D \rangle + \langle \delta \psi_D | H - E | \psi_C^R \rangle = 0, \tag{9.90a}$$

$$\langle \delta \psi_C^R | H - E | \psi_D \rangle + \langle \delta \psi_C^R | H - E | \psi_C^R \rangle = 0, \tag{9.90b}$$

where

$$H''(E) = H - \sum_{l=3}^{\infty} \frac{|H - E|\psi_C^l\rangle\langle\psi_C^l|H - E|}{E_C^l - E}. \tag{9.91}$$

More explicitly, eqs. (9.90a) and (9.90b) consist of four coupled integrodifferential and integral equations. These equations are

$$\left\langle \delta \psi_{Di} \middle| H''(E) - E \middle| \sum_{k=1}^{2} \psi_{Dk} \right\rangle + \left\langle \delta \psi_{Di} \middle| H - E \middle| \sum_{l=1}^{2} a_l \psi_C^l \right\rangle = 0, \tag{9.92a}$$

$$\left\langle \psi_C^m \middle| H - E \middle| \sum_{k=1}^{2} \psi_{Dk} \right\rangle + a_m (E_C^m - E) = 0, \tag{9.92b}$$

with the indices i and m taking on the values 1 and 2. Now by solving eq. (9.92b) for the amplitude a_1 and substituting the resultant expression into eq. (9.92a), the above set of coupled equations can be simplified and the results are

$$\left\langle \delta\psi_{Di} | \hat{H} - E | \sum_{k=1}^{2} \psi_{Dk} \right\rangle + \left\langle \delta\psi_{Di} | H - E | a_2 \psi_C^2 \right\rangle = 0, \quad (i = 1, 2) \tag{9.93a}$$

$$\left\langle \psi_C^2 | H - E | \sum_{k=1}^{2} \psi_{Dk} \right\rangle + a_2 (E_C^2 - E) = 0, \tag{9.93b}$$

with

$$\hat{H}(E) = H''(E) + H_1 = H''(E) - \frac{|H - E|\psi_C^1\rangle\langle\psi_C^1|H - E|}{E_C^1 - E}. \tag{9.94}$$

Because of the way $\hat{H}$ is constructed, one can easily see that it is a hermitian operator.

So far, we have not made any approximations. Now, for simplicity, we shall only consider the special case where the lower resonance, described by ψ_C^1, is mainly a potential resonance with respect to channel 1† and is, in addition, relatively broad compared to the upper resonance which is described by ψ_C^2 and has a comparatively small energy width. With these restrictive assumptions, we can then omit the direct-coupling term $\langle \delta\psi_{Di} | H'' - E | \psi_{Dk} \rangle$ and the term $\langle \delta\psi_{Di} | H_1 | \psi_{Dk} \rangle$ with $i \neq k$ in eq. (9.93a). In this way, the formulation of the problem becomes relatively simple and we shall be able to examine in a rather easy manner the influence of the lower resonance state on the phase shifts in the energy region around the upper resonance state.

With the above simplifications, we obtain from eq. (9.93a) the equation

$$\langle \delta\psi_{Di} | \hat{H} - E | \psi_{Di} \rangle + a_2 \langle \delta\psi_{Di} | H - E | \psi_C^2 \rangle = 0 . \quad (i = 1, 2) \tag{9.95}$$

Together with eq. (9.93b), we can then proceed in exactly the same manner as in the discussion following eqs. (9.64) and (9.65). The result for the S-matrix elements is

$$S_{in} = e^{i(\hat{\delta}_i + \hat{\delta}_n)} \left[\delta_{in} + \frac{i\, 2\pi\, \bar{\gamma}_{2i}\, \bar{\gamma}_{2n}}{E_C^2 - E - \Delta_2(E) - \frac{1}{2} i \sum_{k=1}^{2} \Gamma_{2k}(E)} \right], \tag{9.96}$$

where we have, for simplicity in writing, suppressed the index j for the total angular momentum of the system. In the above equation, the real quantities $\Delta_2(E)$, $\bar{\gamma}_{2k}$, and Γ_{2k} are given by

$$\Delta_2(E) = \langle \psi_C^2 | (H - E) \left(\sum_{k=1}^{2} \hat{G}_{Dk}^P \right) (H - E) | \psi_C^2 \rangle, \tag{9.97}$$

$$\bar{\gamma}_{2k}(E) = i\, e^{-i\hat{\delta}_k(E)} \langle \psi_C^2 | H - E | \hat{\psi}_{Dk}^+(E) \rangle, \tag{9.98}$$

$$\Gamma_{2k}(E) = 2\pi\, \bar{\gamma}_{2k}^2, \tag{9.99}$$

---

† Because of the special nature of ψ_C^1, it will be useful in an actual calculation to project out this function from the open-channel function ψ_{D1}.

where $\hat{G}^{P}_{Dk}$ is the hermitian part of the resolvent operator $\hat{G}^{+}_{Dk}$ obtained by using the solutions of the homogeneous equation

$$\langle \delta\psi_{Dk} | \hat{H}(E) - E^{\beta}_{Dk} | \psi^{\beta}_{Dk} \rangle = 0, \tag{9.100}$$

and $\hat{\psi}^{+}_{Dk}$ is a solution of eq. (9.100) with $E^{\beta}_{Dk} = E$ and is matched as usual to the incoming wave in the kth channel. Also, in eqs. (9.96) and (9.98), the quantities $\hat{\delta}_1$ and $\hat{\delta}_2$ are potential-scattering phase shifts, which can be obtained by examining the functions $\hat{\psi}^{+}_{D1}$ and $\hat{\psi}^{+}_{D2}$ in the asymptotic region.

Because of the presence of the relatively broad resonance described by ψ^{1}_{C}, the potential scattering phase shift $\hat{\delta}_1$ will in particular contain a part which shows a resonance-like behaviour. To see this, we consider the equation

$$\langle \delta\psi_{Di} | \hat{H}(E) - E | \psi_{Di} \rangle = 0 \quad (i = 1, 2) \tag{9.101}$$

in detail. By using eq. (9.94), we can write this equation as

$$\langle \delta\psi_{Di} | H''(E) - E | \psi_{Di} \rangle = \frac{\langle \delta\psi_{Di} | H - E | \psi^{1}_{C} \rangle \langle \psi^{1}_{C} | H - E | \psi_{Di} \rangle}{E^{1}_{C} - E}, \tag{9.102}$$

which, except for trivial differences, has the same form as eq. (9.9). Thus, following the same procedure as that given in subsection 9.2a, we obtain the following expression for the amplitude $\hat{A}_k$ (k = 1, 2) of the outgoing partial wave with total spin j in channel k:

$$\hat{A}_k = e^{2i\hat{\delta}_k} \approx e^{2i\delta''_k(E)} \frac{E^{1}_{C} - E - \Delta_{1k}(E) + \frac{1}{2} i \Gamma_{1k}(E)}{E^{1}_{C} - E - \Delta_{1k}(E) - \frac{1}{2} i \Gamma_{1k}(E)}, \tag{9.103}$$

where

$$\Delta_{1k}(E) = \langle \psi^{1}_{C} | (H - E) G''^{P}_{Dk} (H - E) | \psi^{1}_{C} \rangle, \tag{9.104}$$

$$\Gamma_{1k}(E) = 2\pi | \langle \psi^{1}_{C} | H - E | \psi''^{+}_{Dk} \rangle |^2, \tag{9.105}$$

with G''^{P}_{Dk} being the hermitian part of the resolvent operator G''^{+}_{Dk} obtained by using the solutions of a homogeneous equation containing the operator $H''(E)$. The important point to note here is that $\hat{A}_k$ is a pure phase factor and the phase shift $\delta''_k(E)$ has no resonance-like behaviour in the energy region around the two resonances under consideration.

The expression for $\hat{A}_k$ in eq. (9.103) looks somewhat peculiar; instead of the total level shift and the total level width, the quantities Δ_{1k} and Γ_{1k} appear in the denominator. The reason for this is that, in our derivation, the lower resonance is considered as mainly a potential resonance in channel 1, and hence the coupling of the two channels through the operator H_1 is neglected. Thus, Δ_{12} and Γ_{12} have very small values and the amplitudes $\hat{A}_1$ and $\hat{A}_2$ can be written in good approximation as

$$\hat{A}_1 \approx e^{2i\delta''_1(E)} \frac{E^{1}_{C} - E - \Delta_1(E) + \frac{1}{2} i \Gamma_1(E)}{E^{1}_{C} - E - \Delta_1(E) - \frac{1}{2} i \Gamma_1(E)}, \tag{9.106}$$

$$\hat{A}_2 \approx e^{2i\delta''_2(E)}, \tag{9.107}$$

with $\Delta_1 \approx \Delta_{11}$ and $\Gamma_1 \approx \Gamma_{11}$. Especially for energies not too close to E_C^1, it is expected that these approximate expressions for $\hat{A}_1$ and $\hat{A}_2$, given by eqs. (9.106) and (9.107), should be satisfactory.

Next, we apply the linear approximation, as described in subsection 9.2a, to the level shift and the level width. For this we first define resonance energies E_{1r} and E_{2r} by the equation

$$E_C^i - E_{ir} - \Delta_i(E_{ir}) = 0, \quad (i = 1, 2) \tag{9.108}$$

Then, by making the usual linear expansions of $\Delta_1(E)$ and $\Delta_2(E)$ around E_{1r} and E_{2r} (see eq. (9.21)), we obtain the following expressions for the S-matrix elements in the neighbourhood of the upper resonance:

$$S_{11} = e^{2i\delta_1''(E)} \left(\frac{E - E_{1r} - \frac{i}{2}\Gamma_{1r}}{E - E_{1r} + \frac{i}{2}\Gamma_{1r}} \right) \left[\frac{E - E_{2r} - \frac{i}{2}(\Gamma_{21,r} - \Gamma_{22,r})}{E - E_{2r} + \frac{i}{2}\Gamma_{2r}} \right], \tag{9.109a}$$

$$S_{12} = S_{21} = -e^{i(\delta_1'' + \delta_2'')} \left(\frac{E - E_{1r} - \frac{i}{2}\Gamma_{1r}}{E - E_{1r} + \frac{i}{2}\Gamma_{1r}} \right)^{1/2} \frac{i\bar{\gamma}_{21,r}\,\bar{\gamma}_{22,r}}{E - E_{2r} + \frac{i}{2}\Gamma_{2r}}, \tag{9.109b}$$

$$S_{22} = e^{2i\delta_2''(E)} \frac{E - E_{2r} - \frac{i}{2}(\Gamma_{22,r} - \Gamma_{21,r})}{E - E_{2r} + \frac{i}{2}\Gamma_{2r}}, \tag{9.109c}$$

where

$$\Gamma_{1r} = \left[\frac{\Gamma_1(E)}{1 + \frac{d}{dE}\Delta_1(E)} \right]_{E = E_{1r}}, \tag{9.110a}$$

$$\bar{\gamma}_{2k,r} = \left\{ \frac{\gamma_{2k}}{\left[1 + \frac{d}{dE}\Delta_2(E)\right]^{1/2}} \right\}_{E = E_{2r}}, \quad (k = 1, 2) \tag{9.110b}$$

$$\Gamma_{2k,r} = 2\pi(\bar{\gamma}_{2k,r})^2, \quad (k = 1, 2) \tag{9.110c}$$

$$\Gamma_{2r} = \Gamma_{21,r} + \Gamma_{22,r}. \tag{9.110d}$$

It should be noted that the linear expansion around the upper resonance is a good approximation only when the upper resonance has an appreciably smaller width than the lower resonance.

At a first glance, it might seem as if the function ψ_C^1 influences the S-matrix elements only through the factor

$$e^{2i\hat{\delta}_1(E)} \approx e^{2i\delta_1''(E)} \frac{E - E_{1r} - \frac{i}{2}\Gamma_{1r}}{E - E_{1r} + \frac{i}{2}\Gamma_{1r}}. \tag{9.111}$$

But this is really not so, because $\bar{\gamma}_{21}$, $\bar{\gamma}_{22}$, and Δ_2 in eq. (9.96) have to be calculated with open-channel functions ψ^{β}_{D1} which are solutions of the homogeneous equation (9.100). In this latter equation, the operator $\hat{H}(E)$ appears, which includes the influence of the resonance wave function ψ^1_C. As is seen from eq. (9.14), the solutions of this equation contain a resonance part which describes, in our case, the effect of ψ^1_C on the elastic-scattering wave in channel 1. The appearance of this resonance part has the consequence of enhancing the coupling between the functions ψ^2_C and $\hat{\psi}^+_{D1}$ and thereby increasing the value of Γ_{21}, as follows from eqs. (9.98) and (9.99).

If the couplings of the wave functions ψ^1_C and ψ^2_C with both open channels are of similar order of magnitude, then the mutual coupling of the two channels via ψ^1_C can no longer be neglected in good approximation. In this case, the derivation of the S-matrix elements is more complicated, but one finds again that these elements can be represented in product forms. The major difference here is that, in S_{11} for example, the term within the parentheses is no longer a pure phase factor, as it is in eq. (9.109a).

9.5b. A Specific Example

As a specific example, we consider the two $\frac{5}{2}^-$ levels of ^{7}Li and ^{7}Be in the excitation-energy region between 6 and 8 MeV. We choose this particular example, because the simplifying assumptions made in subsection 9.5a are quite valid; here, the lower $\frac{5}{2}^-$ resonance is mainly a potential resonance with respect to the α + t or the α + ^{3}He channel and the upper resonance is, especially in ^{7}Li, appreciably narrower than the lower resonance. In addition, this example serves to demonstrate in an interesting way the Coulomb effect in mirror nuclei. Because of the difference in Coulomb energy in ^{7}Li and ^{7}Be, the upper resonance in ^{7}Li lies much closer to the n + ^{6}Li threshold than the corresponding resonance in ^{7}Be does to the p + ^{6}Li threshold; consequently, the phase shift in the α + t channel exhibits a distinctly different behaviour from that in the α + ^{3}He channel at energies around the upper resonance, as will be shown explicitly in the following discussion.

We shall now describe very briefly the ansatz for the wave function and the nucleon-nucleon potential used in the calculation.† For definiteness, we shall refer to the ^{7}Li case; the ^{7}Be case can of course be discussed in an entirely similar manner. In the nucleus ^{7}Li, the favoured cluster configurations are the α + t configuration and the n + ^{6}Li configuration. In the former configuration the energetically lowest $\frac{5}{2}^-$ state has $l = 3$ and $S = \frac{1}{2}$, while in the latter configuration the energetically lowest $\frac{5}{2}^-$ state has $l = 1$ and $S = \frac{3}{2}$. Therefore, for the calculation of elastic-scattering and reaction cross sections in the excitation-energy region from 6 to 8 MeV, a two-channel ansatz for the wave function will be made, which contains these two particular cluster configurations. As for the interaction potential, it is clear that one must employ a nucleon-nucleon potential containing a tensor component, because the two configurations in this case differ in orbital angular-momentum quantum number by two units and hence only such a noncentral nucleon-nucleon potential can yield a direct coupling between these configurations.

† For a detailed description of the wave function and the nucleon-nucleon potential employed, see ref. [TE 73].

For a better description of the behaviour of the system in the compound region, one should also add to the wave function bound-structure terms multiplied with linear variational amplitudes to take into account specific distortion effects. However, this has not been done in order to simplify the calculation. Because of this, some undesirable features do arise and these will be mentioned in the discussion below.

Before we present the results of this calculation, we shall first discuss qualitatively why the two $\frac{5}{2}^-$ levels in ^{7}Li have very different widths and why the upper level in ^{7}Li is much narrower than the corresponding level in ^{7}Be, although the widths of the lower $\frac{5}{2}^-$ levels in these two nuclei are nearly the same. As has already been mentioned, the lower $\frac{5}{2}^-$ resonances in ^{7}Li and ^{7}Be are essentially potential resonances with respect to the α + t and the α + ^{3}He channels, respectively. They are shielded from the outside regions by centrifugal barriers, but are well above the respective Coulomb barriers which are rather low in both nuclei. Therefore, the widths of these levels should have rather similar magnitudes, as has indeed been found experimentally. The upper $\frac{5}{2}^-$ level in ^{7}Li has mainly a n + ^{6}Li cluster structure and lies just above the n + ^{6}Li threshold. Consequently, it is deeply imbedded under the centrifugal barrier and is only weakly coupled to the α + t channel through the tensor interaction. Thus, one expects that this level should have a much smaller width than the lower $\frac{5}{2}^-$ level. In ^{7}Be, the situation is somewhat different. Here, due to the additional repulsive p + ^{6}Li Coulomb interaction, the upper $\frac{5}{2}^-$ level lies relatively far away from the p + ^{6}Li threshold. As a result of this, it has a much larger energy width than the corresponding level in ^{7}Li. In addition, it should be noted that the upper $\frac{5}{2}^-$ level in ^{7}Be can also decay into the α + d + p channel. However, the level-broadening effect due to the presence of this extra decay channel is expected to be rather unimportant, because the threshold of this channel is only about 0.1 MeV away from the upper $\frac{5}{2}^-$ level in ^{7}Be.

Now let us go back to the discussion of the two-channel calculation in ^{7}Li. Because of the adoption of a simple wave function and a non-saturating nucleon-nucleon potential, it becomes necessary to adjust the strengths of the central and the spin-orbit parts in each channel in order to obtain nearly correct values for the resonance and the threshold energies. In addition, the strength of the tensor component in the nucleon-nucleon potential has to be chosen such that the width of the upper $\frac{5}{2}^-$ level is correctly obtained. As a consequence of these adjustments, a good agreement between calculated and empirical values for the α + t phase shift δ_1, n + ^{6}Li phase shift δ_2, and the reflection coefficient τ is more or less expected. This is seen in figs. 31 and 32, where the solid curves represent the results of the two-channel calculation as a function of the relative energy E_{R2} in the n + ^{6}Li channel, and the dashed curves show as a comparison the results of a simpler calculation in which the coupling between the α + t and the n + ^{6}Li channels is neglected.

The necessity of making adjustments on the various strengths is certainly an undersirable feature of the calculation, and it is clear that one should perform in the future an improved calculation using a better wave function and a more realistic nucleon-nucleon potential. Meanwhile, however, one can still use this example to study the Coulomb effect in mirror nuclei. For this purpose, the α + ^{3}He phase shift δ_1, the p + ^{6}Li phase shift δ_2, and the reflection coefficient τ are then computed by using the same wave function and the same nucleon-nucleon potential as those used in the ^{7}Li case. The results are shown in figs. 33

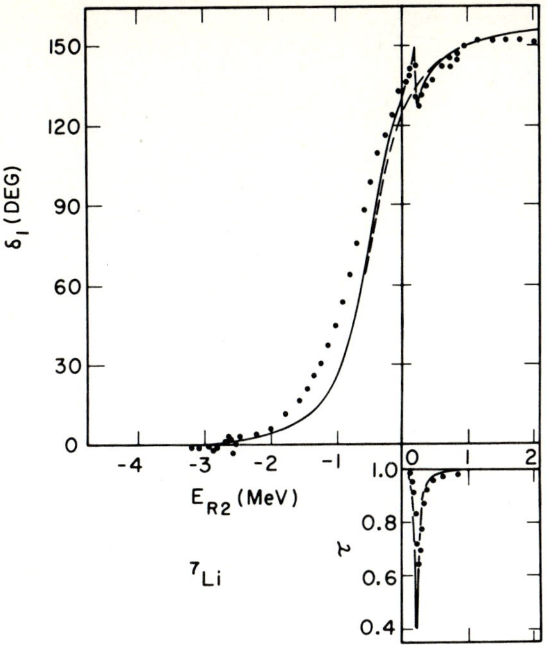

Fig. 31
$\alpha + t$ phase shift δ_1 and reflection coefficient τ as a function of the relative energy E_{R2} in the $n + {}^6Li$ channel. The solid curves represents the result of two-channel calculation, while the dashed curve represents the results of a simpler calculation in which the coupling between the $\alpha + t$ and the $n + {}^6Li$ channels is neglected. Empirical data points are those of ref. [SP 67]. (Adapted from ref. [TE 73].)

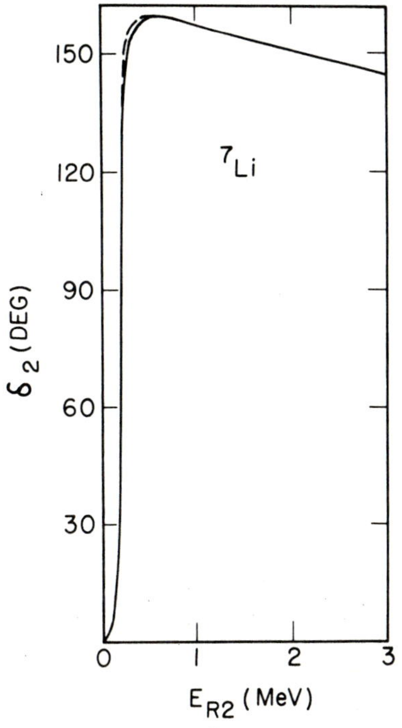

Fig. 32
$n + {}^6Li$ phase shift δ_2 as a function of E_{R2}. The meaning of the solid and dashed curves is the same as that given in the caption of fig. 31. (Adapted from ref. [TE 73].)

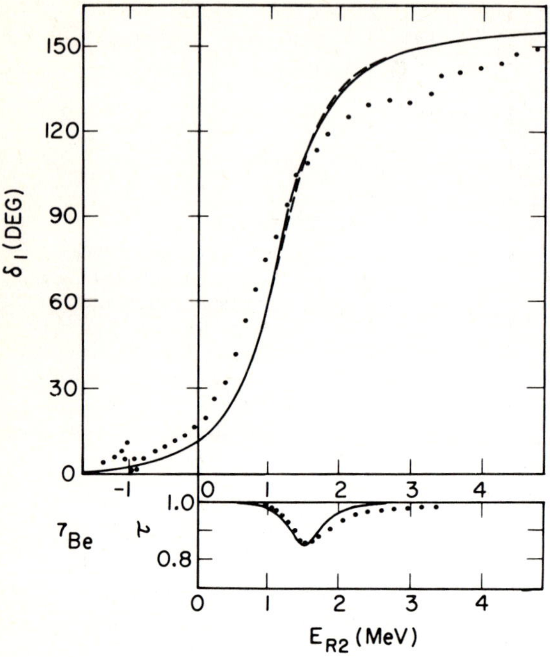

Fig. 33
$\alpha + {}^3$He phase shift δ_1 and reflection coefficient τ as a function of the relative energy E_{R2} in the p + ^{6}Li channel. The solid curves represent the result of a two-channel calculation, while the dashed curve represents the result of a simpler calculation in which the coupling between the $\alpha + {}^3$He and the p + ^{6}Li channels is neglected. Empirical data points are those of ref. [SP 67]. (Adapted from ref. [TE 73].)

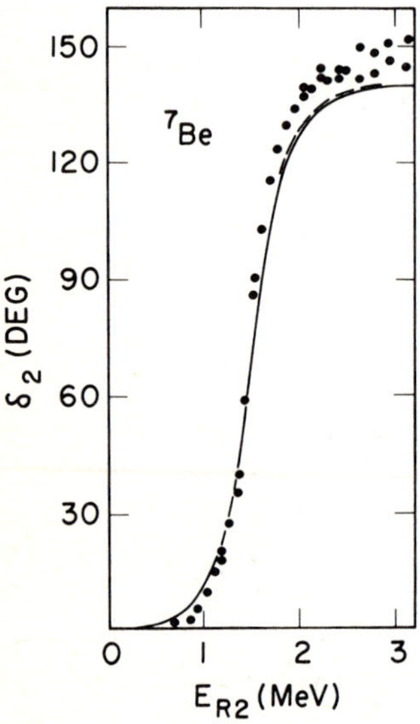

Fig. 34
p + ^{6}Li phase shift δ_2 as a function of E_{R2}. The meaning of the solid and dashed curves is the same as that given in the caption of fig. 33. Empirical data points are those of ref. [PE 69]. (Adapted from ref. [TE 73].)

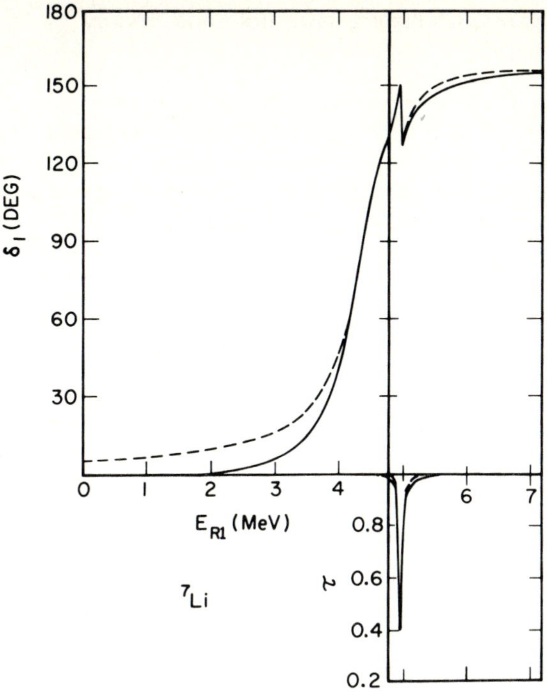

Fig. 35
Comparison of result in ^{7}Li obtained from a two-channel calculation with that obtained by using the resonance formulae of eq. (9.109). (Adapted from ref. [TE 73].)

and 34. Since it is reasonable to expect that the Coulomb interaction will not cause any additional specific distortion of the clusters in the compound region, one anticipates that there will again be a good agreement between calculation and experiment, as is indeed seen from these figures. In particular, it is noted that the upper $\frac{5}{2}^-$ resonance in ^{7}Be has very little influence on the $\alpha + {}^3$He phase shift; this is in contrast to what happens in the ^{7}Li case but in agreement with the empirical finding [SP 67].

The results shown in figs. 31–34 are obtained from two-channel calculations where the direct coupling between the channels is not neglected. In fig. 35, we compare the result (solid curve) obtained from a two-channel calculation in ^{7}Li with that (dashed curve) obtained by using the resonance formulae of eq. (9.109) which are derived under the assumption that the coupling between the α + t and the n + ^{6}Li channels occurs only through the upper $\frac{5}{2}^-$ resonance level. In applying these resonance formulae, the resonance energies of the two $\frac{5}{2}^-$ levels and the various level widths are used as adjustable parameters. From this figure one sees that the agreement between these two calculations is quite satisfactory. The only discrepancies occur near the low- and high-energy ends of these curves and these discrepancies arise mainly as a consequence of the linear-expansion approximation which was used to obtain eq. (9.109) from eq. (9.96). For example, it is clear that, when the energy is below the n + ^{6}Li threshold, the reflection coefficient τ should be equal to one. With the resonance formulae of eq. (9.109), this condition is certainly not exactly fulfilled.

9.6. Behaviour of the Partial Level Width Near a Threshold and Energy-dependent Width Approximation

If a resonance level lies very close to the threshold of a decay channel, then its representation in terms of a Breit-Wigner formula of the type given by eq. (9.17) is not accurate. The reason for this is that the width function $\Gamma(E)$ is sensitively energy-dependent in this case and the approximation represented by eq. (9.22) is no longer valid. In this section, we shall examine this problem in detail for an isolated resonance and show that, by studying the energetical behaviour of this function, a more appropriate resonance formula can be obtained.

The partial-width function $\Gamma_i(E)$ for the decay of an isolated resonance into a channel i is given by

$$\Gamma_i = 2\pi |\gamma_i|^2, \tag{9.112}$$

with

$$\gamma_i = \langle \psi_C^1 | H - E | \psi_{Di}^{\prime +} \rangle. \tag{9.113}$$

In the above equation, the function ψ_C^1 is chosen in such a way that it describes the behaviour of the resonance state in the compound region as well as possible. The function $\psi_{Di}^{\prime +}$ describes potential scattering in channel i; it is constructed with the function ψ_C^1 projected out, such that the potential-scattering phase shift becomes a slowly-varying function of energy and $\psi_{Di}^{\prime +}$ has a very small magnitude in the region of the compound nucleus. Together with the fact that ψ_C^1 is chosen to be an approximate solution of the Schrödinger equation in the compound region, one can then conclude that the contribution to γ_i comes mainly from the surface part of the compound nucleus.

To understand the energetical dependence of $\gamma_i(E)$, one must therefore examine the behaviour of $\psi_{Di}^{\prime +}$ in the surface region of the compound nucleus. Because the resonance is assumed to be deeply imbedded under a high potential barrier, the effective interaction between the clusters, which determines the relative-motion part of $\psi_{Di}^{\prime +}$, contains practically only their mutual Coulomb interaction and a centrifugal term. The form of the relative-motion part inside this potential barrier is thus expected to be nearly independent of the relative energy E_{Ri}; this is so, since in the case considered here, E_{Ri} has a magnitude much smaller than the height of the barrier.† On the other hand, the amplitude A_R of this relative-motion function will depend sensitively on E_{Ri}. This follows from the fact that, for the validity of eq. (9.113), the function $\psi_{Di}^{\prime +}$ must be normalized according to eq. (9.71).

In the case of a pure centrifugal potential, one finds from the behaviour of spherical Bessel functions [AB 64] that, as k_i approaches zero,

$$A_R \propto k_i^{l+\frac{1}{2}}, \tag{9.114}$$

† The relative-motion function inside the barrier falls off in an exponential-like fashion as the relative distance $\hat{R}_i$ becomes smaller.

where l denotes the relative orbital angular momentum between the clusters. If both clusters in the ith channel are charged, then one finds similarly from the behaviour of Coulomb functions [AB 64] that

$$A_R \propto e^{-\pi\eta_i} = e^{-\frac{\pi Z_{i1} Z_{i2} e^2 \mu_i}{\hbar^2 k_i}} \tag{9.115}$$

for small k_i values. It should be noted that A_R in eq. (9.115) is not l-dependent; this comes from the fact that, in the case under consideration here, the wave number k_i is so small that the relative-motion wave function has already fallen to a very small value before the centrifugal barrier becomes effective. Now, using eqs. (9.114) and (9.115), one obtains for small values of k_i

$$\gamma_i = P_l^{1/2}(k_i)\,\tilde{\gamma}_i, \tag{9.116}$$

where

$$\begin{aligned} P_l(k_i) &= k_i^{2l+1} && \text{for } Z_{i1} = 0 \text{ or } Z_{i2} = 0 \\ &= e^{-2\pi\eta_i} && \text{for } Z_{i1} \neq 0 \text{ and } Z_{i2} \neq 0, \end{aligned} \tag{9.117}$$

and $\tilde{\gamma}_i$ is a slowly-varying function of energy. For the validity of eq. (9.116), we should mention again that it is important to have the contribution to γ_i coming mainly from the surface region of the compound nucleus. In our present case where k_i takes on very small values, this condition is certainly well satisfied.

Equation (9.116) shows quantitatively that, even when the resonance state under consideration has a cluster structure corresponding to that of channel i, $\gamma_i(E)$ can become very small if it lies just above the threshold of this channel. In addition, this equation indicates that the Breit-Wigner formula, as given by eqs. (9.75)–(9.77), will become fairly inaccurate in this case, because the assumption that $\gamma_i(E)$ can be considered as constant over the energy width of the resonance level is no longer approximately correct.

An improved resonance formula for the case where the resonance lies very close to an open channel i can be obtained in the following way. One retains the linear expansion for $\Delta(E)$ as given by eq. (9.21) and treats all γ_k with $k \neq i$ as constants over the resonance energy region. But for γ_i only the less restrictive assumption that the part $\tilde{\gamma}_i$ be considered as energy-independent is made. Using this procedure, one obtains for example in the case where there is only one open channel the following S-matrix elements:

$$S_l = e^{2i\delta_l'(E)} \frac{E - E_r - i\pi C_l P_l}{E - E_r + i\pi C_l P_l}, \tag{9.118}$$

with

$$C_l = \left[\frac{|\tilde{\gamma}_i(E)|^2}{1 + \frac{d}{dE}\Delta(E)} \right]_{E = E_r}. \tag{9.119}$$

This resonance formula represents what is known as the energy-dependent width approximation. In the situation being considered here, the use of this formula does yield a better representation of the phase-shift curve determined from experimental results.

Our considerations in this and previous sections show that, when a resonance level happens to lie close to the threshold of a certain open channel, then its property will depend sensitively on its energetical distance from this threshold. This means that, for a reliable calculation, one must try to obtain this energetical distance to within several keV of the experimental value. This is very difficult to do, however, because presently one has to employ relatively simple nucleon-nucleon potentials and adopt various simplifying assumptions in order to make a microscopic study feasible. Therefore, in a practical calculation, it is often necessary to make small adjustments, such as adjusting the strength of the nucleon-nucleon potential, such that this energetical distance can be correctly obtained. Clearly, this is a rather undesirable procedure, but one which seems to be unavoidable at the present stage of development.

10. Resonance Reactions and Isobaric-Spin Mixing

10.1. General Remarks

We have seen in Chapter 9 that, under certain circumstances, the Coulomb interaction can influence very much the properties of resonance levels. In the examples discussed there, this influence comes from the diagonal matrix elements of the Coulomb interaction, which produce an energy shift in the positions of these levels. In this chapter, we shall further investigate the importance of the Coulomb interaction by examining its mixing effect with respect to wave functions of different isobaric spins. Because of the relative weakness of the Coulomb interaction compared with the nuclear interaction, it is of course expected that this isobaric-spin mixing effect can become important only when the levels which are characterized by the same quantum numbers except the isobaric-spin quantum number lie energetically very close to each other.

In section 10.2, we first consider the case where the incoming or elastic channel has a definite value for the isobaric spin.† This will happen when both nuclei in this channel have definite isobaric-spin values and one of them has in addition an isobaric spin $T = 0$. In this case, the mixing of the resonance wave functions will then take place essentially in the region of the compound nucleus. An example where such a mixing effect is important

† For inelastic and rearrangement channels which might also be open, the requirement that they must also possess definite values for the isobaric spin is not necessary.

occurs in the nucleus ^{8}Be. Here the two levels with excitation energies of 16.6 and 16.9 MeV have the same $J^\pi = 2^+$ but no definite isobaric-spin value [MA 65]. As a consequence of the very small difference in the excitation energies, the wave functions describing these two levels in the compound region contain almost equal amounts of $T = 0$ and $T = 1$ components, as will be shown in detail in the following discussion.

Next we consider in section 10.3 the case where the elastic channel in the asymptotic region is described by a mixture of configurations with different isobaric-spin values. This situation arises when both nuclei in this channel have isobaric spins different from zero. The most interesting examples in this category are the analogue resonances found in medium-heavy and heavy nuclei. There the resonance configurations with different isobaric spins in the compound region show a relatively small degree of direct mixing and the coupling between these configurations occurs mainly through the elastic channel.

10.2. Isobaric-Spin Mixing in the Compound Region

We discuss here the case where two resonance levels with the same spin and parity but no well-defined isobaric spin lie near each other and where only the elastic channel with a definite isospin value is open. We shall first derive in subsection 10.2a the relevant phase-shift formula and then consider in subsection 10.2b as a specific example the $\alpha + \alpha$ resonance scattering in the ^{8}Be excitation-energy region around 16 MeV, where the resonance states exhibit an appreciable mixing of $T = 0$ and $T = 1$ configurations.

10.2a. Derivation of a Two-Level Resonance Formula

As in the derivation of the resonance formula for pure elastic-scattering discussed in sections 9.2 and 9.3, we divide the wave function ψ into two parts, i.e.,

$$\psi = \psi_D + \psi_C. \tag{10.1}$$

The function ψ_C is expressed in terms of complete orthonormalized sets of bound-state wave functions $\psi_C^n(T)$, defined in a finite region of configuration space which is larger than but of similar magnitude as that corresponding to the compound nucleus. These functions are constructed as eigenfunctions of the operator $\mathbf{T}^2$ with eigenvalues $T(T+1)$ and are determined in addition from the equation

$$\langle \delta\psi_C(T)|H - E_n(T)|\psi_C^n(T)\rangle = 0. \tag{10.2}$$

In eq. (10.2) it is noted that the variation $\delta\psi_C(T)$, defined in a restricted subspace characterized by a good isospin value, is adopted; this means that we are using here for the construction of $\psi_C^n(T)$ only diagonal elements of the Coulomb interaction with respect to the isobaric spin. We choose to define complete sets of orthonormalized functions in this particular way, because we wish to consider afterwards the mixing of two energetically near-lying states which have the same values of spin and parity but different isobaric-spin quantum numbers.

Using the basis wave functions $\psi_C^n(T)$, we write ψ_C as

$$\psi_C = \psi_C^R + \psi_C^B, \tag{10.3}$$

with

$$\psi_C^R = a_1 \psi_C^1 + a_2 \psi_C^2, \tag{10.4a}$$

$$\psi_C^B = \sum_{n=3}^{\infty} a_n \psi_C^n (T_n), \tag{10.4b}$$

and similarly for the variation $\delta \psi_C$. The functions ψ_C^1 and ψ_C^2 denote the two energetically near-lying states with the same quantum numbers except the isobaric spin, which we wish to consider in more detail. For instance, in the ^{8}Be example to be discussed below, ψ_C^1 will denote a $T = 0$ state and ψ_C^2 a $T = 1$ state. Also, it should be noted that in eq. (10.4b) the summation over n includes wave functions specified by different values of T.

Similar to the derivation given in section 9.3, we eliminate the terms involving ψ_C^B from the projection equation (2.3). The result is

$$\langle \delta \psi_D | H'(E) - E | \psi_D \rangle + \langle \delta \psi_D | H - E | \psi_C^R \rangle = 0, \tag{10.5a}$$

$$\langle \delta \psi_C^R | H - E | \psi_D \rangle + \langle \delta \psi_C^R | H - E | \psi_C^R \rangle = 0, \tag{10.5b}$$

where†

$$H'(E) = H - \sum_{n=3}^{\infty} \frac{|H - E| \bar{\psi}_C^n \rangle \langle \bar{\psi}_C^n | H - E|}{\bar{E}_n - E}, \tag{10.6}$$

with $\bar{\psi}_C^n$ being orthonormalized solutions of the equation

$$\langle \delta \psi_C^B | H - \bar{E}_n | \bar{\psi}_C^n \rangle = 0. \tag{10.7}$$

It should be noted, however, that the functions $\bar{\psi}_C^n$ are not constructed as eigenstates of the operator $\mathbf{T}^2$ and, hence, are not completely identical to the functions $\psi_C^n (T_n)$ with $n \geqslant 3$.

By formally integrating eq. (10.5a) with the help of the open-channel resolvent operator $G_D'^+ (E)$, defined by the equation

$$\langle \delta \psi_D | H'(E) - E | G_D'^+ (E) = \langle \delta \psi_D |, \tag{10.8}$$

we obtain

$$| \psi_D \rangle = | \psi_D'^+ \rangle - G_D'^+ (E) (H - E) | \psi_C^R \rangle. \tag{10.9}$$

Substituting this equation into eq. (10.5b) yields

$$\langle \delta \psi_C^R | H - E | \psi_D'^+ (E) \rangle + \langle \delta \psi_C^R | H - (H - E) G_D'^+ (E) (H - E) - E | \psi_C^R \rangle = 0, \tag{10.10}$$

which represents a system of two inhomogeneous linear equations for the determination of the coefficients a_1 and a_2. If one now performs partial-wave expansions of $\psi_D'^+ (E)$

† In the derivation of eqs. (10.5a) and (10.5b), the Coulomb coupling of wave functions with different isobaric spins contained in ψ_C^B and ψ_C^R is neglected; in other words, we assume that the influence of ψ_C^B on the mixing of ψ_C^1 and ψ_C^2 occurs only indirectly through the open-channel function ψ_D. This is allowed in good approximation, because we are interested in cases where the compound states in ψ_C^B and ψ_C^R which have the same quantum numbers except the isobaric-spin quantum number are well separated from each other. Of course, this is the main reason why we have chosen to separate ψ_C into ψ_C^R and ψ_C^B.

and $G_D^{\prime +}(E)$, and selects the wave which has the same spin j and parity π as the resonances under consideration, then one obtains for the column vector

$$a = \begin{pmatrix} a_1 \\ a_2 \end{pmatrix} \tag{10.11}$$

the following matrix equation:

$$\left(A - i\frac{1}{2}\Gamma\right) a = -\gamma, \tag{10.12}$$

where

$$A = \begin{pmatrix} E_1 - E - \Delta E_{11} & \overline{V}_C - \Delta E_{12} \\ \overline{V}_C - \Delta E_{21} & E_2 - E - \Delta E_{22} \end{pmatrix}, \tag{10.13}$$

$$\Gamma = 2\pi \begin{pmatrix} \gamma_1 \gamma_1^* & \gamma_1 \gamma_2^* \\ \gamma_2 \gamma_1^* & \gamma_2 \gamma_2^* \end{pmatrix}, \tag{10.14}$$

$$\gamma = \begin{pmatrix} \gamma_1 \\ \gamma_2 \end{pmatrix}, \tag{10.15}$$

with

$$E_i = \langle \psi_C^i | H | \psi_C^i \rangle, \quad (i = 1, 2) \tag{10.16}$$

$$\Delta E_{ik} = \langle \psi_C^i | (H - E) G_{Dj}^{\prime P} (H - E) | \psi_C^k \rangle, \quad (i, k = 1, 2) \tag{10.17}$$

$$\overline{V}_C = \langle \psi_C^1 | V_C | \psi_C^2 \rangle, \tag{10.18}$$

$$\gamma_i = \langle \psi_C^i | H - E | \psi_{Dj}^{\prime +}(E) \rangle. \quad (i = 1, 2) \tag{10.19}$$

In eq. (10.13), it is noted that the Coulomb matrix element $\overline{V}_C$ is assumed to be real, which is allowed because the bound-state functions ψ_C^1 and ψ_C^2 can be chosen as real functions. From eq. (10.12), one determines the coefficients a_1 and a_2. Substituting these coefficients into eq. (10.9) and examining the asymptotic behaviour of the resultant expression for ψ_D decomposed into partial waves of total spin j and parity π, one then obtains the phase factor A_j (S-matrix element) for pure elastic-scattering in the incoming channel.

To continue our discussion, we make some further simplifications which are valid in the special example to be studied in the following subsection. Using the convention that the elastic channel and the state ψ_C^1 have the same isobaric spin, we neglect the direct coupling of the elastic channel with the state ψ_C^2 which has a different isospin value; that is, we take into account only the terms ΔE_{11} and γ_1 in eq. (10.12) and set all other ΔE_{ik} and γ_2 in that equation as zero.† These simplifications are certainly allowed if the states under consideration have very good bound structures in the compound region or, in other words, are narrow resonances, and if $\psi_{Dj}^{\prime +}(E)$ has a small amplitude in the compound region and, consequently, describes essentially the phenomenon of

† With these simplifications, the coupling of ψ_C^2 with the elastic channel occurs only through the state ψ_C^1. In this sense, ψ_C^1 acts as a doorway state [FE 74] to the state ψ_C^2.

hard-sphere scattering.† Under these conditions, the spatial overlap between ψ_C^2 and $\psi_{Dj}^{\prime +}(E)$ occurs mainly in the surface region where the Coulomb interaction is much weaker than that in the region of the compound nucleus, and the result of this is that the direct-coupling term involving these two functions becomes quite unimportant.

With the above simplifications, we obtain from eqs. (10.12) and (10.13) the following two equations for the determination of a_1 and a_2:

$$[(E_1 - \Delta E_{11} - E) - i\pi|\gamma_1|^2]\, a_1 + \bar{V}_C\, a_2 = -\gamma_1, \qquad (10.20a)$$

$$\bar{V}_C\, a_1 + (E_2 - E)\, a_2 = 0. \qquad (10.20b)$$

Solving these equations one finds

$$\frac{a_1}{a_2} = -\frac{E_2 - E}{\bar{V}_C} \qquad (10.21)$$

and

$$a_1 = \frac{-\gamma_1 (E_2 - E)}{[E_1 - \Delta E_{11} - E - i\pi|\gamma_1|^2](E_2 - E) - \bar{V}_C^2}. \qquad (10.22)$$

If we now substitute eqs. (10.21) and (10.22) into eq. (10.4a), then we obtain from the asymptotic behaviour of eq. (10.9) the following expression for the amplitude A_j of the outgoing partial wave with total spin j:††

$$A_j = e^{2i\delta_j'(E)} \frac{(E_1 - \Delta E_{11} - E)(E_2 - E) - \bar{V}_C^2 + i\pi|\gamma_1|^2 (E_2 - E)}{(E_1 - \Delta E_{11} - E)(E_2 - E) - \bar{V}_C^2 - i\pi|\gamma_1|^2 (E_2 - E)}, \qquad (10.23)$$

where $e^{2i\delta_j'(E)}$ is the phase factor of the asymptotic outgoing wave without the contribution from the resonance states ψ_C^1 and ψ_C^2. In analogy to eq. (9.56), A_j can also be represented in a product form. The result, after making the usual linear-expansion approximation (see eqs. (9.21), (9.22), (9.58), and (9.59)), is

$$A_j = e^{2i\delta_j'(E)} \frac{\left(E_a - E + i\frac{1}{2}\Gamma_a\right)\left(E_b - E + i\frac{1}{2}\Gamma_b\right)}{\left(E_a - E - i\frac{1}{2}\Gamma_a\right)\left(E_b - E - i\frac{1}{2}\Gamma_b\right)}, \qquad (10.24)$$

where E_a, E_b, Γ_a, and Γ_b are energy-independent quantities. In eq. (10.24), we denote with E_a and Γ_a the resonance energy and energy width of the state which approaches ψ_C^1 in the compound region when $\bar{V}_C$ approaches zero, and with E_b and Γ_b the resonance energy and energy width of the other resonance state.

If the resonances are narrow and do not lie very close to any channel threshold, the quantities $|\gamma_1|^2$ and ΔE_{11} can be considered as approximately energy-independent.††† In

† To satisfy the latter condition as well as possible, one projects out the function ψ_C^1 from ψ_D. As has been discussed in subsection 9.2c, this involves only minor modifications in the formulation. The required changes are only to replace $\psi_{Dj}^{\prime +}$ in eq. (10.19) by $\tilde{\psi}_{Dj}^{\prime +}$ and $G_{Dj}^{\prime P}$ in eq. (10.17) by $\tilde{G}_{Dj}^{\prime P}$.

†† Because we have neglected the direct coupling between the elastic channel and the state ψ_C^2, the term $G_D^{\prime +}(E)(H - E)|a_2 \psi_C^2\rangle$ contained in eq. (10.9) becomes zero.

††† In the ^8Be example to be discussed in subsection 10.2b, a microscopic calculation performed by *Federsel* [FE 74a] using a wave function consisting of two bound states and the $\alpha + \alpha$ channel has indeed shown that $|\gamma_1|^2$ and ΔE_{11} are approximately energy independent.

this case, we can then find simple relationships between the quantities $E_1 - \Delta E_{11}$, E_2, $\overline{V}_C^2$, and $|\gamma_1|^2$ in eq. (10.23) and the observable quantities E_a, E_b, Γ_a, and Γ_b in eq. (10.24). Here we shall not write down all these relationships, but only give expressions for $\overline{V}_C^2$ and the isospin-mixing parameter

$$M = \left(\frac{a_1}{a_2}\right)^2 = \left(\frac{E_2 - E}{\overline{V}_C}\right)^2, \tag{10.25}$$

which determines the degree of mixing of the two states ψ_C^1 and ψ_C^2 in the compound region as a function of the scattering energy. These expressions are

$$\overline{V}_C^2 = \frac{\Gamma_a \Gamma_b}{(\Gamma_a + \Gamma_b)^2}\left[(E_a - E_b)^2 + \frac{1}{4}(\Gamma_a + \Gamma_b)^2\right], \tag{10.26}$$

and

$$M = \frac{[\Gamma_b E_a + \Gamma_a E_b - E(\Gamma_a + \Gamma_b)]^2}{\Gamma_a \Gamma_b [(E_a - E_b)^2 + \frac{1}{4}(\Gamma_a + \Gamma_b)^2]}. \tag{10.27}$$

From eqs. (10.25) and (10.27), one sees that for

$$E = E_2 = \frac{\Gamma_b E_a + \Gamma_a E_b}{\Gamma_a + \Gamma_b} \tag{10.28}$$

which is intermediate between E_a and E_b, M becomes zero. This means that at this particular energy the ψ_C^1 component of the compound state vanishes completely, a result which is certainly only approximately correct because we have neglected the direct coupling between the elastic channel and the state ψ_C^2 via the Coulomb interaction.

At energies equal to E_a and E_b, the isospin-mixing parameter M acquires the following values:

$$M_a = \frac{\Gamma_a}{\Gamma_b} \frac{(E_a - E_b)^2}{(E_a - E_b)^2 + \frac{1}{4}(\Gamma_a + \Gamma_b)^2}, \quad (E = E_a) \tag{10.29a}$$

$$M_b = \frac{\Gamma_b}{\Gamma_a} \frac{(E_a - E_b)^2}{(E_a - E_b)^2 + \frac{1}{4}(\Gamma_a + \Gamma_b)^2}. \quad (E = E_b) \tag{10.29b}$$

Here one sees that for very narrow resonances where Γ_a and Γ_b are both much smaller than $|E_a - E_b|$, there is a complete isobaric-spin mixing ($M_a = M_b = 1$) at these energies when Γ_a equals Γ_b. In this case, it is also noted that the magnitude of the Coulomb matrix element $\overline{V}_C$ is equal to $\frac{1}{2}|E_a - E_b|$.

From these considerations one sees that, in those cases where the approximations made above are valid, one can determine the degree of isobaric-spin mixing of the wave function in the compound region by simply using measured values for the energy and level width of the resonances.

We consider now the limit case where the coupling of the state ψ_C^1 with the elastic channel is also weak. This means that the quantities $|\gamma_1|^2$ and ΔE_{11} appearing in eq. (10.23) have very small values. By comparing eq. (10.23) with eq. (10.24), one can then conclude that the widths Γ_a and Γ_b must also be very small. On the other hand, the ratio Γ_a/Γ_b is not at all small but is given by

$$\frac{\Gamma_a}{\Gamma_b} = -\frac{E_2 - E_a}{E_2 - E_b}$$

$$= \frac{\frac{1}{2}(E_1 - E_2) + \left[\frac{1}{4}(E_1 - E_2)^2 + \bar{V}_C^2\right]^{1/2}}{-\frac{1}{2}(E_1 - E_2) + \left[\frac{1}{4}(E_1 - E_2)^2 + \bar{V}_C^2\right]^{1/2}}. \quad \text{(weak-coupling limit)} \tag{10.30}$$

Using eqs. (10.29a) and (10.29b), one then finds that, in this weak-coupling limit,

$$M_a = \frac{1}{M_b} = \frac{\frac{1}{2}(E_1 - E_2) + \left[\frac{1}{4}(E_1 - E_2)^2 + \bar{V}_C^2\right]^{1/2}}{-\frac{1}{2}(E_1 - E_2) + \left[\frac{1}{4}(E_1 - E_2)^2 + \bar{V}_C^2\right]^{1/2}}. \tag{10.31}$$

As can be shown easily, eq. (10.31) is the same as that which one would obtain if one considers ψ_C^1 and ψ_C^2 as true bound states and neglects from the beginning the coupling to the open channel. This is, of course, to be expected. Therefore, one sees from this consideration that the degree of isobaric-spin mixing depends on the strength of the coupling between ψ_C^1 and the elastic channel, and can deviate quite appreciably from the value given by eq. (10.31) if this coupling is especially strong.

In addition, one can conclude that the degree of isobaric-spin mixing could depend on the reaction process by which the resonance states are excited. For example, if one excites the ^{8}Be states at 16.62 and 16.92 MeV (see the next subsection) by the reaction process ^{9}Be (^{3}He, α) ^{8}Be*, then the degree of isobaric-spin mixing in the compound region could be different from that which one obtains by exciting these states in elastic $\alpha + \alpha$ scattering. This is possible because, in contrast to the $\alpha + \alpha$ case, there is no definite value for the isobaric spin in the incoming channel for the above reaction. To obtain quantitative results about the mixing of states when different reaction processes are used, one must in principle investigate every one of these processes in detail. This can be achieved in the framework of the unified theory presented here by generalizing the trial wave function describing the behaviour of the system. In this section we shall, however, not go further into this discussion, since these generalizations will be rather extensively discussed in following chapters (see especially chapter 12 on direct reactions) where other physical problems are studied.

10.2b. The 16.62 and 16.92 MeV States in ^{8}Be as a Specific Example

As an application of the general formula derived in subsection 10.2a, we consider now the two ^{8}Be states at excitation energies of 16.62 and 16.92 MeV [AJ 74], which have $J^\pi = 2^+$ but no definite value of isobaric spin. Both levels are very narrow compared to the lower-lying 2^+ and 4^+ states, a fact which can be explained in the following qualitative manner. In the excitation-energy region around 16 MeV, one begins to find states in which one α cluster is broken up. These states can be described in good approximation as a coherent mixture of p + ^{7}Li and n + ^{7}Be cluster configurations in the compound region.

For those levels with excitation energies less than about 17 MeV, only the decay into the $\alpha + \alpha$ channel is possible, because the $p + {}^7Li$ and $n + {}^7Be$ channels open at 17.3 and 18.9 MeV, respectively. The spatial overlap of these cluster configurations with the $\alpha + \alpha$ channel is expected to be small; therefore, resonance levels in this particular excitation-energy region should have a rather long lifetime and, consequently, a rather small level width.[†]

With $p + {}^7Li$ and $n + {}^7Be$ cluster configurations, one can form both $T = 0$ and $T = 1$ states. In the compound region, the normalized wave functions, denoted as ψ_C^1 for the $T = 0$ state and ψ_C^2 for the $T = 1$ state, can be written in the following way:

$$\psi_C^1 = \hat{N}_1 \mathcal{A} \{\phi(\mathbf{r}_1, \ldots, \mathbf{r}_8; s_1, \ldots, s_8) [\xi_T(^7Li)\, \pi_8 - \xi_T(^7Be)\, \nu_8]\}, \qquad (10.32a)$$

$$\psi_C^2 = \hat{N}_2 \mathcal{A} \{\phi(\mathbf{r}_1, \ldots, \mathbf{r}_8; s_1, \ldots, s_8) [\xi_T(^7Li)\, \pi_8 + \xi_T(^7Be)\, \nu_8]\}, \qquad (10.32b)$$

where the function ϕ describes the spatial-spin behaviour of the two states, and $\xi_T(^7Li)$ and $\xi_T(^7Be)$ denote the isospin functions of the seven-nucleon system with $T_3 = +\frac{1}{2}$ and $-\frac{1}{2}$, respectively. The quantities $\hat{N}_1$ and $\hat{N}_2$ are normalization constants; they are not equal to each other, because antisymmetrized $p + {}^7Li$ and $n + {}^7Be$ wave functions are not mutually orthogonal. On the other hand, for the levels at 16.62 and 16.92 MeV, the proton and the neutron are only weakly bound to the respective ^7Li and ^7Be cores. This has the consequence that the exchange terms in the normalization integral become much smaller than the direct term and the difference between $\hat{N}_1$ and $\hat{N}_2$ becomes quite small.

We now use eqs. (10.29a) and (10.29b) to determine the degree of isobaric-spin mixing in the 16.62 and 16.92 MeV states of ^8Be. The data which we shall use for the resonance energies and the widths of these two levels are determined from the $^{10}B(d,\alpha)\,^8Be^*$ reaction [NO 71] and not from elastic $\alpha + \alpha$ scattering. Therefore, due to the remark made near the end of subsection 10.2a, one has to be careful in the application of these equations. However, one notes that, as in the $\alpha + \alpha$ channel, the $d + {}^{10}B$ channel has also an isobaric spin equal to zero; this means that the intermediate ^{12}C compound state which decays subsequently into an α particle and an excited ^8Be nucleus has similarly a value of $T = 0$. Consequently, it is reasonable to expect that the values for the resonance energies and level widths of the 16.62 and 16.92 MeV states in ^8Be as determined from the $^{10}B(d,\alpha)\,^8Be^*$ reaction would be very similar to those determined from precision $\alpha + \alpha$ scattering when such measurements become available.

The experimental values for $(E_a - E_b)$, Γ_a, and Γ_b obtained from analyzing the $^{10}B(d,\alpha)\,^8Be^*$ reaction data [NO 71] are

$$\begin{aligned} E_a - E_b &= -290 \text{ keV}, \\ \Gamma_a &= 90 \text{ keV}, \\ \Gamma_b &= 70 \text{ keV}. \end{aligned} \qquad (10.33)$$

[†] Quantitatively, this is expressed by the fact that γ_i given by eq. (10.19) has a small value. In the case discussed in subsection 10.2a where one assumes that only $|\gamma_1|^2$ is different from zero, one obtains by comparing eq. (10.24) with eq. (10.23) the relation

$$|\gamma_1|^2 = \frac{1}{2\pi}(\Gamma_a + \Gamma_b)$$

between $|\gamma_1|^2$ and the widths of the levels.

Substituting these values into eqs. (10.29a) and (10.29b), we find then

$$M_a = 1.20,\qquad M_b = 0.72. \tag{10.34}$$

Equation (10.34) indicates the presence of strong isobaric-spin mixing in both ^{8}Be states under consideration. If there were no isobaric-spin mixing at all, the parameter M would acquire the value 0 or ∞ and the corresponding compound state would have a pure $T = 1$ or a pure $T = 0$ configuration. On the other hand, one can see from eqs. (10.32a) and (10.32b) that, with $\hat{N}_1$ and $\hat{N}_2$ being nearly equal, the compound state will have almost a pure $p + {}^7$Li or $n + {}^7$Be configuration, if M is equal to 1.

From the values of M_a and M_b given in eq. (10.34), one finds that, in the compound region, one of the resonance states under consideration has predominantly a $p + {}^7$Li cluster structure, while the other has predominantly a $n + {}^7$Be cluster structure. Based on the information given by eq. (10.34) alone, one cannot of course decide which of these two states has predominantly a $p + {}^7$Li or a $n + {}^7$Be configuration, because to do this will require the knowledge of not only the magnitude of the ratio a_1/a_2 but also its sign. As can be seen from eqs. (10.32a) and (10.32b), the compound state will have mainly a $p + {}^7$Li configuration if a_1/a_2 has a value close to $+1$ and a $n + {}^7$Be configuration if a_1/a_2 has a value close to -1. To obtain the sign of a_1/a_2, one makes use of eq. (10.21). Since E_2 is known to have a value intermediate between E_a and E_b, this sign is determined once the sign of $\bar{V}_C$ is obtained. Using eqs. (10.18), (10.32a), and (10.32b), one finds that

$$\bar{V}_C = \langle \psi_C^1 | V_C | \psi_C^2 \rangle \approx D\,[E_C(p + {}^7\mathrm{Li}) - E_C(n + {}^7\mathrm{Be})], \tag{10.35}$$

where $E_C(p + {}^7$Li) and $E_C(n + {}^7$Be) denote the Coulomb energies of the $p + {}^7$Li and $n + {}^7$Be configurations, respectively, and D is a positive proportionality constant unimportant for our discussion.† Now, because the electric charge in the $n + {}^7$Be configuration is more concentrated than that in the $p + {}^7$Li configuration, $E_C(n + {}^7$Be) is larger than $E_C(p + {}^7$Li); consequently, $\bar{V}_C$ has a minus value. Therefore, it follows from eq. (10.21) and the argument given above that the 16.62 MeV state has predominantly a $p + {}^7$Li cluster configuration and the 16.92 MeV state has predominantly a $n + {}^7$Be cluster configuration, a conclusion which can of course also be reached by simply examining the general behaviour of the Coulomb energy.

The above discussion also yields a somewhat more general conclusion. It can be seen from eq. (10.21) that, in the energy region where the formulae derived in subsection 10.2a are valid, the wave function in the compound region has predominantly a $p + {}^7$Li structure for $E < E_2$ and a $n + {}^7$Be structure for $E > E_2$.

Finally, it should be mentioned that there is a microscopic calculation performed by *Federsel* [FE 74a], which deals with this particular problem. This calculation was carried out in a similar way as the many examples described in chapter 7. Hence, we shall not discuss it further here, except to say that it does yield satisfactory values for $(E_a - E_b)$, Γ_a, and Γ_b, and show that the approximations made in subsection 10.2a are reasonable.

---

† In deriving eq. (10.35), one notes again that the proton and neutron are weakly bound to their respective cores.

10.3. Isobaric-Spin Mixing in the Incoming Channel

10.3a. Qualitative Description

We shall now discuss resonance reactions where an isobaric-spin mixing occurs already in the incoming channel. As has already been mentioned, this can happen only if both reaction partners have an isobaric spin different from zero. The best studied resonances of this kind are the analogue resonances observed in proton elastic scattering [FO 64]. Here we shall restrict ourselves to an investigation of just these resonances. Furthermore, we shall give only a relatively brief discussion of how the analogue resonances can be described within the framework of the unified nuclear theory discussed in this monograph. For a detailed description of analogue resonances in the case of proton scattering, we refer to the book of *Mahaux* and *Weidenmüller* [MA 69].

Let us first mention briefly the essential features of analogue resonances. Analogue resonances belong to an isospin multiplet; in this multiplet, the state with the largest isobaric-spin component $T_3 = \frac{1}{2}(N-Z)$ forms the ground state or low-excited state of a so-called parent nucleus. For this state the isobaric spin T is equal to T_3 because, for a given T_3 value, this particular value of T yields the energetically most favoured isospin configuration. By carrying out a rotation in the isospin space on this parent state through the use of the lowering operator T^-, multiplet states with the same value of T but smaller T_3 values are produced in neighbouring nuclei. In these multiplet states, the Coulomb energies will be larger than the Coulomb energy in the parent state, because the number of protons is increased. Now, if one considers systems where this increase in Coulomb energy is large enough such that one or more channels become open, then the corresponding analogue state can be observed as a resonance (isobaric analogue resonance) in these channels. Already in the region of the periodic table with $Z \approx 20$, it can be easily estimated

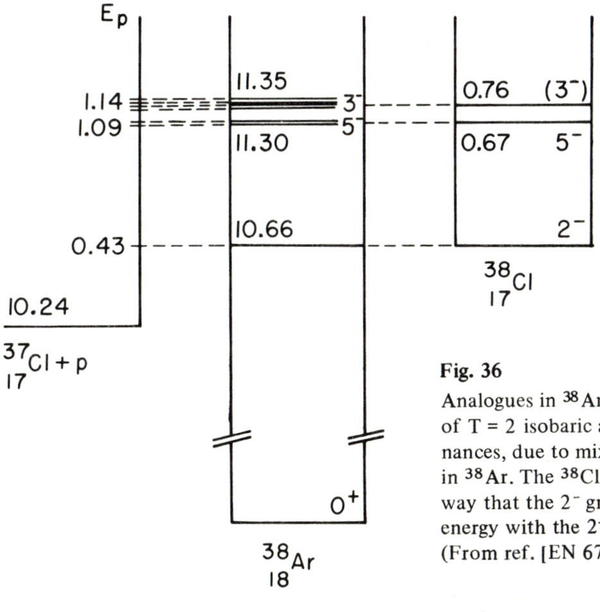

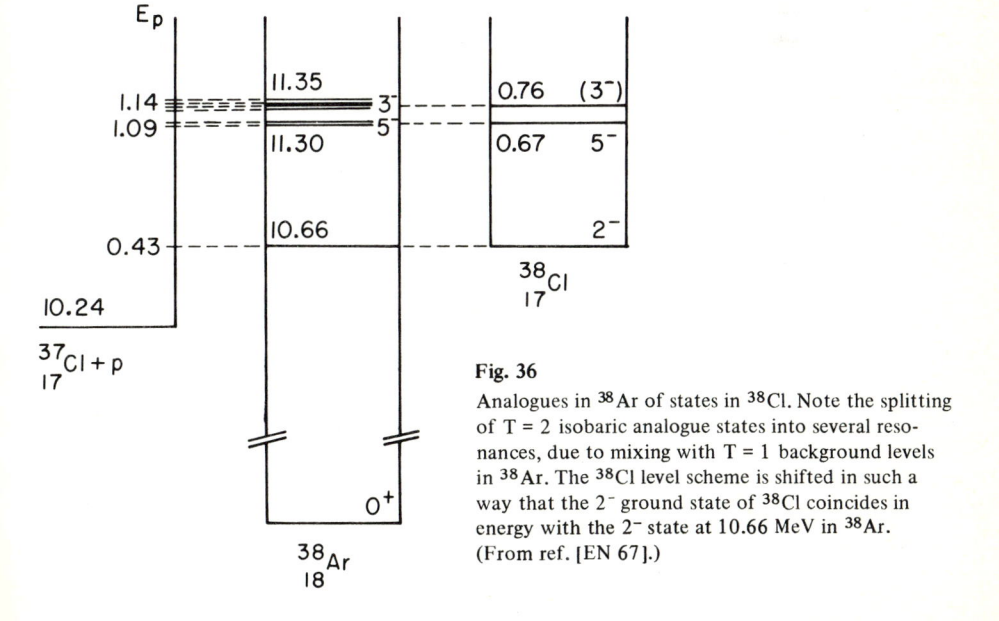

Fig. 36

Analogues in ^{38}Ar of states in ^{38}Cl. Note the splitting of T = 2 isobaric analogue states into several resonances, due to mixing with T = 1 background levels in ^{38}Ar. The ^{38}Cl level scheme is shifted in such a way that the 2^- ground state of ^{38}Cl coincides in energy with the 2^- state at 10.66 MeV in ^{38}Ar. (From ref. [EN 67].)

that this Coulomb-energy increase in going from a parent state with $T_3 = T$ to the analogue state with $T_3 = T - 1$ is so large that the analogue state becomes unstable against nucleon decays and hence can be observed, for instance, in proton elastic scattering.†

As an example of this [EN 67], we discuss qualitatively the ground state of ^{38}Cl and the 10.66 MeV excited state of ^{38}Ar (see fig. 36). These two states with $J^\pi = 2^-$ are members of an isospin quintet (T = 2). The 10.66 MeV state of ^{38}Ar is unstable against proton emission and is the lowest T = 2 state because it is the isobaric analogue of the ground state of the parent nucleus ^{38}Cl. In the excitation-energy region around 11 MeV, the density of T = 2 levels in ^{38}Ar is rather small, which is a consequence of the fact that in the parent ^{38}Cl nucleus the corresponding T = 2 levels lie in a region of low excitation. On the other hand, the density of T = 1 excited states (normal states) in this excitation-energy region is expected to be much higher; the reason for this is that the ground state of ^{38}Ar has T = 1 and, hence, T = 1 states with excitation energies of about 11 MeV are already rather highly excited. The 10.66 MeV analogue state in ^{38}Ar has a rather simple structure, because its corresponding state in the parent nucleus ^{38}Cl can be well represented, for instance, in the shell-model representation by a superposition of a relatively small number of simple configurations. As a consequence, this state should be coupled fairly strongly to the p + ^{37}Cl channel and its decay width should be fairly large.†† This is not the case for the T = 1 levels because, due to their large excitation energies, they possess usually a complicated structure; for instance, in the shell-model representation, they will be represented by a linear superposition of a large number of both simple and complicated configurations. Therefore, the coupling of these states with the p + ^{37}Cl channel is quite small and their decay widths should be significantly smaller than those of the T = 2 states. Other parent states which yield analogue resonances in ^{38}Ar are the 0.67 MeV 5^- and the 0.76 MeV 3^- states in ^{38}Cl. Both parent states have again T = 2. In addition, there exist in ^{38}Ar near these analogue states several T = 1 levels with $J^\pi = 5^-$ and 3^-. These levels can be mixed via the Coulomb interaction with the T = 2 analogue resonances and, therefore, can be readily observed in reactions such as ^{37}Cl(p, γ) ^{38}Ar and ^{37}Cl(p, α) ^{34}S [EN 67].

In heavier nuclei, the situation about analogue resonances is qualitatively similar to that described above. But there is one important difference, which is connected with the fact that the density of excited normal states in the excitation-energy region around the analogue resonance is much larger in a heavier system.††† Therefore, one expects that, when

† In this region of the periodic table, this increase in Coulomb energy is equal to about 6–8 MeV, which is of the order of the separation energy of a nucleon from the nucleus.

†† Isobaric analogue states are usually considered as doorway states belonging to the proton channel in which they are produced. One should note, however, that doorway states are defined especially clearly in terms of particle-hole excitations in the shell-model representation, but not in other cluster representations. Therefore, in general, we do not use this terminology in this monograph.

††† The reason for this is that, with increasing proton number, the excitation energy of the analogue state increases. Therefore, by taking also into account the larger nucleon number in a heavier nucleus, one sees easily that the density of normal states, i. e., states having the same spin and parity as the analogue state but the isospin of the ground state, can become very large in the excitation-energy region around the analogue state.

the nucleon number becomes larger, there will be more interplay between analogue and normal states. However, even though the degree of isobaric-spin mixing between these states can be quite large, the concept of isospin is still a useful one even in heavy nuclei, as has indeed been shown by many experimental studies [BA 67a].

10.3b. Quantitative Formulation in the Case of a Single Open Channel

We present now the quantitative formulation of analogue resonances in the framework of our unified theory. For simplicity, we shall consider mainly the case where only the elastic channel, in which the analogue state is excited, is open. In heavier nuclei this will of course never happen because, in the excitation-energy region where the analogue resonance lies, a number of other channels will also be open. However, many essential conclusions about analogue resonances can already be obtained in this simplified situation. Afterwards, we shall show briefly how one can generalize these considerations if these other open channels are also considered.

We could start our discussion with eq. (9.55) of section 9.3 where a many-level resonance formula for pure elastic-scattering is derived. To see the specific influence of the normal compound states on an analogue state, one would then have to represent the bound-state solutions $\psi_C^{R\lambda}$ of eq. (9.50) as a linear superposition of a function describing the analogue state and a set of functions belonging to a reduced subspace which does not contain the wave function of the analogue state. This is a rather inconvenient procedure; hence, we shall start from the beginning by splitting the wave function in a way which is particularly suited for the purpose of our investigation.

The wave function is written as

$$\psi = \psi_D + b_0 \psi_0 + \psi_C, \tag{10.36}$$

where ψ_D is an open-channel function. The functions ψ_0 and ψ_C are eigenfunctions of the isospin operator with different eigenvalues; they describe the analogue state and the normal states in the compound region, respectively. As discussed above, the analogue state has a relatively simple structure and is, therefore, coupled rather strongly to the open channel. On the other hand, the normal states, which have the same spin and parity as the analogue state, have complicated structures; consequently, their coupling to the open channel is usually quite weak. We choose to split ψ in this particular way, because in so doing the resultant scattering amplitudes or S-matrix elements will exhibit the influence of ψ_0 in an especially clear manner.

By substituting eq. (10.36) into the projection equation (2.3), one obtains

$$\langle \delta \psi_D | H - E | \psi_D \rangle + b_0 \langle \delta \psi_D | H - E | \psi_0 \rangle + \langle \delta \psi_D | H - E | \psi_C \rangle = 0, \tag{10.37a}$$

$$\langle \psi_0 | H - E | \psi_D \rangle + b_0 \langle \psi_0 | H - E | \psi_0 \rangle + \langle \psi_0 | H - E | \psi_C \rangle = 0, \tag{10.37b}$$

$$\langle \delta \psi_C | H - E | \psi_D \rangle + b_0 \langle \delta \psi_C | H - E | \psi_0 \rangle + \langle \delta \psi_C | H - E | \psi_C \rangle = 0. \tag{10.37c}$$

If one makes now a partial-wave expansion for ψ_D, then one can select the wave which has the same spin and parity as ψ_0 and ψ_C. In the following, we shall always work in a particular (j, π) state but, for brevity, omit these indices in the formulation, whenever no confusion may arise.

To find the S-matrix elements from the asymptotic behaviour of ψ_D, we first determine ψ_D as a function of ψ_C and ψ_0. This can be achieved by introducing the resolvent operator G_D^+, defined by the equation

$$\langle \delta \psi_D | H - E | G_D^+ = \langle \delta \psi_D |, \tag{10.38}$$

with outgoing-wave boundary conditions. The result is

$$| \psi_D \rangle = | \psi_D^+ \rangle - b_0 G_D^+ | H - E | \psi_0 \rangle - G_D^+ | H - E | \psi_C \rangle, \tag{10.39}$$

where ψ_D^+ is as usual a solution of the homogeneous equation

$$\langle \delta \psi_D | H - E | \psi_D^+ \rangle = 0, \tag{10.40}$$

and is matched to the incoming wave in the elastic channel. By substituting eq. (10.39) into eqs. (10.37b) and (10.37c), we then obtain the following equations for the determination of b_0 and ψ_C:

$$b_0 \langle \psi_0 | \hat{h} | \psi_0 \rangle + \langle \psi_0 | \hat{h} | \psi_C \rangle = - \langle \psi_0 | H - E | \psi_D^+ \rangle, \tag{10.41a}$$

$$b_0 \langle \delta \psi_C | \hat{h} | \psi_0 \rangle + \langle \delta \psi_C | \hat{h} | \psi_C \rangle = - \langle \delta \psi_C | H - E | \psi_D^+ \rangle, \tag{10.41b}$$

with

$$\hat{h} = (H - E) - (H - E) G_D^+ (H - E). \tag{10.42}$$

Next, we separate G_D^+ into a hermitian and an antihermitian part, and use the hermitian part G_D^P to define a set of orthonormalized basis functions ψ_i $(i = 1, 2, \ldots)$. These functions are solutions of the homogeneous equation

$$\langle \delta \psi_C | H - (H - E) G_D^P (H - E) - E_i (E) | \psi_i \rangle = 0 \tag{10.43}$$

with eigenvalues E_i. In terms of ψ_i, the function ψ_C can be expanded; i.e.,

$$\psi_C = \sum_i b_i \psi_i. \tag{10.44}$$

With this expansion for ψ_C, eqs. (10.41a) and (10.41b) become

$$(E_0 - E - i \pi \gamma_0 \gamma_0^*) b_0 + \sum_i (V_i^* - i \pi \gamma_0 \gamma_i^*) b_i = - \gamma_0, \tag{10.45a}$$

$$(V_k - i \pi \gamma_k \gamma_0^*) b_0 + \sum_i [\delta_{ki} (E_i - E) - i \pi \gamma_k \gamma_i^*] b_i = - \gamma_k, \quad (k = 1, 2, \ldots) \tag{10.45b}$$

where

$$\gamma_0 = \langle \psi_0 | H - E | \psi_D^+ \rangle, \tag{10.46}$$

$$\gamma_i = \langle \psi_i | H - E | \psi_D^+ \rangle, \quad (i = 1, 2, \ldots) \tag{10.47}$$

$$E_0 = \langle \psi_0 | H - (H - E) G_D^P (H - E) | \psi_0 \rangle, \tag{10.48}$$

$$E_i = \langle \psi_i | H - (H - E) G_D^P (H - E) | \psi_i \rangle, \quad (i = 1, 2, \ldots) \tag{10.49}$$

$$V_i = \langle \psi_i | (H - E) - (H - E) G_D^P (H - E) | \psi_i \rangle, \quad (i = 1, 2, \ldots) \tag{10.50}$$

The determination of b_0 and b_i from eqs. (10.45a) and (10.45b) proceeds now in the same way as the determination of b_λ (given by eq. (9.54)) from eq. (9.52). The result is as follows:

$$b_0 = \frac{-\gamma_0 + \left(1 - i\pi \sum_i \frac{\gamma_i \gamma_i^*}{E_i - E}\right)^{-1} \left(\sum_i \frac{V_i^* - i\pi\gamma_i^*\gamma_0}{E_i - E}\gamma_i\right)}{E_0 - E - i\pi\left(\gamma_0 - \sum_i \frac{V_i^*\gamma_i}{E_i - E}\right)\left(\gamma_0^* - \sum_i \frac{V_i\gamma_i^*}{E_i - E}\right)\left(1 - i\pi\sum_i \frac{\gamma_i\gamma_i^*}{E_i - E}\right)^{-1} - \sum_i \frac{V_i V_i^*}{E_i - E}} \quad \ldots (10.51a)$$

$$b_k = -\frac{\gamma_k}{E_k - E}\left(1 - i\pi\sum_i \frac{\gamma_i\gamma_i^*}{E_i - E}\right)^{-1}$$

$$- b_0\left[\frac{i\pi\gamma_k}{E_k - E}\left(\sum_i \frac{V_i - i\pi\gamma_i\gamma_0^*}{E_i - E}\gamma_i^*\right)\left(1 - i\pi\sum_i \frac{\gamma_i\gamma_i^*}{E_i - E}\right)^{-1} \right. \qquad (10.51b)$$

$$\left. + \frac{V_k - i\pi\gamma_k\gamma_0^*}{E_k - E}\right], \qquad (k = 1, 2, \ldots)$$

If one substitutes eqs. (10.44), (10.51a), and (10.51b) into eq. (10.39), then one can obtain from the asymptotic behaviour of eq. (10.39) the S-matrix element for pure elastic-scattering.

Before we do this, we wish to make one remark. At this moment, it might seem as if we had not yet utilized the assumption that only the elastic channel is open. But this is not so, because in separating the operator $\hat{h}$ into a hermitian and an antihermitian part for the derivation of eqs. (10.45a) and (10.45b), we have used the fact that the anti-hermitian part of $\hat{h}$ has the form $-i\pi(H - E)|\psi_D^+\rangle\langle\psi_D^+|(H - E)$. If there are several channels open, then the operator $|\psi_D^+\rangle\langle\psi_D^+|$ has to be replaced by the operator $\Sigma_i|\psi_{Di}^+\rangle\langle\psi_{Di}^+|$, where ψ_{Di}^+ is a wave function describing a reaction initiated through channel i.† This will usually complicate the calculation very much, and we shall come back to this point later.

We return now to the derivation of the S-matrix element. In analogy with our earlier considerations given in sections 9.2 and 9.3, we obtain from the asymptotic behaviour of ψ_D the following expression for this element (i.e., the amplitude of the outgoing wave in the asymptotic region):

$$S = e^{2i\delta}\left[1 - 2\pi i\left(b_0\gamma_0^* + \sum_i b_i\gamma_i^*\right)\right]. \qquad (10.52)$$

† In other words, the ψ_{Di}^+'s are solutions of the same homogeneous equation, but with different asymptotic boundary conditions.

Substituting eqs. (10.51a) and (10.51b) into eq. (10.52) then yields

$$S = e^{2i\delta}\left[\frac{1 + i\pi\sum_i \frac{\gamma_i\gamma_i^*}{E_i - E}}{1 - i\pi\sum_i \frac{\gamma_i\gamma_i^*}{E_i - E}}\right]$$

$$+ 2\pi i \frac{\left(\gamma_0 - \sum_i \frac{V_i^*\gamma_i}{E_i - E}\right)\left(\gamma_0^* - \sum_i \frac{V_i\gamma_i^*}{E_i - E}\right)\left(1 - i\pi\sum_i \frac{\gamma_i\gamma_i^*}{E_i - E}\right)^{-2}}{E_0 - E - i\pi\left(\gamma_0 - \sum_i \frac{V_i^*\gamma_i}{E_i - E}\right)\left(\gamma_0^* - \sum_i \frac{V_i\gamma_i^*}{E_i - E}\right)\left(1 - i\pi\sum_i \frac{\gamma_i\gamma_i^*}{E_i - E}\right)^{-1} - \sum_i \frac{V_i V_i^*}{E_i - E}}$$

... (10.53)

Even though the expression for S looks complicated, it can be shown to be a pure phase factor, as it of course must be. Also, we should mention that the formulae derived so far in this subsection are not restricted in their application to a description of the mutual influence between analogue and normal resonances. They can in fact be applied to any pure-elastic-scattering case where there is a group of close-lying levels in the compound nucleus, coupled to each other through the action of certain components in the nucleon-nucleon potential.

In the special case where the resonance states ψ_i are coupled only indirectly to the elastic-scattering channel via the resonance state ψ_0, the expression for S is much simpler. Under this assumption the quantity γ_i and the term $\langle \psi_i|(H - E)\, G_D^P\, (H - E)|\psi_0\rangle$ in V_i are equal to zero, and one obtains

$$S = e^{2i\delta}\left[\frac{E_0 - E - \sum_i \frac{V_i V_i^*}{E_i - E} + i\pi\gamma_0\gamma_0^*}{E_0 - E - \sum_i \frac{V_i V_i^*}{E_i - E} - i\pi\gamma_0\gamma_0^*}\right]. \tag{10.54}$$

This is eq. (9.3.11) in the book of *Mahaux* and *Weidenmüller* [MA 69], derived in the framework of the shell-model reaction theory for the case of pure elastic-scattering and under the assumption that between the resonance levels and the elastic channel there is only indirect coupling through a doorway state. In our case eq. (10.54), used in conjunction with eqs. (10.46)–(10.50), is of course applicable not only in the case of proton elastic scattering but also in the case when a composite particle is employed as the incident particle. For instance, the example considered in subsection 10.2b, where we discuss $\alpha + \alpha$ scattering by assuming that the direct coupling between the T = 1 compound configuration and the elastic channel can be neglected, is a case belonging to this latter category (see eq. (10.23)).

We now use eq. (10.53) to discuss briefly the properties of analogue resonances. As has already been mentioned, the analogue resonance, in contrast with other resonances, is coupled relatively strongly to the elastic channel.† This means that the energy width $2\pi\gamma_0\gamma_0^*$ is much larger than the energy width $2\pi\gamma_i\gamma_i^*$ $(i = 1, 2, \ldots)$. In addition, it is noted that in heavy nuclei the density of normal states with the same quantum numbers as the analogue state (except isospin) is usually so large that many such normal states are present within the energy width $2\pi|\gamma_0|^2$. Therefore, to see the influence of these narrow but numerous resonances on the scattering cross section in the energy region around the analogue resonance, it is useful to perform an average over an energy interval I centered at E. This interval I should be chosen to be much smaller than the width $2\pi|\gamma_0|^2$, but much larger than both the width $2\pi|\gamma_i|^2$ and the nearest-neighbour spacing between normal states; in other words, many normal states should be contained in the interval I. In performing this average for a certain function F(E), we adopt the following definition:

$$\langle F(E)\rangle_I = \frac{I}{\pi}\int_{-\infty}^{\infty} \frac{F(E')}{(E-E')^2 + I^2}\, dE', \tag{10.55}$$

i.e., we use for the definition of $\langle F(E)\rangle_I$ a Lorentzian distribution as the weight function [BR 59].

Applying the averaging procedure to the elastic-scattering S-matrix element, we obtain

$$\langle S(E)\rangle_I = \frac{I}{\pi}\int_{-\infty}^{\infty} \frac{S(E')}{(E-E')^2 + I^2}\, dE'. \tag{10.56}$$

For the evaluation of the integral involved in the above equation, we have to examine the analytic properties of S(E) in the complex energy plane. It can be shown that S(E) has only poles in the lower half of this plane.†† Furthermore, by neglecting threshold effects, the branch points can be assumed to lie infinitely far away from the energy region of interest. Using these analytic properties, we can then carry out the integration in eq. (10.56) by closing the contour of the integral along a large semi-circle in the upper half of the complex energy plane. The result is

$$\langle S(E)\rangle_I = S(E + iI). \tag{10.57}$$

Thus, to obtain $\langle S(E)\rangle_I$ from eq. (10.53), one simply substitutes E by E + iI at every place in eq. (10.53) where E appears.†††

---

† We assume that the analogue state under consideration is energetically well separated from all other analogue states. Due to the fact that, in the parent nucleus, the corresponding states lie in the low-excitation region, this assumption is usually quite valid.

†† This is indicated, for example, by eqs. (9.17) and (9.55). Especially when one considers narrow resonances and thereby neglects the energy dependence of the shift and width functions, one can see immediately that the poles of these elastic-scattering S-matrix elements lie in the lower half of the complex energy plane. As to the physical reason for this, we shall discuss in a later chapter where we consider some time-dependent problems.

††† We assume that the quantities $\delta(E)$, γ_0, γ_i, E_0, E_i, and V_i vary smoothly with energy and can be considered as constants in the energy region of interest.

To proceed further, we consider a simple "picket-fence" model [MO 67], which is useful for an examination of the essential features of analogue resonances. In this model we assume that, in the energy region of interest,

$$\begin{aligned} \gamma_i &= \gamma, \\ V_i &= V, \\ E_{i+1} - E_i &= d, \end{aligned} \tag{10.58}$$

for all values of i; i. e., we assume that the normal states are uniformly spaced and the coupling matrix elements between the normal states and the analogue state or the elastic channel are equal to one another. Based on these assumptions, we obtain then for a term of the type

$$\sum_i \frac{c_i}{E_i - (E + iI)}, \quad (c_i = \gamma_i \gamma_i^*, \ldots)$$

which appears in the expression for $\langle S(E) \rangle_I$, the following result:

$$\sum_i \frac{c_i}{E_i - (E + iI)} \approx \sum_{n=-\infty}^{\infty} \frac{c}{nd - iI} \approx \int_{-\infty}^{\infty} \frac{c}{d} \frac{dx}{x - i\frac{I}{d}} = \frac{c}{d} i\pi. \tag{10.59}$$

Equation (10.59) shows that, in the limit of a very large level density for the normal states, the sum in eq. (10.59) becomes independent of I. In addition, we note that, even for real c_i, the right side acquires a non-real value. This has the consequence that $\langle S(E) \rangle_I$ is no longer a pure phase factor, in contrast to the quantity S(E). One sees this in an especially clear way by examining again the special case represented by eq. (10.54). By defining the escape width $\Gamma^\uparrow$ as

$$\Gamma^\uparrow = 2\pi \gamma_0 \gamma_0^* \tag{10.60}$$

and the spreading width $\Gamma^\downarrow$ as

$$\Gamma^\downarrow = -2i \sum_i \frac{V_i V_i^*}{E_i - (E + iI)} \approx 2\pi \frac{|V|^2}{d}, \tag{10.61}$$

we obtain

$$\begin{aligned} \langle S(E) \rangle_I &= e^{2i\delta} \frac{E_0 - E + \frac{1}{2} i (\Gamma^\uparrow - \Gamma^\downarrow - 2I)}{E_0 - E - \frac{1}{2} i (\Gamma^\uparrow + \Gamma^\downarrow + 2I)} \\ &\approx e^{2i\delta} \frac{E_0 - E + \frac{1}{2} i (\Gamma^\uparrow - \Gamma^\downarrow)}{E_0 - E - \frac{1}{2} i (\Gamma^\uparrow + \Gamma^\downarrow)}, \end{aligned} \tag{10.62}$$

where we have used the fact that I is much smaller than $\Gamma^\uparrow$. From eq. (10.62), one notes that $|\langle S(E) \rangle_I|$ is smaller than 1; this means that, by performing the averaging procedure, some absorption is introduced into the elastic channel. This absorption is connected with

$\Gamma^{\downarrow}$, which comes from the coupling of ψ_0 to the normal states and describes the transitions from ψ_0 to these states.†

We go back now to the discussion of eq. (10.53) concerning analogue resonances. Using eqs. (10.57) and (10.59), we obtain the following expression for $\langle S(E)\rangle_I$ in the picket-fence model:

$$\langle S(E)\rangle_I \approx e^{2i\delta}\left[\frac{1-\pi^2\frac{|\gamma|^2}{d}}{1+\pi^2\frac{|\gamma|^2}{d}} + 2\pi i\,\frac{\left(\gamma_0 - i\pi\frac{V^*\gamma}{d}\right)\left(\gamma_0^* - i\pi\frac{V\gamma^*}{d}\right)\left(1+\pi^2\frac{|\gamma|^2}{d}\right)^{-2}}{E_0 - E - iI - i\pi\left(\gamma_0 - i\pi\frac{V^*\gamma}{d}\right)\left(\gamma_0^* - i\pi\frac{V\gamma^*}{d}\right)\left(1+\pi^2\frac{|\gamma|^2}{d}\right)^{-1} - i\pi\frac{|V|^2}{d}}\right] \qquad \ldots(10.63)$$

By defining the real quantities

$$Y_C = \pi^2\frac{|\gamma|^2}{d}, \qquad (10.64)$$

$$\Gamma = 2\pi\left[\left(\gamma_0\gamma_0^* - \pi^2\frac{|\gamma V|^2}{d^2}\right)\left(1+\pi^2\frac{|\gamma|^2}{d}\right)^{-1} + \frac{|V|^2}{d}\right], \qquad (10.65)$$

$$\Delta = \pi^2\left(\frac{\gamma_0 V\gamma^* + \gamma_0^* V^*\gamma}{d}\right)\left(1+\pi^2\frac{|\gamma|^2}{d}\right)^{-1}, \qquad (10.66)$$

and the complex quantity

$$\widetilde{\Gamma}_0 = 2\pi\left(\gamma_0 - i\pi\frac{V^*\gamma}{d}\right)\left(\gamma_0^* - i\pi\frac{V\gamma^*}{d}\right)\left(1+\pi^2\frac{|\gamma|^2}{d}\right)^{-2}, \qquad (10.67)$$

eq. (10.63) assumes the form

$$\langle S(E)\rangle_I \approx e^{2i\delta}\left(\frac{1-Y_C}{1+Y_C} + i\,\frac{\widetilde{\Gamma}_0}{E_0 - E - \Delta - i\frac{\Gamma}{2}}\right). \qquad (10.68)$$

In writing eq. (10.68) we have used the relation $I \ll \Gamma$, which comes from the requirement that I is much smaller than $2\pi|\gamma_0|^2$. Here again, one can show that $\langle S(E)\rangle_I$ has a magnitude less than unity, just as in the special case discussed above.

The expression for $\langle S(E)\rangle_I$ given by eq. (10.68) is the same as that derived in the framework of the shell-model reaction theory (see eq. (13.8.2) of ref. [MA 69]). The only point we wish to emphasize again is that the formulae obtained here are valid even in the case where the incident particle is not a single nucleon but a composite particle.

It can be shown from eq. (10.68) that $|\langle S(E)\rangle_I|$ as a function of E shows an asymmetric behaviour around the resonance energy $(E_0 - \Delta)$. This follows from the fact that the

† For an interesting discussion of the physical meaning of $\Gamma^{\uparrow}$ and $\Gamma^{\downarrow}$, see ref. [MA 69].

quantity $\widetilde{\Gamma}_0$ is in general complex. If the condition is such that $\widetilde{\Gamma}_0$ turns out to be purely real, then this asymmetric behaviour will not be present. From eq. (10.67) one sees that this happens when either V or γ becomes zero. In the $\gamma = 0$ case, the normal states are not coupled directly to the incident channel and the lack of asymmetry can be clearly seen from eq. (10.62). In the case where V = 0, the normal states and the analog state are completely uncoupled and again no asymmetry appears. Therefore, in the case of analogue resonances, the asymmetric behaviour of $|\langle S(E)\rangle_I|$ is produced by an intricate interplay between the coupling of the analogue state and the normal states with the incident channel and the mutual coupling of the analogue state with the normal states. By examining eq. (10.50) one notes that, for V to be nonzero, it is not essential that there must be a direct coupling between the analogue and the normal states. Even if $\langle \psi_i | H - E | \psi_0 \rangle$ happens to be negligibly small, the asymmetric behaviour of $|\langle S(E)\rangle_I|$ will still be quite noticeable as long as the term $\langle \psi_i | (H - E) G_D^P (H - E) | \psi_0 \rangle$ has an appreciable magnitude or, in other words, as long as there is a significant amount of indirect coupling between these states via the incident channel.

In reality, the coupling matrix element $\langle \psi_i | H - E | \psi_0 \rangle$, to which only the weak Coulomb interaction contributes, has usually a much smaller magnitude than the coupling matrix element $\langle \psi_i | (H - E) G_D^P (H - E) | \psi_0 \rangle$. This comes from the fact that the incident channel consists of a mixture of isospin eigenstates.† Therefore, due to the fact that the functions $\psi_D^\beta (E)$ (see eqs. (9.11) and (9.16)) appearing in the operator

$$G_D^P = \underset{\beta}{S} P \frac{|\psi_D^\beta (E)\rangle \langle \psi_D^\beta (E)|}{E_D^\beta - E} \tag{10.69}$$

are a mixture of different isospin configurations, cross terms of different isospins will appear in $|\psi_D^\beta (E)\rangle \langle \psi_D^\beta (E)|$, which contribute strongly to the indirect-coupling matrix element $\langle \psi_i | H - E) G_D^P (H - E) | \psi_0 \rangle$. The extreme case where the coupling between the analogue state ψ_0 and the normal states ψ_i occurs only through the operator $(H - E) G_D^P (H - E)$ is commonly referred to as the case of pure external-mixing.

It has already been mentioned several times that the treatment of analogue resonances given here is not restricted to the case involving a single-nucleon channel. In practice, however, it is quite likely that, in elastic-scattering processes, the only composite particle which can be used as an incident particle to study analogue resonances is the nucleus ^{3}He. The reason for this is two-fold. First, for the scattering on a heavy nucleus with $N > Z$, it is necessary to employ a projectile with $T_3 < 0$ in order to reach an analogue state which is characterized by a value of T larger than the value of T_3.†† Second, for projectiles heavier than ^{3}He and with $T_3 < 0$, the compound nucleus becomes so highly excited that analogue resonances, which correspond to low-excited states in the parent nucleus and, therefore, have relatively low excitation energies, can no longer be observed by the process of elastic scattering.

† For example, in proton elastic scattering on a target nucleus having an isospin $T > 0$, it consists of a mixture of $T_> = T + \frac{1}{2}$ and $T_< = T - \frac{1}{2}$ configurations.

†† This statement is made under the assumption that one does not consider complicated rearrangement processes which can occur through the influence of the weak Coulomb interaction.

10.3c. Brief Discussion in the Case of Many Open Channels

Until now, we have considered the case where only the elastic channel is open. Especially in heavy nuclei, this is never the case. There, besides the proton elastic channel, at least neutron channels are open. In this subsection, we shall indicate briefly how one can generalize the considerations of subsection 10.3b to the case where more than one open channel exists.

In the many-channel case, the formulation is in general very complicated. It becomes fairly simple only if one makes the assumption that the direct coupling between different open channels can be neglected (see subsection 9.4a). Under this assumption, which is fortunately justified in good approximation in many instances, the resolvent operator G_D^+ takes on the relatively simple form

$$G_D^+ = \underset{\beta, i}{S} \frac{|\psi_{Di}^\beta\rangle\langle\psi_{Di}^\beta|}{E_{Di}^\beta - E - i\epsilon}, \quad (\epsilon > 0) \tag{10.70}$$

where i is a channel index and the functions ψ_{Di}^β are solutions of the homogeneous equation

$$\langle\delta\psi_{Di}|H - E_{Di}^\beta|\psi_{Di}^\beta\rangle = 0. \quad (i = 1, 2, \ldots) \tag{10.71}$$

With G_D^+ given by eq. (10.70) one can then proceed, starting from eq. (10.39), in the same way as in the single-channel case discussed in the previous subsection.

When the normal resonances overlap each other very strongly, one cannot neglect the direct coupling between different channels. However, just because of the large overlap of these resonances, the formulation is again simplified since the procedure of energy-averaging represented by eq. (10.56) is no longer required.†

The physical consequences in the case of many open channels are essentially the same as those in the case of a single open channel. The main difference is that the energy dependence of the S-matrix element $S_{11}(E)$, which describes elastic scattering in the incident channel, is already smoothed out somewhat by the presence of reaction channels, and the absolute value of $S_{11}(E)$ is already smaller than 1 even before one performs the energy-average over the normal resonances.

In concluding this chapter, we should mention that even though the consideration given here about analogue resonances is quite brief, all the essential ideas are presented. For a more detailed discussion, we refer to, as has been stated before, the book of *Mahaux* and *Weidenmüller* [MA 69].

---

† For a brief discussion of the excitation of analogue resonances in direct reactions, see chapter 12.

11. Optical-Model Potentials for Composite Particles

11.1. General Remarks

The optical model is very successful not only in the description of nucleon-nucleus elastic scattering, but also in the description of the elastic scattering of composite particles by nuclei [HO 71]. In the case of nucleon-nucleus scattering at low energies, the scattering cross section displays isolated resonance peaks and the optical model yields the shape-elastic part of the cross section averaged over a certain energy interval. At higher energies where many reaction channels are open and compound resonances overlap, the cross section becomes slowly varying with energy. Then, the optical model reproduces the full cross section, since the fluctuating part of the cross section is now negligibly small.

For the scattering of composite particles, e. g., α particles or ^{3}He, by medium-heavy and heavy nuclei, the excitation energy of the compound nucleus is already so large at rather low energies that many neutron and proton channels are open and compound resonances overlap strongly. As a consequence, the scattering cross section exhibits a relatively smooth behaviour in its energy variation even in the low-energy region. In addition, it should be mentioned that, in the case of elastic scattering with composite incident particles, the Coulomb barrier is normally so high that, over a large angular region, the scattering cross sections at low energies differ only slightly from the corresponding Rutherford cross sections.

If the target particle is a light nucleus, the situation is different. Here, even when the compound nucleus is rather highly excited, still only few channels are open. In addition, the compound nucleus is often inhibited from decaying into these channels by approximate selection rules. Therefore, in the case of elastic scattering of composite particles by light nuclei, one often observes resonance peaks in the scattering cross section. For example, in the excitation-energy region around 17 MeV in ^{5}He, a resonance peak has been observed in both n + α and d + t elastic scattering.

In section 11.2 we first discuss the optical-model potential in the case where only the elastic channel is open. This discussion follows closely that of *Feshbach* for nucleon-nucleus scattering [FE 58, FE 62], but is equally valid in the case where the incident particles are composite particles. Next, we consider the essential changes which come about when, besides the elastic channel, many other channels are open. Also, in this section, a brief discussion will be given as to the question why a composite particle can penetrate into the target nucleus without being dissolved immediately in the nuclear surface region. This discussion will be useful when one tries to estimate the magnitude of the imaginary part of the optical potential.

Specific examples will be given in section 11.3; there, we shall see that, by making use of the generator-coordinate technique described in subsection 5.2b, even rather large systems can be considered. Finally, in section 11.4, we discuss general features of the effective potential between two clusters, with particular attention being paid to a discussion of its dependence on orbital angular momentum.

11.2. Optical-Model Description of Elastic-Scattering Processes

11.2a. Preliminary Remarks About the Optical-Model Potential

As has been mentioned in section 8.4 we assume, for clarity in discussion, that the interacting nuclei have no spin and no charge. Under this assumption, the elastic-scattering cross section in channel i is given by

$$\sigma_{ii} = \frac{\pi(2l+1)}{k_i^2} |S_{ii}(E) - 1|^2, \tag{11.1}$$

where, for brevity, we have omitted in σ_{ii} and S_{ii} the orbital angular-momentum index l. If we now use the definition given by eq. (10.55) to average σ_{ii} over a small energy interval I which contains many compound resonances, then we obtain

$$\langle \sigma_{ii}(E) \rangle_I = \frac{\pi(2l+1)}{k_i^2} [\langle |S_{ii}(E)|^2 \rangle_I - 2\,\mathrm{Re}\langle S_{ii}(E) \rangle_I + 1]. \tag{11.2}$$

By assuming that the relative kinetic energy of the two clusters in channel i is much larger than I, the relation

$$\langle S_{ii}(E) \rangle_I = S_{ii}(E + iI) \tag{11.3}$$

is again valid. Using this relation, we can write eq. (11.2) as

$$\langle \sigma_{ii}(E) \rangle_I = \sigma_{ii}^{se}(E) + \sigma_{ii}^{ce}(E), \tag{11.4}$$

where $\sigma_{ii}^{se}(E)$ is the shape-elastic cross section given by

$$\sigma_{ii}^{se}(E) = \frac{\pi(2l+1)}{k_i^2} |S_{ii}(E + iI) - 1|^2, \tag{11.5}$$

and $\sigma_{ii}^{ce}(E)$ is the compound-elastic or fluctuation cross section given by

$$\sigma_{ii}^{ce}(E) = \frac{\pi(2l+1)}{k_i^2} [\langle |S_{ii}(E)|^2 \rangle_I - |\langle S_{ii}(E) \rangle_I|^2]. \tag{11.6}$$

From eq. (11.6) one sees that if $S_{ii}(E)$ is approximately energy-independent in the energy interval I, then the fluctuation cross section will be vanishingly small. This is always the case if, in the scattering energy region under consideration, the resonance levels of the compound nucleus overlap each other strongly.

The optical-model potential V_i in channel i is defined as an effective intercluster interaction potential which has a scattering function $\bar{S}_i(E)$ satisfying the condition

$$\bar{S}_i(E) = \langle S_{ii}(E) \rangle_I = S_{ii}(E + iI). \tag{11.7}$$

In other words, it is chosen to yield the shape-elastic part of the energy-averaged scattering cross section. Now, since the S-matrix is unitary and since

$$|\langle S_{ii}(E) \rangle_I| \leqslant \max\{|S_{ii}(E)|\}, \tag{11.8}$$

one finds that

$$|\bar{S}_i(E)| \leqslant 1. \tag{11.9}$$

This means that $\bar{S}_i(E)$ is in general not unitary and V_i is in general not hermitian. As is easily seen, this lack of hermiticity for V_i has two origins: it arises from the energy dependence of $S_{ii}(E)$ caused by the presence of compound resonances and from the existence of reaction channels.

Because $|\bar{S}_i(E)| \leqslant 1$, the optical potential V_i contains in general an imaginary part which describes the absorption of clusters from the incident channel. This imaginary part will vanish only when the equality sign in eq. (11.9) holds. This corresponds to the case where the elastic channel is the only open channel and where the quantity $S_{ii}(E)$ is independent of energy in an interval larger than I.

It should also be mentioned that eq. (11.7) does not determine V_i uniquely. It only determines the asymptotic behaviour of the optical-model wave function, i.e., of the wave function obtained from a two-cluster Schrödinger equation containing the optical-model potential as an interaction potential.

11.2b. Optical-Model Potential for Pure Elastic Scattering

As always, the starting point for our derivation is the projection equation

$$\langle \delta \psi | H - E | \psi \rangle = 0. \tag{11.10}$$

In the present case where the elastic channel (denoted as channel i) is the only open channel, we write

$$\psi = \psi_D + \psi_C, \tag{11.11}$$

where ψ_C describes bound structures in the compound region and ψ_D is an open-channel function, given as in eq. (9.2) by

$$\psi_D = A' \{\hat{\phi}(\tilde{A})\, \hat{\phi}(\tilde{B})\, F(\mathbf{R})\} = \int A' \{\hat{\phi}(\tilde{A})\, \hat{\phi}(\tilde{B})\, \delta(\mathbf{R} - \mathbf{R}'')\}\, F(\mathbf{R}'')\, d\mathbf{R}'', \tag{11.12}$$

with $\mathbf{R} = \mathbf{R}_A - \mathbf{R}_B$. Proceeding now as in subsection 9.2a, we obtain

$$\langle \delta \psi_D | \tilde{H} - E | \psi_D \rangle = 0, \tag{11.13}$$

where

$$\tilde{H} = H - \sum_\alpha \frac{(H - E) |\psi_C^\alpha\rangle \langle \psi_C^\alpha| (H - E)}{E_C^\alpha - E}. \tag{11.14}$$

If we substitute ψ_D of eq. (11.12) and a similar expression for $\delta\psi_D$ into eq. (11.13), we find then

$$\int H(\mathbf{R}', \mathbf{R}'')\, F(\mathbf{R}'')\, d\mathbf{R}'' = 0, \tag{11.15}$$

with

$$H(\mathbf{R}', \mathbf{R}'') = \langle \hat{\phi}(\tilde{A})\, \hat{\phi}(\tilde{B})\, \delta(\mathbf{R} - \mathbf{R}') | \tilde{H} - E | A' \{\hat{\phi}(\tilde{A})\, \hat{\phi}(\tilde{B})\, \delta(\mathbf{R} - \mathbf{R}'')\}\rangle. \tag{11.16}$$

From $H(\mathbf{R}', \mathbf{R}'')$ one can obtain an effective nonlocal but hermitian "microscopic" optical potential $V_i^{\mu}(\mathbf{R}', \mathbf{R}'')$ between the two nuclei A and B. Because of the presence of the factor $(E_C^{\alpha} - E)^{-1}$ in $\widetilde{H}$, this microscopic optical potential is a rapidly varying function of energy and is, therefore, different from the optical potential V_i which yields the energy-averaged elastic-scattering S-matrix element $\overline{S}_i(E)$.†

The optical potential V_i cannot be obtained by simply energy-averaging the microscopic optical potential V_i^{μ} because, in contrast to V_i, V_i^{μ} operates on the function $F(\mathbf{R}'')$ which varies rapidly with energy. Therefore, to derive V_i in the case of pure elastic-scattering, one must adopt another procedure. This procedure will be discussed below, where we shall first perform a partial-wave expansion of the scattering wave function and thereby derive the optical potential in a given orbital angular-momentum state.

In the presence of many compound levels, the wave function ψ_D has the form (see section 9.3) ††

$$|\psi_D(E)\rangle = |\psi_D^+(E)\rangle - \sum_{\lambda=1}^{n_0} G_D^+(E)\, |H - E|\psi_C^{\lambda}\rangle\, b_{\lambda}, \tag{11.17}$$

where n_0 is chosen to be much larger than the number of compound levels contained in the interval I used in the energy-averaging procedure, and b_{λ} is given by

$$b_{\lambda} = \frac{\gamma_{\lambda}}{E_{\lambda} - E}\, \frac{1}{i\pi \displaystyle\sum_{\mu=1}^{n_0} \frac{\gamma_{\mu}\gamma_{\mu}^*}{E_{\mu} - E} - 1}, \tag{11.18}$$

with E_{λ} and γ_{λ} defined by eqs. (9.50) and (9.53), respectively. From the asymptotic behaviour of ψ_D, one obtains the elastic-scattering S-matrix element

$$S_{ii}(E) = e^{2i\delta(E)}\, \frac{1 + i\pi \displaystyle\sum_{\lambda=1}^{n_0} \frac{\gamma_{\lambda}\gamma_{\lambda}^*}{E_{\lambda} - E}}{1 - i\pi \displaystyle\sum_{\lambda=1}^{n_0} \frac{\gamma_{\lambda}\gamma_{\lambda}^*}{E_{\lambda} - E}}. \tag{11.19}$$

† In the case where reaction channels are present (see subsection 11.2c), the situation is somewhat different. There, if the coupling of the compound states to the reaction channels is so strong that the compound states appreciably overlap, V_i^{μ} and V_i will become approximately equal to each other.

†† For simplicity, we omit all nonessential indices, such as the orbital angular-momentum index, the double prime on ψ_D^+, and so on. These indices will be properly written out, only if confusion may otherwise arise.

If we now perform the procedure of energy-averaging, then we find by using eq. (11.3) that the energy-averaged S-matrix element is given by

$$\langle S_{ii}(E)\rangle_I = e^{2i\delta(E)} \frac{1 + i\pi \sum_{\lambda=1}^{n_0} \frac{\gamma_\lambda \gamma_\lambda^*}{E_\lambda - E - iI}}{1 - i\pi \sum_{\lambda=1}^{n_0} \frac{\gamma_\lambda \gamma_\lambda^*}{E_\lambda - E - iI}}. \tag{11.20}$$

In obtaining eq. (11.20), we have again assumed that $\delta(E)$, $\gamma_\lambda(E)$, and $E_\lambda(E)$ have a smooth energy dependence in the energy interval I.

We consider now the case where the compound levels are narrow, i. e., $2\pi|\gamma_\lambda|^2 \ll d$, with d being the average nearest-neighbour spacing. This particular case is chosen for a detailed study, because the influence of compound resonances on the shape-elastic and compound-elastic cross sections can be seen in an especially clear manner. Now, since n_0 is chosen to be much larger than I/d which is itself a large number, we find by using eq. (10.59) that the quantity $-i\pi\,\Sigma\,[\gamma_\lambda\gamma_\lambda^*/(E_\lambda - E - iI)]$ has a similar order of magnitude as the quantity $\pi^2|\gamma_\lambda|^2_{av}/d$, with $|\gamma_\lambda|^2_{av}$ being the average value of $|\gamma_\lambda|^2$ for the compound levels around E. Therefore, in this particular case, $-i\pi\,\Sigma\,[\gamma_\lambda\gamma_\lambda^*/(E_\lambda - E - iI)]$ has a small value and we can write $\langle S_{ii}(E)\rangle_I$ approximately as

$$\begin{aligned}\langle S_{ii}(E)\rangle_I &\approx e^{2i\delta(E)}\left(1 + 2\pi i \sum_{\lambda=1}^{n_0} \frac{\gamma_\lambda\gamma_\lambda^*}{E_\lambda - E - iI}\right)\\ &= e^{2i\delta(E)}\left(1 - \frac{2\pi}{I}\sum_{\lambda=1}^{n_0} g_\lambda(E)\,\gamma_\lambda\gamma_\lambda^*\right)\\ &= e^{2i\delta(E)}\left(1 - \frac{2\pi}{I}\sum_{\lambda=1}^{n_0} g_\lambda(E)\,\langle\psi_D^+|H-E|\psi_C^\lambda\rangle\langle\psi_C^\lambda|H-E|\psi_D^+\rangle\right),\end{aligned} \tag{11.21}$$

where

$$g_\lambda(E) = -\frac{iI}{E_\lambda - E - iI} \tag{11.22}$$

is a smoothly varying function of E and is equal to 1 for $E = E_\lambda$. From eq. (11.21) we can immediately deduce that the desired $\langle S_{ii}(E)\rangle_I$ can be obtained from a wave function $\overline{\psi}_D$ of the following form:

$$|\overline{\psi}_D(E)\rangle = |\psi_D^+(E)\rangle + \frac{i}{I}\sum_\lambda G_D^+(E)|H-E|\psi_C^\lambda\rangle\langle\psi_C^\lambda|H-E|\psi_D^+(E)\rangle\, g_\lambda(E). \tag{11.23}$$

As can be easily seen, this wave function is a formal solution of the equation

$$\langle\delta\overline{\psi}_D|H'' - E|\overline{\psi}_D\rangle - \frac{i}{I}\sum_\lambda \langle\delta\overline{\psi}_D|H-E|\psi_C^\lambda\rangle\langle\psi_C^\lambda|H-E|\psi_D^+\rangle\, g_\lambda(E) = 0, \tag{11.24}$$

where the operator H″ is defined by eq. (9.46) and takes into account the influence of far-lying compound levels.†

To proceed further, we write

$$\langle \psi_C^\lambda | H - E | \psi_D^+ \rangle = \alpha_\lambda \langle \psi_C^\lambda | H - E | \bar{\psi}_D \rangle. \tag{11.25}$$

The constant α_λ can be obtained by substituting eq. (11.23) into eq. (11.25). The result is

$$\alpha_\lambda = \left[1 + \frac{i}{I} \sum_\mu \frac{W_{\lambda\mu} \langle \psi_C^\mu | H - E | \psi_D^+ \rangle g_\mu(E)}{\langle \psi_C^\lambda | H - E | \psi_D^+ \rangle} \right]^{-1}, \tag{11.26}$$

with

$$W_{\lambda\mu} = \langle \psi_C^\lambda | (H - E) \, G_D^+ \, (H - E) | \psi_C^\mu \rangle. \tag{11.27}$$

If we now substitute eq. (11.25) into eq. (11.24), then we obtain

$$\langle \delta \bar{\psi}_D | \bar{H} - E | \bar{\psi}_D \rangle = 0, \tag{11.28}$$

where

$$\bar{H} = H'' + U, \tag{11.29}$$

with

$$U = -\frac{i}{I} \sum_\lambda (H - E) | \psi_C^\lambda \rangle \, \alpha_\lambda \, g_\lambda \, \langle \psi_C^\lambda | (H - E). \tag{11.30}$$

We see that U is a nonhermitian operator. Consequently, the effective Hamiltonian $\bar{H}$ is also nonhermitian, as is to be expected from our earlier discussion.††

The optical potential V_l or its spatial representative $V_l(R', R'')$ in a state with relative orbital angular momentum l can be obtained from $\bar{H}$ in the following way. We substitute

$$\begin{aligned} \bar{\psi}_D &= A' \left\{ \hat{\phi}(\widetilde{A}) \, \hat{\phi}(\widetilde{B}) \, \frac{1}{R} f_l(R) \, P_l(\cos\theta) \right\} \\ &= \int A' \{ \hat{\phi}(\widetilde{A}) \, \hat{\phi}(\widetilde{B}) \, \delta(\mathbf{R} - \mathbf{R}'') \, P_l(\cos\theta'') \, R'' \} \, f_l(R'') \, dR'' \, d\Omega'' \end{aligned} \tag{11.31}$$

and

$$\begin{aligned} \delta \bar{\psi}_D &= A' \left\{ \hat{\phi}(\widetilde{A}) \, \hat{\phi}(\widetilde{B}) \, \frac{1}{R} \delta f_l(R) \, P_l(\cos\theta) \right\} \\ &= \int A' \{ \hat{\phi}(\widetilde{A}) \, \hat{\phi}(\widetilde{B}) \, \delta(\mathbf{R} - \mathbf{R}') \, P_l(\cos\theta') \, R' \} \, \delta f_l(R') \, dR' \, d\Omega' \end{aligned} \tag{11.32}$$

† It should be noted that the variations $\delta \psi_D$ and $\delta \bar{\psi}_D$ define the same subspace.

†† It should be mentioned that the effective Hamiltonian $\bar{H}$ and the optical potential derived from it are not unique. For example, the expression for the optical potential will depend on how much one projects out the compound structures from the open-channel function. Because of this, even the optical-model relative-motion function is not uniquely defined in the region of strong interaction. On the other hand, as has already been mentioned, the asymptotic behaviour of the relative-motion function is always uniquely determined.

into eq. (11.28). This yields the integrodifferential equation

$$\int \bar{H}_l (R', R'')\, f_l (R'')\, dR'' = 0, \tag{11.33}$$

with

$$\bar{H}_l = \int \langle \hat{\phi}(\tilde{A})\, \hat{\phi}(\tilde{B})\, \delta(\mathbf{R} - \mathbf{R}')\, R'\, P_l(\cos\theta')\, |\bar{H} - E|\, A' \{\hat{\phi}(\tilde{A})\, \hat{\phi}(\tilde{B}) \times \delta(\mathbf{R} - \mathbf{R}'')\, R''\, P_l(\cos\theta'')\}\rangle\, d\Omega'\, d\Omega''. \quad \ldots (11.34)$$

More explicitly, eq. (11.33) can be written in the following familiar form:

$$-\frac{\hbar^2}{2\mu}\left[\frac{d^2}{dR'^2} - \frac{l(l+1)}{R'^2}\right] f_l(R') + \int_0^\infty V_l(R', R'')\, f_l(R'')\, dR'' = E_R\, f_l(R'), \tag{11.35}$$

where

$$V_l(R', R'') = V_{Dl}(R')\, \delta(R' - R'') + k_l(R', R''), \tag{11.36}$$

with $V_{Dl}(R')$ being a complex direct potential and $k_l(R', R'')$ being a nonhermitian kernel. Thus, we see that, because of the indistinguishability of the nucleons, the optical potential between two nuclei is in general a complicated nonlocal (and energy-dependent) potential.

To see more clearly the influence of compound states, we examine $\bar{\psi}_D$ and $\langle S_{ii}(E)\rangle_I$ in more detail. Within the approximation in which the expression for $\langle S_{ii}(E)\rangle_I$ given by eq. (11.21) was obtained, we can set α_λ in eq. (11.30) as 1. This is allowed because, as is seen from eqs. (11.26) and (11.27), the neglected term in α_λ is of the order of $\pi^2 |\gamma_\lambda|^2_{av}/d$ which is small compared to unity. With this simplification, eq. (11.28) becomes

$$\langle \delta \bar{\psi}_D | H'' - \left[\frac{i}{I} \sum_\lambda (H - E)|\psi_C^\lambda\rangle\, g_\lambda\, \langle \psi_C^\lambda|(H - E)\right] - E|\bar{\psi}_D\rangle = 0. \tag{11.37}$$

We now consider the solution of the above equation in the first Born approximation. Then, by introducing the resolvent operator G_D^+ for the equation $\langle \delta \bar{\psi}_D | H'' - E|\bar{\psi}_D\rangle = 0$, we find

$$|\bar{\psi}_D\rangle = |\psi_D^+\rangle + \frac{i}{I} \sum_\lambda G_D^+ |H - E|\psi_C^\lambda\rangle\, g_\lambda\, \langle \psi_C^\lambda |H - E|\psi_D^+\rangle. \tag{11.38}$$

From the asymptotic behaviour of $\bar{\psi}_D$, the energy-averaged elastic-scattering S-matrix element $\langle S_{ii}(E)\rangle_I$ can be obtained. The result is

$$\langle S_{ii}(E)\rangle_I = e^{2i\delta(E)} \left(1 - \frac{2\pi}{I} \sum_\lambda g_\lambda\, \langle \psi_D^+ |H - E|\psi_C^\lambda\rangle \langle \psi_C^\lambda |H - E|\psi_D^+\rangle\right), \quad \ldots (11.39)$$

which is the same as that given by eq. (11.21). This must be so of course because eq. (11.21), which was the starting point for the derivation of the operator U, is correct to first order in $\pi^2 |\gamma_\lambda|^2_{av}/d$.†

† For the following consideration, we could have started immediately from eq. (11.21). However, it is sometimes interesting to show how one gets back to the original assumption from which one has started.

For further consideration we shall adopt, for simplicity in discussion, the picket-fence model as represented by eq. (10.58); that is, we shall assume $\gamma_\lambda = \gamma$ and $E_{\lambda+1} - E_\lambda = d$ for all values of λ. In this model, the expression for $\langle S_{ii}(E)\rangle_I$ is reduced to

$$\langle S_{ii}(E)\rangle_I = e^{2i\delta(E)}\left(1 - 2\pi^2 \frac{|\gamma|^2}{d}\right), \tag{11.40}$$

and the absorption cross section in the lth orbital angular-momentum state is given by

$$\sigma_{abs} \approx \frac{\pi(2l+1)}{k_i^2}\, 4\pi^2 \frac{|\gamma|^2}{d}. \tag{11.41}$$

Thus one sees that the absorption probability of an incident particle is proportional to the quantity $|\gamma|^2/d$ for the compound levels.

It might seem fictitious to speak of an absorption probability when the absorption cross section has its origin in the energy-averaging procedure. That this is not so may be demonstrated in the following way [FR 55a]. We consider an incoming wave packet which, due to its finite spatial extension, has an energy spread considerably larger than the energy width of the sharp compound level being excited in the scattering process. Then it can be shown, by employing the time-dependent projection equation (2.1) for instance, that the scattered wave packet may turn out to be considerably more extended compared to the incoming wave packet. The extension of this scattered wave packet is of the order of $(\hbar/2\pi|\gamma|^2)\,v$, with v being the relative velocity of the interacting particles. This indicates that, at an intermediate time, a compound nucleus of lifetime $\Delta\tau = \hbar/2\pi|\gamma|^2$ may be formed, which represents an absorption from the viewpoint of the incoming wave packet.

From eq. (11.40) we can compute the corresponding shape-elastic and compound-elastic cross sections. The result for the compound-elastic cross section is, for example,

$$\sigma_{ii}^{ce} \approx \frac{\pi(2l+1)}{k_i^2}\, 4\pi^2 \frac{|\gamma|^2}{d}. \tag{11.42}$$

This means that, at least in the case of narrow resonances and pure elastic-scattering, the cross section σ_{ii}^{ce} is defined in such a way that it describes the absorption of incoming particles into the compound levels.

In the case where the compound levels are not very narrow, one cannot use eq. (11.21) for $\langle S_{ii}(E)\rangle_I$ as the starting point for the derivation of the optical potential, but has to use eq. (11.20). However, the procedure to construct the operator U in the effective Hamiltonian $\overline{H}$ is completely analogous to that described above. One realizes easily that in this case one obtains

$$U = -\frac{i}{I}\sum_\lambda \frac{(H-E)|\psi_C^\lambda\rangle\, \overline{g}_\lambda\, \langle\psi_C^\lambda|(H-E)}{1 + \frac{i}{I}\sum_\mu \frac{W_{\lambda\mu}\langle\psi_C^\mu|H-E|\psi_D^+\rangle\, \overline{g}_\mu(E)}{\langle\psi_C^\lambda|H-E|\psi_D^+\rangle}}, \tag{11.43}$$

with

$$\overline{g}_\lambda(E) = \frac{iI}{E_\lambda - E - iI}\left(i\pi \sum_{\mu=1}^{n_0} \frac{\gamma_\mu\gamma_\mu^*}{E_\mu - E - iI} - 1\right)^{-1} \tag{11.44}$$

which reduces to $g_\lambda(E)$ in the limit case involving narrow levels.

We consider now briefly the other limit case where the compound levels overlap very strongly. In this case, the quantity $-i\pi \sum_\lambda [\gamma_\lambda \gamma_\lambda^* /(E_\lambda - E - iI)]$ has a value much larger than 1 and one obtains from eq. (11.20)

$$\langle S_{ii}(E)\rangle_I \approx -e^{2i\delta(E)}. \tag{11.45}$$

From this expression for $\langle S_{ii}(E)\rangle_I$, we can infer some of the properties of the corresponding operator U. Clearly, this operator must be such a nonlocal operator that it yields an additional phase shift of $\pi/2$ over the phase shift associated with $\psi_D^+(E)$. But, since $\langle S_{ii}(E)\rangle_I$ is still a pure phase factor,† the compound-elastic cross section is equal to zero, indicating that U does not describe on the average any absorption into the compound states. This is of course understandable, because the combined width of these levels is now so large that a compound nucleus having a relatively long lifetime can no longer be produced.

The above limit case which involves broad levels and pure elastic-scattering does not occur in nature, because the strong overlap of resonance levels is always created by transitions to many open inelastic and reaction channels. However, even though this limit case is not realistic, it does indicate that, in the case of strongly overlapping levels, the energy-averaging procedure does not need to be carried out and the compound-elastic cross section is equal to zero.

11.2c. Optical-Model Potential in the Presence of Reaction Channels

In the presence of inelastic and reaction channels, the effective Hamiltonian $\overline{H}$ (see eq. (11.29)) can be constructed in a similar way as that described in subsection 11.2b. This construction is, however, rather complicated if one does not consider the case where the compound levels strongly overlap. Since in practical applications the case with strongly overlapping resonances is in fact the most important one, we shall for simplicity restrict ourselves to a consideration of this relatively simple problem in this subsection.

To treat this problem we separate ψ into three parts, i. e.,

$$\psi = \psi_1 + \psi_2 + \psi_3, \tag{11.46}$$

where ψ_1 describes the incoming or elastic channel, ψ_2 describes all other open channels, and ψ_3 describes the bound structures of the compound system. Using again the projection equation (2.3), we obtain then as the starting point for our consideration the following set of equations:

$$\langle \delta\psi_1 | H - E | \psi_1\rangle + \langle \delta\psi_1 | H - E | \psi_2\rangle + \langle \delta\psi_1 | H - E | \psi_3\rangle = 0, \tag{11.47a}$$

$$\langle \delta\psi_2 | H - E | \psi_1\rangle + \langle \delta\psi_2 | H - E | \psi_2\rangle + \langle \delta\psi_2 | H - E | \psi_3\rangle = 0, \tag{11.47b}$$

$$\langle \delta\psi_3 | H - E | \psi_1\rangle + \langle \delta\psi_3 | H - E | \psi_2\rangle + \langle \delta\psi_3 | H - E | \psi_3\rangle = 0. \tag{11.47c}$$

To see the influence of ψ_2 and ψ_3 on the incoming channel, we formally eliminate them in the usual way. From eq. (11.47b) it follows that

$$|\psi_2\rangle = -G_2^+(H - E)|\psi_1 + \psi_3\rangle, \tag{11.48}$$

† This does not mean that U must necessarily be a hermitian operator.

where G_2^+ is the resolvent operator for the equation

$$\langle \delta \psi_2 | H - E | \psi_2 \rangle = 0 \tag{11.49}$$

with outgoing-wave boundary conditions.† Substituting eq. (11.48) into eq. (11.47c) yields

$$\langle \delta \psi_3 | H_3 - E | \psi_1 \rangle + \langle \delta \psi_3 | H_3 - E | \psi_3 \rangle = 0, \tag{11.50}$$

with

$$H_3 = H - (H - E)\, G_2^+ (H - E). \tag{11.51}$$

The formal solution of eq. (11.50) is

$$|\psi_3\rangle = -\, G_3 (H_3 - E) |\psi_1\rangle, \tag{11.52}$$

where G_3 is a resolvent operator, defined by the equation

$$\langle \delta \psi_3 | H_3 - E | G_3 = \langle \delta \psi_3 | . \tag{11.53}$$

Using eqs. (11.48) and (11.52), we can proceed to construct the effective Hamiltonian $\bar{H}$. However, before we do this, we shall first discuss in some detail the structure of the operator G_3.

We start this discussion by showing that, because of the presence of G_2^+, the operator H_3 is a nonhermitian operator. For this we write G_2^+ in the following spectral representation:

$$G_2^+ = \underset{\beta}{S} \frac{|\psi_2^\beta\rangle \langle \psi_2^\beta|}{E_2^\beta - E - i\epsilon} = \underset{\beta}{S}\, P \frac{1}{E_2^\beta - E} |\psi_2^\beta\rangle \langle \psi_2^\beta| + i\pi |\psi_2^+(E)\rangle \langle \psi_2^+(E)| . \tag{11.54}$$

In writing eq. (11.54), we have assumed that only one channel besides the elastic channel is open. This is only done for simplicity in discussion. When there are more than one reaction channel open, then the second term in eq. (11.54) has to be replaced by the operator $i\pi \, \Sigma_i |\psi_{2i}^+\rangle \langle \psi_{2i}^+|$, with the index i denoting the various independent orthogonal solutions of the homogeneous equation (11.49) and, similarly, the summation over β has to include all the independent solutions at each of the energy E_2^β. Now, by substituting eq. (11.54) into eq. (11.51), we obtain

$$H_3 = H - \underset{\beta}{S}\, P \frac{(H-E)|\psi_2^\beta\rangle \langle \psi_2^\beta|(H-E)}{E_2^\beta - E} - i\pi (H-E)|\psi_2^+(E)\rangle \langle \psi_2^+(E)|(H-E), \tag{11.55}$$

where the first two terms on the right side represent the hermitian part of H_3 and the third term represents the antihermitian part.

Because H_3 is nonhermitian, the spectral representation of the resolvent operator G_3 is constructed from solutions of the following homogeneous equations:

$$\langle \delta \psi_3 | H_3 - E_3^\beta | \psi_3^\beta \rangle = 0, \tag{11.56a}$$

$$\langle \delta \psi_3 | H_3^+ - \widetilde{E}_3^\beta | \widetilde{\psi}_3^\beta \rangle = 0. \tag{11.56b}$$

† Here again, we work in a state with a given total angular momentum.

In these equations, the eigenvalues E_3^β and $\widetilde{E}_3^\beta$ are complex conjugates of each other, i. e.,

$$E_3^\beta = (\widetilde{E}_3^\beta)^*, \tag{11.57}$$

and the eigenfunctions ψ_3^β and $\widetilde{\psi}_3^\beta$ satisfy the orthonormality and completeness relations

$$\langle \widetilde{\psi}_3^\beta | \psi_3^{\beta'} \rangle = \delta_{\beta\beta'}, \tag{11.58a}$$

$$\sum_\beta | \psi_3^\beta \rangle \langle \widetilde{\psi}_3^\beta | = 1, \tag{11.58b}$$

with 1 being a unit operator in the subspace spanned by $\delta \psi_3$.

To prove the relations given by eqs. (11.57), (11.58a), and (11.58b), one expands the bound-state solutions ψ_3^β and $\widetilde{\psi}_3^\beta$ in terms of a complete orthonormal set of functions ψ_{3m} in the subspace defined by $\delta \psi_3$; i. e., one writes

$$\psi_3^\beta = \sum_m b_m^\beta \psi_{3m}, \tag{11.59a}$$

$$\widetilde{\psi}_3^\beta = \sum_m \widetilde{b}_m^\beta \psi_{3m}. \tag{11.59b}$$

Such an orthonormal set can be chosen, for example, as consisting of eigenfunctions of the hermitian part H_3^h of the operator H_3; in other words, one can choose ψ_{3m} as solutions of the equation

$$\langle \delta \psi_3 | H_3^h - E_{3m} | \psi_{3m} \rangle = 0. \tag{11.60}$$

Now, by denoting the matrix elements of the nonhermitian operator H_3 in this representation as $H_{3,nm}$, then one obtains from eqs. (11.56a) and (11.56b) the following set of equations:

$$\sum_m [H_{3,nm} - E_3^\beta \delta_{nm}] b_m^\beta = 0, \tag{11.61a}$$

$$\sum_n (\widetilde{b}_n^\beta)^* [H_{3,nm} - (\widetilde{E}_3^\beta)^* \delta_{nm}] = 0. \tag{11.61b}$$

From these equations, one sees that E_3^β and $(\widetilde{E}_3^\beta)^*$ obey the same secular equation and, therefore, the relation given by eq. (11.57) follows. Also, it can be shown easily from eqs. (11.61a) and (11.61b) that, with proper normalization, the relations

$$\sum_m (\widetilde{b}_m^\beta)^* b_m^{\beta'} = \delta_{\beta\beta'} \tag{11.62}$$

and

$$\sum_\beta b_m^\beta (\widetilde{b}_n^\beta)^* = \delta_{mn} \tag{11.63}$$

hold. With these relations, one realizes immediately that eqs. (11.58a) and (11.58b) are valid.

After these considerations, it is clear that the spectral representation of G_3 is

$$G_3 = \sum_\beta \frac{|\psi_3^\beta\rangle\langle\tilde{\psi}_3^\beta|}{E_3^\beta - E}. \tag{11.64}$$

To understand the meaning of G_3, let us consider the special situation where the resonances do not overlap strongly. In this situation, the antihermitian part H_3^a of H_3 can be considered as a small perturbation and the complex energy eigenvalue E_3^m is approximately given by

$$E_3^m = E_{3m} + \langle\psi_{3m}|H_3^a|\psi_{3m}\rangle. \tag{11.65}$$

Using the expression for H_3^a in the case where only one reaction channel is present (see eq. (11.55)), we obtain

$$E_3^m = E_{3m} - i\pi|\gamma_m|^2, \tag{11.66}$$

with

$$\gamma_m = \langle\psi_{3m}|H - E|\psi_2^+(E)\rangle. \tag{11.67}$$

Equation (11.66) can be easily generalized to the case where more than one reaction channel exists and where the direct coupling between the reaction channels is neglected. The result is

$$E_3^m = E_{3m} - i\pi\sum_k |\gamma_{mk}|^2, \tag{11.68}$$

with k being an index labelling the various reaction channels. Thus, one sees that, in the situation involving compound levels which overlap very little, the imaginary part of E_3^m is proportional to the level width resulting from the coupling to reaction channels and the resolvent operator G_3 becomes

$$G_3 = \sum_m \frac{|\psi_{3m}\rangle\langle\psi_{3m}|}{E_{3m} - E - i\pi\sum_k |\gamma_{mk}|^2} \tag{11.69}$$

in first-order approximation.

For our considerations the important point to note is that, because of the coupling of the compound resonances to the reaction channels, the energy eigenvalues E_3^β are complex quantities. If the resonance levels of the compound nucleus are so dense that the mean nearest-neighbour spacing is much smaller than the average magnitude of the imaginary parts of E_3^β, then the resolvent operator G_3 becomes a smoothly varying function of the energy. This situation exists usually at high scattering energies where the density of the compound states is very large and where many reaction channels are open, with the consequence that the levels strongly overlap one another. As has already been mentioned at the beginning of this subsection, this is the situation which we shall, for simplicity, adhere to in the following discussion.

We continue now our discussion on the formal elimination of ψ_2 and ψ_3 from eq. (11.47a). By substituting eq. (11.52) into eq. (11.48), we find

$$|\psi_2\rangle = -G_2^+|(H-E) - (H-E)G_3(H-E) + (H-E)G_3(H-E)G_2^+(H-E)|\psi_1\rangle. \tag{11.70}$$

Using eq. (11.52) and eq. (11.70), we obtain finally from eq. (11.47a) the following equation

$$\langle \delta \psi_1 | \overline{H} - E | \psi_1 \rangle = 0, \tag{11.71}$$

where

$$\overline{H} = H + U_1, \tag{11.72}$$

with

$$U_1 = -[(H-E)\, G_2^+ (H-E) + (H-E)\, G_3 (H-E) - (H-E)\, G_2^+ (H-E)\, G_3 (H-E) - (H-E)\, G_3 (H-E)\, G_2^+ (H-E) + (H-E)\, G_2^+ (H-E)\, G_3 (H-E)\, G_2^+ (H-E)]. \tag{11.73}$$

In a way similar to that described in subsection 11.2b (see the discussion after eq. (11.30)), we can now derive from $\overline{H}$ an angular-momentum dependent, nonhermitian, microscopic optical potential $V_1^\mu (R', R'')$ in the incoming channel. Since both G_2^+ and G_3 vary smoothly with energy,† the operator $\overline{H}$ is a smooth function of energy and the procedure of energy-averaging the S-matrix element is not required. Consequently, in the special case under consideration here where the compound resonances overlap strongly, the optical potential $V_1 (R', R'')$ is nearly identical to the microscopic optical potential $V_1^\mu (R', R'')$ obtained from $\overline{H}$.

The operator U_1 describes the coupling of the elastic channel with itself via transitions to bound structures and reaction channels and, as is seen from eq. (11.73), has a rather complicated structure. Therefore, it is in general not possible to obtain this operator explicitly and for practical calculations one must devise a semi-phenomenological method in which the influence of U_1 is taken into account in a much simplified manner. To see how this can be done, we study qualitatively the effect of U_1 by dividing it into a hermitian part U_1^h and an antihermitian part U_1^a. The hermitian part U_1^h is mainly responsible for the mutual specific distortion of the clusters in the incoming channel; its effect will be small in comparison with the effect of the operator (H – E), if the clusters have relatively low compressibilities and if the coupling between the elastic channel and the reaction channels is weak enough such that the reaction cross section is not too large compared with the elastic-scattering cross section.†† As for the antihermitian part U_1^a, it describes mainly the absorption of clusters from the incoming channel or, in other words, transitions from the incoming channel to other channels during the time when the clusters penetrate each other.

In view of the above discussion, it is reasonable in many instances to adopt a semi-phenomenological procedure [BR 71] which works in the following way. One calculates at first the real part††† of the optical potential from the equation

$$\langle \delta \psi_1 | H - E | \psi_1 \rangle = 0. \tag{11.74}$$

† We need $G_2^+ (E)$ in an energy region where the spectrum of E_2^β is continuous.

†† We should emphasize that the smooth energy dependence of G_3 is a consequence of the coupling between the bound structures and the reaction channels.

††† In an actual calculation, one commonly makes a minor phenomenological adjustment of the real part in order to compensate approximately for the effect of U_1^h not taken into account in this procedure.

To this real part one then adds a phenomenological imaginary part which can be either in the form of an antihermitian kernel or simply in the form of an imaginary local potential.

For simplicity in calculation, the imaginary part of the optical potential is usually chosen as a local, sometimes l-dependent [TH 72a], potential. At the present moment, this is certainly a reasonable choice to make since, as has been mentioned, the optical potential is in any case not uniquely defined by the elastic-scattering S-matrix element. In fact, it is not even uniquely defined when the relative-motion function $f_l(R')$ in the elastic channel is given. For instance, from eq. (11.35) which contains an optical potential $V_l(R', R'')$ given by eq. (11.36), one can solve for the function $f_l(R')$. But, using this function, one can in turn construct an equivalent local, energy- and l-dependent potential

$$V_l^*(R') = \frac{1}{f_l(R')} \int_0^\infty V_l(R', R'')\, f_l(R'')\, dR''. \tag{11.75}$$

In section 11.4, we shall come back to this potential and discuss some of its interesting properties.

11.2d. Mean Free Path of a Cluster in a Target Nucleus

In subsections 11.2b and 11.2c, we have discussed how one can construct, at least in principle, optical potentials in the elastic channel for composite bombarding particles. In addition, we have outlined a semi-phenomenological method to carry out practical optical-model calculations. This method is particularly useful when the imaginary part of the optical potential is not too large or, in other words, when the bombarding particle can penetrate appreciably into the target nucleus without being instantly absorbed in the surface region. Otherwise, the scattering will be similar to diffraction by a black disc and may be more simply treated by using approximate methods based on diffraction theory.

In the case of nucleon scattering by a medium-heavy or heavy nucleus, the depth of the imaginary part of the optical potential is generally around 5–10 MeV. For an incoming nucleon of about 50 MeV, this means that it has a mean free path of the order of the diameter of the target nucleus or a lifetime in the target nucleus of about 10^{-22} sec.[†] Because of this relatively long mean free path, one can observe in nucleon-nucleus scattering processes angular distributions which are different from the diffraction pattern resulting from nucleon scattering by a black disc. Similarly, of course, one can account for the appearance of optical giant resonances [BA 52] which occur if the lifetime of the incoming nucleon in the target nucleus is of the order of or longer than its traversal time through the nucleus [WI 61].

As is seen from eq. (11.73), an actual calculation of the imaginary part of the optical potential is very difficult because, in principle, one has to take into account all open channels explicitly. However, by describing the target nucleus as a Fermi gas, one can

† This is the average length of time which a nucleon can stay inside a target nucleus without losing its energy by collisions with other nucleons.

make in the case of nucleon-nucleus scattering approximate calculations [GR 68] which do yield results in fair agreement with the values of the imaginary-potential depth mentioned above.

If one considers now the scattering of composite particles by nuclei, then the situation at first looks quite different. For example, take the case of α-nucleus scattering. When the α particle touches the nuclear surface, the average kinetic energy E_{kin} of each constituent nucleon in the compound system is about 20 MeV, an estimate which follows from a simple consideration using the Fermi gas model. Thus, we can say that the time required for an exchange of a nucleon in the α particle with a nucleon in the target nucleus is of the order of

$$\tau \approx \frac{\frac{R_\alpha}{2}}{(2\,E_{kin}/M)^{1/2}} \approx 10^{-23}\ \text{sec.} \qquad (11.76)$$

If the time given above were a correct estimate of the lifetime of the α particle in the target nucleus, as was often argued some years ago, then one could conclude that the lifetime of an α particle is considerably shorter than its time of traversal across the nucleus and the possibility for the occurrence of any prominent resonance effect, as for instance the observation with composite bombarding particle of optical giant resonances, can be precluded.†

In the above discussion, the very important influence of the indistinguishability of the nucleons has, however, not been taken into account. One cannot simply use an equation of the form given by eq. (11.76) to obtain an estimate of the lifetime of a composite particle in a target nucleus. Based on considerations in the preceding chapters, the lifetime of a nuclear state must be derived by using a totally antisymmetrized wave function, such as a cluster wave function of the type

$$\psi = A' \{\hat{\phi}_B\, \hat{\phi}_T\, F(\mathbf{R}_B - \mathbf{R}_T)\}, \qquad (11.77)$$

where $\hat{\phi}_B$ and $\hat{\phi}_T$ describe the internal degrees of freedom of the bombarding and target clusters, respectively, while $F(\mathbf{R}_B - \mathbf{R}_T)$ denotes as usual the relative-motion function. Because of the totally anti-symmetric nature of ψ indicated by the operator A', the exchange effects considered in eq. (11.76), which occur in about 10^{-23} sec, do not determine the lifetime of nuclear clusters (i.e., the time before the cluster structure described by eq. (11.77) is destroyed). Each such exchange of nucleons merely introduces a minus sign for the wave function and, therefore, do not alter the correlation structure of the cluster wave function.††

The inclusion of exchange effects through the use of a totally antisymmetric wave function often results in a considerable increase of the lifetime of a cluster within nuclear matter over that given by eq. (11.76). As extreme examples, the low-lying states of ^{16}O

---

† The inclusion of centrifugal barrier effects leaves this conclusion essentially unchanged, as can be seen easily.

†† Because nucleons are indistinguishable, such exchange effects are of course unobservable. To consider that eq. (11.76) correctly estimates the lifetime of an α cluster within a larger nucleus is to assume that the behaviour of a nuclear system can be properly described by using a wave function of the form (11.77) without the antisymmetrization operator A'.

and the low-energy $\alpha + \alpha$ scattering states considered earlier have an α-cluster structure which persists indefinitely in first approximation.†

If the lifetime of the bombarding cluster in the nucleus is prolonged to the extent that it is about the same as, or longer than, the traversal time across the nucleus, then we would expect optical phenomena (other than diffraction phenomena appearing in the scattering by black discs), as for instance giant resonance phenomena, to occur.

Similar to the nucleon case, one can make a rough estimation of the lifetime of a composite cluster in nuclear matter by using the Fermi gas model. For an α particle of bombarding energy around 10 MeV, such an estimation [WI 61] leads to a lifetime in nuclear matter of about 2 to 3×10^{-23} sec, which corresponds to a mean free path of about 2 to 3 fm.††

The above estimate for the mean free path of an α cluster should be a fairly reasonable one when the target nucleus is quite heavy. Therefore, α clusters cannot penetrate very deeply into heavy target nuclei; this means that, in heavy nuclei, the α clusters are not destroyed in a short time only if there are configurations in which they stay mainly in the nuclear surface region where the nucleon density is relatively small. These configurations are the ones where the orbital angular momentum between the α cluster and the target nucleus is large and are important, for example, in considerations on α-particle transfer reactions.

When the target nucleus is relatively light ($A \lesssim 40$), the mean free path for an α cluster is likely to be somewhat longer than the estimate of 2–3 fm given above. The reason for this is that, due to conservation laws in connection with total spin and parity, the probability of exciting certain energy levels in the compound nucleus is quite small. In other words, the number of compound levels which can be effectively excited is appreciably smaller than that actually present. This reduction effect, which is particularly important in a lighter system, is not taken into account when the Fermi gas model is used to estimate the cluster lifetime. Therefore, in the case of α-particle scattering by a light nucleus, the chance that the α particle can pass through the nucleus is quite appreciable and optical resonance phenomena are expected to appear.

Similar considerations can also be made when other composite particles, such as ^{3}He, ^{12}C, ^{16}O and so on, are used as incident particles. Here again, it turns out that these clusters can exist for an appreciable length of time if they stay in the nuclear surface region. This happens when the relative orbital angular momentum between the cluster and the target nucleus is large, and is the situation which very often exists in heavy-ion transfer reactions.

11.3. Specific Examples

In this section, we shall discuss some examples to illustrate the semi-phenomenological procedure suggested in subsection 11.2c for the study of elastic-scattering processes. In

† This statement needs to be somewhat modified, if specific distortion effects are taken into consideration.

†† The velocity of an α cluster in nuclear matter is approximately the same as that of a nucleon in nuclear matter, because the α cluster moves in an average potential which is about 4 times deeper than that for a nucleon [WI 65].

these examples, the hermitian part of the optical potential is calculated from eq. (11.74) which describes the elastic scattering of the incident particle without taking into account the influence of compound states and other open channels. To this hermitian part, a phenomenological antihermitian part represented by a purely local imaginary potential, is then added. By adjusting the parameters contained in this imaginary potential, one attempts to fit as well as possible the experimental results on differential cross section, polarization, and so on. As the examples given below show, this simple procedure does work fairly well, in the sense that all essential features of the elastic-scattering data can be satisfactorily reproduced.

11.3a. ^{3}He + α Scattering

The scattering of ^{3}He by α particles at low energies of 1.7 and 4.975 MeV is discussed in subsection 7.3a. In this subsection where we consider scattering at higher energies [FU 75], an imaginary potential $\mathrm{i}\,W_l(R)$ is introduced into the single-channel formulation of section 5.5 (see eq. (5.102)), in order to take approximate account of open reaction channels. This imaginary potential is taken to be of the form

$$W_l(R) = [1 + C_I(-1)^l]\,\widetilde{W}(R) \tag{11.78}$$

which has an odd-even l-dependence. We choose $W_l(R)$ in this particular way, since it was found in an analysis of p + α scattering [TH 72a] that the inclusion of such a feature can yield an improved agreement with the experimental differential cross-section and polarization results.† As for the choice of the absorptive potential $\widetilde{W}(R)$, we use the information learned from a study of the $\alpha + \alpha$ scattering problem [BR 71]. There we found that the complex phase shifts determined empirically by *Darriulat et al.* [DA 65] can be fitted rather well by using

$$\widetilde{W}(R) = -W_0 \left\{ \frac{1}{1 + e^{(R - R_W)/a_W}} + \frac{4\,e^{(R - R_W)/a_W}}{[1 + e^{(R - R_W)/a_W}]^2} \right\}, \tag{11.79}$$

which contains both a volume and a surface term. The geometry parameters used were $R_W = 3.0$ fm and $a_W = 0.5$ fm. These parameters are chosen such that the rms radius of the volume part of $\widetilde{W}(R)$ is approximately equal to the rms radius of the direct nuclear-central potential. Since it seems reasonable to expect that the absorption mechanism in the ^{3}He + α case should be similar to that in the $\alpha + \alpha$ case, we shall also use eq. (11.79) for $\widetilde{W}(R)$, and since the ^{3}He nucleus is somewhat larger than the α particle, we choose

$$\begin{aligned} R_W &= 3.2 \text{ fm}, \\ a_W &= 0.5 \text{ fm}. \end{aligned} \tag{11.80}$$

With the geometry parameters fixed in this way, there are only two adjustable parameters C_I and W_0 left, which will then be varied to yield a best fit to the experimental data for ^{3}He + α scattering at each of the energies considered.

The parameters (C_I, W_0) are determined to be (− 0.75, 1.15 MeV), (− 0.60, 2.50 MeV), and (− 0.45, 3.15 MeV) at energies of 15.95, 24.36, and 37.00 MeV, respectively. They have relatively large uncertainties, which results in the fact that the total reaction cross

† For a discussion on the rationale of choosing such an imaginary potential, see ref. [TH 72a].

sections at these energies can only be determined to within a rather large range of 300–500 mb. Also, it is noted that the magnitude of W_0 is appreciably smaller than the corresponding value for the scattering of ^{3}He by heavy nuclei. The reason for this is that, at the same scattering energy, the number of open channels in a heavy system is much larger than that in a light system.

A comparison between calculated and experimental differential cross sections at the above-mentioned energies is shown in fig. 37. Here it is seen that, except for minor discrepancies, the agreement is quite satisfactory. In particular, one notes that, at 37.00 MeV, the rise in experimental cross section at angles larger than 100° is well reproduced. This is a consequence of the fact that, in this calculation, a totally anti-symmetrized wave function has been employed [TA 71] and, therefore, exchange processes [TH 71] are properly taken into account.

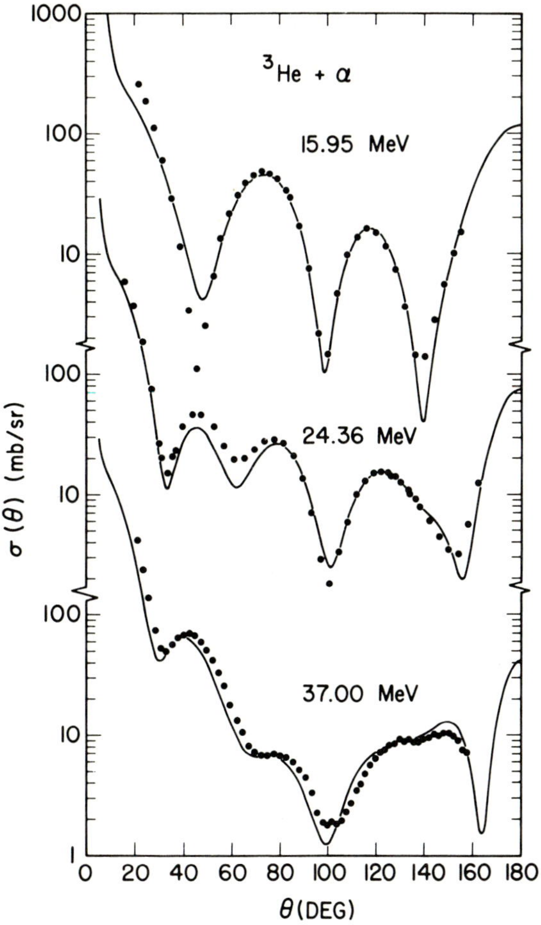

Fig. 37
Comparison of calculated differential cross sections for ^{3}He + α scattering with experimental data at 15.95, 24.36, and 37.00 MeV. The data at 15.95 MeV are those of ref. [JA 70], at 24.36 MeV are those of ref. [SC 69], and at 37.00 MeV are those of ref. [FE 71].

In conclusion, it is seen from the results discussed in section 5.5, subsection 7.3a, and this subsection that the behaviour of the ^{3}He + α system can be well described by employing a wave function of the form given by eq. (5.100). The only defect, albeit a relatively minor one in this particular case, is the lack of consideration of the specific distortion effect. This effect can be properly incorporated into the formulation by using the procedure described for d + α scattering in subsection 7.3c, but such a calculation has so far not yet been performed.

11.3b. p + ^{16}O Scattering

In the single-channel approximation [TH 75], the wave function for the p + ^{16}O system is written as

$$\psi = A \left\{ \tilde{\phi}(^{16}\tilde{O})\, \pi_1 \sum_{J=\frac{1}{2}}^{\infty} \sum_{l=J-\frac{1}{2}}^{J+\frac{1}{2}} \frac{1}{R} f_{Jl}(R)\, Y^{\frac{1}{2}}_{Jl\frac{1}{2}} \right\}, \qquad (11.81)$$

where π_1 is an isospin function for the proton, $\tilde{\phi}\,(^{16}\tilde{O})$ is the ^{16}O-wave function given by eq. (7.70), and $Y^{\frac{1}{2}}_{Jl\frac{1}{2}}$ is a normalized spin-angle function. The nucleon-nucleon potential used in this calculation has the form given by eq. (5.98), with the exchange-mixture parameter u and the spin-orbit strength determined in a way as discussed below.

The integrodifferential equation satisfied by $f_{Jl}(R)$ is obtained by using the projection equation (2.3) and the generator-coordinate technique described in subsections 5.2b and 7.2a; it is

$$\left\{ \frac{\hbar^2}{2\mu} \left[\frac{d^2}{dR^2} - \frac{l(l+1)}{R^2} \right] + E_R - V_N(R) - V_C(R) - \eta_{Jl} V_{so}(R) \right\} f_{Jl}(R)$$

$$= \int_0^{\infty} k_l(R, R')\, f_{Jl}(R')\, dR', \qquad (11.82)$$

where η_{Jl} is given by eq. (5.103). In computing the spin-orbit and Coulomb contributions, we have for simplicity omitted the exchange part by neglecting the antisymmetrization between the incident proton and the nucleons in the target nucleus. In addition, since the result of a previous N + α calculation [RE 70] was found to be insensitive to the choice of the spin-orbit range parameter λ, we have computed V_{so} in the zero-range approximation, i. e., by letting λ approach infinity.

The exchange-mixture parameter u in the nucleon-nucleon potential is adjusted to yield the experimental n + ^{16}O separation energy in the first excited $\frac{1}{2}^+$ state in ^{17}O [AJ 71]. The resultant value of u is 0.825, which is reasonably close to the value required in a similar calculation on $\alpha + \alpha$ scattering [BR 71]. With u fixed in this way, the spin-orbit strength is then varied to reproduce the observed value of about 5 MeV for the $d_{5/2} - d_{3/2}$ splitting [AJ 71]. The $\frac{5}{2}^+$ state so calculated lies at 0.74 MeV below the $\frac{1}{2}^+$ state mentioned above; this agrees quite well with the experimental result.

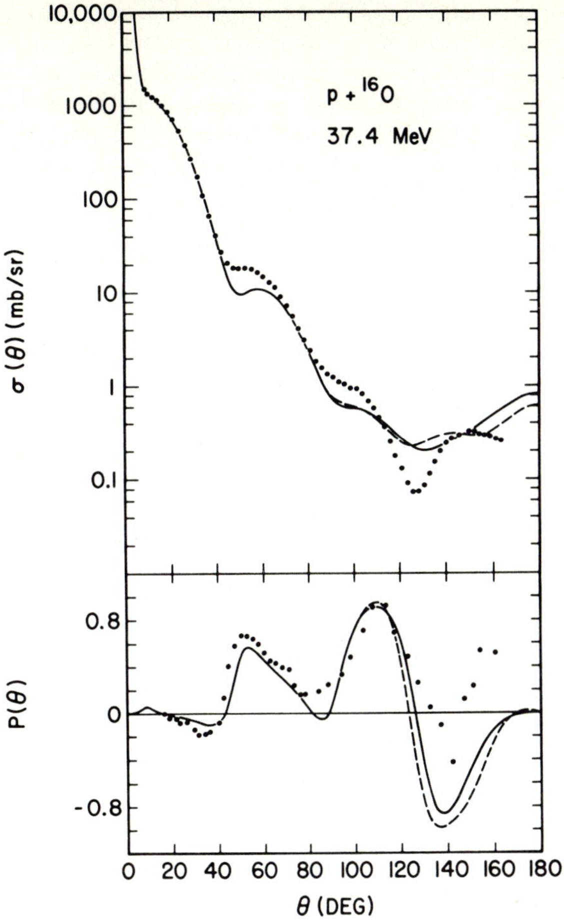

Fig. 38
Comparison of calculated and experimental results for p + ^{16}O scattering at 37.4 MeV. The experimental data are those quoted in ref. [VA 69]. (From ref. [TH 75].)

A comparison between calculated (solid curves) and experimental p + ^{16}O differential cross sections and polarizations at E_R = 37.4 MeV is shown in fig. 38. To take reaction effects into account, we have replaced V_N (R) in eq. (11.82) by V_N (R) + i$\widetilde{W}$(R), with $\widetilde{W}$(R) being the imaginary potential obtained by van Oers and Cameron [VA 69] in an optical-model analysis of the p + ^{16}O scattering data. As is seen, the agreement is quite satisfactory at angles up to about 130°. Beyond 130°, the calculation does correctly predict a rise in the cross section, but a detailed agreement is not obtained.†

In an optical-model study of p + ^{40}Ca scattering [GR 72], it was found that the incorporation of an odd-even *l*-dependence into the imaginary potential can significantly

† The calculated value for the total reaction cross section is 452 mb, which agrees well with the experimental value of 440 ± 35 mb [CA 67].

improve the agreement with experiment at large angles. To see if a similar feature might also be required here, a calculation was made with $\widetilde{W}(r)$ multiplied by a factor $[1 + C_I (-1)^l]$ as in eq. (11.78). The result obtained with $C_I = 0.03$ is shown by the dashed curves in fig. 38. Here one sees that, with a value of C_I similar in magnitude to that used in $p + {}^{40}Ca$ scattering, there is indeed a noticeable improvement in the shape of the cross-section curve at backward angles.

To see the effect of the antisymmetrization operator A' (see eq. (6.3)), we have also made a calculation in which the kernel k_l and the spin-orbit potential V_{so} in eq. (11.82) are set as zero but the direct potential $V_N(R)$ is multiplied by a factor C_a. The results for $p + {}^{16}O$ scattering at 30 MeV are shown in fig. 39, where the dashed and dot-dashed curves represent results for unantisymmetrized calculations with $C_a = 1.0$ and 1.4, respectively, while the solid curve represents the result of the fully-antisymmetrized calculation. Here it is seen that there is an over-all agreement between the antisymmetrized calculation and the unantisymmetrized calculation using $C_a = 1.4$, which indicates that the antisymmetrization operator A' does have an important effect. At extreme backward angles, there is a discrepancy of a factor of about 2.5. Even though this discrepancy may seem quite large, one should note that in $He^3 + \alpha$ scattering at 44.5 MeV where a similar comparison has been made [TA 71], the cross sections at backward angles differ by a factor of almost 1000. Therefore, we conclude that, for $p + {}^{16}O$ scattering at relatively high energies, the probability amplitude for the pickup of a ${}^{15}N$ cluster by the incident proton, which contributes mainly to cross sections in the backward directions, is rather small; this is indeed to be expected from an intuitive viewpoint, because a ${}^{15}N$ cluster is many times heavier than a proton.†

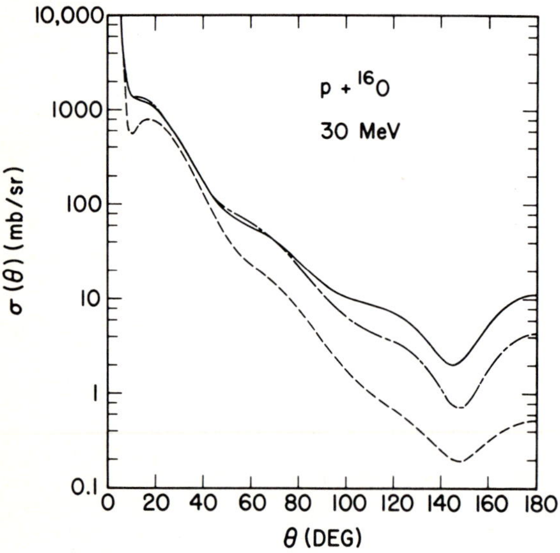

Fig. 39
Comparison of antisymmetrized (solid curve) and unantisymmetrized (dashed and dot-dashed curves) calculations. The unantisymmetrized calculation refers to the case where an equivalent local potential ($1.0\ V_N$ or $1.4\ V_N$) is used as the nuclear interaction between the proton and the ${}^{16}O$ nucleus. (From ref. [TH 75].)

† For a detailed study of the importance of exchange processes in an analogous case of $N + \alpha$ scattering, see ref. [TH 71].

11.3c. $\alpha + {}^{16}O$ Scattering

The formulation of the $\alpha + {}^{16}O$ problem was given in subsection 7.3e. In that subsection it was shown that, at an energy of 14.7 MeV where the reaction cross section is not expected to be large, a good agreement between calculated and experimental differential cross sections can be obtained. Here we consider the scattering at a higher energy of 19.2 MeV and take reaction effects approximately into account by adding into the formulation an imaginary potential of Woods-Saxon form with a depth of 5 MeV. The results are shown in fig. 40, where the dashed and solid curves are obtained with and without the imaginary potential, respectively. From this figure, it is again seen that the agreement between calculation and experiment is rather satisfactory when absorption is considered. The depth of the imaginary potential is somewhat smaller than that required for α-scattering by heavy nuclei. This is connected with the fact that ${}^{16}O$ in its ground state has a very stable structure.

From fig. 40, one sees also that especially the cross sections at backward angles are reduced by the inclusion of reaction effects. This is associated with the fact that, with the resultant imaginary potential, partial waves of low-l values are appreciably absorbed, while partial waves corresponding to grazing collisions are not. Indeed, such features are to be expected of the imaginary potential because, as was discussed in subsection 11.2d, an α cluster has a relatively long lifetime only when it stays mainly in the surface region of the nucleus.

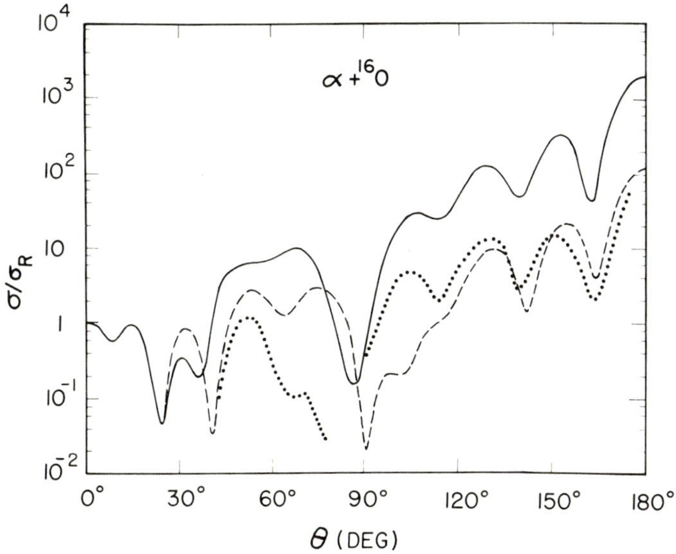

Fig. 40. Comparison of calculated and experimental differential cross sections for $\alpha + {}^{16}O$ scattering at 19.2 MeV. The dashed and solid curves represent results obtained with and without the imaginary potential, respectively. The experimental data, represented by solid dots, are those of ref. [OE 72]. (Adapted from ref. [SU 72].)

The α + ^{16}O scattering calculation can also be used to explain some general features of α-particle scattering by nuclei. It was found experimentally [OE 72] that, when α particles are scattered by target nuclei ($A \lesssim 50$) consisting of closed shells or closed shells plus a few holes, the differential cross sections at backward angles exhibit a rapid rise. On the other hand, for target nuclei having a few nucleons outside of closed shells, the backward rise becomes much smaller or even vanishes almost entirely. To understand these phenomena, we must first give general reasons as to why large backward peaks in the differential cross sections can occur. Afterwards, we have then to discuss under what circumstances these backward peaks become small or even completely disappear.

The discussion can be facilitated by considering specifically α + ^{16}O scattering which does show large cross sections at backward angles (see figs. 18 and 40). By examining the behaviour of the partial-wave scattering amplitudes, one can easily associate the appearance of large backward cross sections with the finding that the effective interaction between the α and the ^{16}O clusters depends on the parity of the relative-motion function.† To understand this qualitatively, we consider the low excited states of the compound nucleus ^{20}Ne in the oscillator-model representation. Based on energetical arguments discussed previously, one can conclude that these states have predominantly a α + ^{16}O cluster structure. The lowest α-cluster states of ^{20}Ne have positive parity and can be described approximately in the oscillator shell-model representation as having a configuration of a closed shell plus two protons and two neutrons in the 2s–1d shell. In this configuration the relative-motion function between the clusters possesses 8 oscillator quanta of excitation and the relative orbital angular momentum can be equal to 0, 2, 4, 6, and 8. States with relative orbital angular-momentum values larger than 8 cannot occur, if one restricts these four nucleons to the 2s–1d shell. The next higher α-cluster states are obtained by lifting one nucleon into the next oscillator shell; this results in 9 oscillator quanta of excitation in the relative motion and negative-parity α-cluster states with $l = 1, 3, 5, 7$, and 9. One can proceed in this way and obtain further group of positive-parity states in an even higher excitation-energy region and so on. The essential point to note is that, in this approximation, the states of even-l and odd-l values are bunched together in different energy regions. Certainly, one will find in reality large deviations from this simple picture, especially since scattering states, not bound states, are being considered. However, it is reasonable to expect that the actual energy spectrum will exhibit some features of this odd-even characteristic; i.e., one expects that in one excitation-energy region many broad potential-resonance states with even-l values appear, while in another excitation-energy region many broad potential-resonance states with odd-l values appear. This is important for our discussion, since it is well known that a bunching of scattering states with exclusively odd-l or even-l values will produce an angular distribution for elastic scattering which is symmetrical with respect to 90° in the centre-of-mass system. Of course, due to the large energy overlap between these broad odd-l and even-l resonances, this symmetrical behaviour around 90° will be appreciably perturbed, but some effects will remain, which are responsible for the occurrence of large backward peaks in α + ^{16}O scattering.

---

† In other words, the effective interaction in even-l states is distinctly different from that in odd-l states. In section 11.4, we shall further discuss this odd-even effect in detail.

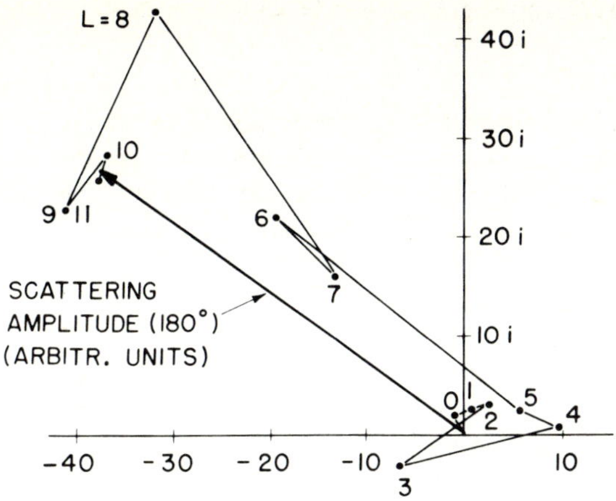

Fig. 41
Contributions of various partial waves to the $\alpha + {}^{16}O$ scattering amplitude in the backward direction. (Adapted from ref. [SU 72].)

The above argument is indeed supported by our quantitative study of $\alpha + {}^{16}O$ scattering. In fig. 41, we show at 19.2 MeV the contribution of various partial waves to the total scattering amplitude at $\theta = 180°$. Here it is seen that the main contribution does come from even-l partial waves with $l = 6$ and 8.

As should be quite clear, the above explanation for the appearance of large backward peaks in $\alpha + {}^{16}O$ scattering is quite general and can be applied to other cases of α-scattering on closed-shell nuclei. Thus our task now is to discuss why, for target nuclei having a few nucleons outside of closed shells, the backward peaks become comparatively small or sometimes even nearly nonexistent. For this we refer again to fig. 40. There one sees that, by the inclusion of transitions to other open channels through the addition of a phenomenological imaginary potential, the differential cross sections, especially at backward angles, are reduced to a large extent. This means that, in those cases where the transition probabilities to other open channels are large, the backward peaks in the elastic-scattering angular distributions will usually be quite small. This is indeed the case for α-scattering on ${}^{18}O$ and ${}^{44}Ca$ [OE 72], where the target nuclei contain relatively loosely bound nucleons and, therefore, the total reaction cross sections are expected to be fairly large. On the other hand, for target nuclei such as ${}^{14}C$, ${}^{15}N$, and ${}^{39}K$, the backward cross sections for α-particle scattering should be significantly larger, because these target nuclei have relatively stable structure; this has in fact been found to be the case experimentally [OE 72].

There is another reason which may also be partially responsible for the appearance of smaller backward peaks when the target nuclei contain loosely bound nucleons. This comes from the fact that, in such cases, the effective potential between the incident and the target nuclei has a rather large surface diffuseness. As a result of this, the backward cross sections will become comparatively small, because it is known that the probabilities of scattering into backward directions are somewhat reduced when the effective potential is smoothed out in the surface region.

11.4. Features of Effective Local Potentials between Nuclei

11.4a. Wave-Function-Equivalent Local Potentials

In subsection 11.2c it was shown that, for every elastic-scattering relative-motion function $f_l(R')$ which is determined from the integrodifferential equation (11.35), one can construct an equivalent local potential $V_l^*(R')$ which yields exactly the same relative-motion function. Here we shall discuss briefly the features of such wave-function-equivalent local potentials, because potentials similar to these are used frequently in phenomenological optical-model calculations.

In fig. 42 effective potentials $V_l^*(R')$ for $^3He + \alpha$ scattering at energies of 1.7 and 18.0 MeV are shown [BR 68a]. To obtain these potentials, we have used a single-channel formulation by neglecting all transitions to other open channels. The consequence of this is that V_l^* becomes a purely real potential which describes no absorption of particles from the incident channel. As is noted from fig. 42, the effective potentials $V_l^*(R')$ possess a number of singularities. These arise from the fact that a wave function obtained from a nonlocal potential can have nonzero second derivatives at values of R′ where the wave

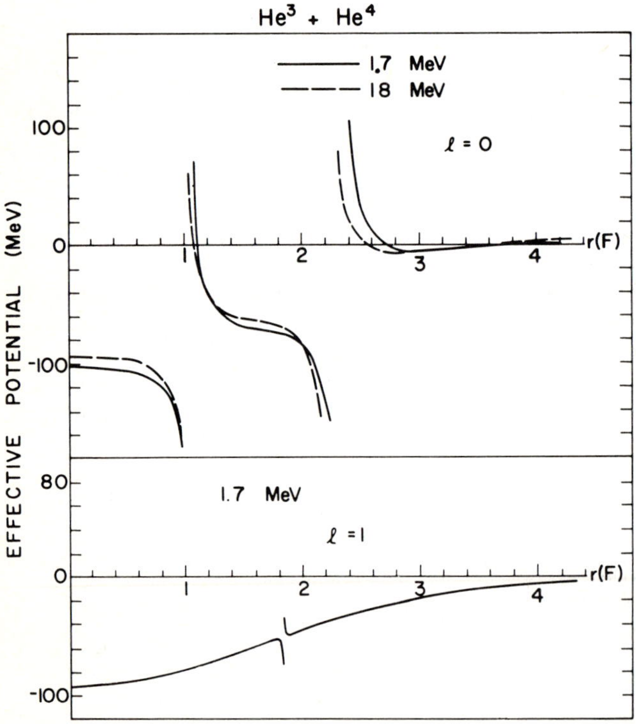

Fig. 42. Effective potentials $V_l^*(R')$ for $l = 0$ and 1 in the $^3He + \alpha$ system. (From ref. [BR 68a].)

function is itself equal to zero.† If absorption had been considered, these singularities would no longer be present because the elastic-scattering wave function $f_l(R')$ would be complex and, in general, one would not expect the real and imaginary parts to vanish simultaneously. On the other hand, even with absorption taken into account, relatively rapid changes in $V_l^*(R')$ would still occur in the nuclear surface region and these changes would be increasingly damped for increasing absorption probabilities. This is demonstrated in fig. 43, where the real parts of the effective potentials $V_0^*(R')$ for $\alpha + \alpha$ scattering at 32 MeV are shown for various values of the strength of the absorptive potential [BR 71]. From this figure, one indeed notes that the inclusion of absorption does smooth the potential curve to a significant extent.

The singularities or rapid variations of the effective potentials will usually not have a very strong influence on the scattering, because at low energies the local reduced wavelength of the relative motion will be large compared to the spatial extent of the singularity and at high energies the occurrence of many reaction channels will appreciably smooth the singularities. Therefore, one expects that, in most cases, the effective potential $V_l^*(R')$ can be represented by a smoothed potential which closely resembles $V_l^*(R')$ but is free from narrow singularities or rapid variations. This is indeed the type of potential commonly used in phenomenological optical-model studies. However, there is still the possibility that in special cases rather rapidly varying potentials may have to be chosen in order to describe the experimental scattering data and the use of such potentials may have significant consequences especially in distorted-wave calculations.

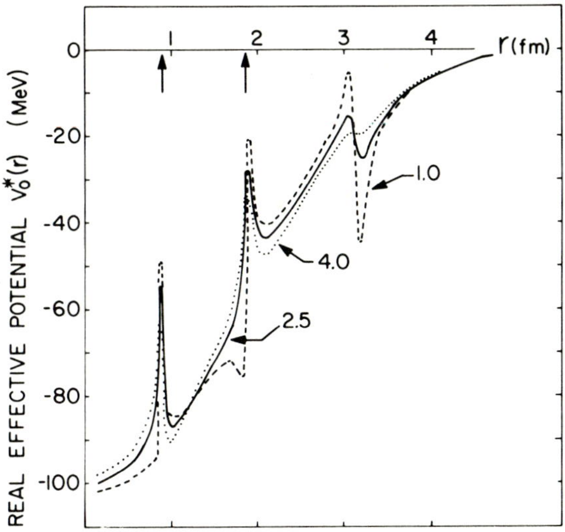

Fig. 43
The real part of the effective potential $V_0^*(R')$ for the $\alpha + \alpha$ system at 32 MeV. The three curves are labeled by the absorptive strength W_0 in MeV (see ref. [BR 71]). The arrows indicate the positions of the first two, no-absorption singularities. (From ref. [BR 71].)

† It can be shown that no singularities can occur in nuclear matter because of the translational invariance of the interaction, but they seem to be a general feature of the nuclear surface region where the wavelength of the relative motion changes rapidly.

11.4b. Phase-Equivalent Local Potentials

Microscopic resonating-group calculations of the type illustrated in sections 7.3 and 11.3 have been performed in many nuclear systems. From the results of these calculations, one finds that the effective potential $\widetilde{V}$ between nuclei, which yields nearly the phase shifts of the corresponding resonating-group calculation, has the following properties:

(i) It has an odd-even feature; that is, the interaction in even orbital angular-momentum states is distinctly different from that in odd orbital angular momentum states [BR 68a, TA 69, GR 71, TA 71].

(ii) The nature of the odd-even l-dependence may depend on the value of the channel spin [CH 73b, BR 74].

(iii) Its short-range part contains a repulsive core; this particular feature is necessary in order to avoid the occurrence of spurious bound states [SW 67, BR 74a, KO 75].

In the following paragraphs, we shall discuss these properties in detail. In addition, we shall show that an effective potential, which is constructed in a very simple manner but does contain these features, can reproduce the resonating-group result to a satisfactory extent.

11.4b (i). Odd-Even l-Dependence

To demonstrate the odd-even l-dependence, we use ^{3}H + α scattering as an example. By using the formulation given in section 5.5, but with the spin-orbit and exchange Coulomb potentials omitted for simplicity, we obtain the following integrodifferential equation for the radial function $f_l(R')$:

$$\left\{\frac{\hbar^2}{2\mu}\left[\frac{d^2}{dR'^2}-\frac{l(l+1)}{R'^2}\right]+E_R-V_D(R')-V_C(R')\right\}f_l(R') \\ =\int_0^\infty k_l(R',R'')\,f_l(R'')\,dR'', \tag{11.83}$$

where V_D is the direct nuclear potential. In fig. 44, the values of δ_l calculated with a Serber nucleon-nucleon potential are shown as a function of l at an energy of 40 MeV. These values are shown as dots and crosses with a connecting dashed line as a visual aid. In this figure, it is apparent that the phases possess an odd-even feature wherein the odd-l and even-l phases show distinctly different decreasing trends with l.[†]

To obtain a connection between the odd-even behaviour in δ_l and the effective local interaction between the ^{3}H and α clusters, we solve the equation

$$\left\{\frac{\hbar^2}{2\mu}\left[\frac{d^2}{dR'^2}-\frac{l(l+1)}{R'^2}\right]+E_R-\widetilde{V}_l(R')-V_C(R')\right\}\widetilde{f}_l(R')=0, \tag{11.84}$$

which contains an effective potential $\widetilde{V}_l(R')$. The results with $\widetilde{V}_l = 0.9\,V_D$, $1.1\,V_D$, and $1.3\,V_D$ are shown by the solid and dot-dashed lines in fig. 44. Here one sees that the odd-l

† It should be mentioned that *Giamati et al.* [GI 63] have noticed an odd-even feature in their analysis of p + α scattering.

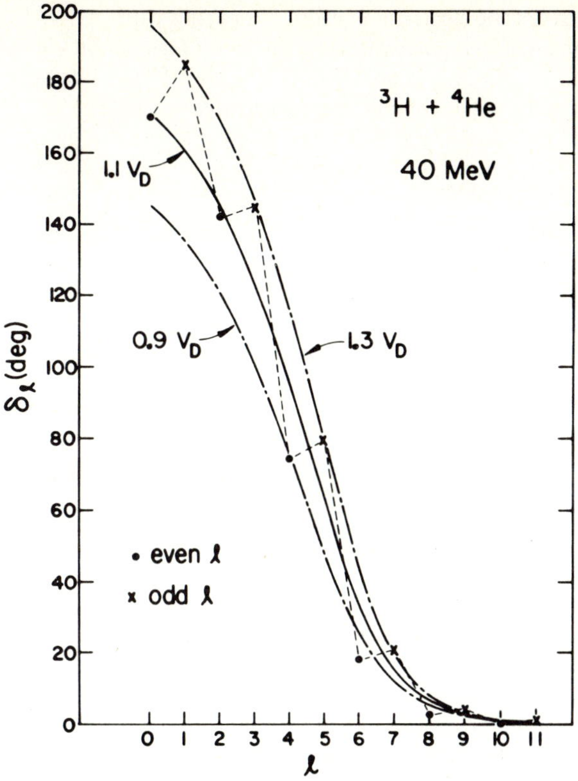

Fig. 44
Phase shift δ_l as a function of the orbital angular momentum. The dots and crosses represent the values obtained from eq. (11.83), while the solid and dot-dashed lines represent the values obtained from eq. (11.84) with $\widetilde{V}_l = 0.9\ V_D$, $1.1\ V_D$, and $1.3\ V_D$.

and even-l resonating-group phase shifts are fairly well reproduced by using effective potentials of $1.3\ V_D$ and $0.9\ V_D$, respectively, while the curve representing the result with $\widetilde{V}_l = 1.1\ V_D$ cuts right across the dashed line. From this we conclude that, to a fair approximation, it is possible to represent the interaction between the ^{3}H and α clusters by an effective potential of the form

$$\widetilde{V}_l = (V_D + V_a) + (-1)^l V_b, \tag{11.85}$$

with

$$V_a = 0.1\ V_D, \quad V_b = -0.2\ V_D. \tag{11.86}$$

The fact that V_a and V_b are quite different from zero indicates that, even at this relatively high energy of 40 MeV, the effects of antisymmetrization which gives rise to the kernel function k_l in eq. (11.83) are very important and cannot at all be neglected.

The odd-even behaviour can be qualitatively understood by using the bunching-of-states argument given in subsection 11.3c. Equivalently, one can also explain it in the following manner. In the oscillator picture for the ^{3}H + α system, the number of oscillator quanta for the relative motion between the clusters is equal to 3 for states with $l = 1$ and 3, and 4 for states with $l = 0, 2$, and 4. Thus, the two clusters are closer together on the average in odd-l states than in even-l states. As a consequence, the odd-l effective interaction will be stronger than the even-l effective interaction in this particular case. Similar argument can also be applied to other systems. For example, in the $d + \alpha$ case, the even-l effective interaction is expected to be stronger, which has indeed been verified by an explicit resonating-group calculation [TH 73].

For a more quantitative understanding of the odd-even features, we consider the case of $n + \alpha$ scattering [TH 71] discussed in subsection 7.2a, which has a relatively simple structure for the kernel function. What we shall do is to compute the scattering amplitude in the Born approximation, and then construct an equivalent local potential between the neutron and the α cluster, which yields exactly the same scattering amplitude in this approximation.

The Born scattering amplitude for the direct potential V_D is given by

$$f_D = -\frac{\mu}{2\pi\hbar^2}\int e^{-i\mathbf{k}_f\cdot\mathbf{R}'}\, V_D(R')\, e^{i\mathbf{k}_i\cdot\mathbf{R}'}\, d\mathbf{R}', \tag{11.87}$$

with $\mathbf{k}_i$ and $\mathbf{k}_f$ being the initial and final propagation vectors. By using the explicit form of V_D given in eq. (7.24) and performing the integration, one obtains

$$f_D = \frac{\mu}{2\pi\hbar^2} V_0 (4w - m + 2b - 2h) \left(\frac{\pi}{\kappa}\right)^{3/2} \exp\left[-\frac{4a + 3\kappa}{4a\kappa} k^2 \sin^2\frac{\theta}{2}\right], \tag{11.88}$$

where $k = |\mathbf{k}_i| = |\mathbf{k}_f|$, and θ is the scattering angle in the center-of-mass system. Similarly, the scattering amplitudes for the kernel terms can be calculated with the equation

$$f_i = -\frac{\mu}{2\pi\hbar^2}\int e^{-i\mathbf{k}_f\cdot\mathbf{R}'}\, K_i(\mathbf{R}', \mathbf{R}'')\, e^{i\mathbf{k}_i\cdot\mathbf{R}''}\, d\mathbf{R}'\, d\mathbf{R}''. \tag{11.89}$$

Using the expressions for K_i ($i = 1, 2, 3$) given in eqs. (7.26)–(7.28), one can easily carry out the integration for f_i. For our present purpose, it will be convenient (see ref. [TH 71] for details) to rewrite f_K, which is a sum of f_1, f_2, and f_3, as

$$f_K = f_1' + f_2' + f_3', \tag{11.90}$$

with

$$f_1' = \frac{\mu}{2\pi\hbar^2} V_0 (-w + 4m - 2b + 2h) \left(\frac{16\pi}{3a + 16\kappa}\right)^{3/2} \times \exp\left[-\frac{25}{4(3a + 16\kappa)} k^2\right] \exp\left[-\frac{4(3\kappa - a)}{a(3a + 16\kappa)} k^2 \sin^2\frac{\theta}{2}\right], \tag{11.91}$$

$$f_2' = -2\frac{\mu}{2\pi\hbar^2} V_0 (3w + 3m) \left(\frac{16\pi}{3a + 10\kappa}\right)^{3/2} \times \exp\left[-\frac{9(a + 2\kappa)}{4a(3a + 10\kappa)} k^2\right] \exp\left[-\frac{4(a + 2\kappa)}{a(3a + 10\kappa)} k^2 \cos^2\frac{\theta}{2}\right], \tag{11.92}$$

$$f_3' = -\frac{\mu}{2\pi\hbar^2}\left(\frac{16\pi}{3a}\right)^{3/2}\left\{E_\alpha - V_0(-3w-3m)\left(\frac{a}{a+2\kappa}\right)^{3/2} - e^2\left(\frac{2a}{\pi}\right)^{1/2}\right.$$
$$\left. - \frac{\hbar^2}{2M}\left[3a+\frac{3}{4}k^2+\frac{4}{3}k^2\cos^2\frac{\theta}{2}\right]\right\}\exp\left(-\frac{3}{4a}k^2\right)\exp\left(-\frac{4}{3a}k^2\cos^2\frac{\theta}{2}\right) \qquad (11.93)$$

From the above equations, it can be easily seen that the Born amplitudes f_D and f_1' are peaked in the forward direction, while the Born amplitudes f_2' and f_3' are peaked in the backward direction.

It is rather simple to construct equivalent local potentials which yield the amplitudes f_1', f_2', and f_3' in the Born approximation. Recognizing the fact that, for smoothly varying potentials at relatively high energies, the Born scattering amplitude for a Wigner-type potential is large mainly in forward directions, while the Born scattering amplitude for a Majorana-type potential is large mainly in backward directions, we can easily find these equivalent potentials; these will be labeled as V_1, $V_2 P^R$, and $V_3 P^R$, respectively, and are given by

$$V_1 = -V_0(-w+4m-2b+2h)\left(\frac{4a}{3\kappa-a}\right)^{3/2}$$
$$\times \exp\left[-\frac{25}{4(3a+16\kappa)}k^2\right]\exp\left[-\frac{a(3a+16\kappa)}{4(3\kappa-a)}\mathbf{R}'^2\right], \qquad (11.94)$$

$$V_2 P^R = 2V_0(3w+3m)\left(\frac{4a}{a+2\kappa}\right)^{3/2}$$
$$\times \exp\left[-\frac{9(a+2\kappa)}{4a(3a+10\kappa)}k^2\right]\exp\left[-\frac{a(3a+10\kappa)}{4(a+2\kappa)}\mathbf{R}'^2\right]P^R, \qquad (11.95)$$

$$V_3 P^R = 8\left[E_\alpha - V_0(-3w-3m)\left(\frac{a}{a+2\kappa}\right)^{3/2} - e^2\left(\frac{2a}{\pi}\right)^{1/2}\right.$$
$$\left. - \frac{\hbar^2}{2M}\left(\frac{9}{2}a+\frac{3}{4}k^2-\frac{3}{4}a^2 R'^2\right)\right]\exp\left(-\frac{3}{4a}k^2\right)\exp\left(-\frac{3a}{4}\mathbf{R}'^2\right)P^R, \qquad (11.96)$$

where P^R is a Majorana operator exchanging the position coordinates of the neutron and the α particle. The total equivalent local interaction is, therefore,

$$\widetilde{V} = (V_D + V_1) + (V_2 + V_3)P^R, \qquad (11.97)$$

which contains a Majorana component, indicating that it takes on different values depending upon whether the relative orbital angular momentum between the neutron and the α particle is even or odd.

The $n + \alpha$ study shows clearly that the odd-even l-dependence in the effective interaction arises as a consequence of the antisymmetrization procedure. If the wave function is not antisymmetrized with respect to the interchange of the incident and target nucleons, then, with the nucleon-nucleon potential of eq. (7.5), the nonlocal interaction represented by the kernel function will vanish and the only nuclear interaction between the clusters will

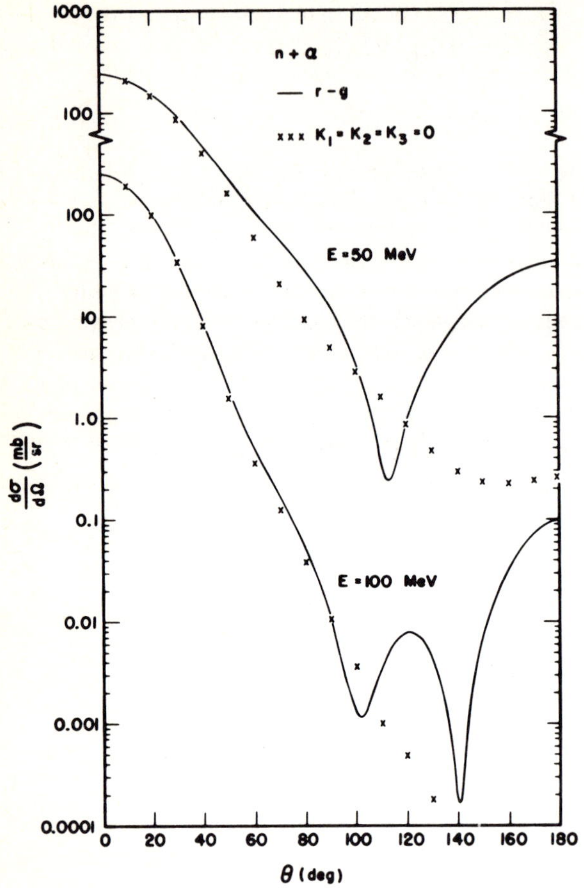

Fig. 45
Comparison of differential cross sections obtained with the resonating-group calculation and differential cross sections obtained using the direct potential V_D only. (From ref. [TH 71].)

be the direct potential V_D which has no odd-even dependence. The result of this is that the behaviour of the differential cross section in the backward angular region can no longer be satisfactorily explained. This is shown in fig. 45, where a comparison is made between the resonating-group result and the result of a calculation in which the kernel term is set as zero. From this figure it can be seen that the most prominent feature of this comparison is indeed the lack of a rise in the differential cross section at backward angles for the case where only V_D is included.

11.4b (ii). Channel-Spin Dependence

Because of the Pauli principle, the nature of the odd-even dependence can be different in different channel-spin states. To illustrate this, we show in fig. 46 the phase shifts as a function of the orbital angular momentum in both the $^3H + ^3He$ system and the $d + ^3He$ system [BR 74]. From this figure one finds that in the $S = \frac{3}{2}$ state of the $d + ^3He$ system the even-l effective interaction is stronger, whereas in the $S = \frac{1}{2}$ state of the d + ^{3}He system the odd-l effective interaction is stronger. For the ^{3}H + ^{3}He system, the situation is different; here it is seen that the effective interaction is stronger in even-l states, regardless of the channel spin of the system.

The above finding can be easily explained by using the argument given in subsection 11.4b (i) concerning the strength of the effective interaction in odd-l or even-l states. In the $S = \frac{3}{2}$ state of the d + ^{3}He system, the neutron in the deuteron cluster cannot penetrate into the 1s-neutron hole of the ^{3}He cluster. This causes the ^{3}He cluster to act toward the deuteron as a spin-isospin saturated substructure, with the consequence that the d + ^{3}He system in this channel-spin state behaves very much like the d + α system. In the $S = \frac{1}{2}$ state, however, the neutron can penetrate into the 1s-neutron hole of ^{3}He to form a cluster similar to a diffuse α particle; thus, in this channel-spin state, the d + ^{3}He system has a similarity to the p + α system and even more so to the more diffuse ^{3}He + α system [CH 73b].

An interesting way to demonstrate the different behavior in the two channel-spin states is shown in fig. 47. Here we depict at 15 MeV the d + ^{3}He partial differential scattering cross sections $\sigma_4(\theta)$ and $\sigma_2(\theta)$ in the states with $S = \frac{3}{2}$ and $S = \frac{1}{2}$, respectively. As a comparison, we show also the differential cross sections for d + α and ^{3}He + α scattering at the same energy. From this figure one sees clearly that the shape of the σ_4 curve is similar to that of the d + α curve and the shape of the σ_2 curve is similar to that of the ^{3}He + α curve, a result which is in full accord with the discussion given in the preceding paragraph.

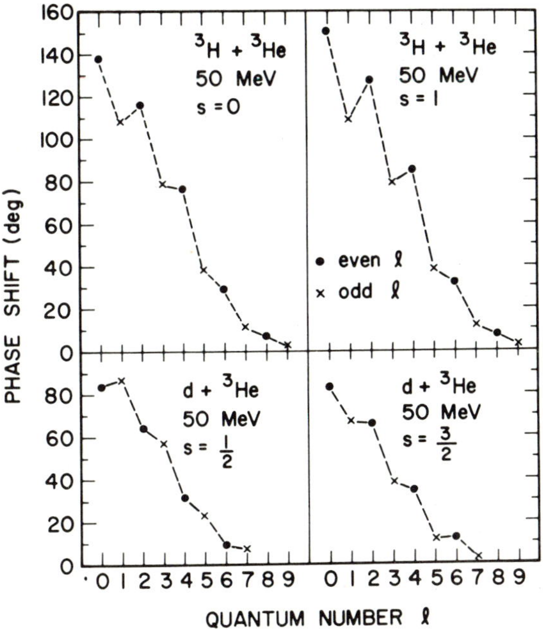

Fig. 46
Phase shifts as a function of the orbital angular momentum l and the channel spin S in both the ^{3}H + ^{3}He system and the d + ^{3}He system. (From ref. [BR 74].)

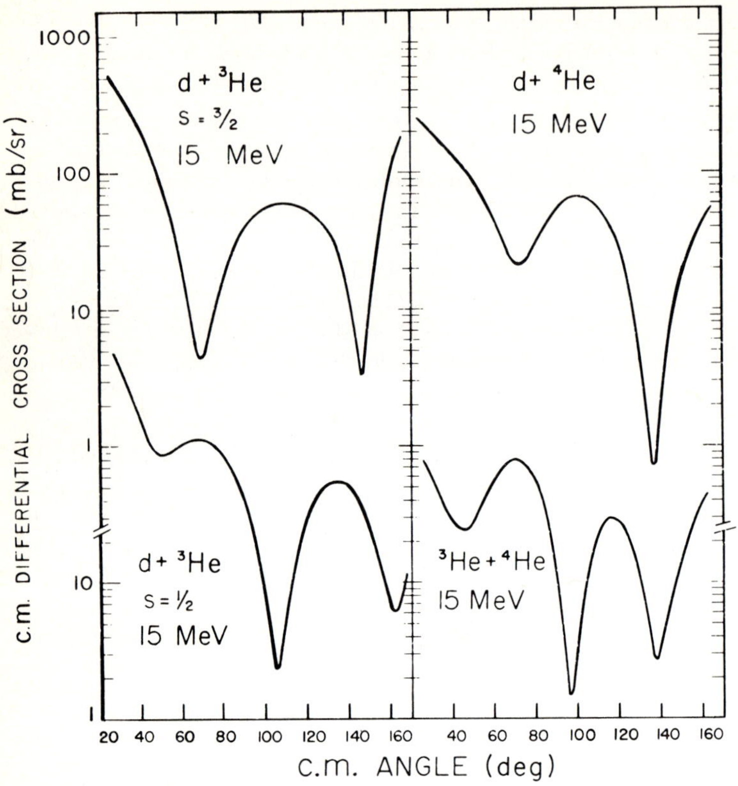

Fig. 47. Comparison of quartet (S = 3/2) and doublet (S = 1/2) cross sections for d + ^{3}He scattering with cross sections for d + α and ^{3}He + α scattering, all at 15 MeV. (From ref. [CH 73b].)

In an analogous manner one can understand the situation in the ^{3}H + ^{3}He system. Here in both channel-spin states, a nucleon from one of the clusters can penetrate into the 1s-hole of the other cluster. Therefore, it is expected not only that the ^{3}H + ^{3}He interaction should be rather insensitive to the channel spin, but also that the system should behave similarly to the d + α system and to the d + ^{3}He system in its $S = \frac{3}{2}$ state. As is seen from fig. 46, this is indeed the case.

It is conceivable that the spin effects described here will be important in many other scattering problems, such as the d + ^{15}N scattering problem. This means that if one wishes to conduct a simple, macroscopic examination of these problems, then according to the discussion presented in this subsection, one may have to employ effective interactions between nuclei which have not only an odd-even l-dependence but also a dependence on channel spin.

11.4b (iii). Pauli Repulsive Core

In the above discussion, it is shown that one of the consequences of the Pauli principle is to introduce an odd-even l-dependence into the effective interaction between two clusters. Here we shall show that, as another consequence of this principle, the effective interaction contains also a repulsive core (Pauli repulsive core) in its short-range part. To see this, we consider again the $\alpha + \alpha$ system as an example and make use of the conclusion reached in section 6.2 concerning the difference in behaviour between antisymmetrized and unantisymmetrized wave functions.

Based on the oscillator model for the $\alpha + \alpha$ system in $l = 0, 2, 4$ states,† it is argued in section 6.2 that, if antisymmetrization is not carried out, the function $g_n(R)$ appearing in eq. (6.5) should preferably be chosen as

$$g_n(R) = R^4 \exp(-bR^2), \tag{11.98}$$

rather than that given by eq. (6.6). This means that a reasonable requirement for $\widetilde{V}$ is that it produce, in the low angular-momentum states of the $\alpha + \alpha$ system with $l \leqslant 4$, a relative-motion function which has a radial part proportional to R^4 for small intercluster distances. As can be seen easily, this requirement can be simply met if one assumes that, in the effective potential $\widetilde{V}$, there is an l-dependent repulsive component which has for $l \leqslant 2$ the form

$$\begin{aligned} V_R(R) &= \frac{\hbar^2}{2\mu R^2}\,[N(N+1) - l(l+1)], \qquad (R < R_{Cl}) \\ &= 0, \qquad\qquad\qquad\qquad\qquad\qquad\quad (R > R_{Cl}) \end{aligned} \tag{11.99}$$

with $N = 4$ and is equal to zero for $l > 2$. It should be noted that, in the expression for V_R, two cutoff radii R_{C0} and R_{C2} have been introduced; this is necessary, because the Pauli principle is not expected to be effective when the separation distance between the clusters is large.

To examine whether the inclusion of the Pauli repulsive core in the effective potential $\widetilde{V}$ is useful or not, we have analyzed the phase-shift data [CH 75, NI 58, TO 63, WE 64] for $\alpha + \alpha$ scattering in the center-of-mass energy range from 2 to 15 MeV [BR 74a]. The form of $\widetilde{V}$ used is

$$\widetilde{V} = V_R + V_A, \tag{11.100}$$

with V_R given by eq. (11.99) and V_A obtained by folding Gaussian α-particle densities together with the direct part of a nucleon-nucleon potential of Yukawa form (see ref. [CH 75] for details). With this potential, the Yukawa parameters are adjusted to fit the phase shift δ_4. The excellent fit obtained is shown in fig. 48. The resultant potential V_A obtained in this way is then added to V_R, and the cutoff radii R_{C0} and R_{C2} are varied to obtain least-squared fits to δ_0 and δ_2. The values of R_{C0} and R_{C2} obtained are equal

† As has been shown by explicit resonating-group calculations described in subsections 7.3a and 7.3b, it is reasonable to use the oscillator model for a discussion of the behaviour of the relative-motion functions in the region where the clusters penetrate each other strongly.

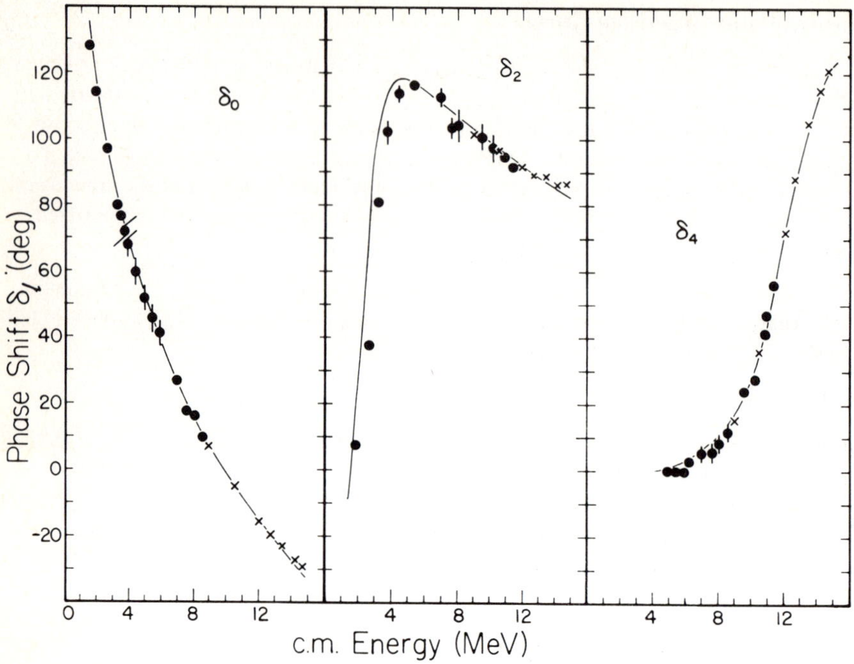

Fig. 48. $\alpha + \alpha$ phase shifts δ_0, δ_2, and δ_4 as a function of the scattering energy. The solid curves show results of the calculation using an effective potential. (From ref. [BR 74a].)

to 2.82 and 2.78 fm, respectively, and the fits are also shown in fig. 48. It is interesting that both R_{C0} and R_{C2} turn out to be about equal to the sum of the rms radii of the two α clusters, which is of course rather to be expected. Furthermore, since R_{C0} and R_{C2} are nearly equal, we find that the R^4-requirement contained in V_R automatically yields the strength of the repulsion in the $l = 2$ state as equal to about 0.7 of the in the $l = 0$ state, a feature found empirically by Ali and Bodmer [AL 66].

11.4b (iv). An Odd-Even Model for Elastic Scattering

In this subsection we shall show that an odd-even model, in which an effective potential incorporating all the features discussed above is employed, can satisfactorily describe the behaviour of a scattering system. This will be demonstrated in the $^3H + \alpha$ case by comparing the differential cross-section result obtained from this model with that obtained from a resonating-group calculation [PA 75].

In the odd-even model, the effective potential $\widetilde{V}$ between the ^3H and α clusters has the form of eq. (11.100). The repulsive part V_R is chosen as given by eq. (11.99) for $l \leqslant 2$ and is set equal to zero for $l > 2$. The value of N is taken to be 4 for $l = 0$ and 2, and 3 for $l = 1$. The part V_A is chosen to have an odd-even l-dependence; it is given by

$$V_A = \left(\frac{c_e + c_o}{2} + \frac{c_e - c_o}{2} P^R \right) V_D, \tag{11.101}$$

with V_D being the direct nuclear potential appearing in eq. (11.83). The adjustable parameters contained in the potential $\widetilde{V}$ are the odd-even parameters c_o and c_e, and the cutoff radii R_{Cl}. These parameters are varied at each energy to achieve a best fit with the results of a $^3H + \alpha$ resonating-group calculation obtained with a Serber interaction and with the exchange-Coulomb potential turned off.

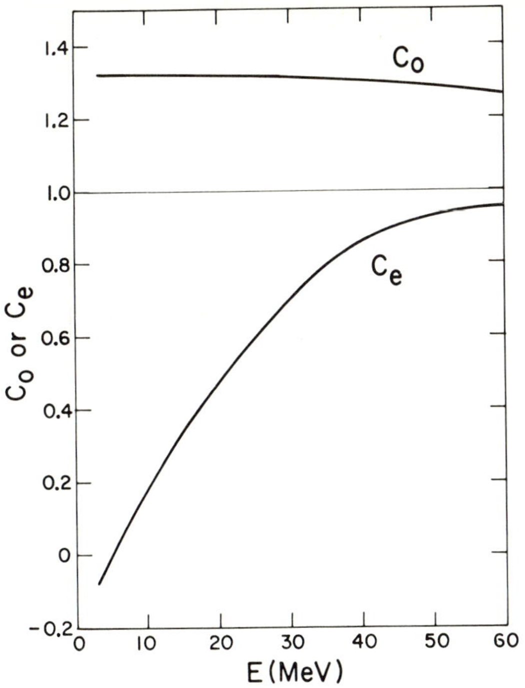

Fig. 49
Odd-even parameters c_o and c_e as a function of the scattering energy in the $^3H + \alpha$ case. (From ref. [PA 75].)

The calculation with $\widetilde{V}$ shows that the cutoff radius R_{Cl} can be chosen as both energy- and l-independent. A good choice is $R_{Cl} = 3.05$ fm, which should be compared with the corresponding value of about 2.80 fm obtained in the $\alpha + \alpha$ case. This shows again that the cutoff radius is about equal to the sum of the rms radii of the clusters involved, indicating that the repulsive effect arising from the Pauli principle is important only when the clusters begin to overlap.

The behaviour of c_o and c_e is depicted in fig. 49. Here one notes the following features: (a) Both c_o and c_e approach 1 as the energy becomes large; this is expected, because the Pauli principle should be less important at higher energies. (b) c_o is larger than c_e. This feature is not a universal one for the effective potential; for example, in the $d + \alpha$ case, c_e will be larger than c_o. (c) The parameter c_o is nearly energy-independent over a wide energy range, whereas the parameter c_e varies rapidly with energy.

Differential cross sections are calculated at energies from 3 to 40 MeV. Because of the near constancy of c_o, we have simply used $c_o = 1.319$ in this whole range. The results at

10 and 25 MeV are illustrated in fig. 50, where the cross sections calculated with $\widetilde{V}$ (dashed curves) are compared with those from the resonating-group study (solid curves). As is seen, the agreement between these calculations is fairly good, which indicates that the odd-even model with Pauli repulsive core is satisfactory in this problem.

In conclusion, we feel that, even though this model is certainly not as refined as the resonating-group method, it is simple enough for a systematic analysis of many systems and can thereby yield further understanding about the importance of the Pauli principle in nuclear problems.

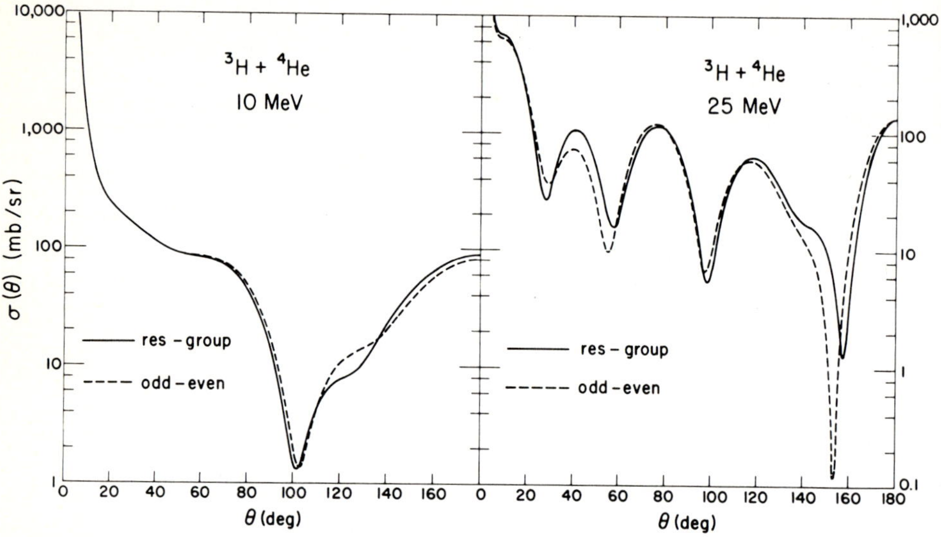

Fig. 50. Comparison of cross sections calculated with the resonating-group method (solid curves) and the odd-even model (dashed curves). (From ref. [PA 75].)

12. Direct Reactions

12.1. General Remarks

Direct reactions constitute a powerful tool in the investigation of nuclear properties. However, since for many-nucleon systems even reasonably accurate solutions of the Schrödinger equation, which satisfy the Pauli principle, are very difficult to obtain, analyses of experimental data have commonly been based in a restrictive manner on a variety of phenomenological models for the reaction mechanisms [TO 61, AU 70]. In our opinion, the most serious shortcoming of all these models lies in the fact that one does not know to what extent the requirement of the Pauli principle is satisfied.

In this chapter, we shall discuss how direct-reaction processes can be described in the framework of the unified theory presented in this monograph. Because in this theory fully antisymmetric wave functions are employed, the influence of the Pauli principle is taken correctly into account. Hence, one can not only study the relationship between this principle and the various phenomenological direct-reaction mechanisms, but also answer the important question as to the circumstances under which the antisymmetrization of the wave function can be partially or even totally omitted.

We shall derive the most important formulae of the direct-reaction theory in section 12.2. In section 12.3, we discuss two specific examples, namely, the rearrangement reactions $\alpha(p, d)\,{}^3He$ and ${}^6Li\,(p, {}^3He)\,\alpha$. These reactions will be studied at relatively high energies in both the incoming and outgoing channels, in order to satisfy the basic assumptions which enter into the derivation of the general formulae. Then, in section 12.4, we shall learn by means of these same reactions the influence of the Pauli principle. In particular, we shall study the consequences if certain exchange terms in the nuclear wave function are neglected in the calculation.

12.2. Derivation of the General Formulae

We start again from the projection equation

$$\langle \delta\psi | H - E | \psi \rangle = 0. \tag{12.1}$$

In direct-reaction processes, one is interested in transitions from the incoming channel to a specific reaction channel. Therefore, we write ψ this time as

$$\psi = \psi_1 + \psi_2 + \psi_R , \tag{12.2}$$

where ψ_1 describes the incoming channel, ψ_2 describes the reaction channel in which one is interested, and ψ_R describes all other open channels and all bound structures of the compound system. Using eq. (12.2) and the equation

$$\delta\psi = \delta\psi_1 + \delta\psi_2 + \delta\psi_R , \tag{12.3}$$

one obtains from eq. (12.1) the following set of coupled equations:

$$\langle \delta \psi_1 | H - E | \psi_1 \rangle + \langle \delta \psi_1 | H - E | \psi_2 \rangle + \langle \delta \psi_1 | H - E | \psi_R \rangle = 0, \tag{12.4a}$$

$$\langle \delta \psi_2 | H - E | \psi_1 \rangle + \langle \delta \psi_2 | H - E | \psi_2 \rangle + \langle \delta \psi_2 | H - E | \psi_R \rangle = 0, \tag{12.4b}$$

$$\langle \delta \psi_R | H - E | \psi_1 \rangle + \langle \delta \psi_R | H - E | \psi_2 \rangle + \langle \delta \psi_R | H - E | \psi_R \rangle = 0. \tag{12.4c}$$

Equations (12.4a)–(12.4c) constitute a set of an infinite number of coupled integro-differential and integral equations. Therefore, to solve these equations one must make certain approximations. In the following, we shall first discuss the different types of approximation which can be made, depending upon the energy region which one is concerned with.

At low bombarding energies, there are only one or a few reaction channels open and the density of compound configurations is also quite small. Bearing in mind further that, especially for low excitation energies, the influence of the Pauli principle can effectively reduce the number of independent configurations necessary to yield an adequate description of the behaviour of the system in the compound region, one can expect to obtain reasonably satisfactory results by taking into account only a small number of coupled equations. Then, it should be possible to solve the problem numerically, especially for lighter nuclear systems. An example for this is the often-mentioned $\alpha(p, d)\,^3He$ reaction in the region of about 22–25 MeV proton bombarding energy. In this case, the calculation is relatively simple because one essentially needs to consider only the $p + \alpha$ and the $d + {}^3He$ open channels plus a small number of bound configurations.

In the medium-energy region, the situation is different because the number of open channels and compound configurations increases rapidly. Here the behaviour of the system will, in general, be dominated by the presence of the relatively dense-lying bound structures. As a consequence, the number of coupled equations which must be taken into account is quite large and the calculation becomes often extremely complicated.

A simpler situation exists again at energies much higher than the Coulomb barriers between the fragments. Due to the large excitation energy involved, compound resonances become quite broad and overlap each other strongly. As a result of this, one may not need to consider ψ_R in eq. (12.2) explicitly, but rather can take its influence into approximate account by introducing into the formulation either nonlocal potentials having a non-hermitian part or complex local potentials having a smooth energy-dependence. The experimental fact that, at high bombarding energies, the angular distribution of the reaction products often shows a rather slow dependence on the energy can be considered as an indication of the validity of this type of approximation. In this chapter, we shall be concerned mainly with this particular situation.

We go back now to the treatment of eqs. (12.4a)–(12.4c). First, we eliminate ψ_R formally from these equations by introducing the resolvent operator G_R^+ which is defined by the equation

$$\langle \delta \psi_R | H - E | G_R^+ = \langle \delta \psi_R | . \tag{12.5}$$

This resolvent operator has the spectral representation

$$G_R^+ (E) = \mathop{S}_{\beta} \frac{| \psi_R^\beta \rangle \langle \psi_R^\beta |}{E_R^\beta - E - i\epsilon}, \tag{12.6}$$

where the ψ_R^β are solutions of the homogeneous equation

$$\langle \delta \psi_R | H - E_R^\beta | \psi_R^\beta \rangle = 0. \tag{12.7}$$

Using this operator, we can then reduce eqs. (12.4a)–(12.4c) to the following set of equations:

$$\langle \delta \psi_1 | H' - E | \psi_1 \rangle + \langle \delta \psi_1 | H' - E | \psi_2 \rangle = 0, \tag{12.8a}$$

$$\langle \delta \psi_2 | H' - E | \psi_1 \rangle + \langle \delta \psi_2 | H' - E | \psi_2 \rangle = 0. \tag{12.8b}$$

where

$$H' = H + W_R, \tag{12.9}$$

with

$$W_R = -(H - E)\, G_R^+ \,(H - E). \tag{12.10}$$

From the way the energy-dependent operator W_R is constructed, one can easily see that it describes the influence of the bound structures and open channels in the subspace spanned by $\delta \psi_R$ on the elastic channel and the reaction channel under consideration. Because of the presence of open channels, G_R^+ is in general nonhermitian and so is W_R. Furthermore, since we shall later need G_R^+ in an energy region where many channels are open and where no sharp resonances exist, G_R^+ (E) and consequently W_R (E) will usually show a rather smooth energy dependence in this energy region.†

Equations (12.8a) and (12.8b) can be used as the starting point to discuss two-channel calculations and the various approximations usually made in these calculations. Since direct reactions are commonly described as two-channel processes, these equations should of course also be appropriate for the discussion of such reactions.

Direct reactions usually take place in an energy region where many channels are open and where compound resonances overlap each other strongly. As discussed above, this has the consequence that W_R and therefore H' show a smooth energy dependence and it is not necessary to perform an energy-averaging procedure over resonances. In connection with this, one can also see that there are two simplifications which can be made in direct-reaction calculations. First, one can neglect in eq. (12.8a) the "feedback" term

† To understand better the properties of G_R^+ and W_R; we write ψ_R as

$$\psi_R = \psi_{R1} + \psi_{R2},$$

where ψ_{R1} describes the open-channel functions besides ψ_1 and ψ_2, and ψ_{R2} describes the bound structures in the compound region. By introducing resolvent operators G_{R1}^+ and G_{R2} belonging to subspaces spanned by $\delta \psi_{R1}$ and $\delta \psi_{R2}$, respectively, we obtain easily that

$$G_R^+ = G_{R1}^+ + G_{R2} - G_{R1}^+ (H - E) G_{R2} - G_{R2} (H - E) G_{R1}^+$$
$$+ G_{R1}^+ (H - E) G_{R2} (H - E) G_{R1}^+ = G_{R1}^+ + [1 - G_{R1}^+ (H - E)] G_{R2} [1 - (H - E) G_{R1}^+].$$

Using this form for G_R^+, we find then that W_R has a structure similar to U_1 of eq. (11.73). Therefore, utilizing the same argument as given in subsection 11.2c, we can conclude that both G_R^+ and W_R have a smooth energy-dependence in the energy region of interest.

$\langle \delta \psi_1 | H' - E | \psi_2 \rangle$ which represents the influence of the second channel (reaction channel) on the first one (elastic channel). Second, one needs to take into account only direct transitions from channel 1 to channel 2; in other words, one can neglect in eq. (12.8b) the term $\langle \delta \psi_2 | W_R | \psi_1 \rangle$ and use $\langle \delta \psi_2 | H - E | \psi_1 \rangle$ instead of $\langle \delta \psi_2 | H' - E | \psi_1 \rangle$.

The first simplification or the weak-coupling approximation results in the so-called distorted-wave Born approximation in direct-reaction theories. This simplification is justified, because at relatively high energies the Pauli principle is less effective in reducing the difference between different cluster structures and because the wave-number difference between the initial and the reaction channels becomes larger as the energy increases.† Both of these effects have the consequence of reducing the coupling between two open channels. As for the omission of the term $\langle \delta \psi_2 | W_R | \psi_1 \rangle$, it is allowed in good approximation because this term represents the transition from channel 1 to channel 2 via compound structures and other open channels. At high energies where the open channels are not effectively shielded by Coulomb and centrifugal barriers, it is certainly much more probable that, when the compound nucleus is excited, it will decay into these many other open channels rather than make a transition into the particular reaction channel under consideration.

With these simplifications, eqs. (12.8a) and (12.8b) become

$$\langle \delta \psi_1 | H' - E | \psi_1 \rangle = 0, \tag{12.11a}$$

$$\langle \delta \psi_2 | H - E | \psi_1 \rangle + \langle \delta \psi_2 | H' - E | \psi_2 \rangle = 0. \tag{12.11b}$$

Now, we introduce the resolvent operator G_2^+ defined by the equation

$$\langle \delta \psi_2 | H' - E | G_2^+ = \langle \delta \psi_2 | . \tag{12.12}$$

Because of the nonhermiticity of the operator H', the spectral representation of G_2^+ requires the solutions of two adjoint homogeneous equations

$$\langle \delta \psi_2 | H' - E_2^\beta | \psi_2^\beta \rangle = 0, \tag{12.13a}$$

$$\langle \widetilde{\psi}_2^\beta | H' - E_2^\beta | \delta \psi_2 \rangle = 0. \tag{12.13b}$$

It should be mentioned that the solutions ψ_2^β and $\widetilde{\psi}_2^{\beta *}$ are in general not complex conjugate to each other, but the equations which define them have, as explained in subsection 11.2c, the same complex eigenvalue-spectrum. In contrast to the solutions of eqs. (11.56a) and (11.56b) given in subsection 11.2c, eqs. (12.13a) and (12.13b) have also continuous eigenvalues. These continuous eigenvalues which we need must be real because, if the clusters in channel 2 are well separated, then ψ_2^β and $\widetilde{\psi}_2^\beta$ must describe the behaviour of two noninteracting clusters in channel 2.

---

† The wave-number difference at high energies is approximately given by

$$\Delta k = k_1 - k_2 \approx k_1 \, [1 - (\mu_2/\mu_1)^{1/2}],$$

with μ_1 and μ_2 being the reduced mass in the initial and reaction channels, respectively.

Similar to the discussion given in subsection 11.2c, it can be shown that ψ_2^β and $\widetilde{\psi}_2^\beta$ satisfy the orthonormality and completeness relations

$$\langle \widetilde{\psi}_2^\beta | \psi_2^{\beta'} \rangle = \delta_{\beta\beta'}, \tag{12.14a}$$

$$\underset{\beta}{\mathrm{S}} | \psi_2^\beta \rangle \langle \widetilde{\psi}_2^\beta | = 1, \tag{12.14b}$$

with 1 being a unit operator in the subspace spanned by $\delta\psi_2$. Now, by employing these adjoint solutions, one can then write G_2^+ as

$$G_2^+ = \underset{\beta}{\mathrm{S}} \frac{|\psi_2^\beta\rangle\langle\widetilde{\psi}_2^\beta|}{E_2^\beta - E - i\epsilon}, \tag{12.15}$$

where the factor $i\epsilon$ in the denominator again ensures that G_2^+ satisfy outgoing-wave boundary conditions.

Using the operator G_2^+, we formally integrate eq. (12.11b) to obtain

$$|\psi_2\rangle = -G_2^+ (H - E)|\psi_1\rangle. \tag{12.16}$$

Here the function ψ_1 is a solution of eq. (12.11a); it describes the elastic scattering and absorption of the clusters in the incoming channel. If we now decompose G_2^+, ψ_1, and ψ_2 into partial waves of given total angular momentum j and parity π, then by examining the asymptotic behaviour of ψ_{2j} we obtain the following expression for the transition or reaction amplitude $A_{j,12}$:

$$A_{j,12} \propto \left(\frac{v_1}{v_2}\right)^{1/2} \langle \widetilde{\psi}_{2j}(E) | H - E | \psi_{1j} \rangle, \tag{12.17}$$

where v_1 and v_2 denote the relative velocity of the clusters in channels 1 and 2, respectively, and $\langle \widetilde{\psi}_{2j}(E)|$ denotes the solution of eq. (12.13b) with $E_2^\beta = E$. Thus, we see that the amplitude $A_{j,12}$, which we derive here in the weak-coupling approximation, corresponds to the direct-reaction amplitude in the framework of the distorted-wave Born approximation.

Equation (12.17) suggests the following procedure for a semi-phenomenological direct-reaction calculation. One calculates at first the real parts of the optical potentials for the relative-motion functions in channel 1 and channel 2 from the equations

$$\langle \delta\psi_1 | H - E | \psi_1 \rangle = 0, \tag{12.18a}$$

$$\langle \delta\psi_2 | H - E | \psi_2 \rangle = 0. \tag{12.18b}$$

To these real parts one then adds phenomenological imaginary parts to describe the absorption of clusters in these channels for elastic-scattering processes. These phenomenological imaginary parts may be in the form of either an antihermitian kernel or simply an imaginary local potential. The reaction amplitude is then calculated from eq. (12.17). As is easily seen, this procedure is relatively simple, but does contain a problem of over-counting. This is associated with the fact that, in determining the imaginary parts of the optical potentials, the mutual influence of the two channels should not be taken into account, according to the discussion given above which results in eq. (12.17). However, this over-counting problem should not lead to any serious consequences, because at high energies there are many reaction channels open.

In phenomenological direct-reaction calculations, the antisymmetrization of the wave function is usually neglected completely or only taken partially into account. This is in general undesirable for the following reasons. First, with antisymmetrization properly considered, all phenomenological direct-reaction processes, such as knockout process, heavy-particle pickup process, and so on, are automatically taken into account and there is no need to introduce these processes into the calculation in an ad hoc manner, as is commonly done in phenomenological studies. Second, as has been discussed especially in subsection 11.2d, the procedure of antisymmetrization is necessary to explain the fact that an incident cluster can penetrate appreciably into the surface region of the target nucleus. Since the absolute value of the transition matrix element appearing in eq. (12.17) can especially depend sensitively upon how deeply the incident particle penetrates, one can easily see that the neglect of antisymmetrization will create serious problems in most cases.

However, there are situations under which a partial antisymmetrization can still yield satisfactory results. For example, at high bombarding energies and when one is only interested in calculating the differential reaction cross section over a limited angular region, say, the forward angular region, then one can in fact omit, to a good approximation, most of the exchange terms between the incident and the target nuclei. In section 12.4, we shall study the $\alpha(p, d)\,^3He$ and the $^6Li(p, {}^3He)\,\alpha$ reactions in detail in order to illustrate the validity of this type of simplification.

We discuss briefly now how one can refine and generalize the considerations given above. For instance, it can happen that, even at relatively high excitation energies, broad potential resonances exist in either channel 1 or channel 2, or both channels. In the case where such resonances occur in channel 2, it becomes then a poor approximation to neglect the feedback term $\langle \delta\psi_1 | H' - E | \psi_2 \rangle$ in eq. (12.8a). A much better approximation would be to omit only the coupling terms $\langle \delta\psi_2 | W_R | \psi_1 \rangle$ and $\langle \delta\psi_1 | W_R | \psi_2 \rangle$ in eqs. (12.8a) and (12.8b), and solve the following set of more correct equations:

$$\langle \delta\psi_1 | H' - E | \psi_1 \rangle + \langle \delta\psi_1 | H - E | \psi_2 \rangle = 0, \qquad (12.19a)$$

$$\langle \delta\psi_2 | H - E | \psi_1 \rangle + \langle \delta\psi_2 | H' - E | \psi_2 \rangle = 0. \qquad (12.19b)$$

By adopting again the semi-phenomenological procedure of introducing imaginary potentials into the formulation, one can solve these coupled equations to obtain the S-matrix elements. As in the case of the weak-coupling approximation, a calculation of this type can be performed with reasonable effort if the nuclear system under consideration is not too heavy.†

† To solve eqs. (12.19a) and (12.19b) is not much more difficult than to solve eqs. (12.11a) and (12.11b). Therefore, one should in general adopt the coupled-reaction-channel approach of using eqs. (12.19a) and (12.19b), rather than the weak-coupling approach of using eqs. (12.11a) and (12.11b).

If, in the energy region of interest, broad potential resonances exist in an open channel (channel 3) other than channel 1 or channel 2, then it is useful to take into account this open channel explicitly. Therefore, one writes now the wave function as

$$\psi = \psi_1 + \psi_2 + \psi_3 + \psi_R, \tag{12.20}$$

and obtains, instead of eqs. (12.19a) and (12.19b), the following set of coupled equations:

$$\langle \delta \psi_1 | H' - E | \psi_1 \rangle + \langle \delta \psi_1 | H - E | \psi_2 \rangle + \langle \delta \psi_1 | H - E | \psi_3 \rangle = 0, \tag{12.21a}$$

$$\langle \delta \psi_2 | H - E | \psi_1 \rangle + \langle \delta \psi_2 | H' - E | \psi_2 \rangle + \langle \delta \psi_2 | H - E | \psi_3 \rangle = 0, \tag{12.21b}$$

$$\langle \delta \psi_3 | H - E | \psi_1 \rangle + \langle \delta \psi_3 | H - E | \psi_2 \rangle + \langle \delta \psi_3 | H' - E | \psi_3 \rangle = 0, \tag{12.21c}$$

where the operator H' contains the influence of all bound structures and open channels other than channels 1, 2, and 3 on these three channels. For lighter nuclear systems, also this type of calculation can be carried out with reasonable effort [HA 74].

Another situation which frequently arises is when the target and residual nuclei have excited states which may be rather strongly excited in a direct-reaction process and must, therefore, be explicitly taken into consideration. Examples of this type occur when the nuclear system under investigation contains nuclei which have low-lying vibrational or rotational states. To carry out an accurate microscopic calculation in such a situation is of course very complicated, especially because of the large number of nucleons involved. At present, there exist only phenomenological calculations (so-called multi-step or coupled-channel Born-approximation calculations) which combine aspects of both the coupled-channel approach and the distorted-wave Born approximation, and which do yield reasonably satisfactory results (see, for example, refs. [OL 72, OL 73, OL 75, OS 76]).

Finally, before we proceed to the next section, we wish to consider briefly another interesting case which concerns the excitation of analogue resonances in direct reactions. As has been discussed in section 10.3, relatively sharp analogue resonances can exist even in the region of rather high excitation, because these resonances are shielded from the elastic channel by a high Coulomb barrier and transitions to other open channels can occur only through the weak Coulomb interaction. Clearly, the presence of these analogue resonances will have an important influence upon the behaviour of the direct-reaction cross sections.

Examples of this kind involve analogue resonances of relatively high excitation energy observed in (p, n), (^{3}He, n), (^{3}He, t), and so on reactions. For these cases, in contrast to the case discussed in section 10.3, the coupling of the complicated resonances to the reaction channels having the same isobaric spin as these complicated resonances is so large that these resonances overlap each other strongly, with the consequence that the energy-averaging procedure does not need to be carried out. This simplifies the formulation very much, as can be seen easily from the discussion given in section 10.3.

To calculate the reaction amplitude one can use again eq. (12.17), if one assumes that no potential resonances exist in the reaction channels. But this time one makes for ψ_1 the ansatz

$$\psi_1 = \psi_{1D} + a\,\psi_A, \tag{12.22}$$

where ψ_{1D} is the wave function of the elastic channel with mixed isobaric spin and ψ_A describes the analogue state with good isobaric spin. By using eq. (12.11a), one obtains then

$$a = -\frac{\langle \psi_A | H' - E | \psi_{1D}^+ \rangle}{E_A - E - \langle \psi_A | (H' - E)\, G_{1D}^+ \,(H' - E) | \psi_A \rangle} \tag{12.23}$$

and

$$|\psi_{1D}\rangle = |\psi_{1D}^+\rangle - a\, G_{1D}^+ \,|H' - E|\psi_A\rangle, \tag{12.24}$$

where ψ_{1D}^+ and the resolvent operator G_{1D}^+ come from, in the usual way, the nonhermitian operator H' of eq. (12.9). The quantity

$$E_A = \langle \psi_A | H' | \psi_A \rangle \tag{12.25}$$

is a complex energy expectation value; it is approximately equal to $\langle \psi_A | H | \psi_A \rangle$, because the operator H' has its contribution only from intermediate states which have an isobaric-spin quantum number different from that of ψ_A. For the same reason, one can usually, as a good approximation, replace also the operator $(H' - E)$ appearing in eqs. (12.23) and (12.24) by the simpler operator $(H - E)$.

If one substitutes eqs. (12.22)–(12.24) into eq. (12.17), then one obtains (with the index j omitted)

$$S_{12} = \left(\frac{v_2}{v_1}\right)^{1/2} A_{12} \propto \Big[\langle \widetilde{\psi}_2 | H - E | \psi_{1D}^+ \rangle - \langle \widetilde{\psi}_2 | (H - E) - (H - E)\, G_{1D}^+ \,(H' - E) | \psi_A \rangle$$
$$\times \frac{\langle \psi_A | H' - E | \psi_{1D}^+ \rangle}{E_A - E - \langle \psi_A | (H' - E)\, G_{1D}^+ \,(H' - E) | \psi_A \rangle}\Big]. \tag{12.26}$$

The first term on the right side describes the direct coupling between the elastic channel with mixed isobaric spin and the reaction channel under consideration, while the second term describes the direct and indirect influence of the analogue resonance on the reaction amplitude. If one separates now S_{12} into a real and an imaginary part, and takes into account as energy dependence only the linear energy dependence in the real part of the denominator $[E_A - E - \langle \psi_A | (H' - E)\, G_{1D}^+ \,(H' - E) | \psi_A \rangle]$, then one finds that, similar to the discussion given in subsection 10.3b concerning the asymmetry of the elastic-scattering S-matrix element, the asymmetry of $|S_{12}|^2$ is again produced by an interplay between the direct and indirect coupling of the incident channel with the considered reaction channel.

Another example, where the isobaric analogue resonance influences the direct-reaction amplitude, is the rearrangement reaction ^{207}Pb (d, p) ^{208}Pb (ground state). In this example, the isobaric analogue state occurs again in the compound nucleus,† with its parent state in the nucleus ^{209}Pb. But, in contrast to the case just considered above, here the elastic channel has a good isobaric spin which is different from the isobaric spin of the analogue resonance and the reaction channel under consideration has an isobaric-spin mixing. Since the analogue resonance is coupled this time strongly to the considered reaction channel,

† One can similarly treat cases where the residual nucleus has isobaric analogue resonances (see ref. [SC 75]).

one can no longer use the simplified equations (12.11a) and (12.11b) for the calculation of the reaction cross section in the energy region around the analogue resonance. Rather, one has to use the more accurate equations (12.19a) and (12.19b) which include the feedback of channel 2 on channel 1. As for the wave function, one again adopts the form given in eq. (12.2); however, one adds now the analogue state with a linear variational amplitude to the reaction-channel function ψ_2, rather than to the elastic-channel function ψ_1 as in eq. (12.22).

12.3. Specific Examples

In the preceding section, we have outlined the general formulation of direct-reaction calculations within the framework of the unified theory, and discussed various approximations which could be made under different circumstances. Here we shall consider two specific examples, namely, the ^{3}He (d, p) α [CH 74] and the ^{6}Li (p, ^{3}He) α [HU 72, SC 74] reaction calculations. These calculations will be performed in what we shall call the antisymmetrized direct-reaction (ADR) approximation; that is, we shall use eqs. (12.19a) and (12.19b), and take other open channels crudely into account by introducing phenomenological imaginary potentials into the coupled-channel formulation.

12.3a. ^{3}He (d, p) α Reaction

The ^{3}He (d, p) α reaction is the inverse of the α (p, d) ^{3}He reaction discussed in subsection 7.3f, where we were mainly concerned with the resonance excitation of the relatively low-lying $\frac{3}{2}^+$ state in the compound nucleus ^{5}Li. Here, we are instead interested in the behaviour of this reaction at higher energies where a number of reaction channels are open and, consequently, sharp resonances are no longer present.

For simplicity, we use the purely central nucleon-nucleon potential of eqs. (5.81)–(5.83). Because of this, the spin-$\frac{3}{2}$ channel of the d + ^{3}He system is not coupled to the p + α channel and, therefore, will not be discussed here (see ref. [CH 73b]). In the channel spin $\frac{1}{2}$ state, the wave function is assumed as

$$\begin{aligned}\psi = &\mathcal{A}\,\{N_f^{-1/2}\,\phi_h(123)\,\phi_d(45)\,F(\mathbf{R}_h - \mathbf{R}_d)\,\xi_f(s,t)\}\\ &+ \mathcal{A}\,\{N_g^{-1/2}\,\phi_\alpha(1234)\,G(\mathbf{R}_\alpha - \mathbf{R}_p)\,\xi_g(s,t)\},\end{aligned} \tag{12.27}$$

where ξ_f and ξ_g denote appropriate spin-isospin functions, and the quantities $1/\sqrt{N_f}$ and $1/\sqrt{N_g}$ are normalization factors (see subsection 7.2b). The functions ϕ_d, ϕ_h, and ϕ_α describe the spatial behaviour of the deuteron cluster, the ^{3}He cluster, and the α cluster, respectively, with the arguments in these functions indicating which nucleons are involved in these clusters; such a labeling is permissible, because the wave function ψ is totally antisymmetrized. The functions $F(\mathbf{R}_h - \mathbf{R}_d)$ and $G(\mathbf{R}_\alpha - \mathbf{R}_p)$ are linear variational functions, describing the relative motion of the clusters in each channel. Finally, as a convention, we have used the subscripts f and g to refer to the d + ^{3}He and p + α channels, respectively.

By using the projection equation (12.1), we obtain the following coupled integrodifferential equations for $F(\mathbf{R}'_f)$ and $G(\mathbf{R}'_g)$:

$$\left[\frac{\hbar^2}{2\mu_f}\nabla^2_{R'_f} + E_f - V_{Nf}(\mathbf{R}'_f) - V_{Cf}(\mathbf{R}'_f)\right] F(\mathbf{R}'_f)$$
$$= \int K_{ff}(\mathbf{R}'_f, \mathbf{R}''_f)\, F(\mathbf{R}''_f)\, d\mathbf{R}''_f + \int K_{fg}(\mathbf{R}'_f, \mathbf{R}''_g)\, G(\mathbf{R}''_g)\, d\mathbf{R}''_g, \quad (12.28a)$$

$$\left[\frac{\hbar^2}{2\mu_g}\nabla^2_{R'_g} + E_g - V_{Ng}(\mathbf{R}'_g) - V_{Cg}(\mathbf{R}'_g)\right] G(\mathbf{R}'_g)$$
$$= \int K_{gf}(\mathbf{R}'_g, \mathbf{R}''_f)\, F(\mathbf{R}''_f)\, d\mathbf{R}''_f + \int K_{gg}(\mathbf{R}'_g, \mathbf{R}''_g)\, G(\mathbf{R}''_g)\, d\mathbf{R}''_g, \quad (12.28b)$$

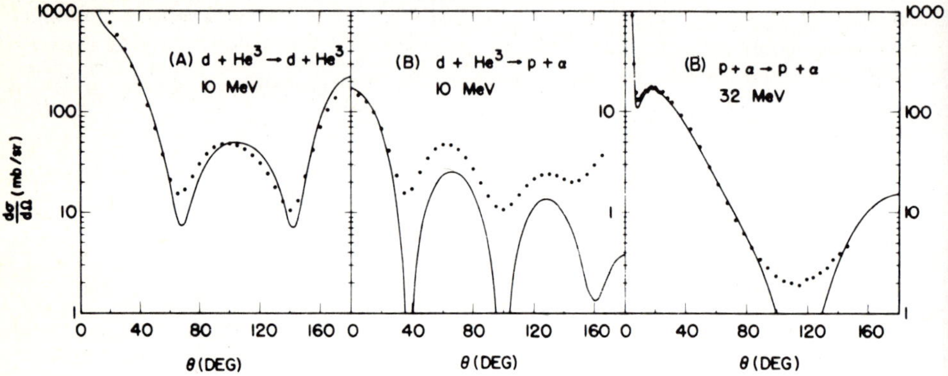

Fig. 51. Comparison of calculated and experimental results for differential scattering and reaction cross sections in the five-nucleon system.

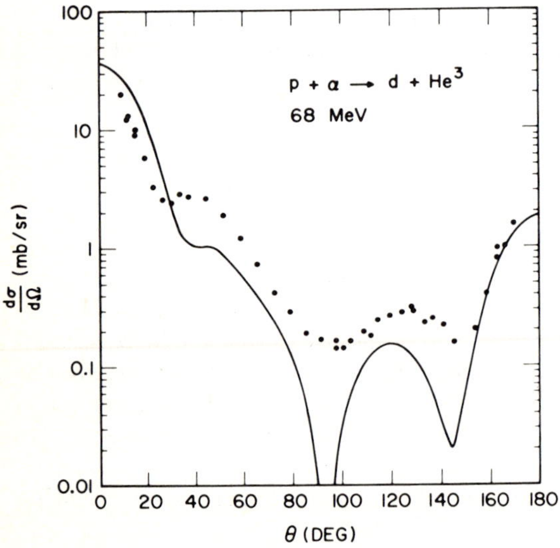

Fig. 52
Comparison of calculated and experimental differential reaction cross sections for the $\alpha(p, d)^3He$ reaction at E_g = 68 MeV. (From ref. [CH 74].)

where E_f and E_g denote, respectively, the relative energies of the clusters in the $d + {}^3He$ and $p + \alpha$ channels, and the other symbols are explained in ref. [CH 74]. These equations are then solved to yield the S-matrix and, subsequently, the differential scattering and reaction cross sections.

In order to account approximately for the many-body breakup channels not considered explicitly, we have again added phenomenological imaginary potentials into the formulation. This is done by simply replacing V_{Nf} by $V_{Nf} + i W_f$ and V_{Ng} by $V_{Ng} + i W_g$ in eqs. (12.28a) and (12.28b), and a similar replacement for the direct nuclear potential in the channel-spin $\frac{3}{2}$, $d + {}^3He$ channel. Typical results are shown in fig. 51 for the ${}^3He(d, p)\alpha$ reaction at $E_f = 10$ MeV,† and in fig. 52 for the $\alpha(p, d){}^3He$ reaction at $E_g = 68$ MeV. From these figures it is seen that there is a fairly good agreement with experimental data [BR 57a, KI 72, VO 74] over a rather wide range of energies. The presence of deep minima and the slight underestimate in the reaction cross section are likely due to the fact that, in these calculations, the nucleon-nucleon potential employed has no noncentral components.

We have also examined [CH 74a] the $\alpha(p, d)\,{}^3He$ reaction in the weak-coupling approximation or the distorted-wave Born approximation (DWBA). As is indicated by

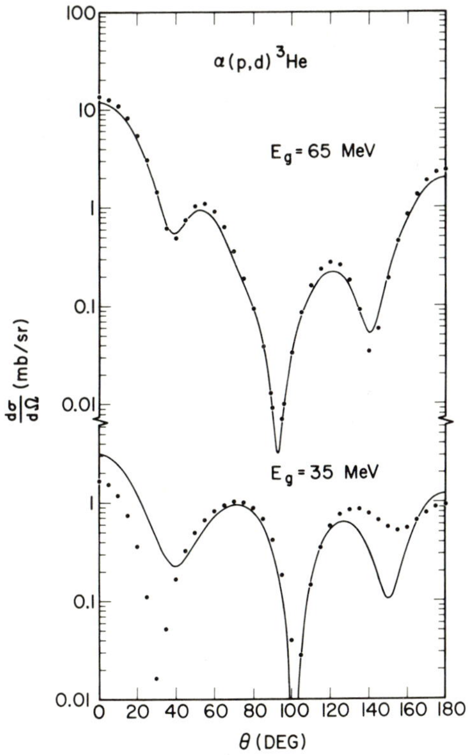

Fig. 53. Differential reaction cross sections for the $\alpha(p, d){}^3He$ reaction at $E_g = 65$ and 35 MeV. The solid dots show the results obtained by solving eqs. (12.28a) and (12.28b) with no further approximation, while the solid curves show the results obtained using the DWBA. (From ref. [CH 74a].)

† The value of E_g corresponding to $E_f = 10$ MeV is 30.45 MeV.

the discussion given in section 12.2, this is accomplished by solving eqs. (12.28a) and (12.28b) with K_{gf} (but not K_{fg}) set equal to zero. In fig. 53, a comparison between the differential reaction cross sections computed by solving eqs. (12.28a) and (12.28b) with no approximation (solid dots) and by solving these equations in the DWBA (solid curves) is shown. One can see that at a relatively high energy of E_g equal to 65 MeV, the DWBA is quite valid. At E_g = 35 MeV, the DWBA still reproduces the features of the full calculation, but the magnitude is somewhat incorrect.

12.3b. ^{6}Li (p, ^{3}He) α Reaction

As another example of an ADR calculation, we consider the ^{6}Li (p, ^{3}He) α reaction [HU 72, SC 74]. Here also, we employ for simplicity a purely central nucleon-nucleon potential; consequently, the coupling of the two channels is again effected only through channel spin $\frac{1}{2}$ states. In these states, the wave function is written as

$$\psi = \psi_f + \psi_g, \tag{12.29}$$

where

$$\psi_f = A\,\{N_f^{-1/2}\,\phi_{Li}(1234;56)\,F(\mathbf{R}_P - \mathbf{R}_{Li})\,\xi_f(s,t)\}, \tag{12.30}$$

and

$$\psi_g = A\,\{N_g^{-1/2}\,\phi_\alpha(1234)\,\phi_h(567)\,G(\mathbf{R}_h - \mathbf{R}_\alpha)\,\xi_g(s,t)\}, \tag{12.31}$$

with the subscripts f and g referring to the p + ^{6}Li and ^{3}He + α channels, respectively. The function ϕ_{Li} describes the spatial behaviour of the ^{6}Li cluster in its ground state; it is chosen to have a form suggested by the cluster model, namely, a form describing the relative oscillation of a deuteron and an α cluster.†

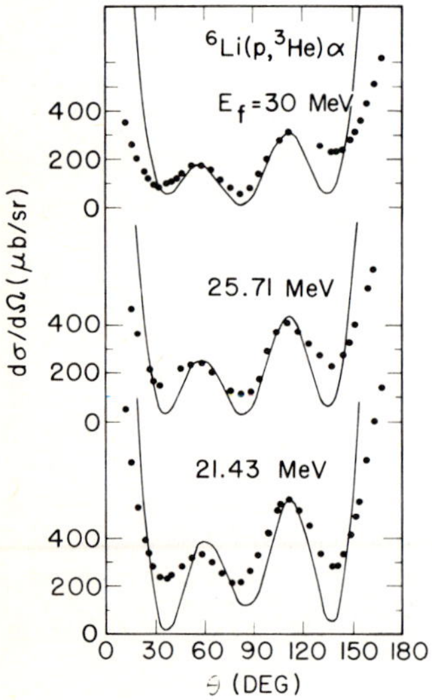

Fig. 54
Comparison of calculated and experimental differential reaction cross sections for the ^{6}Li(p, ^{3}He) α reaction. (Adapted from ref. [SC 74].)

† For the explicit form of the nucleon-nucleon potential employed and the cluster internal functions, see ref. [HU 72].

Substituting the wave function of eqs. (12.29)–(12.31) into the projection equation (12.1) leads again to coupled integrodifferential equations of the form given by eqs. (12.28a) and (12.28b). To these latter equations, we add then phenomenological imaginary potentials which are adjusted to fit as well as possible the peaks of the experimental differential reaction cross sections. The results at $E_f = 21.43, 25.71$, and 30.0 MeV are shown in fig. 54, where one sees that the calculation does reproduce the main features of the experimental data [SC 74]. The systematic deviation of the calculated cross sections from the experimental results at forward ($\lesssim 30^\circ$) and backward ($\gtrsim 150^\circ$) angles is likely a consequence of the fact that, in these calculations, a relatively simple $d + \alpha$ relative-motion function has been employed.

The DWBA has also been examined for this reaction. Here again, it was found that, at sufficiently high energies, the DWBA calculation does yield rather accurate results.

12.4. Influence of the Pauli Principle on Direct-Reactions

The examples in section 12.3 show that, at relatively high energies, coupled-channel ADR calculations can yield results which are in satisfactory agreement with experimental data. These examples can, therefore, be used to examine the connection between the Pauli principle and the various direct-reaction processes commonly employed in phenomenological studies. Specifically, what we wish to show in this section is that, because of the use of a totally antisymmetrized wave function, phenomenological direct-reaction processes, such as the pickup process and so on, are automatically contained in the ADR calculation.

The procedure we use is to first show an association between a direct-reaction process and certain terms in the antisymmetric wave function or, equivalently, certain terms in the coupling kernels K_{fg} and K_{gf} of eqs. (12.28a) and (12.28b). By comparing the result of the ADR calculation with that obtained from a calculation in which only these certain terms in K_{fg} and K_{gf} are included, we can then obtain a better understanding about the meaning of this particular direct-reaction process. In addition, through this type of investigation, we shall be able to answer the question concerning the situation under which a partial antisymmetrization of the wave function is permissible. This is important, because the use of fully antisymmetrized wave functions will be rather difficult in reaction calculations involving medium or heavy-weight nuclei.

Before we proceed further, we shall first discuss by means of the $^6Li\,(p, {}^3He)\,\alpha$ example, the types of direct-reaction processes which are involved. This is a particularly interesting example because, as has been discussed in subsection 3.5c, the lowest states of ^6Li can be described equivalently in the oscillator-limit case as having either a $d + \alpha$ cluster structure or a $t + {}^3He$ cluster structure.

In the $d + \alpha$ cluster representation for ^6Li, the direct-reaction processes involved are as follows:

(A) Deuteron-pickup and α-knockout process. Here the incoming proton picks up a deuteron cluster, thereby forming an outgoing ^3He particle.

(B) Single-particle exchange process. The incoming proton is exchanged with one of the protons in the α cluster. The exchanged proton is then "transferred" to the deuteron cluster, thus forming an outgoing ^{3}He particle. This process is sometimes referred to in the literature as a heavy-particle stripping process.

Similarly, one can consider the target nucleus ^{6}Li in the t + ^{3}He cluster representation. Here the direct-reaction processes involved are then the following:

(C) Triton-pickup and ^{3}He-knockout process. The incoming proton picks up a triton cluster and forms in this way an outgoing α particle.

(D) Single-particle exchange process. The incoming proton is exchanged with one of the protons in the ^{3}He cluster. The exchanged proton is "transferred" to the triton cluster, thus forming an outgoing α particle. This process is again sometimes referred to as a heavy-particle stripping process.

It is well-known that, in a realistic description of the ^{6}Li ground state, the d + α cluster structure is predominant. In our discussion here, we shall however employ wave functions in the oscillator-limit case where the equivalence of the d + α and the t + ^{3}He cluster description of the ^{6}Li ground state is complete. The reason for this is that, in this section, our main aim is to illustrate the possible implications arising from the use of different cluster representations for the target nucleus and, therefore, the adoption of completely equivalent target wave functions is necessary.

Finally, we wish to mention one important point. In the ^{6}Li (p, ^{3}He) α reaction, the Q-value is large and positive. As a consequence of this, the ^{3}He and α clusters always have, except in states of large relative orbital angular momentum, an appreciable chance of being close to each other. This means that the probability for the reaction to take place when the deuteron and the α clusters are in close proximity is quite appreciable. Thus, in this example, we shall always use a completely antisymmetric wave function with respect to the nucleons which make up the target nucleus because, as has often been mentioned, the correlations between the nucleons are substantially affected by the antisymmetrization procedure if the clusters overlap each other strongly. This is in contrast with many phenomenological direct-reaction calculations where a complete antisymmetrization of the target wave function is frequently not carried out.

In subsection 12.4a, we shall first make an examination of the effect of the Pauli principle in the ^{6}Li (p, ^{3}He) α reaction using the plane-wave Born approximation (PWBA). Then, in subsection 12.4b, we shall consider both the α(p, d) ^{3}He and the ^{6}Li (p, ^{3}He) α reactions with the coupled-channel formulation in order to make certain that the general conclusions reached by the simple PWBA treatment are essentially correct.

12.4a. Study of Direct-Reaction Mechanisms in the Plane-Wave Born Approximation

The reaction amplitude in the plane-wave Born approximation is simply given by

$$A_{fg}\,(\text{PWBA}) = c_P \langle \mathcal{A}\, \{\phi_\alpha (1234)\, \phi_h (567)\, \xi_g\, e^{i\mathbf{k}_g \cdot \mathbf{R}_g}\}\, | H - E |\, \phi_{Li} (1234; 56)\, \xi_f\, e^{i\mathbf{k}_f \cdot \mathbf{R}_f} \qquad (12.32)$$

with

$$\mathbf{R}_f = \mathbf{r}_7 - \frac{1}{6}(\mathbf{r}_1 + \mathbf{r}_2 + \mathbf{r}_3 + \mathbf{r}_4 + \mathbf{r}_5 + \mathbf{r}_6), \tag{12.33a}$$

$$\mathbf{R}_g = \frac{1}{3}(\mathbf{r}_5 + \mathbf{r}_6 + \mathbf{r}_7) - \frac{1}{4}(\mathbf{r}_1 + \mathbf{r}_2 + \mathbf{r}_3 + \mathbf{r}_4), \tag{12.33b}$$

and c_P being a constant factor, unimportant for our present consideration. In eq. (12.32), it is noted that the argument of the ^{6}Li wave function is written in the manner (1234; 56) to indicate that ^{6}Li is being considered as having a d + α cluster structure.

For our further discussion, it is appropriate to write A in the form

$$A = A_6(1 - P_{17} - P_{27} - P_{37} - P_{47} - P_{57} - P_{67}). \tag{12.34}$$

In the above equation, P_{ij} is an operator which exchanges the spatial, spin, and isospin coordinates of nucleons i and j, and A_6 is the antisymmetrization operator for nucleons 1 to 6 which make up the target nucleus ^{6}Li in the unantisymmetrized wave function for the initial channel, as denoted in eq. (12.32).

With eq. (12.34) we can separate A_{fg} into two parts

$$A_{fg} = A_{fg}^A + A_{fg}^B, \tag{12.35}$$

where

$$A_{fg}^A = c_P \langle A_6(1 - P_{57} - P_{67})\{\phi_\alpha(1234)\,\phi_h(567)\,\xi_g\, e^{i\mathbf{k}_g \cdot \mathbf{R}_g}\} \\ |H - E|\,\phi_{Li}(1234;56)\,\xi_f\, e^{i\mathbf{k}_f \cdot \mathbf{R}_f}\rangle, \tag{12.36a}$$

$$A_{fg}^B = c_P \langle A_6(-P_{17} - P_{27} - P_{37} - P_{47})\{\phi_\alpha(1234)\,\phi_h(567)\,\xi_g\, e^{i\mathbf{k}_g \cdot \mathbf{R}_g}\} \\ |H - E|\,\phi_{Li}(1234;56)\,\xi_f\, e^{i\mathbf{k}_f \cdot \mathbf{R}_f}\rangle. \tag{12.36b}$$

From the form of eq. (12.36a), one can see that the incoming proton comes out in the outgoing ^{3}He particle. Therefore, A_{fg}^A is the Born amplitude for the deuteron-pickup and α-knockout process (process A) mentioned in section 12.4. Similarly, it is easy to see that A_{fg}^B is the Born amplitude for the single-particle exchange process (process B).

Let us discuss now the behaviour of the amplitude A_{fg}^A. For this we consider first the quantity

$$A_{fg,d}^A = c_P \langle A_6(1 - P_{57} - P_{67})\{\phi_\alpha(1234)\,\phi_h(567)\,\xi_g\, e^{i\mathbf{k}_g \cdot \mathbf{R}_g}\} \\ |H_d - E|\,\phi_{Li}(1234;56)\,\xi_f\, e^{i\mathbf{k}_f \cdot \mathbf{R}_f}\rangle, \tag{12.37}$$

which is a part of A_{fg}^A, obtained by omitting in eq. (12.36a) all terms of the Hamiltonian H which contain Majorana space-exchange operators. In $A_{fg,d}^A$ there are permutations which do not exchange nucleons between the two clusters in the final channel. The corresponding integrals have then the common factor

$$\Lambda_d(\mathbf{k}_f, \mathbf{k}_g; \mathbf{r}_1, \ldots, \mathbf{r}_7) = e^{i\mathbf{k}_f \cdot \mathbf{R}_f - i\mathbf{k}_g \cdot \mathbf{R}_g} \tag{12.38}$$

which mainly determines their magnitudes as a function of the angle between $\mathbf{k}_f$ and $\mathbf{k}_g$. This is an important factor, because at high energies the wavelengths corresponding to k_f and k_g are much shorter than the wavelengths corresponding to the internal motion of the nucleons inside the clusters and the range of the nucleon-nucleon potential. Thus, if for a certain combination of $\mathbf{k}_f$ and $\mathbf{k}_g$ the factor Λ_d oscillates rapidly as a function of the nucleon spatial coordinates, the corresponding integrals in the amplitude $A^A_{fg,d}$ will have a small magnitude. On the other hand, if the values of $\mathbf{k}_f$ and $\mathbf{k}_g$ are such that Λ_d is not too rapidly varying, then the magnitudes of the corresponding integrals in $A^A_{fg,d}$ will be considerably larger.

At high energies, the wave numbers k_f and k_g satisfy the approximate relation

$$k_g \approx \sqrt{2}\, k_f = \sqrt{2}\, k. \tag{12.39}$$

By using eqs. (12.33a) and (12.33b), and defining further unit vectors $\mathbf{n}_f$ and $\mathbf{n}_g$ in the directions of $\mathbf{k}_f$ and $\mathbf{k}_g$, respectively, we can write Λ_d as

$$\begin{aligned}\Lambda_d = \exp\Big[& ik\left(\mathbf{n}_f - \frac{\sqrt{2}}{3}\,\mathbf{n}_g\right)\cdot \mathbf{r}_7 - ik\left(\frac{1}{6}\,\mathbf{n}_f + \frac{\sqrt{2}}{3}\,\mathbf{n}_g\right)\cdot(\mathbf{r}_5 + \mathbf{r}_6) \\ & - ik\left(\frac{1}{6}\,\mathbf{n}_f - \frac{\sqrt{2}}{4}\,\mathbf{n}_g\right)\cdot(\mathbf{r}_1 + \mathbf{r}_2 + \mathbf{r}_3 + \mathbf{r}_4)\Big] .\end{aligned} \tag{12.40}$$

A close examination of this expression shows that Λ_d is more rapidly varying when $\mathbf{n}_f$ and $\mathbf{n}_g$ are antiparallel than when $\mathbf{n}_f$ and $\mathbf{n}_g$ are parallel. Similarly, one can easily see that the other integrals in $A^A_{fg,d}$, which, due to the presence of the operator A_6, involve exchanges of nucleons between the two clusters in the final channel, exhibits essentially the same angular behaviour at high energies. This means that, as far as the amplitude $A^A_{fg,d}$ is concerned, there is a maximum probability for the ^{3}He particle to come out in the forward direction and a minimum probability in the backward direction. These two particular directions correspond, of course, to a minimum and maximum momentum-change for the incident proton, respectively.

The quantity

$$\begin{aligned}A^A_{fg,x} = c_p \langle & A_6\,(1 - P_{57} - P_{67})\,\{\phi_\alpha(1234)\,\phi_h(567)\,\xi_g\, e^{i\mathbf{k}_g\cdot\mathbf{R}_g}\} \\ & |H_x|\,\phi_{Li}(1234;56)\,\xi_f\, e^{i\mathbf{k}_f\cdot\mathbf{R}_f}\rangle ,\end{aligned} \tag{12.41}$$

where H_x is an operator which represents the part of the Hamiltonian H containing Majorana space-exchange operators, can be studied in a similar manner. For example, the space exchange of nucleon 1 with nucleon 7 will result in the appearance of a factor

$$\begin{aligned}\Lambda_x = \exp\Big[& ik\left(\mathbf{n}_f + \frac{\sqrt{2}}{4}\,\mathbf{n}_g\right)\cdot \mathbf{r}_1 - ik\left(\frac{1}{6}\,\mathbf{n}_f + \frac{\sqrt{2}}{3}\,\mathbf{n}_g\right)\cdot(\mathbf{r}_5 + \mathbf{r}_6 + \mathbf{r}_7) \\ & - ik\left(\frac{1}{6}\,\mathbf{n}_f - \frac{\sqrt{2}}{4}\,\mathbf{n}_g\right)\cdot(\mathbf{r}_2 + \mathbf{r}_3 + \mathbf{r}_4)\Big]\end{aligned} \tag{12.42}$$

in some of the terms. An examination of this and similar factors reveals that, for this particular space exchange, a new feature arises; that is, these factors will give rise to a backward maximum in the ^{3}He angular distribution. At high energies, this backward maximum will, however, be quite small compared to the forward maximum coming from non-space-exchange contributions, as can be shown by a more detailed investigation of the two amplitudes $A^A_{fg,d}$ and $A^A_{fg,x}$.†

Thus, in conclusion, we find that process A gives rise to both forward and backward peaks in the ^{3}He angular distribution. These peaks can be conveniently regarded as arising from the deuteron-pickup process and the α-knockout process, respectively. We should emphasize again, however, that at high energies the backward peak from this process is small compared with the forward peak and, hence, does not need to be seriously considered in most discussions.

The amplitude A^B_{fg}, which represents the single-particle exchange process B, can be examined in a very similar way. Here the result shows that this process also yields both forward and backward peaks in the ^{3}He angular distribution, but with the backward peak having a much larger magnitude.

Next, we study the case where the target nucleus ^{6}Li is considered in the t + ^{3}He cluster representation. Here it is convenient to write the amplitude A_{fg} (PWBA) as

$$A_{fg}\,(\mathrm{PWBA}) = c_P \langle \mathcal{A} \{\phi_h(123)\,\phi_\alpha(4567)\,\xi_g\, e^{i\mathbf{k}_g\cdot\bar{\mathbf{R}}_g}\} \\ |H - E|\,\phi_{Li}(123;456)\,\xi_f\, e^{i\mathbf{k}_f\cdot\bar{\mathbf{R}}_f}\rangle, \tag{12.43}$$

with

$$\bar{\mathbf{R}}_f = \mathbf{R}_f = \mathbf{r}_7 - \frac{1}{6}(\mathbf{r}_1 + \mathbf{r}_2 + \mathbf{r}_3 + \mathbf{r}_4 + \mathbf{r}_5 + \mathbf{r}_6), \tag{12.43a}$$

$$\bar{\mathbf{R}}_g = \frac{1}{3}(\mathbf{r}_1 + \mathbf{r}_2 + \mathbf{r}_3) - \frac{1}{4}(\mathbf{r}_4 + \mathbf{r}_5 + \mathbf{r}_6 + \mathbf{r}_7). \tag{12.43b}$$

Using eq. (12.34) we can again separate A_{fg} into two parts, i.e.,

$$A_{fg} = A^C_{fg} + A^D_{fg}, \tag{12.44}$$

where

$$A^C_{fg} = c_P \langle \mathcal{A}_6 (1 - P_{47} - P_{57} - P_{67}) \{\phi_h(123)\,\phi_\alpha(4567)\,\xi_g\, e^{i\mathbf{k}_g\cdot\bar{\mathbf{R}}_g}\} \\ |H - E|\,\phi_{Li}(123;456)\,\xi_f\, e^{i\mathbf{k}_f\cdot\bar{\mathbf{R}}_f}\rangle, \tag{12.45a}$$

$$A^D_{fg} = c_P \langle \mathcal{A}_6 (-P_{17} - P_{27} - P_{37}) \{\phi_h(123)\,\phi_\alpha(4567)\,\xi_g\, e^{i\mathbf{k}_g\cdot\bar{\mathbf{R}}_g}\} \\ |H - E|\,\phi_{Li}(123;456)\,\xi_f\, e^{i\mathbf{k}_f\cdot\bar{\mathbf{R}}_f}\rangle. \tag{12.45b}$$

As is easily seen, A^C_{fg} is the Born amplitude for the triton-pickup and ^{3}He-knockout process (process C), while A^D_{fg} is the Born amplitude for the single-particle exchange process (process D).

---

† A quantitative PWBA treatment has also been carried out for the α(p, d) ^{3}He problem [CH 75a], and similar conclusions have been reached.

An analysis similar to that described above can now be performed to determine the behaviour of A^C_{fg} and A^D_{fg}. The result shows that, in the ^{3}He angular distribution, process C produces mainly a backward peak, while process D produces mainly a forward peak.

Our considerations using two different cluster representations for ^{6}Li show that there is a certain degree of equivalence between the pickup process in one representation and the exchange process in the other (complementary) representation. From the discussion given above, one can easily understand that this equivalence should become quite exact at high energies and when the target wave function is constructed in the oscillator shell-model limit. As one deviates from the oscillator limit and chooses for the description of ^{6}Li a more realistic cluster wave function, then the degree of this equivalence will be reduced. But even then, one can still expect that our general conclusions concerning the appearance of forward and backward peaks for the various direct-reaction processes will remain essentially correct.

Finally, we wish to mention that the general aspects of our considerations will remain true in reactions involving heavier nuclei. Due to the larger mass of the target nucleus, the oscillatory nature of the factor Λ_d in eq. (12.38) will be more pronounced. As a consequence of this, one anticipates that our discussion in regard to the contributions from different direct-reaction processes should become even more valid in a heavier system.

12.4b. Study of Direct-Reaction Mechanisms with the Coupled-Channel Formulation

The considerations given in the preceding subsection are more or less of qualitative nature. In this subsection, we shall quantitatively verify the results obtained there by means of coupled-channel calculations. In addition, we wish to examine how valid these results are at lower energies. The problems which will be considered here are the α(p, d) ^{3}He and ^{6}Li(p, ^{3}He) α reactions. For the latter reaction, we shall also explicitly show the equivalence mentioned above concerning the general behaviour of direct-reaction processes in different target representations.

12.4b (i). α(p, d) ^{3}He Reaction

The formulation of the α(p, d) ^{3}He reaction is given in subsection 12.3a.† By expressing the operator A as

$$A = A_4(1 - P_{15} - P_{25} - P_{35} - P_{45}), \qquad (12.46)$$

where A_4 is the antisymmetrization operator for the α cluster, we can divide it into two parts in an analogous way as in eqs. (12.36a) and (12.36b). With this division, the coupling kernel K_{fg} can then be written as

$$K_{fg} = K^A_{fg} + K^B_{fg}, \qquad (12.47)$$

† It should be noted that here the p + α channel (labelled by g) is chosen as the incident channel.

where

$$K_{fg}^{A}(\mathbf{R}_f', \mathbf{R}_g'') = (N_f N_g)^{-1/2} \langle A_4 (1 - P_{45}) \{\phi_h(123)\, \phi_d(45)\, \xi_f\, \delta(\mathbf{R}_f - \mathbf{R}_f')\} |H - E| \phi_\alpha(1234)\, \xi_g\, \delta(\mathbf{R}_g - \mathbf{R}_g'')\rangle, \quad (12.48a)$$

$$K_{fg}^{B}(\mathbf{R}_f', \mathbf{R}_g'') = (N_f N_g)^{-1/2} \langle A_4 (-P_{15} - P_{25} - P_{35}) \{\phi_h(123)\, \phi_d(45)\, \xi_f\, \delta(\mathbf{R}_f - \mathbf{R}_f')\} |H - E| \phi_\alpha(1234)\, \xi_g\, \delta(\mathbf{R}_g - \mathbf{R}_g'')\rangle, \quad (12.48b)$$

with

$$\mathbf{R}_f = \frac{1}{3}(\mathbf{r}_1 + \mathbf{r}_2 + \mathbf{r}_3) - \frac{1}{2}(\mathbf{r}_4 + \mathbf{r}_5), \quad (12.49a)$$

$$\mathbf{R}_g = \frac{1}{4}(\mathbf{r}_1 + \mathbf{r}_2 + \mathbf{r}_3 + \mathbf{r}_4) - \mathbf{r}_5. \quad (12.49b)$$

One can clearly see from the forms of eqs. (12.48a) and (12.48b) that K_{fg}^A corresponds to a process in which the incident proton is contained in the outgoing deuteron, while K_{fg}^B corresponds to a process in which the incident proton is contained in the outgoing ^{3}He. In other words, K_{fg}^A describes a one-nucleon pickup process (process A), while K_{fg}^B describes a single-nucleon exchange process (process B) or a two-nucleon pickup process (process C).†

Since the purpose of this study is to gain an understanding of direct-reaction mechanisms, we shall simplify the calculation of subsection 12.3a in order to bring out the important features in a clear manner. These simplifications are: (i) a single-Gaussian

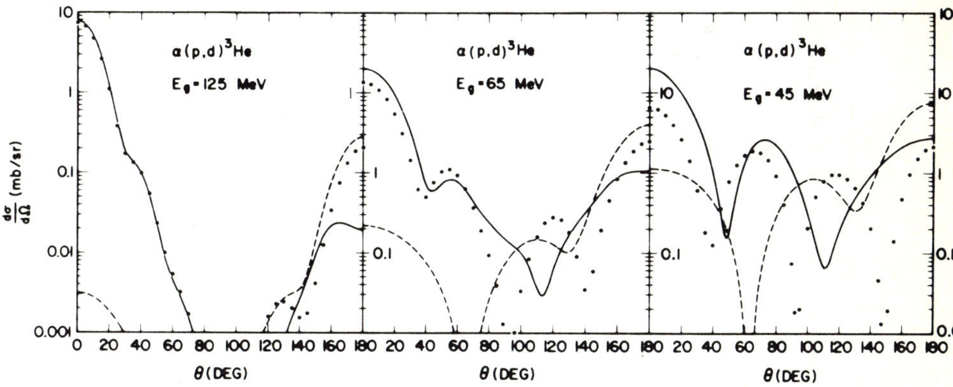

Fig. 55. Differential reaction cross sections for the α(p, d) ^{3}He reaction at E_g = 125 MeV, 65 MeV, and 45 MeV. The solid dots show the results obtained using the full coupling kernel K_{fg}, the solid curves show the results obtained with process A, and the dashed curves show the results obtained with process B. (From ref. [CH 74a].)

† Due to the particular choice of ϕ_α (see eq. (5.85)), processes B and C are entirely equivalent in this problem.

deuteron function which yields the same rms radius as the three-Gaussian function employed in the calculation of subsection 12.3a, is used, (ii) a simplified nucleon-nucleon potential with a Serber exchange mixture is employed; this potential is given by eqs. (7.5) and (7.57)–(7.60), and (iii) the Coulomb interaction is neglected. These simplifications do not change the essential features of the results described in subsection 12.3a, but they do allow for a more straightforward interpretation.

Differential reaction cross sections at E_g = 125, 65, and 45 MeV [CH 74a] are shown in fig. 55. In this figure, the solid dots represent the results obtained using the full coupling kernel K_{fg}, while the solid and dashed curves represent the results obtained when K_{fg}^B and K_{fg}^A are respectively set equal to zero.† Here it is seen that, at a rather high energy of E_g = 125 MeV, process A yields a large forward peak and a much smaller backward peak, while process B yields a large backward peak and a much smaller forward peak. This is in agreement with the PWBA result of subsection 12.4a. Also, one sees that, at this energy, the cross section in the forward angular region is explained very well by process A (i. e., K_{fg}^A alone is important), while the cross section in the backward angular region is explained satisfactorily by process B (i. e., K_{fg}^B alone is important). At lower energies, this separation into angular regions where either process A or process B alone can describe the reaction becomes less clear. At E_g = 65 MeV, for example, the use of a single direct-reaction mechanism can still account for the differential reaction cross sections in the forward ($\lesssim 70°$) and backward ($\gtrsim 140°$) angular regions in a fairly satisfactory manner, although a lack of detailed agreement does begin to appear. In fact, it can be seen that, at E_g = 45 MeV, the contributions from both processes become quite comparable at almost all angles; therefore, interference effects are important and the reaction is rather poorly described, even in a limited angular region, by a single process.

In conclusion, it is seen from this study that exchange effects are quite important in the α(p, d) ^{3}He reaction and a proper treatment of these effects by the use of a totally antisymmetrized wave function is in general necessary for a satisfactory description of the differential reaction cross section over the entire angular region. However, at sufficiently high energies, we do find that a simpler description in terms of a single direct-reaction mechanism in either the forward or the backward angular region does yield adequate results for this reaction.

12.4b (ii). ^{6}Li (p, ^{3}He) α Reaction

The procedure of studying the roles played by various direct-reaction mechanisms in the ^{6}Li (p, ^{3}He) α reaction [HU 72] is similar to that given above for the α(p, d) ^{3}He reaction. That is, we write the coupling kernel K_{fg} as a sum of K_{fg}^A and K_{fg}^B which correspond to processes A and B, respectively, or as a sum of K_{fg}^C and K_{fg}^D which correspond to processes C and D, respectively. By setting these kernel terms successively equal to zero and comparing the result with that obtained using the full coupling kernel K_{fg} (ADR calculation), one can then study the importance of each of these direct-reaction processes individually.

† Unless otherwise stated, when we say that K_{fg}^A (K_{fg}^B) is set equal to zero, we imply that the hermitian conjugate kernel K_{gf}^A (K_{gf}^B) is also set equal to zero.

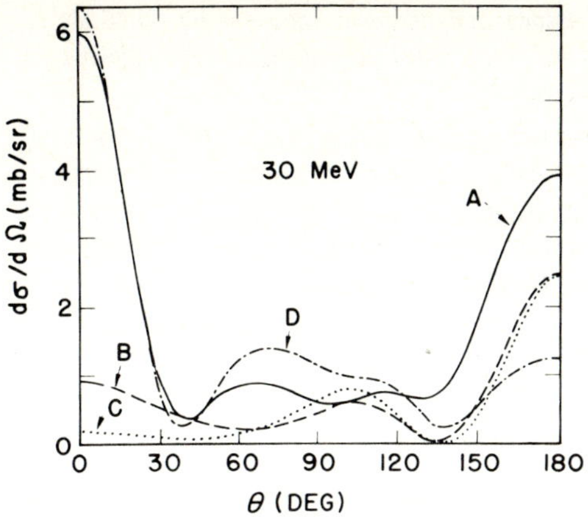

Fig. 56
Comparison of ^{6}Li (p, ^{3}He) α differential reaction cross sections obtained with processes A, B, C, and D at E_f = 30 MeV. (Adapted from ref. [HU 72].)

First, we discuss the result of our investigation regarding the equivalence of processes A and D, or processes B and C. As was mentioned in subsection 12.4a, a complete equivalence of these processes is approached only at the high-energy limit. From our calculation, we have found that this equivalence is indeed nonexistent at E_f = 10 MeV and at lower energies. This means that at relatively low energies the reaction amplitudes for processes A and B, which are defined when the target nucleus ^{6}Li is described in the d + α cluster representation, have to be given as a linear superposition of reaction amplitudes for processes C and D, which are defined when the target nucleus ^{6}Li is described in the t + ^{3}He cluster representation.

However, already at an energy of E_f = 30 MeV, this equivalence of processes A and D or B and C begins to hold fairly well. In fig. 56 it is shown that for processes A and D this equivalence exists especially in the ^{3}He forward directions and for processes B and C especially in the ^{3}He backward directions. With increasing energy, the equivalence of

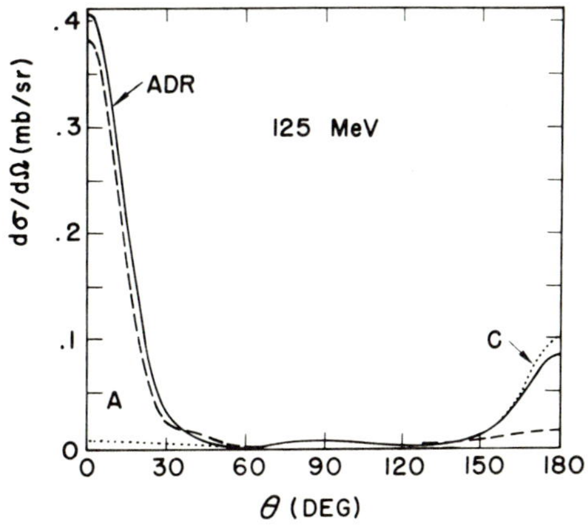

Fig. 57
Comparison of ^{6}Li(p, ^{3}He) α differential reaction cross sections obtained with processes A and C and the ADR calculation at E_f = 125 MeV. (Adapted from ref. [HU 72].)

these direct-reaction processes is found to be accomplished even more; at $E_f = 125$ MeV, the angular distributions obtained with processes A and B can practically no longer be distinguished from those obtained with processes D and C.

At high energies we find also that it is possible to describe the ^{6}Li (p, ^{3}He) α reaction in either the forward or the backward angular region by means of a single direct-reaction process. This is illustrated at $E_f = 125$ MeV in fig. 57, where one sees that processes A (or D) and C (or B) yield an adequate description of the ADR result at forward and backward angles, respectively. When one goes to lower energies, then such a simplified description in terms of a single reaction mechanism is again found to become less adequate, exactly as has been learned in the α (p, d) ^{3}He study described in the preceding subsection.

12.5. Concluding Remarks

In section 12.2, we have presented the general formulation of the direct-reaction theory and discussed various approximations which could be made in different circumstances. Then, we have applied the coupled-reaction-channel approach discussed in that section to two realistic examples, namely, the ^{3}He (d, p) α and the ^{6}Li (p, ^{3}He) α reactions, and showed that a satisfactory agreement with experiment can be obtained in both cases. These examples were then used to examine the physical background of various phenomenological direct-raction mechanisms. As a result of this examination, it was found that the usual description of direct reactions as pickup process, single-particle exchange process, and so on is not unequivocal. These designations will depend on the particular cluster representation which one chooses to describe the target nucleus, and there are situations under which some of these direct-reaction mechanisms are very nearly equivalent.

When the energy is high, it was shown that the differential reaction cross section in either the forward or the backward angular region can be adequately described by a single direct-reaction process. This means that, at high energies, a complete antisymmetrization between the nucleons in the incident and target clusters may not need to be carried out. At lower energies where most phenomenological direct-reaction analyses have been performed, we have found, however, that exchange effects are generally important and the use of a single direct-reaction mechanism may not be adequate even in a limited angular region. It should be noted that in these phenomenological analyses the differential reaction cross section was usually normalized by fitting some experimental peaks in the forward and backward directions. By such a procedure, it was frequently possible to obtain fairly reasonable agreement with experimental data. However, in view of our finding that in the low- and medium-energy region interference effects between different direct-reaction processes are important, one must conclude that the interpretation of the results from these analyses in terms of the considered reaction processes may need to be carefully examined.

In appendix D we shall briefly discuss, by means of the ^{6}Li (p, ^{3}He) α reaction, the connection between the direct-reaction theory commonly used in phenomenological analyses and the considerations presented here. This discussion will be useful, because it brings into view the approximations which are inherent in these phenomenological calculations and shows how one can modify these calculations for better results.

13. Some Considerations About Heavy-Ion Transfer Reactions

13.1. General Remarks

In recent years, a great deal of theoretical and experimental interest has been concentrated on heavy-ion reactions [BR 56, BR 56a] as a tool to study nuclear structure and reaction mechanisms.† In this chapter, we shall therefore briefly show how a theoretical description of these reactions can be included in the unified theory of this monograph. What we shall do here is not to cover this field in any complete manner, but rather to concentrate on some crucial points, in connection with the Pauli principle, which are essential for a satisfactory description of such reactions from a microscopic viewpoint.

First, we shall briefly summarize some of the main characteristics of heavy-ion reactions [AN 70]. These are as follows:

(a) The condition $kR_i \gg 1$, with R_i being the nuclear interaction radius, is always satisfied. This means that a large number of partial waves is involved in the process and, consequently, the motion of the scattering partners outside the nuclear-interaction region can be well described by classical orbits. Thus, for scattering energies much higher than the Coulomb barrier, the angle θ_C up to which the differential cross section has a Coulomb behaviour can be calculated classically and is given by

$$\theta_C \approx 2 \tan^{-1} (\eta/L), \qquad (13.1)$$

where η is the Sommerfeld parameter and L is the orbital-angular-momentum quantum number of the highest partial wave which feels the nuclear interaction. For $\theta > \theta_C$, the differential cross section generally decreases more rapidly than that calculated with the Rutherford formula. However, a deviation from this general rule can occur when the process of elastic transfer has a large probability. An example of this is the scattering of α particles by ^6Li, where the transfer of a loosely bound deuteron cluster between the α-cluster cores does yield large scattering cross sections at backward angles.

(b) Because of the large number of protons involved, the Coulomb interaction has generally a larger influence than the nuclear interaction.†† This is especially so for reactions occurring at energies well below the Coulomb barrier. In this latter situation, some nuclear-interaction terms can in fact be neglected, thus simplifying the calculation considerably.

---

† By heavy-ion physics, one generally intends to mean physics of all processes in which both incident and target nuclei have a mass number greater than about four.

†† One interesting point should also be mentioned. In heavy-ion scattering involving nuclei containing a large number of nucleons ($A \gtrsim 100$), Coulomb-deformation effects can become quite important. These deformation effects can, however, be handled in a good approximation by using classical means, as has especially been shown by *Greiner* and his collaborators (see, e.g., ref. [EI 72]).

(c) Heavy ions have very short mean free paths in nuclear matter, so that reactions at relatively low energies occur mainly in the peripheral region; this makes detailed knowledge of the bound-state wave function inside the nucleus unnecessary.

(d) The mass and the angular momentum of the transferred cluster are respectively much smaller than the masses of the interacting nuclei and the maximum angular momentum involved in the process.

As has been discussed [AN 70], the above features allow the use of certain simplifications in the evaluation of transition amplitudes both for scattering and for transfer reactions. Here we shall concentrate mainly on one problem in connection with these simplifications, namely, the influence of the Pauli principle. The reason why we choose to consider this particular problem is quite obvious. For heavy-ion reactions involving medium or heavy-weight nuclei with $A \gtrsim 20$, microscopic calculations employing totally antisymmetrized wave functions are very difficult to perform. Therefore, it is important to know the types of antisymmetrization terms which may be omitted from the calculation without making serious errors.

For the purpose mentioned above, we shall study two specific examples, namely, the elastic scattering of α particles by ^{6}Li and the ^{6}Li(p, ^{3}He) α reaction. These particular examples are chosen, because the nuclear systems involved are large enough that certain of their features are representative of those occurring in heavy-ion reactions, yet at the same time are small enough that antisymmetrization can be completely carried out. With these examples, we can then compare results of fully antisymmetrized calculations with those from calculations where we omit certain exchange terms which, from physical arguments, are not expected to have much contribution. In this way, we can gain information about the importance of various types of exchange terms and, thereby, learn to perform approximate but reasonably accurate calculations in more complicated systems.

In addition, we shall employ the $\alpha + {}^6$Li example to study further the odd-even feature in the effective potential between nuclei and derive, under certain conditions, an equation which is similar to that used in the LCNO (linear combination of nuclear orbitals) model introduced by *von Oertzen* [VO 70]. This is useful, because the LCNO model has been fairly successful for a semi-phenomenological description of heavy-ion scattering processes where elastic transfer is important.

13.2. Specific Examples to Study the Influence of Antisymmetrization

In this section, we study the influence of antisymmetrization by examining two specific examples, namely, the $\alpha + {}^6$Li elastic scattering and the ^{6}Li(p, ^{3}He) α reaction. But before we present the details of these examinations, we shall first make some statements concerning the types of simplifications which are expected to be generally valid in heavy-ion problems. For this we consider the interaction of a heavy ion C_1 with a target nucleus composing of a core C_2 and a group D of loosely bound nucleons. At low bombarding energies below the Coulomb barrier, the relative-motion function between the target and the projectile will be vanishingly small at small intercluster distances; consequently, exchange terms in the wave function describing the interchange of one, two, or more nucleons between C_1 and C_2 should not contribute appreciably to the scattering amplitude. Certainly, for the loosely bound nucleons, their antisymmetrization with the nucleons of the cores C_1 and C_2 will in general be more important. In addition, it should be noted that, again as a result of the large separation between C_1 and C_2 in this particular energy region, the nuclear interaction between these cores can usually be omitted and only the long-ranged Coulomb interaction needs to be taken into consideration.

As the bombarding energy increases, the cores will eventually come into proximity of each other. Here, however, due to transitions into other channels, low partial waves will almost completely be absorbed from the elastic channel and the transfer channel under consideration; these partial waves can, therefore, be taken into account in a relatively simple manner by using, e. g., absorptive potentials. Higher partial waves, which contribute significantly to elastic and transfer processes, are now under the influence of high centrifugal and Coulomb barriers. Because of this, one can expect that for these higher partial waves the same class of exchange terms as that mentioned above can be neglected from the calculation.

With the omission of these exchange terms, the reduction in computational effort is quite appreciable. In the following, we shall show explicitly by means of the above-mentioned specific examples that, under appropriate conditions, these omissions are in fact reasonable and could lead to feasible calculations in even heavier systems.

13.2a. $\alpha + {}^6$Li Elastic Scattering at Low Energies

Low energy $\alpha + {}^6$Li elastic scattering can be considered not only as an example for heavy-ion elastic scattering, but also as an example for transfer reactions in which nucleons in one of the colliding particles are transferred to the other colliding particle. Using the $d + \alpha$ representation for ^{6}Li, we have in this example two identical α clusters (α_1 and α_2) as the cores C_1 and C_2, and a deuteron cluster which is transferred between these α clusters.

In the low-energy region we choose, for energetical reasons, the following trial wave function for the $\alpha + {}^6$Li system:

$$\psi = A' \{\hat{\phi}(\tilde{\alpha}_1)\, \hat{\phi}(\widetilde{\mathrm{Li}}_2)\, F(\mathbf{R})\}, \qquad (13.2)$$

where

$$\hat{\phi}(\widetilde{\mathrm{Li}}_2) = A' \{\hat{\phi}(\tilde{\mathrm{d}})\, \hat{\phi}(\tilde{\alpha}_2)\, \chi(\boldsymbol{\rho})\}. \qquad (13.3)$$

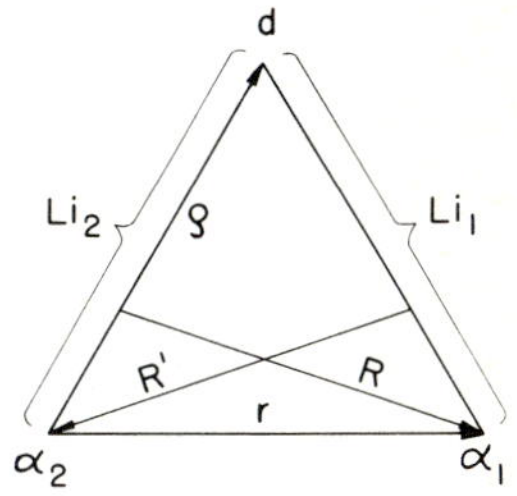

Fig. 58.
Definition of various spatial vectors used in the $\alpha + {}^6$Li system.

In the above equations, the various symbols are explained in fig. 58 and A' is an operator which antisymmetrizes the wave function with respect to nucleons in different clusters. The functions $\hat{\phi}(\tilde{\alpha}_1)$, $\hat{\phi}(\tilde{\alpha}_2)$, and $\hat{\phi}(\tilde{\mathrm{d}})$ are antisymmetrized internal functions for the clusters involved; they are chosen to have simple Gaussian spatial dependence for simplicity in calculation. The function $\chi(\boldsymbol{\rho})$ describes the relative motion between the deuteron and the α_2 clusters, and is also chosen to have a Gaussian form.† The linear variational function $F(\mathbf{R})$ is again determined from the projection equation

$$\langle \delta \psi | H - E | \psi \rangle = 0. \tag{13.4}$$

In the Hamiltonian H, the nucleon-nucleon potential used is the one given by eqs. (7.5) and (7.57)–(7.59) with y = 0.94. For this system a completely antisymmetrized calculation [CL 75a, RE 75] is still feasible, since the number of nucleons involved is not yet too large.

To see the influence of antisymmetrization, we systematically drop various exchange terms from the wave function of eq. (13.2). Thus, a second calculation was performed where the antisymmetrization between the deuteron cluster and the α cores was carried out, but where all terms containing one-, two-, or three-particle exchanges between the cores are omitted. Of course, because the two cores α_1 and α_2 are really identical, it is necessary to include the 4-particle $\alpha - \alpha$ exchange term in the calculation in order to preserve the Bose symmetry in this problem. With this approximation (approximation I), the wave function has then the following simpler form:

$$\psi_I = \hat{\phi}(\tilde{\alpha}_1)\,\hat{\phi}(\widetilde{\mathrm{Li}}_2)\,F(\mathbf{R}) + \hat{\phi}(\tilde{\alpha}_2)\,\hat{\phi}(\widetilde{\mathrm{Li}}_1)\,F(\mathbf{R}'), \tag{13.5}$$

where

$$\hat{\phi}(\widetilde{\mathrm{Li}}_1) = A'\,\{\hat{\phi}(\tilde{\mathrm{d}})\,\hat{\phi}(\tilde{\alpha}_1)\,\chi(\boldsymbol{\rho} - \mathbf{r})\}, \tag{13.6}$$

and $\mathbf{R}'$ is defined again in fig. 58. According to the discussion given above, this approximation should yield reasonably accurate results at energies below the Coulomb plus centrifugal barrier.

In the case where the target nucleus shows a strong $C_2 + D$ clustering and/or the residual nucleus in the transfer channel shows a strong $C_1 + D$ clustering (i.e., the cluster D is weakly bound to one or both cores), then it is possible that one can omit even those terms where the nucleons in the cluster D are exchanged with the nucleons in the core C_2 and/or the core C_1. To see if this is a valid low-energy approximation (approximation II) in the $\alpha + {}^6\mathrm{Li}$ case where the deuteron cluster is indeed loosely bound, we have performed a further calculation in which only the 4-particle $\alpha - \alpha$ exchange term is taken into consideration.

The results of the calculations [CL 75, CL 75a] are presented in fig. 59, where the nuclear phase shift δ_l is shown as a function of the relative energy E_R. In this figure, the solid curves represent the results of the fully antisymmetrized calculation, while the dashed and dotted curves represent the results obtained with approximation I and approximation II respectively. The arrows indicate the heights of the Coulomb-plus-centrifugal barriers, calculated with a reasonable value of 5.2 fm for the nuclear interaction radius.

With the fully antisymmetrized wave function of eq. (13.2), the calculation predicts the existence of both an $l = 0$ bound state with $E_R = -2.1$ MeV and an $l = 2$ resonance state in the low-energy region (see fig. 59). This means that, from the viewpoint of the

† Explicit expressions for $\hat{\phi}(\tilde{\mathrm{d}})$, $\hat{\phi}(\tilde{\alpha}_1)$, $\hat{\phi}(\tilde{\alpha}_2)$, and $\chi(\boldsymbol{\rho})$ are given in ref. [CL 75].

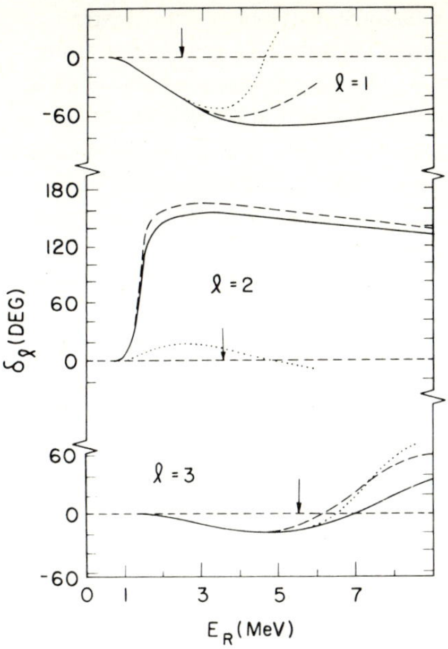

Fig. 59.
$\alpha + {}^6Li$ nuclear phase shift δ_l as a function of the relative energy E_R. The solid curves represent the results of the fully antisymmetrized calculation, while the dashed and dotted curves represent the results obtained with approximation I and approximation II, respectively. (Adapted from ref. [CL 75].)

cluster model, the T = 0 energy spectrum of ^{10}B should consist, as its lowest states, of one 3^+ state, one 2^+ state, and two 1^+ states. Experimentally [AJ 66], one finds that this is indeed the case. The calculated positions of these states are, however, too high on the energy scale – a discrepancy which comes from the fact that, in our present calculation involving an easily compressible ^6Li cluster, specific distortion effects have not been accounted for (see subsection 7.3c).†

To determine the validity of approximations I and II, we compare the phase-shift results exhibited in fig. 59. For the odd-*l* phase shifts one sees that, at energies somewhat below the Coulomb-plus-centrifugal barrier, there is a good agreement between the fully antisymmetrized calculation and the calculations where exchange terms are partially omitted.†† When the energy is above the barrier, the situation is, however, quite different. Here the colliding particles can penetrate each other, and a fully antisymmetrized calculation has, in general, to be performed in order to obtain accurate results.

In the case of even-*l* phase shifts, a more careful interpretation has to be made. For the $l = 2$ partial wave at energies below the Coulomb-plus-centrifugal barrier, it is seen from fig. 59 that the phase shifts calculated with fully and partially (especially approximation II) antisymmetrized wave functions can be quite different. The reason for this is that,

† Specific distortion effects can be crudely taken into account [TH 73] by adjusting the parameter y of eq. (7.59). A recent calculation by *Clement* [CL 75b] shows indeed that, if y is increased from 0.94 to 1.10, the agreement between calculated and experimental level positions is greatly improved.

†† The calculation also verifies that, at energies below the barrier, one can also neglect in good approximation the short-ranged nuclear interaction between the α cores.

as mentioned above, a ^{10}B compound state, which has essentially an $\alpha + {}^6Li$ cluster structure, exists in the energy region below the barrier. Therefore, at $\alpha + {}^6Li$ scattering energies near the resonance energy of this state, the colliding particles can come close to each other and no exchange terms can reasonably be omitted from the calculation.

The above-mentioned discrepancy between fully and partially antisymmetrized calculations will, in general, not occur in heavy-ion scattering where the heavy ions involved contain appreciably more than 12 nucleons. This is so, because compound states which can be strongly excited from the incident channel (i. e., potential resonances) are not expected to exist below the Coulomb-plus-centrifugal barrier. In cases involving not-too-heavy colliding particles (for instance, $^{12}C + {}^{12}C$ scattering), such states can, however, appear for a certain time as pre-equilibrium or doorway states before they decay into compound-nucleus configurations. For these cases, a partial antisymmetrization of the wave function may therefore be inadequate and can be employed only upon a careful examination of the level structure of the compound nucleus.

To demonstrate that low-energy $\alpha + {}^6Li$ elastic scattering can also be considered as an example for transfer reactions, we study this problem with approximation I (see eq. (13.5)) in a somewhat different manner. Instead of using the one-channel wave function ψ_I which contains one variational function $F(\mathbf{R})$, one can also use for this problem a two-channel function ψ_I', given by

$$\psi_I' = \hat{\phi}(\tilde{\alpha}_1)\,\hat{\phi}(\widetilde{Li}_2)\,F(\mathbf{R}) + \hat{\phi}(\tilde{\alpha}_2)\,\hat{\phi}(\widetilde{Li}_1)\,G(\mathbf{R}'), \tag{13.7}$$

which contains two independent variational functions $F(\mathbf{R})$ and $G(\mathbf{R}')$. By solving for $F(\mathbf{R})$ and $G(\mathbf{R}')$ in the usual way, one can subsequently calculate all observable magnitudes as, for instance, the $\alpha + {}^6Li$ elastic-scattering cross section and so on. For these calculations, it is of course necessary to add the scattering and transfer amplitudes coherently, because the two channels are in reality indistinguishable. In this way, one can easily see that one obtains with the two-channel formulation the same results for the observable quantitites as with the one-channel formulation.

One useful consequence of treating $\alpha + {}^6Li$ scattering as a two-channel problem is as follows. From this treatment one can readily understand that the oscillatory behaviour of the differential scattering cross section comes as a result of the interference between direct and transfer processes.

13.2b. $^6Li\,(p, {}^3He)\,\alpha$ Reaction in States of Large Orbital Angular Momentum

The $\alpha + {}^6Li$ elastic-scattering problem is a rather artificial example of a transfer reaction, since the Q-value involved is equal to zero. In real transfer reactions the Q-values are, however, always nonzero, with the consequence that the antisymmetrization of the wave function may, in general, have to be carried out in a rather complete manner. To see this, let us consider a reaction where the Q-value is large and positive. For such a reaction, the relative energy of the nuclei in the transfer channel can be quite large, even when the relative energy in the incident channel has a small value. In states of low orbital angular momentum, the residual and the outgoing nuclei can therefore penetrate each other and no simplifications with respect to antisymmetrization could be made. On the

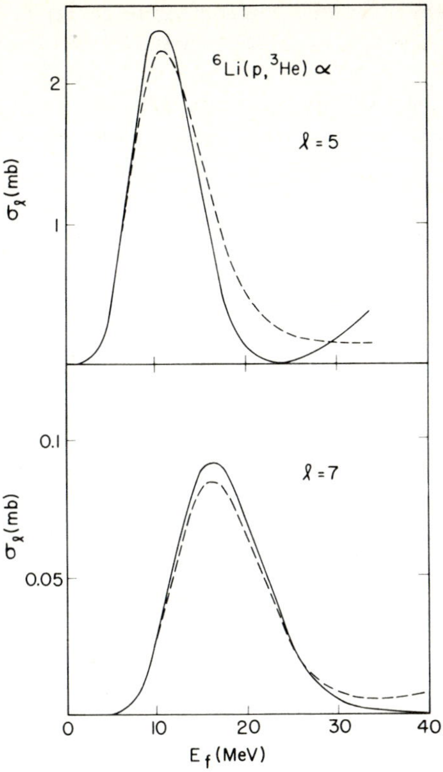

Fig. 60
Partial reaction cross section σ_l as a function of E_f for the ^6Li(p, ^3He) α reaction. The solid and dashed curves represent the results obtained with full and partial antisymmetrization, respectively. (Adapted from ref. [CL 75].)

other hand, in states of large orbital angular momentum, the energies in both the incident and the transfer channels may turn out to be smaller than the corresponding Coulomb-plus-centrifugal barrier heights and, according to arguments given above, some of the exchange terms may then be reasonably omitted from the calculation.

To verify the above statements, we study again the ^6Li (p, ^3He) α reaction which was considered also in subsection 12.3b. In this reaction, the Q-value is about 4 MeV which is already considerably larger than the Coulomb barrier of about 1.5 MeV in the $^3He + \alpha$ channel, calculated with a nuclear interaction radius equal to 4.0 fm.

The results [CL 75] for the partial reaction cross section σ_l with $l = 5$ and 7 are shown in fig. 60 as a function of E_f, the relative energy of the nuclei in the p + ^6Li channel. In this figure, the solid curves represent the results obtained with complete antisymmetrization, while the dashed curves represent the results obtained when the antisymmetrization between the incoming proton and the ^6Li target and between the outgoing ^3He and α clusters is neglected. Here one sees that, in the $l = 5$ case, the two curves begin to deviate significantly from each other at E_f equal to 15 MeV. This corresponds to a value for E_g, the relative energy of the clusters in the $^3He + \alpha$ channel, equal to about 19 MeV, a value which is somewhat smaller than the Coulomb-plus-centrifugal barrier of about 24 MeV in the ^{3}He + α channel. In the case with $l = 7$, the situation is similar. Here the deviation occurs at about 30 MeV for E_f, which corresponds to an E_g-value of about 34 MeV, again somewhat smaller than the ^{3}He + α Coulomb-plus-centrifugal barrier height of 44 MeV. These results confirm, therefore, our prediction, based on physical arguments, that the antisymmetrization between the interacting nuclei can be neglected if these nuclei in both the initial and the final channels do not penetrate each other.†, ††

13.3. Further Discussion of the Odd-Even Feature in the Effective Potential between Nuclei

In subsection 11.4b we have presented evidence to indicate that, at relatively high energies, the effective potential between nuclei should contain an odd-even orbital-angular-momentum dependence. Here we shall show explicitly by means of the $\alpha + {}^6$Li elastic-scattering example that, under conditions which are usually well satisfied in heavy-ion scattering at subcoulomb energies, a single-channel microscopic formulation of the scattering problem does lead to a one-body equation for the intercluster relative motion which involves an effective local potential having such a dependence on the orbital angular momentum and which is similar to the equation used in the LCNO (linear combination of nuclear orbitals) model of *von Oertzen* [VO 70].

Our present consideration is for heavy-ion scattering at energies close to or below the Coulomb barrier and where potential resonances of the compound nucleus do not exist. Under these conditions, it was shown in subsection 13.2a that, in the $\alpha + {}^6$Li case, the scattering can be adequately described by using approximation II where only the 4-particle $\alpha - \alpha$ exchange term is included in the calculation.

Another approximation which we shall use is to assume that the mass of the transferred cluster is much smaller than the masses of the core nuclei. In the $\alpha + {}^6$Li case, this condition is clearly not well satisfied. However, in scattering problems where rather heavy nuclei are involved (e.g., ^{16}O + ^{17}O scattering), this is certainly a reasonable approximation, because the number of loosely bound nucleons which can be transferred at subcoulomb energies is usually much smaller than the number of nucleons in the cores.

With these approximations, the spatial vectors $\mathbf{R}$ and $\mathbf{R}'$ can be written as $\mathbf{r}$ and $-\mathbf{r}$, respectively (see fig. 58), and the wave function describing the $\alpha + {}^6$Li system is simply given by

$$\psi = \hat{\phi}(\tilde{\alpha}_1)\ \{\hat{\phi}(\tilde{d})\,\hat{\phi}(\tilde{\alpha}_2)\,\chi(\boldsymbol{\rho})\}\ F(\mathbf{r}) + \hat{\phi}(\tilde{\alpha}_2)\ \{\hat{\phi}(\tilde{d})\,\hat{\phi}(\tilde{\alpha}_1)\,\chi(\boldsymbol{\rho}-\mathbf{r})\}\ F(-\mathbf{r}), \tag{13.8}$$

where the expressions inside the curly brackets represent ^{6}Li target functions obtained by omitting antisymmetrization with respect to nucleons in the α and deuteron clusters. Next, we introduce the relative momentum $\mathbf{p}_\rho$ between the deuteron cluster and the

† In this situation, one can then also neglect the short-ranged nuclear interaction between the cores which are the α particle and the proton in this example.

†† For a consideration of the problems, discussed in section 13.2, from a phenomenological coupled channel approach see e.g. ref. [BA 75] and references given there.

cluster α_2 and the relative momentum $\mathbf{p}_r$ between the two α clusters, and write the Hamiltonian approximately as

$$H \approx H_{\alpha_1} + H_{\alpha_2} + H_d + \frac{1}{2\mu_\rho} p_\rho^2 + \frac{1}{2\mu_r} p_r^2 + V_{\alpha_1,\alpha_2} + V_{d,\alpha_1} + V_{d,\alpha_2}, \tag{13.9}$$

where H_{α_1}, H_{α_2}, and H_d are the internal Hamiltonians of the clusters involved, and μ_ρ and μ_r represent the reduced masses of the $d + \alpha$ and $\alpha + \alpha$ systems, respectively. The quantity V_{α_1,α_2} represents a sum of two-nucleon potentials between nucleons in the cluster α_1 and nucleons in the cluster α_2, and the quantities V_{d,α_1} and V_{d,α_2} are defined in a similar way. In writing eq. (13.9), we have again made use of the condition that the mass of the transferred cluster is small compared to the masses of the cores.

To proceed further, we use again the projection equation (13.4). By carrying out the summation over all spin and isospin coordinates and the integration over all cluster internal spatial coordinates, we obtain, after some algebraic manipulation, the equation†

$$\left[T_r - E_r + \frac{J(\mathbf{r}) \pm K(\mathbf{r})}{1 \pm S(\mathbf{r})}\right] F_\pm(\mathbf{r}) = 0 \tag{13.10}$$

satisfied by the functions $F_\pm(\mathbf{r})$, defined as

$$F_\pm(\mathbf{r}) = F(\mathbf{r}) \pm F(-\mathbf{r}). \tag{13.11}$$

In eq. (13.10), the definitions of $S(\mathbf{r})$, $J(\mathbf{r})$, and $K(\mathbf{r})$ are as follows:

$$S(\mathbf{r}) = N_\rho \langle \chi(\boldsymbol{\rho}) | \chi(\boldsymbol{\rho} - \mathbf{r}) \rangle, \tag{13.12a}$$

$$J(\mathbf{r}) = N_\rho \langle \chi(\boldsymbol{\rho}) | \bar{V}_{d\alpha}(\boldsymbol{\rho} - \mathbf{r}) | \chi(\boldsymbol{\rho}) \rangle + \bar{V}_{\alpha\alpha}(\mathbf{r}), \tag{13.12b}$$

$$\begin{aligned} K(\mathbf{r}) = {} & T_r S(\mathbf{r}) - S(\mathbf{r}) T_r + S(\mathbf{r}) \bar{V}_{\alpha\alpha}(\mathbf{r}) \\ & + N_\rho \langle \chi(\boldsymbol{\rho}) | T_\rho + \bar{V}_{d\alpha}(\boldsymbol{\rho}) + \bar{V}_{d\alpha}(\boldsymbol{\rho} - \mathbf{r}) - E_\rho | \chi(\boldsymbol{\rho} - \mathbf{r}) \rangle, \end{aligned} \tag{13.12c}$$

where

$$N_\rho = \langle \chi(\boldsymbol{\rho}) | \chi(\boldsymbol{\rho}) \rangle^{-1} \tag{13.13}$$

and the Dirac brackets indicate an integration over the spatial variable $\boldsymbol{\rho}$. Also, in the above equations, T_r and T_ρ denote, respectively, the kinetic-energy operators for the $\alpha + \alpha$ and $d + \alpha_2$ relative motions, and E_r and E_ρ denote, respectively, the relative energies of the clusters in the $\alpha + {}^6Li$ system and in the nucleus ${}^6Li$. The quantities $\bar{V}_{\alpha\alpha}$ and $\bar{V}_{d\alpha}$ represent the effective interactions between the relevant clusters; they are obtained from V_{α_1,α_2} and V_{d,α_2} by summing and integrating over appropriate cluster internal coordinates. It should be noted that, because the α and ${}^6Li$ clusters do not

† For a derivation of this equation from a slightly different and more general viewpoint, see ref. [CL 74]. In this derivation, the wave function for the $\alpha + {}^6Li$ system is written as $\psi = \hat{\phi}(\widetilde{\alpha}_1)\, \hat{\phi}(\widetilde{Li}_2)\, F(\mathbf{r}) + \hat{\phi}(\widetilde{\alpha}_2)\, \hat{\phi}(\widetilde{Li}_1)\, F(-\mathbf{r})$, with $\hat{\phi}(\widetilde{Li})$ being the normalized (antisymmetrized) ground-state eigenfunction of the ${}^6Li$ Hamiltonian.

strongly penetrate each other at subcoulomb energies, the nuclear part of $\bar{V}_{\alpha\alpha}$ can be omitted and only the long-ranged Coulomb part needs to be taken into consideration.†

The function $F_+(\mathbf{r})$ contains only even-l partial waves, while the function $F_-(\mathbf{r})$ contains only odd-l partial waves.†† Recognizing this, one can easily see from eq. (13.10) that the effective potential between the α and ^{6}Li clusters has, as expected, an odd-even dependence on the orbital angular momentum. Indeed, this is the type of effective potential which has been employed in the LCNO model [BO 72a, VO 70] for a fairly successful analysis of the experimental data on heavy-ion scattering.

The utility of eq. (13.10) in treating heavy-ion scattering problems has not been fully examined at this moment. A preliminary test [CL 74] comparing calculated and experimental $\alpha + {}^6$Li differential cross sections at subcoulomb energies did show some promise in using this simplified approach. However, this is not a truly conclusive test, because the conditions under which eq. (13.10) is derived are not well satisfied in the $\alpha + {}^6$Li case. For a more definitive conclusion about the usefulness of this equation, it will be necessary to consider scattering problems such as ^{16}O + ^{17}O scattering and so on, where the approximations made in this derivation are much more valid.

13.4. Concluding Remarks

The considerations about the $\alpha + {}^6$Li scattering and the ^{6}Li(p, ^{3}He)α reaction at low energies can be extended to other elastic-scattering and transfer-reaction problems involving heavier nuclei. Here we list some examples which can be calculated in a rather straightforward fashion if one neglects the antisymmetrization between nucleons in different cores, the short-ranged nuclear interaction of the cores, and specific distortion effects. These examples are:

(a) Low-energy elastic-scattering processes, such as ^{16}O + ^{17}O scattering, ^{16}O + ^{18}F scattering, and ^{16}O + ^{20}Ne scattering. In these examples, the outside nucleons belong to the 2s − 1d shell in a shell-model description and the elastic-scattering channel is identical with the transfer channel as in the $\alpha + {}^6$Li case.

(b) Low-energy transfer reactions, such as ^{6}Li(^{16}O, ^{18}F)α reaction, ^{20}Ne(^{12}C, ^{16}O)^{16}O reaction, ^{7}Li(^{16}O, ^{19}F)α reaction, and ^{7}Be(^{16}O, ^{20}Ne)^{3}He reaction.

---

† We should also mention that the expression for $K(\mathbf{r})$ may be somewhat simplified. Because at subcoulomb energies the functions $F_\pm(\mathbf{r})$ are small at small intercluster distances and because, as a consequence of the weak binding between the deuteron and the α clusters in ^{6}Li, the function $S(\mathbf{r})$ is rather long-ranged, the influence of the operator $[T_r S(\mathbf{r}) - S(\mathbf{r}) T_r]$ on $F_\pm(\mathbf{r})$ may be neglected as compared with the influence of T_r on $F_\pm(\mathbf{r})$. Further, by choosing $\chi(\boldsymbol{\rho})$ to satisfy the equation $[T_\rho + \bar{V}_{d\alpha}(\boldsymbol{\rho}) - E_\rho]\,\chi(\boldsymbol{\rho}) = 0$ at least in the region where the deuteron and α clusters do not strongly overlap, we can then write $K(\mathbf{r})$ approximately as

$$K(\mathbf{r}) \approx S(\mathbf{r})\,\bar{V}_{\alpha\alpha}(\mathbf{r}) + N_\rho \langle \chi(\boldsymbol{\rho}) | \bar{V}_{d\alpha}(\boldsymbol{\rho} - \mathbf{r}) | \chi(\boldsymbol{\rho} - \mathbf{r}) \rangle .$$

†† It is noted from eq. (13.10) that even-l and odd-l partial waves do not couple with each other. This comes from the fact that in our derivation the approximate Hamiltonian of eq. (13.9) is used and, consequently, the effective operators which act on F_+ and F_- depend only on p_r^2 and $|\mathbf{r}|$.

The reason why these processes can be calculated rather straightforwardly is as follows. With the above-mentioned simplifications, the potential and kernel terms appearing in the corresponding coupled integrodifferential equations will mainly be the same as those belonging to simpler processes, as can be easily seen from the derivation, given in section 13.3, of the LCNO-type equation for $\alpha + {}^6Li$ scattering. For example, in the ${}^7Be({}^{16}O, {}^{20}Ne)\,{}^3He$ reaction where an α cluster is transferred, one needs, besides the Coulomb terms, mainly nuclear potential and kernel terms for the $\alpha + {}^3He$ and $\alpha + {}^{16}O$ systems and these terms are already known at this moment [MA 75, SU 72, TA 63].

Specific distortion effects, caused by the interaction between the loosely bound nucleons and the cores and by the relatively strong Coulomb interaction commonly present in heavy-ion reactions, can be taken into account in the usual manner. As was shown in subsection 4.2b, this can be achieved by adding to the trial wave function bound-state terms with linear variational amplitudes.

In heavy-ion transfer reactions with relatively large Q-values, it frequently happens that the residual nucleus is formed initially in a pre-equilibrium or doorway state which is not an eigenstate for the nucleus.† But due to the fact that in this pre-equilibrium state the relative orbital angular momentum between the transferred cluster and the core is usually quite large (in other words, the cluster is captured at the surface of the core), this state has very often a rather long lifetime ($\tau > 10^{-22}$ sec, see subsection 11.2d). Therefore, in such reactions, the residual nucleus in this pre-equilibrium state can be treated in a good approximation as stable and appropriate omission of exchange terms can correspondingly be considered in the calculation of various probability amplitudes.

Finally, we have shown by making reasonable approximations that, at subcoulomb energies, the effective potential between nuclei has an odd-even dependence on the orbital angular momentum. Together with the discussion given in subsection 11.4b, we are therefore of the opinion that, for a phenomenological study of the scattering data, the optical potential used should contain such a feature. In this respect, it is interesting to note that a number of phenomenological analyses [BO 72a, FR 72, VO 70, VO 74] at both high and low energies has shown that potentials of this type are indeed required in order to yield a satisfactory agreement with experimental results.

† This means that the pre-equilibrium or doorway state is represented by a superposition of a large number of compound-nucleus states.

14. Collective States

14.1. General Remarks

Our discussions have so far been mainly concentrated on nuclear problems dealing with scattering and reactions. To show that our theory is equally well suited for studying the physical properties of nuclear bound systems, we shall give in this chapter a microscopic description of rotational states. We choose to illustrate with this particular type of collective states simply for definiteness; as will be seen, our considerations can certainly be applied similarly also to other collective phenomena, such as nuclear vibrational excitations.

In section 14.2 we start with a discussion of rotational states in even-even nuclei with K = 0, where K is the projection of the angular momentum on the symmetry axis of the intrinsic wave function. Then in section 14.3 we generalize the considerations to include rotational states of odd-even and odd-odd nuclei, and also of even-even nuclei with K-values different from zero. Some general arguments are then given in section 14.4 concerning the circumstances under which nuclear rotational states appear; these arguments will again be based on the projection equation

$$\langle \delta \psi | H - E | \psi \rangle = 0, \tag{14.1}$$

which has been employed in all our previous discussions.

Electromagnetic transitions between rotational levels are briefly discussed in section 14.5. In section 14.6 we discuss the relationship between our and other descriptions of nuclear rotational states, and in section 14.7 we describe a method appropriate for quantitative studies of collective states in medium-heavy and heavy nuclei. Numerical examples are presented in section 14.8. Finally, in section 14.9, concluding remarks are made.

14.2. Rotational States of Even-Even Nuclei with K = 0

To discuss the essential properties of the energetically lowest rotational states in deformed even-even nuclei, we shall start by considering a specific example which is concerned with the three lowest states in the nucleus ^{8}Be. In this case, typical attempts [PE 56, WH 37] to explain the observed rotational energy spectrum have been to treat this nucleus as a rigid rotator; that is, one writes the wave functions for these states as

$$\hat{\psi}_{lm} = \hat{\phi}(\widetilde{\alpha}_1)\, \hat{\phi}(\widetilde{\alpha}_2)\, g(R)\, Y_{lm}(\theta, \varphi), \tag{14.2}$$

where $\hat{\phi}(\widetilde{\alpha}_1)$ and $\hat{\phi}(\widetilde{\alpha}_2)$ are antisymmetric functions describing the behaviour of the two α particles. The function $g(R)$ specifies the dependence on the radial distance $R = |\mathbf{R}_1 - \mathbf{R}_2|$ and is chosen, in the rigid-rotator model, to be the same for all angular-momentum values. In eq. (14.2), the important point to note is that the antisymmetrization is carried out within the α particles but not between them. If one now computes energy expectation values with these rigid-rotator functions, then one finds that the potential energies in all rotational states are identical because, with the wave function of eq. (14.2), the average distances between the nucleons do not depend on l. On the other hand, the kinetic energy increases with increasing l, thus giving rise to the characteristic energy spectrum

$$E_l = E_0 + \frac{\hbar^2}{2J} l(l+1) \tag{14.3}$$

where J denotes the moment of inertia.

A correct explanation of these levels must take into account the antisymmetrization between the α particles; i.e., the α particles must be treated as α clusters. In all our previous considerations about the lowest ^{8}Be levels, we have of course always done this. Here what we wish to do is to study these levels from a somewhat different viewpoint, namely, to see in what respect these levels can be considered as belonging to a rotational band. This study will be useful, because it will give us insight as to the general structure of rotational states in even-even nuclei with K = 0.

For this purpose, let us use the following oscillator cluster functions for ^{8}Be:

$$\psi_{lm} = A' \{\hat{\phi}(\tilde{\alpha}_1)\,\hat{\phi}(\tilde{\alpha}_2)\,\chi_{lm}(\mathbf{R})\}, \tag{14.4}$$

where $\hat{\phi}$ has a spatial part given by eq. (5.9a) and

$$\chi_{lm}(\mathbf{R}) = (R^4 + b_l R^2 + c_l)\exp(-aR^2)\,Y_{lm}(\theta, \varphi), \tag{14.5}$$

with b_l and c_l being l-dependent constants. Because the wave function ψ_{lm} is now totally antisymmetrized, the parts with factors $b_l R^2$ and c_l are of no consequence; they vanish upon antisymmetrization, since they correspond to relative motions between the α clusters which contain less than 4 quanta of excitation (see the discussion given in section 3.4). By comparing eqs. (14.2) and (14.4), one sees then that, if the rigid-rotator functions of eq. (14.2) with $g(R) = R^4 \exp(-aR^2)$ were totally antisymmetrized, they would be identical to the corresponding oscillator cluster functions of eq. (14.4). Despite this apparent similarity, the totally antisymmetrized cluster functions lead to a completely different explanation of the energy differences between the levels. Now the levels with $l = 0$, 2, and 4 all have the same kinetic energy but are split by differing potential energy [WI 59a]. Hence, the failure to antisymmetrize between the clusters in the rigid-rotator model shifts the cause of energy splitting completely from the potential energy to the kinetic energy.† In the oscillator limit for ^{8}Be, this shifting takes place in such a way that the ratio of rotational energy difference $(E_4 - E_0)/(E_2 - E_0) = 10/3$ is preserved.

For a more precise investigation of the change in nature of the energy spectrum caused by the transition from distinguishable to indistinguishable particles, we introduce angular parameter-coordinates†† $\omega' = (\theta', \varphi')$ and rewrite the antisymmetrized ^{8}Be wave function in the following way:

$$\psi_{lm} = \int \psi_i(\tilde{\mathbf{r}}_1, \ldots, \tilde{\mathbf{r}}_8; \theta', \varphi')\, Y_{lm}(\theta', \varphi')\, d\omega', \tag{14.6}$$

† This type of antisymmetrization effect is a very general one. We shall discuss it further later.

†† The introduction of angular parameter coordinates enables us to define the symmetry axis of the intrinsic wave function independently of the nucleon coordinates (see subsection 3.7b).

where ψ_i is an intrinsic wave function defined as

$$\psi_i = A' \{\hat{\phi}(\tilde{\alpha}_1)\,\hat{\phi}(\tilde{\alpha}_2)\,g(R)\,\delta(\omega - \omega')\}. \tag{14.7}$$

In the above equation, the function g(R) can be any normalizable function; i. e., we shall now consider cases which are more general than the oscillator limit case. Also, in eq. (14.7) the delta function $\delta(\omega - \omega')$ is given by

$$\delta(\omega - \omega') = \frac{\delta(\theta - \theta')\,\delta(\varphi - \varphi')}{\sin\theta} = \sum_{\lambda\mu} Y_{\lambda\mu}(\theta, \varphi)\, Y^*_{\lambda\mu}(\theta', \varphi'). \tag{14.8}$$

In the case where $\omega' = 0$, eq. (14.8) is further reduced to†

$$\delta(\omega) = \frac{1}{2\pi}\frac{\delta(\theta)}{\sin\theta} = \sum_{\lambda}\left(\frac{2\lambda + 1}{4\pi}\right)^{1/2} Y_{\lambda 0}(\theta, \varphi). \tag{14.9}$$

From the form of eq. (14.7), one sees that ψ_i describes a configuration which is rotationally symmetric around a symmetry axis specified by the angles θ' and φ'. This means that the wave function ψ_{lm} of eq. (14.6) is now written as a superposition of these intrinsic functions which differ only in the directions of their symmetry axes, with their amplitudes given by the function $Y_{lm}(\theta', \varphi')$.

For further considerations, it is useful to rewrite eq. (14.6) in still another way. For this we introduce the operator [ED 60]

$$R(\Omega') = \exp(-i\varphi' L_z/\hbar)\,\exp(-i\theta' L_y/\hbar) \tag{14.10}$$

for a finite rotation through the Euler angles

$$\Omega' = (-\varphi', -\theta', 0). \tag{14.11}$$

In eq. (14.10), L_y and L_z are the y- and z-components of the total orbital angular-momentum operator, respectively. As can be shown [ED 60], the operator $R(\Omega')$ rotates a scalar function depending on the nucleon spatial coordinates first around the y-axis by an angle θ' and then around the z-axis by an angle φ'. In terms of this operator, we see that [KR 73]

$$\delta(\omega - \omega') = R(\Omega')\,\delta(\omega). \tag{14.12}$$

Further, by making use of the fact that $\hat{\phi}(\tilde{\alpha}_1)$ and $\hat{\phi}(\tilde{\alpha}_2)$ are spherically symmetric, we obtain

$$\psi_i = R(\Omega')\,\psi_{i,0}, \tag{14.13}$$

---

† The function $\delta(\theta)$ is defined in such a way that

$$\int_0^{\theta_0} f(\theta)\,\delta(\theta)\,d\theta = f(0)$$

for $0 < \theta_0 \leqslant \pi$.

where

$$\psi_{i,0} = A' \{\hat{\phi}(\tilde{\alpha}_1)\, \hat{\phi}(\tilde{\alpha}_2)\, g(R)\, \delta(\omega)\} \tag{14.14}$$

is an intrinsic wave function which is rotationally symmetric around the z-axis. Using eqs. (14.6), (14.13), and (14.14), we can finally write ψ_{lm} as

$$\psi_{lm} = \frac{1}{2\pi} \left(\frac{2l+1}{4\pi}\right)^{1/2} \int d\Omega'\, D^{l*}_{m0}(\Omega')\, R(\Omega')\, \psi_{i,0}, \tag{14.15}$$

where we have used the relation

$$Y_{lm}(\theta', \varphi') = \left(\frac{2l+1}{4\pi}\right)^{1/2} D^{l*}_{m0}(-\varphi', -\theta', 0), \tag{14.16}$$

with D^{l}_{m0} being the Wigner D-function with K = 0 as given by *Edmonds* [ED 60]. Equation (14.15) shows that ψ_{lm} may be considered as obtained from an intrinsic function $\psi_{i,0}$ by making an appropriate angular-momentum projection.

The wave functions of eq. (14.15) can now be used to investigate the structure of ^{8}Be rotational states by computing various expectation values and transition matrix elements. But before we discuss the properties of these states, we shall first mention a generalization of the above consideration in ^{8}Be to include other even-even nuclei with K = 0. This generalization can be achieved by simply generalizing the form of the intrinsic wave function $\psi_{i,0}$. Thus, what one does is to choose an appropriate many-nucleon wave function $\psi_{i,0}$ which is an eigenfunction of the operator $J_z = L_z + S_z$ with an eigenvalue $\hbar K = 0$,† i.e.,

$$J_z |\psi_{i,0}\rangle = 0\, |\psi_{i,0}\rangle. \tag{14.17}$$

For example, $\psi_{i,0}$ can be a two-cluster state of the form

$$\psi_{i,0} = A' \{\hat{\phi}(\tilde{A})\, \hat{\phi}(\tilde{B})\, g(R)\, \delta(\omega)\}, \tag{14.18}$$

where the clusters A and B have no internal angular momentum. More generally, one can write $\psi_{i,0}$ as a superposition of two-cluster states with K = 0, i.e.,

$$\psi_{i,0} = \sum_j A' \{\hat{\phi}(\tilde{A}_j)\, \hat{\phi}(\tilde{B}_j)\, g_j(R_j)\, \delta(\omega_j)\}. \tag{14.19}$$

Using these intrinsic wave functions, one can then obtain again rotational wave functions with K = 0 of the form given by eq. (14.15), with $R(\Omega')$ assuming the more general form

$$R(\Omega') = \exp(-i\varphi' J_z/\hbar)\, \exp(-i\theta' J_y/\hbar), \tag{14.20}$$

where J_y and J_z are, respectively, the y- and z-components of the total angular-momentum operator.

† Note that $\psi_{i,0}$ is usually not an eigenfunction of the operator $\mathbf{J}^2$.

We shall now discuss how to compute energy-expectation values in rotational states. For this we mention first some properties of the Young operator C^J_{MK} (see, e. g., ref. [KR 73]), defined as

$$C^J_{MK} = \frac{2J+1}{8\pi^2} \int d\Omega\, D^{J*}_{MK}(\Omega)\, R(\Omega). \tag{14.21}$$

In the above equation, D^J_{MK} is the Wigner D-function whose K-value is generally different from zero and $R(\Omega)$ is the rotation operator given by

$$R(\Omega) = \exp(i\alpha J_z/\hbar)\exp(i\beta J_y/\hbar)\exp(i\gamma J_z/\hbar), \tag{14.22}$$

with α, β, and γ being the Euler angles. As has been frequently shown, the operation of $R(\Omega)$ on a wave function describes the rotation of this wave function first around the z-axis by an angle $-\gamma$, then around the y-axis by an angle $-\beta$, and finally around the z-axis by an angle $-\alpha$. Also, with $R(\Omega)$ given by eq. (14.22), the Wigner D-function D^J_{MK} takes on the form

$$D^J_{MK}(\alpha, \beta, \gamma) = \exp(iM\alpha)\, d^J_{MK}(\beta)\exp(iK\gamma), \tag{14.23}$$

with $d^J_{MK}(\beta)$ being the reduced d-function extensively discussed in many references [ED 60].

From an intrinsic function $\psi_{i,K}$ which is an eigenfunction of J_z with eigenvalue $\hbar K$, one can now form the projected state

$$\psi_{JM}(K) = C^J_{MK}\, \psi_{i,K}. \tag{14.24}$$

By using the fact that under rotations the Young operator transforms according to the equation

$$R(\Omega)\, C^J_{MK} = \sum_{M'} C^J_{M'K}\, D^J_{M'M}(\Omega), \tag{14.25}$$

one can show that $\psi_{JM}(K)$ is an eigenstate of the operators $\mathbf{J}^2$ and J_z with eigenvalues $\hbar^2 J(J+1)$ and $\hbar M$, respectively.

In addition, the Young operator satisfies the relations

$$C^J_{MK} = C^{J+}_{KM} \tag{14.26}$$

and

$$C^J_{MK} \sum_{J', M'} a_{J'M'}\, \psi_{J'M'} = a_{JK}\, \psi_{JM}. \tag{14.27}$$

From eq. (14.27) one can further show that the operator equation

$$C^J_{KM}\, C^J_{MK'} = C^J_{KK'} \tag{14.28}$$

is satisfied.

Using the above-mentioned properties of the Young operator, we now compute in even-even nuclei the energy expectation values in K = 0 rotational states which are described by wave functions of the form†

$$\psi_{JM}(0) = C^J_{M0}\,\psi_{i,0} = \frac{2J+1}{8\pi^2}\int d\Omega'\, D^{J^*}_{M0}(\Omega')\,R(\Omega')\,\psi_{i,0}, \tag{14.29}$$

with $R(\Omega')$ given by eq. (14.20). These expectation values are

$$E_{JM}(0) = \frac{\langle \psi_{JM}(0)|H|\psi_{JM}(0)\rangle}{\langle \psi_{JM}(0)|\psi_{JM}(0)\rangle}, \tag{14.30}$$

with

$$H = \sum_i T_i + \sum_{i<j} V_{ij} - T_C. \tag{14.31}$$

Let us first consider the denominator in eq. (14.30), which is

$$N_{JM}(0) = \langle \psi_{JM}(0)|\psi_{JM}(0)\rangle. \tag{14.32}$$

By employing eqs. (14.26) and (14.28), we obtain

$$N_{JM}(0) = \langle \psi_{i,0}|C^J_{0M}\,C^J_{M0}|\psi_{i,0}\rangle = \langle \psi_{i,0}|C^J_{00}|\psi_{i,0}\rangle. \tag{14.33}$$

Using eq. (14.21) then yields

$$\begin{aligned} N_{JM}(0) &= \frac{2J+1}{8\pi^2}\int d\Omega\,\langle \psi_{i,0}|D^{J^*}_{00}(\Omega)\,R(\Omega)|\psi_{i,0}\rangle \\ &= \frac{2J+1}{2}\int_0^\pi d\beta\,\sin\beta\,d^J_{00}(\beta)\,\langle \psi_{i,0}|e^{i\beta J_y/\hbar}|\psi_{i,0}\rangle, \end{aligned} \tag{14.34}$$

where we have used eq. (14.17) and the fact that the d-function is real.

Equation (14.34) is interesting, because the part of the integrand

$$n(\beta) = \langle \psi_{i,0}|e^{i\beta J_y/\hbar}|\psi_{i,0}\rangle \tag{14.35}$$

depends only on the angle β, but not on the angles α and γ. Thus, because of the presence of the rotation operator $e^{i\beta J_y/\hbar}$, $n(\beta)$ represents an overlap integral between an intrinsic wave function $\psi_{i,0}$ whose symmetry axis is the z-axis and another intrinsic wave function whose symmetry axis is oriented at an angle β with respect to the z-axis. For the computation of $n(\beta)$, it is of course necessary to perform integrations and summations over all nucleon coordinates of the system.

In a similar way, one can also calculate the quantity

$$H_{JM}(0) = \langle \psi_{JM}(0)|H|\psi_{JM}(0)\rangle. \tag{14.36}$$

† For simplicity in later discussion, we have chosen the numerical multiplicative factor in eq. (14.29) to be different from that in eq. (14.15).

If one uses the fact that the many-nucleon Hamiltonian H is rotationally invariant and, therefore, commutes with C^J_{KM}, then one obtains

$$H_{JM}(0) = \langle \psi_{i,0} | H C^J_{00} | \psi_{i,0} \rangle = \frac{2J+1}{2} \int_0^\pi d\beta \sin\beta \, d^J_{00}(\beta) \langle \psi_{i,0} | H e^{i\beta J_y/\hbar} | \psi_{i,0} \rangle . \qquad \ldots (14.37)$$

Similar to the meaning of $n(\beta)$, the quantity

$$h(\beta) = \langle \psi_{i,0} | H e^{i\beta J_y/\hbar} | \psi_{i,0} \rangle , \qquad (14.38)$$

which occurs in the integrand in eq. (14.37), represents an energy-overlapping integral of two intrinsic wave functions which differ only in the directions of their symmetry axes.

We could have obtained the results represented by eqs. (14.34) and (14.37) in a more direct way by using immediately K = 0 rotational wave functions of the type given by eqs. (14.6) and (14.7) [WI 59a]. But then the generalization of these results to rotational states of odd-even nuclei and so on would not be as straightforward as with the derivation presented here.

We go back now to ^{8}Be as a specific case in studying the properties of the rotational energy spectrum. In this two-cluster example, $\psi_{i,0}$ is given by eq. (14.14) and the normalization overlapping integral $n(\beta)$ is given by

$$n(\beta) = \langle \psi_{i,0} | \psi_i \rangle = \langle A' \{ \hat\phi(\tilde\alpha_1) \hat\phi(\tilde\alpha_2) g(R) \delta(\omega) \} | A' \{ \hat\phi(\tilde\alpha_1) \hat\phi(\tilde\alpha_2) g(R) \delta(\omega - \omega') \} \rangle , \qquad \ldots (14.39)$$

where $\omega' = (\theta', \varphi')$ is equal to $(-\beta, -\alpha)$ when expressed in terms of the Euler angles.

First, we shall study the case where the two clusters are considered as distinguishable, i.e., the antisymmetrization between the two α clusters is neglected (rigid-rotator model). In this case, the symmetry axis can be defined as the line joining the center-of-mass points of the two α particles and, consequently, ω is a uniquely defined function of the nucleon spatial coordinates. As a result of this, two intrinsic functions which differ in the directions of their symmetry axes are orthogonal to each other, i.e.,

$$\begin{aligned} n(\beta) &= \left[\int_0^\infty |g(R)|^2 R^2 dR \right] \frac{\delta(\beta)}{2\pi \sin\beta} \\ &= \left[\int_0^\infty |g(R)|^2 R^2 dR \right] \sum_\lambda \frac{2\lambda+1}{4\pi} P_\lambda(\cos\beta), \end{aligned} \qquad (14.40)$$

where we have assumed that the α-particle wave functions are properly normalized. By further substituting eq. (14.40) into eq. (14.34), we then obtain

$$N_{JM}(0) = \frac{2J+1}{4\pi} \int_0^\infty |g(R)|^2 R^2 dR . \qquad (14.41)$$

Thus one sees that the only J-dependence of $N_{JM}(0)$ is contained in the factor $(2J+1)$.

In a similar way, one can compute $H_{JM}(0)$ in the present case where one does not antisymmetrize between the α clusters. One realizes easily that in $h(\beta)$ the terms which come from the rotationally invariant two-nucleon potentials† contain a δ-function dependence on β. As a consequence, the part of $H_{JM}(0)$ resulting from two-nucleon potentials has again as its J-dependence only the factor $(2J+1)$.

The situation is different with the kinetic energy. Here the operator which represents the centrifugal energy of the $\alpha+\alpha$ system has the form

$$T_{cent} = \frac{1}{2\mu_R R^2} \mathbf{L}^2, \qquad (14.42)$$

where μ_R is the $\alpha+\alpha$ reduced mass. By applying this operator on the intrinsic function ψ_i, one obtains

$$T_{cent}\,\psi_i = \frac{\hbar^2}{2\mu_R R^2} \hat{\phi}(\tilde{\alpha}_1)\,\hat{\phi}(\tilde{\alpha}_2)\,g(R) \sum_{\lambda\mu} \lambda(\lambda+1)\,Y_{\lambda\mu}(\theta,\varphi)\,Y^*_{\lambda\mu}(-\beta,-\alpha). \qquad \ldots(14.43)$$

Using eq. (14.43), one finds that the part of $h(\beta)$ resulting from T_{cent} is given by

$$h_{cent}(\beta) = \left[\frac{\hbar^2}{2\mu_R}\int_0^\infty |g(R)|^2\,dR\right] \sum_\lambda \lambda(\lambda+1)\,\frac{2\lambda+1}{4\pi}\,P_\lambda(\cos\beta). \qquad (14.44)$$

The corresponding part of $H_{JM}(0)$ is then

$$H_{JM}^{cent}(0) = \frac{2J+1}{4\pi}\,J(J+1)\,\frac{\hbar^2}{2\mu_R}\int_0^\infty |g(R)|^2\,dR. \qquad (14.45)$$

The other terms in $H_{JM}(0)$ coming from the rest of the kinetic-energy operator can be similarly considered and the result is that they have as their J-dependence only the factor $(2J+1)$. Therefore, by using eq. (14.30), one finds that

$$E_{JM}(0) = E_0 + \frac{\hbar^2}{2\mu_R}\left\langle\frac{1}{R^2}\right\rangle J(J+1), \qquad (14.46)$$

where

$$\left\langle\frac{1}{R^2}\right\rangle = \int_0^\infty |g(R)|^2\,dR \Big/ \int_0^\infty |g(R)|^2\,R^2\,dR. \qquad (14.47)$$

Thus, it is seen from this discussion that, when the antisymmetrization between the α clusters is not carried out, the potential energy has no J-dependence and the J-dependence of $E_{JM}(0)$ is contained entirely in the rotational kinetic energy.

† For clarity in discussion, our argument is based on the use of a two-nucleon potential having no velocity-dependence and no Majorana component. This restriction is certainly not necessary for our later consideration where totally antisymmetrized wave functions are used.

The situation changes completely if one fully antisymmetrizes the rotational wave function. Here the complication arises because every permutation of the nucleons redefines the center-of-mass coordinates of the two α clusters. As a consequence, two intrinsic wave functions $\psi_i(\tilde{r}_1, \ldots, \tilde{r}_8; \theta', \varphi')$ and $\psi_i(\tilde{r}_1, \ldots, \tilde{r}_8; \theta'', \varphi'')$ are no longer necessarily orthogonal to each other if (θ', φ') and (θ'', φ'') are not equal to each other. This means that both $n(\beta)$ and the part of $h(\beta)$ resulting from two-nucleon potentials have a more complicated dependence on β than that indicated by eq. (14.40). Therefore, even the potential energy of the rotational state now becomes a function of J, in contrast to the case with unantisymmetrized wave functions.

In the oscillator limit case where the wave function is given by eqs. (14.4) and (14.5), the result is especially interesting. Here a calculation [WI 59a] employing a relatively simple central nucleon-nucleon potential shows that the first three levels with J = 0, 2, and 4 follow the energetical pattern of rotational states. Now, however, the kinetic energies of these three levels are all the same, while the potential energy has a J-dependence. In addition, it was shown that the rotational energy spectrum breaks off for $J > 4$; in other words, energies of states with $J > 4$ no longer satisfy the simple $J(J+1)$ rule. This is of course intimately connected with the fact that, with ^8Be having a $(1s)^4\,(1p)^4$ configuration, no α-cluster states with $J > 4$ can be formed.

If one assumes that the function $g(R)$ is so long-ranged that the resultant wave function describes a highly deformed ^8Be nucleus, then a well-formed rotational spectrum will again appear. In this case, it is easy to see that $n(\beta)$ and $h(\beta)$ will be strongly peaked around $\beta = 0$ and $\beta = \pi$. This must be so, because the overlap between two antisymmetrized intrinsic wave functions of this type should be rather small if their symmetry axes are not nearly parallel to each other. Therefore, in calculating $N_{JM}(0)$ and $H_{JM}(0)$, one can expand $d^J_{00}(\beta)$ in a power series of β near $\beta = 0$ and $\beta = \pi$ [ED 60, VE 63], i.e.,

$$d^J_{00}(\beta) = 1 - \frac{1}{4} J(J+1)\beta^2 + \ldots, \tag{14.48a}$$

$$d^J_{00}(\pi - \beta) = (-1)^J \left[1 - \frac{1}{4} J(J+1)\beta^2 + \ldots\right], \tag{14.48b}$$

and use only the first two terms in these expansions. In this way, one can easily see that to a good approximation $E_{JM}(0)$ will follow a rotational pattern in this particular case.

As has already been mentioned very often, the Bose character of the α cluster demands that α-cluster states in ^8Be have only even parity. Because of this, $\psi_{i,0}$ must also have even parity and one can readily show that the quantities $N_{JM}(0)$ and $H_{JM}(0)$ are equal to zero for odd values of J.† The reason for this is as follows. Because the α cluster has zero internal angular momentum, the operation of $e^{i\pi J_y/\hbar}$ or $e^{-i\pi J_y/\hbar}$ on $\psi_{i,0}$ is equivalent to a parity operation.†† Therefore, since $n(\beta)$ and $h(\beta)$ are, respectively, equal to $n(-\beta)$ and

† For the $\alpha + \alpha$ system, this discussion is of course superfluous. However, we present it here, simply because it can be extended to cases where the clusters involved are not identical.

†† The parity operation on $\psi_{i,0}$ is defined as

$$P\psi_{i,0}(r_1, \ldots, r_8; s, t) = \psi_{i,0}(-r_1, \ldots, -r_8; s, t).$$

$h(-\beta)$, one finds that $n(\beta) = n(\pi - \beta)$ and $h(\beta) = h(\pi - \beta)$. By further using the fact that $d^J_{00}(\beta) = (-1)^J d^J_{00}(\pi - \beta)$, one obtains then the consequence that $N_{JM}(0)$ and $H_{JM}(0)$ become zero for odd J-values.

After this slight diversion, we go back to the ^{8}Be discussion. Our consideration shows that, if ^{8}Be is sufficiently deformed, there will appear a $J(J + 1)$-type rotational energy spectrum which does not break off after a few states. This is in agreement with the prediction of the rigid-rotator model. There is, however, one major difference which is worth mentioning. With totally antisymmetrized wave functions it is found that the energy differences between various levels come from differences in both kinetic and potential energies, rather than from differences in kinetic energy alone as in the rigid-rotator model.†

One realizes without further comments that the general results which we have obtained for the ^{8}Be rotational wave functions are similarly valid for the $K = 0$ rotational wave functions of other even-even nuclei. For instance, in relatively light even-even nuclei one obtains in the oscillator limit, where the intrinsic deformations are comparatively small, also rotational spectra which generally follow the $J(J + 1)$ rule but break off after a certain number of states. An example of this is the nucleus ^{20}Ne in the oscillator limit. There the low-lying even parity $\alpha + {}^{16}$O cluster states with $J^\pi = 0^+, 2^+$, and so on are regularly spaced, but the regular pattern breaks off after the state with $J^\pi = 8^+$. The reason for this break-off is again that, with 4 nucleons in the 2s–1d oscillator shell, $\alpha + {}^{16}$O cluster states with $J > 8$ cannot be formed.

We wish to mention that, in contrast to the ^{8}Be energy spectrum, the ^{20}Ne energy spectrum does not have an exact $J(J + 1)$ energetical structure. Even in the oscillator limit, it is only for 1p-shell nuclei that one obtains by using a nucleon-nucleon potential without noncentral components an exact $J(J + 1)$-type energy sequence which persists for a certain finite number of states, as in the ^{8}Be case (see, e.g., ref. [FE 37]).

Using the same argument as presented above, one can show that, for $K = 0$ rotational wave functions of appreciably deformed even-even nuclei, one obtains always a $J(J + 1)$-type rotational spectrum because here again $n(\beta)$ and $h(\beta)$ are strongly peaked around $\beta = 0$ and $\beta = \pi$. In fact, for heavy nuclei already a relatively small deformation is sufficient to produce a strong peaking in $n(\beta)$ and $h(\beta)$ and, consequently, a well-formed rotational structure in the energy spectrum. To see this, we consider intrinsic wave functions which are constructed as antisymmetrized products of single-particle wave functions in deformed potential wells. For two such intrinsic wave functions, the overlap integral will be large only when their symmetry axes are nearly parallel to each other. This can be seen in the following way. The overlap of two single-particle wave functions for the same nucleon but in different potential wells will become smaller as the angle β between the symmetry axes increases [PE 57]. But any reduction in the overlap of two single-particle orbitals in different intrinsic wave functions will be raised, roughly speaking, to the power of the number of nucleons which contribute to the deformation of the considered nucleus

---

† For very large deformations, the energy differences will be mainly kinetic in origin. This is so because, for such ^{8}Be configurations, the antisymmetrization between the α clusters yields relatively minor consequences.

and, hence, will lead to a greatly reduced value for the complete overlap integral. Of course, the procedure of antisymmetrization will increase somewhat the overlap of two intrinsic wave functions with different symmetry axes and therefore, decrease the peaking of $n(\beta)$ and $h(\beta)$.† However, it can still be expected that, in general, the conclusion reached above should be essentially valid.

The above argument shows, therefore, that the peaking of $n(\beta)$ and $h(\beta)$ is most likely to be pronounced when the atomic weight A is large and when the nucleus has its last shell roughly half filled so as to produce a large deformation [BO 52, BO 53, PE 57, YO 57]. For closed shell nuclei the peaking of $n(\beta)$ and $h(\beta)$ vanishes completely due to antisymmetrization and, consequently, these nuclei do not have low-lying rotational states.

Our discussion about the peaking of $n(\beta)$ and $h(\beta)$ is based on intrinsic wave functions constructed with single-particle orbitals in deformed potentials. It is clear, however, that the conclusion reached should remain valid if the intrinsic wave functions are represented by cluster wave functions or superpositions of cluster wave functions. This must be so because, as has often been mentioned, the antisymmetrization procedure reduces to a large extent the difference between different representations.

As our consideration has shown, already a relatively small deformation can lead to a well-formed rotational band in a heavy nucleus. But this does not mean that the energy differences between rotational levels come mainly from differences in kinetic energy, as in the case of a highly deformed ^8Be nucleus discussed above. Because of the small deformation, all clusters in the nucleus can penetrate each other appreciably, with the consequence that antisymmetrization effects between the clusters are important and need always to be properly considered for the low-lying rotational states in such a heavy nucleus.

In contrast with the situation in ^8Be, low-lying $K = 0$ rotational bands of both positive and negative parities can very often exist in even-even nuclei. This comes from the fact that, from an intrinsic state $\psi_{i,0}$, one can in general project out states with any angular momentum J. There are, however, certain restrictions which limit the possible combinations of spin and parity. To see this, let us consider the intrinsic function $\psi_{i,0}$ of eq. (14.18). As is quite evident, this intrinsic function for different clusters A and B does not have a definite parity but can be expressed as the sum of a positive-parity state $\psi_{i,0}^+$ and a negative-parity state $\psi_{i,0}^-$ by writing the function $\delta(\omega)$ as

$$\delta(\omega) = \sum_{\lambda\ \text{even}} \left(\frac{2\lambda+1}{4\pi}\right)^{1/2} Y_{\lambda 0}(\theta,\varphi) + \sum_{\lambda\ \text{odd}} \left(\frac{2\lambda+1}{4\pi}\right)^{1/2} Y_{\lambda 0}(\theta,\varphi). \qquad (14.49)$$

† A striking example for the increase in overlap between two intrinsic wave functions when antisymmetrization is fully considered is the ground state of ^{16}O which can be described as having an $\alpha + {}^{12}C$ cluster structure. In the oscillator limit, the intrinsic wave function $\psi_{i,0}$, if not antisymmetrized, describes the behaviour of a nucleus which is relatively strongly deformed. However, after a complete antisymmetrization of $\psi_{i,0}$, this deformation will completely vanish, and $n(\beta)$ and $h(\beta)$ will no longer be peaked.

With $\psi_{i,0}$ separated in this way, it is then easy to see that by projection one can obtain states with $J^\pi = 0^+, 1^-, 2^+, 3^-$, and so on, a result which is typical of two-cluster configurations where the clusters involved have no internal angular momentum. Further restrictions may arise from the Pauli principle. For example, in the case of α-cluster states in ^{8}Be, the Pauli principle requires that the relative-motion part of the wave function be symmetric with respect to four-nucleon exchange. This means that the negative-parity part of $\psi_{i,0}$ vanishes by antisymmetrization. Thus, one can choose from the beginning an intrinsic wave function which definitely has positive parity, thereby resulting in the occurrence of only α-cluster states with $J^\pi = 0^+, 2^+$, and so on.

The important point of the above discussion is that restrictions on possible J^π values may be traced to symmetry properties contained in the intrinsic wave function (see, e.g., ref. [KR 69a]). Therefore, in constructing the intrinsic wave function, one must always try to properly take both dynamical structure and symmetry properties into consideration.

In even-even nuclei one expects that states of even and odd angular momenta should, in general, be projected from intrinsic wave functions of rather different structure. For instance, if one considers a pure two-cluster case where both clusters have the same intrinsic parity and no internal spin, then the intrinsic wave function takes on the form

$$\psi^+_{i,0} = A'\left\{ \hat\phi(\tilde{A})\,\hat\phi(\tilde{B})\,g_+(R) \sum_{\lambda\ \mathrm{even}} \left(\frac{2\lambda+1}{4\pi}\right)^{1/2} Y_{\lambda 0}(\theta,\varphi)\right\} \qquad (14.50)$$

for states with $J^\pi = 0^+, 2^+, 4^+$, and so on, and the form

$$\psi^-_{i,0} = A'\left\{ \hat\phi(\tilde{A})\,\hat\phi(\tilde{B})\,g_-(R) \sum_{\lambda\ \mathrm{odd}} \left(\frac{2\lambda+1}{4\pi}\right)^{1/2} Y_{\lambda 0}(\theta,\varphi)\right\} \qquad (14.51)$$

for states with $J^\pi = 1^-, 3^-, 5^-$, and so on.

It is well known that the ground-state band of an even-even nucleus consists of states with even angular momentum and positive parity.† This band is therefore obtained by projection from an intrinsic function $\psi^+_{i,0}$. The band containing states with odd angular momentum will start at a certain excited state; it is obtained by projection from an intrinsic function $\psi^-_{i,0}$. An example is provided by the nucleus ^{20}Ne where, in addition to the positive-parity ground-state rotational band, there exists also a negative-parity band starting with the 5.6 MeV 1^- level (see, e.g., ref. [HO 68]). Similar to the states in the positive-parity band, the states in the negative-parity band can also be approximately described as having an $\alpha + {}^{16}$O cluster structure; i.e., the intrinsic wave functions for both bands can be chosen as those given by eqs. (14.50) and (14.51), with A and B representing the α and the ^{16}O clusters, respectively.

In the oscillator cluster model, the states in the ^{20}Ne positive-parity and negative-parity bands are $\alpha + {}^{16}$O cluster states where the relative motions between the clusters

---

† Because of the Pauli principle and the fact that the nucleon-nucleon potential is on the average attractive, it is energetically favoured to pair off the protons and the neutrons.

contain 8 and 9 oscillator quanta of energy, respectively. This indicates that in this model the functions $g_+(R)$ and $g_-(R)$ in eqs. (14.50) and (14.51) are the radial parts of different harmonic-oscillator wave functions and are, therefore, quite distinct from each other.

It is clear that the above discussion concerning even- and odd- parity bands is quite general and not restricted only to the case involving $\psi_{i,0}$. For instance, one can also apply it to any intrinsic function $\psi_{i,K}$ where K is the angular-momentum projection on the symmetry axis and can have a value different from zero. It should be noted, however, that the association of positive parity with even angular-momentum value and negative parity with odd angular-momentum value does not hold in general, but is true only under special circumstances discussed above where nucleons are paired off or where the clusters involved have no internal spin.

In addition to the above considerations about the energetical behaviour of rotational levels in even-even nuclei, one can also investigate the properties of electromagnetic transitions between these levels. This latter type of investigation should shed more light on the structure of rotational states and we shall discuss it in section 14.5.

14.3. Generalization of Rotational Wave Functions

We now generalize the formulation of section 14.2 to include rotational states of odd-even and odd-odd nuclei, and of even-even nuclei with K-value not equal to zero. This generalization is necessary, since for these cases one can certainly no longer assume that like nucleons are paired off or that the clusters involved have no internal spin.

For the present discussion, we start with an antisymmetrized intrinsic wave function $\psi_{i,K}$ which is an eigenfunction of the operator J_z with eigenvalue $\hbar K$, i.e.,

$$J_z\,\psi_{i,K} = \hbar K\,\psi_{i,K}. \tag{14.52}$$

Using this intrinsic function, we can then obtain rotational wave functions $\psi_{JM}(K)$ by utilizing eq. (14.24) which contains the Young operator C^J_{MK} defined in eq. (14.21).

Before we discuss the general properties of these rotational states, we shall first show by a specific example how such an intrinsic wave function can be explicitly constructed. In this example, we consider a nucleus which is composed of two α clusters and an extra nucleon. The α clusters are assumed to be unexcited and the extra nucleon is assumed to rotate around the symmetry axis of the two α clusters with an angular-momentum component K along this axis. By introducing Jacobi coordinates†

$$\begin{aligned} \mathbf{R}_C &= \mathbf{R}_{\alpha_1} - \mathbf{R}_{\alpha_2}, \\ \mathbf{R}_N &= \frac{1}{2}(\mathbf{R}_{\alpha_1} + \mathbf{R}_{\alpha_2}) - \mathbf{r}_N, \end{aligned} \tag{14.53}$$

† The Jacobi coordinate system, together with the definition of orbital angular-momentum operators and so on, is discussed in subsection 3.7a.

with $\mathbf{R}_{\alpha_1}$ and $\mathbf{R}_{\alpha_2}$ being the center-of-mass coordinates of the α clusters and $\mathbf{r}_N$ being the position vector of the extra nucleon, and by orienting the symmetry axis in the direction of the z-axis, we can then write the intrinsic wave function as

$$\psi_{i,K} = \mathcal{A}\,\{\hat{\phi}(\tilde{\alpha}_1)\,\hat{\phi}(\tilde{\alpha}_2)\,g(R_C)\,\delta(\omega_C)\,\phi_K\}. \tag{14.54}$$

In the above equation the function ϕ_K, which depends on the spatial coordinates R_N, θ_N, and φ_N and the spin coordinate s_N, may be chosen either as

$$\phi_K(\mathbf{R}_N, s_N) = G(R_N)\,\Theta(\theta_N)\,e^{i(K-\frac{1}{2})\varphi_N}\,\alpha(s_N) \tag{14.55}$$

or as

$$\phi_K(\mathbf{R}_N, s_N) = G(R_N)\,\Theta(\theta_N)\,e^{i(K+\frac{1}{2})\varphi_N}\,\beta(s_N). \tag{14.56}$$

With either of these forms, one can easily see that ϕ_K satisfies the equation

$$J_{Nz}\,\phi_K = \hbar K\,\phi_K, \tag{14.57}$$

where J_{Nz} is the z-component of the angular-momentum operator $\mathbf{J}_N = \mathbf{L}_N + \mathbf{S}_N$, with $\mathbf{S}_N$ being the spin-angular-momentum operator of the extra nucleon and $\mathbf{L}_N$ being the orbital-angular-momentum operator corresponding to the coordinate $\mathbf{R}_N$. We do not demand of course that ϕ_K be also an eigenfunction of the operator $\mathbf{J}_N^2$; indeed, the function ϕ_K of eq. (14.56) or eq. (14.57) does not possess this property.

It is easy to see that the intrinsic function $\psi_{i,K}$ of eq. (14.54) does meet the required condition. Because the α clusters have no internal angular momentum, one finds that

$$J_z\,\psi_{i,K} = \mathcal{A}\,\{\hat{\phi}(\tilde{\alpha}_1)\,\hat{\phi}(\tilde{\alpha}_2)\,(L_{Cz} + J_{Nz})\,g(R_C)\,\delta(\omega_C)\,\phi_K\}, \tag{14.58}$$

where L_{Cz} is the z-component of the relative orbital angular momentum between the two α clusters. In eq. (14.58), it is noted that

$$L_{Cz}\,g(R_C)\,\delta(\omega_C) = 0, \tag{14.59}$$

a result which follows immediately from the use of eq. (14.9). Utilizing this result and eq. (14.57), one sees then that eq. (14.52) is indeed satisfied.

The applicability of the intrinsic function of eq. (14.54) is somewhat limited, because the resultant projected wave functions $\psi_{JM}(K)$ tend to have rather special properties. This can be seen in the following way. For $K > \frac{1}{2}$, the quadrupole moment obtained with the ground-state wave function $\psi_{KK}(K)$ is in general positive. The reason is that the extra nucleon rotates around a prolate core formed by the two α clusters, and the negative single-particle quadrupole moment of the extra nucleon is usually smaller (for K not too large) than the positive core quadrupole moment. We should mention, of course, that the antisymmetrization of the wave function usually reduces the magnitude of the quadrupole moment because it tends, as has often been pointed out, to reduce the strong correlations which are present between nucleons before antisymmetrization. In fact, there are situations where even the sign of the quadrupole moment can be changed by anti-symmetrization. Therefore, one must always be rather careful in using qualitative arguments based on an unantisymmetrized wave function. In the present case, however, as long as the function $g(R_C)$ is chosen such that the mutual penetration of the two α clusters is not too strong, the problem will remain that one can generally obtain only a positive value for the ground-state quadrupole moment when the intrinsic function of eq. (14.54) is used.

The above-mentioned problem can be remedied by generalizing the form of the intrinsic wave function. This can be achieved, for instance, by replacing the function $\delta(\omega_C)$ in eq. (14.54) by a function $\lambda(\theta_C)$ and thereby writing $\psi_{i,K}$ as

$$\psi_{i,K} = A\{\hat{\phi}(\tilde{\alpha}_1)\hat{\phi}(\tilde{\alpha}_2)\, g(R_C)\,\lambda(\theta_C)\,\phi_K\}. \qquad (14.60)$$

As for the meaning of $\lambda(\theta_C)$, it can be considered as a factor which roughly determines the angular distribution of the core matter-density with respect to the z-axis.

If $\lambda(\theta_C)$ has a large magnitude only when θ_C is around $\pi/2$, then the core will be oblate in shape and the quadrupole moment will acquire a negative value. Thus one sees that, depending upon the specific choice of $\lambda(\theta_C)$, the ground-state quadrupole moment can now be either positive or negative.

The above procedure of constructing intrinsic wave functions with a given value of K can easily be generalized. For instance, one can construct in a similar manner $\psi_{i,K}$'s for which the core itself has an angular-momentum projection different from zero along the symmetry axis[†] and for which several extra nucleons rotate around this axis. Also, by introducing additional Jacobi coordinates, one can construct intrinsic functions in cases where the core consists of three or more clusters. Furthermore, by linearly superposing different $\psi_{i,K}$ with the same K-value, one obtains new intrinsic wave functions having this value of K. In this way, it is possible to construct very flexible intrinsic functions which lead to angular-momentum states $\psi_{JM}(K)$ of the most general nature.[††]

The intrinsic wave function $\psi_{i,K}$ is not uniquely defined. This can be seen by expanding $\psi_{i,K}$ in terms of angular-momentum eigenstates, i.e.,

$$\psi_{i,K} = \sum_J a_J\, \psi_{JK}. \qquad (14.61)$$

Because the operator $R(\Omega)$ commutes with $\mathbf{J}^2$, the effect of rotation is to mix angular-momentum eigenstates with different M values but not eigenstates with different J values. Likewise, the application of the Young operator, which is a linear superposition of rotation operators, to the intrinsic function $\psi_{i,K}$ mixes only eigenstates with different M values but not eigenstates with different J values. This has the following consequence: if in eq. (14.61) one uses a different set of a_J values, thus defining a new intrinsic

---

† We emphasize that the symmetry axis of $\psi_{i,K}$ is always the z-axis of the fixed coordinate system, because $\psi_{i,K}$ is the intrinsic function which will be acted upon afterwards by the rotation operator $R(\Omega)$ (see eqs. (14.21) and (14.24)).

†† It should be mentioned that K cannot be considered as a quantum number of the system. This is associated with the fact that it is not possible to define a symmetric operator whose eigenvalue is K, when applied to an angular-momentum eigenfunction constructed from an intrinsic function $\psi_{i,K}$.

function, then the application of the Young operator to this new function will yield a projected wave function $\psi_{JM}(K)$ which is the same, except for a physically irrelevant normalization factor, as the one obtained from the original intrinsic function.† Thus, there are many intrinsic wave functions which are entirely equivalent to each other. In section 14.4, we shall discuss the question of choosing the optimal one for the construction of rotational wave functions.

From eq. (14.61) and the above remarks, one sees also that the largest J value which can appear in a rotational band is the largest J value which is contained in $\psi_{i,K}$. Especially for deformed medium- and heavy-weight nuclei, the maximum J value contained in $\psi_{i,K}$ is rather large. Therefore, the rotational band-structure is usually destroyed by centrifugal and coriolis effects long before this maximum J-value is reached.

As is well known, the value of the total angular momentum J has to be larger than or equal to the magnitude of K. Here we mention two examples in which the total angular momentum of the lowest state in a rotational band is larger than $|K|$. The first example is the negative-parity band in ^{20}Ne starting with the 5.6 MeV 1^- level [HO 68]; this band with K = 0 was already briefly discussed in the previous section. The second example is the negative-parity band in ^{226}Ra which consists of states with $J^\pi = 1^-, 3^-$, and 5^-, with the lowest 1^- state having an excitation energy of 253 keV [AL 56]. For this band, the K-value of the intrinsic wave function is again zero, a result which was determined experimentally from the observed relative intensities in the γ-decay of these states.

With the generalized wave functions $\psi_{JM}(K)$ which describe the behaviour of rotational states in odd-even nuclei and so on, one can calculate the energies of these states in a completely analogous manner as that described in section 14.2 for the K = 0 rotational states in even-even nuclei. In this calculation, one obtains equations for $N_{JM}(K)$, $H_{JM}(K)$, $n(\beta)$, and $h(\beta)$ which are similar to those for the corresponding quantities in the K = 0 case given by eqs. (14.34), (14.35), (14.37), and (14.38), except that the functions $\psi_{i,0}$ and $d^J_{00}(\beta)$ are replaced by the functions $\psi_{i,K}$ and $d^J_{KK}(\beta)$, respectively.

For medium-heavy and heavy deformed nuclei, $n(\beta)$ and $h(\beta)$ are again strongly peaked around $\beta = 0$ and $\beta = \pi$. Therefore, for the approximate calculation of $N_{JM}(K)$ and $H_{JM}(K)$, one can expand $d^J_{KK}(\beta)$ in a power series of β around $\beta = 0$ and $\beta = \pi$, and keep only the necessary lower-order terms. That is, one uses the following approximate expressions for $d^J_{KK}(\beta)$ [VE 63]:

$$
\begin{aligned}
d^J_{KK}(\beta) &\approx 1 - \frac{1}{4}[J(J+1) - K^2]\beta^2, && \text{for any K} \\
d^J_{KK}(\pi-\beta) &\approx (-1)^J - \frac{1}{4}(-1)^J J(J+1)\beta^2, && \text{for K} = 0 \\
d^J_{KK}(\pi-\beta) &\approx -\frac{1}{2}(-1)^{J+\frac{1}{2}}\left(J+\frac{1}{2}\right)\beta, && \text{for K} = \frac{1}{2} \\
d^J_{KK}(\pi-\beta) &\approx -\frac{1}{8}(-1)^J J(J+1)\beta^2, && \text{for K} = 1
\end{aligned}
\tag{14.62}
$$

† This can also be shown by using eq. (14.27).

and so on. In general, the expansion of $d^J_{KK}(\pi - \beta)$ for arbitrary K-value starts with a term which is $|2K|^{th}$ power in β.

Because the Hamiltonian H is invariant under reflection, it is useful to introduce again a positive-parity intrinsic function $\psi^+_{i,K}$ and a negative-parity intrinsic function $\psi^-_{i,K}$. Each of these intrinsic functions will generate a different rotational band, and the two bands will usually not fit energetically into the same rotational pattern. The reason for this is, as has already been discussed in section 14.2 for the ^{20}Ne case, that the nucleon correlation structure in a positive-parity state is quite different from that in a negative-parity state. As a consequence, the functions $n(\beta)$ and $h(\beta)$ are in general parity-dependent. Especially for $h(\beta)$ this parity dependence can be rather large, since the nucleon-nucleon potential has a strong short-ranged part.

We have seen that, in even-even nuclei with K = 0, positive-parity rotational bands contain even angular-momentum states and negative-parity rotational bands contain odd angular-momentum states. In odd-odd nuclei with K = 0, the situation is, however, quite different; here a band having a definite parity will normally contain states of both even and odd J-values. The reason for this is that, in an odd-odd nucleus, the spin angular momentum is very often different from zero and, therefore, the total angular momentum is composed of both an orbital part and a spin part. This has the consequence that a specification of the parity does not usually exclude either even or odd values of J.† The same situation can arise in an excited even-even nucleus if the excitation energy is so large that the intrinsic S = 0 configuration of this nucleus can be broken up. If K = 0 rotational bands exist for such configurations where S is nonzero, then one cannot expect the occurrence of a definite connection between parity and angular-momentum values.

The explicit derivation of the rotational structure of $E_{JM}(K)$, using eqs. (14.30)–(14.38) generalized to the case of arbitrary K-values, is very much similar to the derivation given by *Verhaar* [VE 63, VE 63a, VE 64] who has continued the type of investigation started by *Peierls* and *Yoccoz* [PE 57, YO 57]. Hence, we shall not present here the details of this derivation, but only summarize the essential results.

For intrinsic functions $\psi_{i,K}$ which yield a strong peaking in $n(\beta)$ and $h(\beta)$, the results are as follows:

(a) One obtains the usual $J(J+1)$ structure for rotational bands, except in the special case where $K = \frac{1}{2}$.

(b) If $K = \frac{1}{2}$, then since $d^J_{\frac{1}{2}\frac{1}{2}}(\pi - \beta)$ starts with a term linear in β, one obtains the well-known decoupling effect for states with this special value of K (see eq. (21) of ref. [VE 63]).

(c) For K = 0 bands in odd-odd nuclei, the theory predicts an odd-even shift; that is, in a rotational band having a definite parity, the positions of even angular-momentum states are shifted relative to the positions of odd angular-momentum states. In addition, the moments of inertia associated with these even and odd angular-momentum states are generally different (see eq. (20) of ref. [VE 63]).

† In such cases, the operation of $e^{\pm i\pi J_y/\hbar}$ on $\psi_{i,K}$ is not equivalent to a parity operation.

(d) If the number of outside nucleons is small compared with the number of nucleons which contribute to the deformation of the core, the moment of inertia changes smoothly as one goes from one nucleus to the neighbouring one. Only if the core wave function changes drastically at such a transition is this not the case.

(e) The intrinsic functions $\psi_{i,K}$ and $\psi_{i,-K}$ are usually degenerate in energy. As a consequence, one must use, for the generation of rotational wave functions, an intrinsic function $\psi_{i,|K|}$ which is a linear superposition of $\psi_{i,K}$ and $\psi_{i,-K}$. Because of the symmetry properties [ED 60] possessed by the function $d^J_{K,-K}(\beta)$ which appear in the mixed terms of N_{JM} and H_{JM}, this modification does not cause any essential change in the results listed above as items (a)–(d).

14.4. Energetical Preference of Rotational Configurations

In the previous two sections, we have described the way to construct rotational wave functions $\psi_{JM}(K)$ and presented a rather detailed discussion of the properties of such wave functions. Here we shall study the question as to the circumstances under which these wave functions can be considered as good approximate solutions of the many-particle Schrödinger equation.

To discuss this question, we employ energetical arguments based on the projection equation (14.1) or, in the present case of dealing with bound-state configurations, the variational equation

$$\delta \langle \psi | H - E | \psi \rangle = 0. \qquad (14.63)$$

For bound-state problems, it is of course well known that eqs. (14.1) and (14.63) are entirely equivalent. Equation (14.63) is, however, somewhat more convenient for our present considerations; it expresses the fact that a wave function ψ_i describing a bound state of the nucleus is determined by minimizing the quantity $\langle \psi_i | H | \psi_i \rangle / \langle \psi_i | \psi_i \rangle$ subject to the condition that ψ_i be orthogonal to all energetically lower bound states possessing the same set of quantum numbers. In other words, eq. (14.63) determines the energetically most favoured configuration under the above-mentioned orthogonality restriction.

We now use the property of eq. (14.63) to determine whether rotational wave functions $\psi_{JM}(K)$ of eq. (14.24) represent in good approximation the solutions of the Schrödinger equation. What we wish to show is that there are situations in which, from the viewpoint of a variational calculation employing linear variational parameters, these rotational wave functions are not mixed very much with other wave functions characterized by the same set of quantum numbers, such as spin, parity, and so on.† For this we utilize eq. (2.12) which was discussed in chapter 2 but will be mentioned here again for convenience. If there are two orthonormalized functions ψ_a and ψ_b with energy expectation values E_a and E_b, then one can find from eq. (14.63) two solutions

$$\psi_1 = a_1 \psi_a + b_1 \psi_b, \qquad (14.64a)$$

$$\psi_2 = a_2 \psi_a + b_2 \psi_b, \qquad (14.64b)$$

† As has already been pointed out before, the use of linear variational parameters guarantees that the solutions of eq. (14.1) or eq. (14.63) are automatically orthogonal to each other.

which have energy expectation values E_1 and E_2, respectively. If the mixing is not too large such that E_1 is close to E_a and E_2 is close to E_b, then one obtains the relation

$$\left|\frac{b_1}{a_1}\right| \approx \left|\frac{a_2}{b_2}\right| \approx \left|\frac{\langle\psi_a|H|\psi_b\rangle}{E_a - E_b}\right|. \tag{14.65}$$

Equation (14.65) indicates that two states are only slightly mixed if their energy-overlapping integral is much smaller than their energy difference.† Thus, if one can present arguments to show that, under proper circumstances, rotational wave functions of eq. (14.24) do yield small energy-overlapping integrals in conjunction with wave functions of other correlation nature, then they must already be good representatives of the solutions of the Schrödinger equation.

Because the nuclear force is short-ranged and on the average attractive, energetically favoured nuclear substructures or clusters are present which very often cause strong prolate or oblate intrinsic deformation in the nuclear ground state. The Pauli principle, which acts to prevent a complete penetration of the clusters, is largely responsible for this situation which, as has already been discussed several times before, is especially pronounced in light nuclei such as ^6Li, ^7Be, ^8Be, ^9Be, ^{19}F, ^{20}Ne, etc. In heavier nuclei with partially filled highest shells, the clusters formed by nucleons in these highest shells can further cause the core to be deformed. The result of this is that heavy nuclei of such shell-model configurations have especially large intrinsic deformations, as has indeed been observed experimentally.

For the nuclei mentioned above, intrinsic deformations are energetically favoured not only in the ground state but also in the lower excited states. To express this fact mathematically, one uses for the description of these states wave functions given by eq. (14.24), with the intrinsic function chosen, for instance, to have the form of eq. (14.18) or eq. (14.19) which contains the feature of cluster formation.

We should mention that one can obtain the wave functions of eq. (14.24) also from the variational equation (14.63).†† For this one adopts as trial functions linear superpositions of intrinsic functions $R(\Omega)\,\psi_{i,K}$ and uses their amplitudes, which depend on Euler angles $\Omega = (\alpha, \beta, \gamma)$, as variational amplitudes.

Other states having the same quantum numbers as the rotational states usually have much higher excitation energies of the order of a few MeV, because to form these states one has to break up energetically favoured correlations, such as cluster correlation, pairing correlation, and so on. Furthermore, the energy-overlap integral $|\langle\psi_a|H|\psi_b\rangle|$

† The approximate relation given by eq. (14.65) does not change appreciably if the wave functions ψ_a and ψ_b are not completely orthogonal to each other. Also, it remains approximately valid if more than two wave functions are mixed with one another, as long as the mixing does not become too large.

†† A derivation of this kind has already been given by *Peierls* and *Yoccoz* [PE 57, YO 57]. Also, it is interesting to mention that the mixing of energetically degenerate intrinsic functions $\psi_{i,K}$ and $\psi_{i,-K}$, mentioned at the end of section 14.3, can be obtained from this variational equation. Because of the symmetry property of the overlap matrix elements N_{JM} and H_{JM} with respect to the transformation from K to - K, these intrinsic functions are always mixed with amplitudes of the same absolute value.

between a rotational state and a state of other configuration will usually be rather small for the following reasons. First, breaking up the energetically favoured correlations in a rotational state will drastically change its character. Especially for medium-heavy and heavy nuclei where many nucleons contribute to the deformation, this will result in a small magnitude for the energy-overlap integral. Second, because of the short-range nature of the nuclear force, the value of an energy-overlap integral tends to be small when the states involved have rather different structures. Therefore, it follows from eq. (14.65) that in medium-heavy and heavy nuclei, rotational states with excitation energies less than about 1 MeV are not mixed appreciably with states of other configurations. Consequently, these nuclei have usually well-formed rotational bands up to rather large values of J.†

The above considerations are especially applicable to intrinsically deformed even-even nuclei. Therefore, let us consider these nuclei further and add some remarks to the discussion given in section 14.2. In that section, it was mentioned that in even-even nuclei there exist low-lying positive- and negative-parity bands containing even- and odd-angular momentum states, respectively. That the ground-state bands of these nuclei have always positive parity is, as mentioned above, due to the Pauli principle and the fact that the nuclear force has a short-ranged and attractive nature. As an example for a lighter system, we have mentioned the nucleus ^{20}Ne. Here a simple way to explain this particular behaviour is that, in the oscillator limit, the relative motion between the α and the ^{16}O clusters has 8 oscillator quanta of excitation for the positive-parity ground-state band, compared with 9 for the lowest negative-parity rotational band. Consequently, the negative-parity band is expected to lie higher on the energy scale than the ground-state band, as has indeed been observed experimentally. For medium-heavy and heavy even-even deformed nuclei, there is an equally simple explanation. In the j–j coupling shell model, the energetically favoured configuration is that even numbers of protons and neutrons occupy the same subshell and these protons and neutrons form pairs with angular-momentum projection equal to zero on the symmetry axis. This type of pairing correlation is particularly favourable, because then the spatial overlap between the single-particle wave functions becomes the largest and the potential energy attains thereby a maximum negative value.

The discussion given above shows that, for the ground-state bands in medium-heavy and heavy nuclei, one must construct positive-parity intrinsic functions $\psi_{i,K}$ which take pair correlations fully into account. For these intrinsic functions which have K = 0, the operation $e^{\pm i\pi J_y/\hbar}$ is again equivalent to a parity operation. Therefore, analogous to our discussion in the ^{8}Be case, it follows that the angular-momentum projected states obtained from such intrinsic functions can have only even values of J. An example for this is the nucleus ^{226}Ra which has not only a positive-parity ground-state band but also a negative-parity band which contains only states with odd angular-momentum values.

In odd-even and odd-odd nuclei, the situation to have well-formed rotational bands is usually not as favourable as that in even-even nuclei. The reason for this is that the K-value of the intrinsic wave function can be changed simply by changing the spin projection of

† In medium-heavy and heavy deformed nuclei, the spacing between neighbouring rotational levels is of the order of 100 keV.

the outside nucleons on the symmetry axis. This, however, does not break up any energetically strongly favoured correlations. Therefore, in these nuclei, the mixing of states with different K-values can be quite appreciable. As can be shown by using the equations for $N_{JM}(K)$ and $H_{JM}(K)$, generalized to the case where such K-mixing is taken into account, this mixing usually changes appreciably the rotational nature of the energy spectrum.

If the odd-even and odd-odd nuclei happen to be very strongly deformed and if the K-value of the outside nucleons is large, then the difference between the energy-expectation values of states with different values of K becomes relatively large. In such cases (see, e.g., ref. [BO 54]), the K-mixing is small and the rotational energy spectrum can be preserved quite well up to an excitation energy of several hundred keV or more.

At high excitation energies, it can happen that, by breaking up energetically favoured correlations, new strongly deformed nuclear states, e.g., other α-cluster or quartet states [AR 70, DA 71], can be formed. These states can then serve as the band heads of new rotational bands. That these bands have very often good rotational structure even though their excitation energies are rather high ($\gtrsim$ 10 MeV) can be attributed to the fact that the states involved have special structure and, consequently, their energy-overlap integrals with neighbouring states are usually quite small.

In light nuclei the occurrence of states with good rotational structure is much rarer than that in medium-heavy and heavy nuclei. The reasons for this are as follows:

(a) For a deformed odd-even or odd-odd nucleus, the coupling of the outside nucleons with the core is usually weaker for a light nucleus than for a heavy one, because of the smaller number of nucleons which form the nuclear core of the former. Therefore, the degree of K-mixing in a light deformed nucleus is expected to be considerably larger than that in a heavy deformed nucleus. This has the consequence that well-formed rotational bands scarcely ever occur in light odd-even or odd-odd nuclei.

(b) Again because of the smaller number of nucleons in the cores of deformed light nuclei, their moments of inertia are usually relatively small. Thus, successive rotational levels in these nuclei have large energy separations, and centrifugal and coriolis effects, which can be described by the mixing of states with the same quantum numbers, play an important role. In addition, the large energy spacing has the consequence that already for rotational levels of relatively low J-values, states of other configurations but with the same quantum numbers can appear nearby. These states can mix appreciably with the rotational states because the number of nucleons involved is fairly small and, consequently, the overlap between these and the rotational wave functions is expected to be significantly larger than that in a heavier nucleus (see, e.g., refs. [PE 57, WI 59a, YO 57]).

(c) In a deformed light nucleus, the peaking of the functions $n(\beta)$ and $h(\beta)$ is usually not as pronounced as that in a deformed heavy nucleus. The result of this is that, in the power-series expansion of $d^J_{KK}(\beta)$, higher-power terms become also important. This again disturbs the rotational nature of the energy spectrum. Therefore, even if a set of wave functions in a deformed light nucleus has a good rotational structure, i.e., if the intrinsic function $\psi_{i,K}$ is nearly J-independent, the energy spectrum of the corresponding rotational band may very often deviate appreciably from the $J(J+1)$ rule (for $K \neq \frac{1}{2}$).

An example for this is the positive-parity rotational band in ^{20}Ne when considered in the oscillator limit (see section 14.2).

In light even-even nuclei it sometimes happens that, in the region of high excitation, a strongly deformed J = 0 state appears. Such a state can then serve as the band head of a rotational band with a very good rotational spectrum because, as a result of its large deformation, the functions $n(\beta)$ and $h(\beta)$ can be strongly peaked even if the number of nucleons contributing to the intrinsic deformation is relatively small. The rotational band in ^{24}Mg which starts at a level with an excitation energy of 11.75 MeV [MO 56] is an example for this.

We should discuss one more point. As was mentioned in section 14.3, the intrinsic wave function $\psi_{i,K}$ is not uniquely defined. In approximate calculations where one retains only the lower-order terms of β in the expansion of $d^J_{KK}(\beta)$, one should in principle always choose intrinsic functions which yield the strongest peaking in the normalization overlap integral $n(\beta)$ and especially in the energy overlap integral $h(\beta)$. In should be noted, however, that the choice of the intrinsic function is not a very critical one. Even if one uses an intrinsic function which does not quite yield the strongest peaking in these integrals, the values $E_{JM}(K)$ will still be correct to first order because the small errors made in the numerator and denominator of eq. (14.30) will approximately compensate each other. This must be so, of course, since the exact values of E_{JM} will be the same, no matter which one of the equivalent intrinsic wave functions is used.

14.5. Electromagnetic Transitions between Rotational Levels

In this section, we study the properties of electromagnetic transitions between rotational states of appreciably deformed nuclei. This is a useful study, since it will give us more understanding concerning the effect of antisymmetrization on rotational states.

For transitions of type σ (electric or magnetic) and multipole order λ between two states of the same rotational band, the reduced transition probability has the form

$$B(\sigma\lambda; J_i \to J_f) = \frac{1}{2J_i+1} \sum_{\mu} \sum_{M_i} \sum_{M_f} \frac{|\langle \psi_{J_f M_f}(K) | O^{\sigma}_{\lambda\mu} | \psi_{J_i M_i}(K) \rangle|^2}{\langle \psi_{J_f M_f}(K) | \psi_{J_f M_f}(K) \rangle \langle \psi_{J_i M_i}(K) | \psi_{J_i M_i}(K) \rangle} \qquad \ldots (14.66)$$

where J_i and J_f denote the angular momenta of the initial and final states, respectively. The quantity $O^{\sigma}_{\lambda\mu}$ is the multipole operator; it is given, for instance, in the electric quadrupole (E2) case by

$$O^{e}_{2\mu} = Q_{2\mu} = \sum_{i=1}^{A} \bar{r}_i^2 \, Y^*_{2\mu}(\bar{\theta}_i, \bar{\varphi}_i) \frac{1}{2}(1 - \tau_{i3}), \qquad (14.67)$$

with $(\bar{r}_i, \bar{\theta}_i, \bar{\varphi}_i)$ specifying the position of the ith nucleon with respect to the center of mass of the nucleus. By using now the Wigner-Eckart theorem, we obtain after some algebraic manipulation the following approximate expression in the case of appreciably deformed intrinsic wave functions:†

$$B(\sigma\lambda; J_i \rightarrow J_f) = (J_f, K, \lambda, 0 | J_f, \lambda, J_i, K)^2 \frac{2J_f + 1}{2J_i + 1} |M_\lambda^\sigma|^2, \qquad (14.68)$$

where

$$M_\lambda^\sigma = \frac{\int \sin\beta \, d\beta \langle \psi_{i,K} | O_{\lambda 0}^\sigma \, e^{i\beta J_y/\hbar} | \psi_{i,K} \rangle}{\int \sin\beta \, d\beta \langle \psi_{i,K} | e^{i\beta J_y/\hbar} | \psi_{i,K} \rangle} \qquad (14.69)$$

and $(J_f, K, \lambda, 0 | J_f, \lambda, J_i, K)$ is a Clebsch-Gordan or vector-coupling coefficient written in the notation of *Edmonds* [ED 60].

Equations (14.68) and (14.69) yield the intensity rule of the collective model [BO 53, KE 59] for transitions between states of the same rotational band without K-mixing. However, the magnitude of the reduced transition probability calculated from these equations is not the same as that obtained in the collective model, as will be seen from the specific example to be discussed below.

As a specific example, we consider the reduced E2 transition probability $B(E2; J + 2 \rightarrow J)$ in a deformed even-even nucleus with $K = 0$. Here the J-dependent part of $B(E2; J + 2 \rightarrow J)$ becomes

$$(J, 0, 2, 0 | J, 2, J + 2, 0)^2 \frac{2J + 1}{2J + 5} = \frac{3}{2} \frac{(J + 1)(J + 2)}{(2J + 3)(2J + 5)}, \qquad (14.70)$$

which is exactly the same as that from the collective model [BO 53] for such transitions. However, it is only in the case where $\psi_{i,0}$ consists of a single two-cluster configuration and where the clusters are assumed as distinguishable (i.e., the antisymmetrization between the clusters is neglected) that one obtains for the reduced transition probability the collective-model expression

$$B_{coll}(E2; J + 2 \rightarrow J) = \frac{15}{32\pi} Q_0^2 \frac{(J + 1)(J + 2)}{(2J + 3)(2J + 5)}, \qquad (14.71)$$

where Q_0 is an intrinsic quadrupole moment defined as

$$Q_0 = \frac{\left\langle \hat{\psi}_{i,0} \left| \sum_{i=1}^{A} (3\bar{z}_i^2 - \bar{r}_i^2) \frac{1}{2}(1 - \tau_{i3}) \right| \hat{\psi}_{i,0} \right\rangle}{\langle \hat{\psi}_{i,0} | \hat{\psi}_{i,0} \rangle}, \qquad (14.72)$$

with $\hat{\psi}_{i,0}$ being an unantisymmetrized intrinsic wave function. The reason for this is that only under these assumptions can one treat two intrinsic wave functions with different symmetry axes (i.e., $\beta \neq 0$) as orthogonal to each other (see the discussion given in section 14.2).

The value of B(E2) obtained with eqs. (14.68) and (14.69) is usually smaller than the corresponding value in the collective model. This can be understood by observing that the

† For a discussion of this derivation, see ref. [KR 73].

procedure of antisymmetrization generally reduces, as has often been mentioned, these effects which are associated with strong nuclear correlations.

In deriving eqs. (14.68) and (14.69), we have assumed that all overlapping integrals are so strongly peaked that the reduced d-function which occurs in the integrands may be approximated by only the leading term in its series expansion. For appreciably deformed medium-heavy and heavy nuclei, this is of course a valid assumption. But one has to be careful in dealing with relatively light nuclei. In these nuclei, the peaking of the overlapping integrals may not be strong enough to justify the adoption of such an assumption. For example, in the nucleus ^{8}Be where in the oscillator limit the three lowest levels have an energetically pure rotational structure, the ratio $B(E2; 4 \to 2)/B(E2; 2 \to 0)$ deviates quite appreciably from that derived from eqs. (14.68) and (14.69) [WI 59a].

The above considerations can be extended to include higher-order corrections connected with K-mixing or the mixing between rotational and other configurations, and to electromagnetic transitions between states in rotational bands specified by different values of K. These extensions are, however, not expected to yield much further information concerning the effect of antisymmetrization on rotational states. Therefore, we shall not discuss them here but refer to the paper by Verhaar [VE 64] on these subjects.

14.6. Relationship with other Descriptions of Nuclear Rotational States

We now discuss the relationship between the rotational wave functions discussed in this chapter and other rotational wave functions, namely, the wave functions in the unified model [BO 52, MO 57] and those of *Peierls, Yoccoz,* and *Verhaar* [PE 57, VE 63, VE 63a, VE 64, YO 57].

Let us first go to the unified model and consider, for simplicity, K = 0 states in an axially symmetric even-even nucleus. In this case, the wave function has the form

$$\psi_u = \left(\frac{2J+1}{8\pi^2}\right)^{1/2} X_K(\tilde{r}')\, D^J_{MK}(\Omega), \tag{14.73}$$

where $X_K(\tilde{r}')$ is an intrinsic function which depends on a set of coordinates $\tilde{r}'$ defined with respect to the body-fixed frame. For K = 0, one can use the relation between the Wigner D-functions and the spherical harmonics Y_{JM} [ED 60] to write ψ_u as

$$\psi_u = (-1)^M (2\pi)^{-1/2} X_0(\tilde{r}')\, Y_{JM}(\theta, \varphi). \tag{14.74}$$

By comparing the above equation with eq. (14.2), we see that the unantisymmetrized ^{8}Be wave function or, in fact, any unantisymmetrized two-cluster wave function has the form of eq. (14.74). Indeed, the reason for this similarity is that, when the antisymmetrization between the clusters is neglected, one can construct a symmetry axis of the intrinsic wave function as the line which joins the uniquely defined center-of-mass points of the two clusters.

If one carries out the antisymmetrization completely such that the wave function has the form of eq. (14.4), then the similarity with the wave function of the unified model vanishes. Now, there is the complication that every permutation contained in the anti-symmetrization procedure gives rise to new definitions for the cluster internal spatial coordinates $\bar{\mathbf{r}}_i$ and cluster relative coordinate **R** (see eq. (3.9)) in terms of the nucleon spatial coordinates. The consequence of this is that cluster center-of-mass coordinates can no longer be uniquely defined. Therefore, one must introduce parameter coordinates in order to define the symmetry axis of the intrinsic function and the rotational wave function has to be written in an integral representation of the form given by eq. (14.6) or eq. (14.29).† One result of this is that the energy differences between rotational states are now also partially potential in origin, in contrast to the situation in the unantisymmetrized case where these energy differences come entirely from differences in kinetic energy.

In the case where the intrinsic wave function consists of a superposition of different two-cluster configurations, it can be realized easily that, even if one does not antisymmetrize between the clusters, the symmetry axis cannot be defined uniquely in terms of nucleon spatial coordinates. Therefore, we conclude from our discussion that the rotational wave functions discussed in this chapter become similar to those of the unified model only under the following rather special circumstances: (a) if the rotational states can be described by single two-cluster configurations and (b) if the antisymmetrization between the two clusters can be considered as unimportant.

The rotational wave functions used by *Peierls, Yoccoz,* and *Verhaar* are much more similar to those we introduce here. The difference is that they construct their intrinsic wave function $\psi_{i,K}$ in terms of single-particle orbitals in a rotationally symmetric deformed potential well. This has the disadvantage that the symmetry axis of the intrinsic function oscillates around the symmetry axis of the deformed well. If one projects out from this intrinsic function eigenstates of the total angular momentum, then these projected wave functions will not be translationally invariant.†† Therefore, the intrinsic wave functions of *Peierls et al.* are not very suitable in constructing rotational functions for nuclei with small deformations or for light nuclei where the peaking of the overlapping integrals is not very strong. In our way of constructing rotational wave functions, there is no problem with the total center-of-mass oscillation, because the center-of-mass motion is always split off in a clear manner from the very beginning.

The main advantage of our method is that it allows the construction of translationally invariant rotational wave functions of the most general kind and yields a natural interpretation of the difference in behaviour between rotational energy spectra in light and heavy nuclei.

The mathematical problems which appear in the method of *Peierls et al.* are largely identical with those which we are confronted with. Consequently, we could simply take over many of their results. Therefore, instead of presenting here all mathematical details, we have frequently referred to the papers of these authors.

---

† Because of antisymmetrization, one has to be careful about the meaning of K. In the unified model, K represents the projection of the total angular momentum on the body-fixed symmetry axis. In our case, K is not a physical observable (see for this the second footnote on p. 290).

†† As to how one can achieve translational invariance and obtain thereby momentum-eigenstates, see sect. 14.7.

14.7. Construction of Intrinsic Wave Functions for Quantitative Studies of Collective States in Medium-Heavy and Heavy Nuclei

To show the connection between rotational states in light and heavy nuclei, we have employed until now cluster wave functions for the construction of intrinsic states $\psi_{i,K}$. For the quantitative study of collective states in medium-heavy and heavy nuclei, it is however more convenient to use another, principally equivalent, formulation which has been adopted especially by Mang, Faessler, and others (see, e. g., refs. [BA 73] and [FA 76], and references quoted therein). In this section, we shall briefly discuss this formulation and indicate how it fits into the framework of the unified nuclear theory presented here.

This formulation is closely related to the method of Peierls *et al.* mentioned in section 14.6. By utilizing single-particle wave functions in a rotationally symmetric, deformed Nilsson potential specified by a mean radius R and a deformation parameter β, one constructs a totally antisymmetric intrinsic wave function $\psi_{i,K}$ with angular-momentum component K along the symmetry axis and a given parity. To add further flexibility into the formulation, one can introduce additional parameters Δ_i $(i = 1, \ldots, s)$, e.g., continuous or discrete parameters Δ_p and Δ_n which describe the degree of breaking up of proton and neutron pairing correlations (antipairing effect) due to centrifugal and Coriolis effects [MO 60]. In this way of construction, $\psi_{i,K}$ is therefore a function of these parameters R, β, and Δ_i, in addition to all the nucleon coordinates. Applying now the Young operator C^J_{MK} (see eqs. (14.21) and (14.24)) to $\psi_{i,K}$, one then obtains an eigenfunction for the angular-momentum operators. The total wave function with projected angular-momentum values J and M for the description of collective states has thus the following generalized Hill-Wheeler form† [HI 53]:

$$\psi_{JM}(K) = \int F^J_{MK}(R, \beta, \Delta_1, \ldots, \Delta_s)\, C^J_{MK}\, \psi_{i,K}(R, \beta, \Delta_1, \ldots, \Delta_s; \vec{r}_1, \ldots, \vec{r}_A) \times dR\, d\beta\, d\Delta_1 \ldots d\Delta_s. \tag{14.75}$$

If one wishes to consider also the case where K-mixing is important, then it is necessary to use a linear superposition of these wave functions; that is, one uses now

$$\psi_{JM} = \sum_K \psi_{JM}(K) \tag{14.76}$$

with $|K| \leqslant J$. By mixing wave functions of different K values, one can, for instance, describe deviations of the intrinsic wave function from rotational symmetry, which is equivalent to the introduction of the parameter γ in the collective model [BO 53].††

---

† The parameters $R, \beta, \Delta_1, \ldots, \Delta_s$ are commonly called the generator coordinates.

†† One can introduce the parameter γ directly into the intrinsic wave function ψ_i and the corresponding variational amplitude F^J_M.

The wave functions $\psi_{JM}(K)$ and ψ_{JM} are not translationally invariant. To achieve translational invariance one uses, instead of the intrinsic function $\psi_{i,K}$ in eq. (14.75), but another intrinsic function $\widetilde{\psi}_{i,K}$ given by

$$\widetilde{\psi}_{i,K} = \psi_{i,K}(R, \beta, \Delta_1, \ldots, \Delta_s; \mathbf{r}_1, \ldots, \mathbf{r}_A)\, \delta(\mathbf{R}_C) \qquad (14.77)$$

with

$$\mathbf{R}_C = \frac{1}{A} \sum_{m=1}^{A} \mathbf{r}_m . \qquad (14.78)$$

As is evident, $\widetilde{\psi}_{i,K}$ represents a wave function describing a system with its center-of-mass fixed at the origin.[†] To perform the integrations required for the evaluation of the matrix elements, one writes then

$$\delta(\mathbf{R}_C) = \left(\frac{1}{2\pi}\right)^3 \int e^{i\mathbf{S}\cdot\mathbf{R}_C}\, d\mathbf{S} \qquad (14.79)$$

and utilizes the generator-coordinate technique discussed in subsection 5.2b.

To determine the superposition amplitudes $F^J_{MK}(R, \beta, \Delta_1, \ldots, \Delta_s)$, one substitutes eq. (14.76) into the projection equation (14.1). In this way, one obtains a coupled set of generalized Hill-Wheeler integral equations [HI 53] for the determination of these amplitudes. For medium-heavy and heavy nuclei, it is usually a good approximation to simplify the calculation by simply adopting the intrinsic function $\psi_{i,K}$ rather than the intrinsic function $\widetilde{\psi}_{i,K}$ because, even with the generator-coordinate technique of subsection 5.2b, the computation with $\widetilde{\psi}_{i,K}$ can be quite laborious.

For pure rotational states, no K-mixing is present and the superposition amplitude F^J_{MK} becomes independent of J and M. This means that, in going from one rotational state to the next, the centrifugal and Coriolis forces do not change the relative importance of the intrinsic function in its dependence on the generator coordinates $R, \beta, \Delta_1, \ldots, \Delta_s$ (see, e. g., ref. [FA 76] and references given there). Under these specific circumstances, one can then easily see that all considerations given in sections 14.1–14.6 remain unchanged.

14.8. Specific Examples

14.8a. $\alpha + {}^{16}O$ Cluster States in ${}^{20}Ne$

As a first example, we consider the positive- and negative-parity K = 0 rotational states in ${}^{20}Ne$ which have an $\alpha + {}^{16}O$ cluster structure. In this case, the binding and resonance energies in these states can be determined from an $\alpha + {}^{16}O$ resonating-group calculation which was formulated in subsection 7.3e. To obtain the resonance energy of a state, we calculated the phase shift as a function of the relative energy E_R between the clusters

† We should mention that the singularity of $\delta(\mathbf{R}_C)$ does not create any difficulty because, in the A-nucleon Hamiltonian, the center-of-mass kinetic-energy operator is subtracted off (see eq. (4.6)).

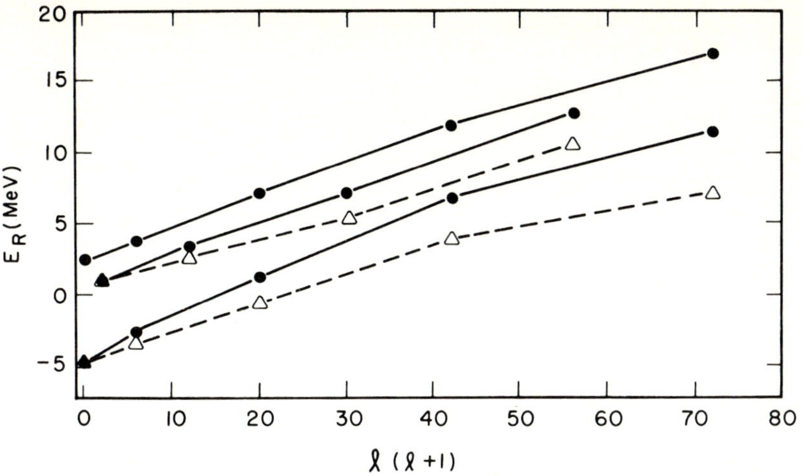

Fig. 61. Comparison of calculated (solid dots) and experimental (open triangles) binding and resonance energies for the α + ^{16}O cluster states in ^{20}Ne. (Adapted from ref. [SU 72].)

and then performed a Breit-Wigner type single-level analysis. The results [SU 72] are shown in fig. 61 by solid dots, with connecting lines drawn through them as a visual aid. From this figure it is seen that the negative-parity band is higher on the energy scale than the positive-parity ground-state band, in agreement with the discussion given in preceding sections.

Experimental results [HO 68] for the binding and resonance energies are shown by open triangles also in fig. 61. As is seen, the agreement between calculation and experiment is fairly satisfactory, but the experimental energy differences between adjacent rotational levels are all somewhat smaller than the calculated ones. The reason for this discrepancy is probably that, in this calculation, a nonsaturating nucleon-nucleon potential is used and the radius-change effect of the α cluster, mentioned in section 3.7, has not been taken into consideration.

14.8b. Rotational States in ^{22}Ne

For the nucleus ^{22}Ne, a detailed microscopic calculation to compute the energies of rotational levels has been carried out by *Goeke et al.* [GO 73]. In that calculation which fits into the general scheme of section 14.7, a projected Hartree-Fock-Bogoliubov (PHB) formalism was used, with the intrinsic wave function containing a variational parameter which measures the degree of pairing correlation. The value of this variational parameter is then determined in each rotational state by minimizing the expectation value of the Hamiltonian calculated with respect to the projected wave function. In this way, it was found that this parameter is state-dependent, indicating that the degree of pairing correlation is not the same in every state of the ^{22}Ne rotational band.

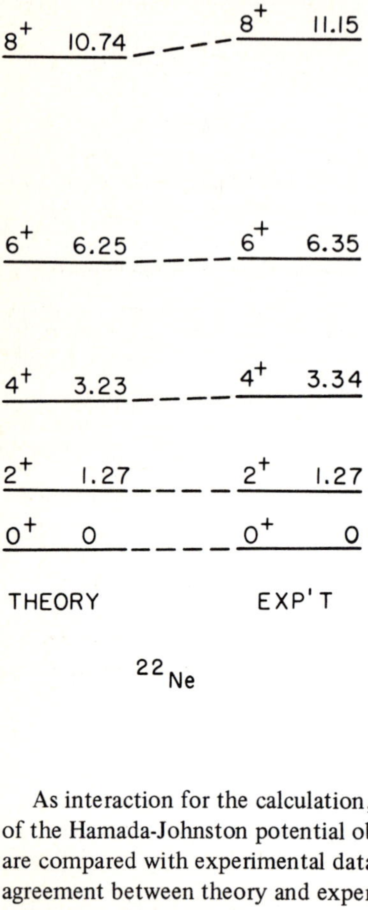

Fig. 62
Comparison of calculated and experimental rotational spectra of ^{22}Ne. (Adapted from ref. [GO 73].)

As interaction for the calculation, these authors used the effective G-matrix elements of the Hamada-Johnston potential obtained by *Barrett et al.* [BA 70, BA 71]. The results are compared with experimental data [DA 70, KU 67] in fig. 62. Here one sees that the agreement between theory and experiment is indeed quite good. In particular, the calculation shows that the moment of inertia increases mostly with increasing excitation energy (centrifugal stretching), which is in agreement with experimental observation.

The wave functions used in this calculation are not translationally invariant, but this can be remedied by using the procedure described in section 14.7. In addition, they are not eigenfunctions of the proton and neutron number operators. This comes from the fact that the intrinsic wave functions are constructed within the field-theoretical PHB formalism in which the proton and neutron numbers are not conserved. By projecting out wave functions corresponding to definite proton and neutron numbers from the intrinsic functions and then utilizing the projection equation (14.1), this defect also can be remedied. We should mention, however, that to take these corrections into account will cause a considerable increase in the amount of computational effort. Therefore, as a first approximation, one may simply choose to forego these corrections. Since ^{22}Ne is already relatively heavy, it is expected that especially the omission of the center-of-mass correction will result in only minor consequences.

14.8c. Backbending

With the type of formalism given in section 14.7, the phenomenon of backbending can also be quantitatively explained. Backbending means that, at a certain critical spin value, the effective moment of inertia, e. g. for elements in the rare-earth region such as ^{162}Er, begins to increase strongly with increasing excitation energy.† This phenomenon occurs as a consequence of the overlap of two rotational bands. In one of these bands the nucleons are mainly paired, while in the other band two $i_{13/2}$ neutrons are unpaired but strongly aligned along the axis of rotation. These two bands are coupled to each other due to the Coriolis force. As the rotational energy increases, this coupling becomes more important; consequently, the angular momentum of the rotational state can increase appreciably with only a moderate increase in the rotational energy, thus resulting in a large change in the effective moment of inertia.

By introducing into $\psi_{i,K}$ parameters which determine the relative importance of pairing and alignment, one can study the backbending phenomenon in a quantitative manner ([FA 76] and references contained therein) by using the procedure described in section 14.7. Such calculations show, for instance, that in the backbending region the deformation of the nucleus remains more or less constant; this means that, with the pair-breaking and alignment occurring in this region, no appreciable change in the deformation of the nucleus is effected. It should be mentioned that, in contrast to the calculation described in subsection 14.8b, it is important for calculations on backbending to always use intrinsic wave functions which correspond to definite proton and neutron numbers. On the other hand, because of the large mass involved, the approximation of not separating off the total center-of-mass motion may still be adopted.

14.9. Concluding Remarks

The main purpose of this chapter is to show that rotational states in light and heavy nuclei have essentially the same microscopic structure and, therefore, can be treated from a unified viewpoint. To show this, we have utilized cluster wave functions and Hill-Wheeler type functions to construct translationally invariant rotational wave functions of the most general kind, and introduced parameter and generator coordinates which enable us to separate collective variables from nucleon coordinates.

As has frequently been pointed out, the procedure of antisymmetrization reduces the differences between different representations. Therefore, the Hill-Wheeler functions of the Hartree-Fock-Bogoliubov type mentioned earlier may be quite similar to wave functions which exhibit explicitly the feature of cluster formation. In this respect, it is interesting to note that antisymmetrization also reduces the amount of intrinsic nuclear deformation. Because of this, the peaking of the overlapping integrals becomes smaller, with the consequence that the effective moment of inertia is reduced compared to the moment of inertia of a rigid body. For medium-heavy and heavy nuclei, this reduction is often by a factor of 2 or more. On the other hand, as has just been discussed, the effective

† The origin of the name "backbending" was the observation that, in the high-spin region of certain elements, the excitation energy increases as the angular frequency of rotation decreases.

moment of inertia can increase appreciably if the internal structure of the nucleus in a certain excitation-energy region is changed. An example for this is the observed large change of this quantity in the backbending region where nucleons lose pairing correlations but become strongly aligned.

Using similar considerations, one can also investigate the microscopic structure of vibrational and other collective excitations of the nucleus, including their mutual coupling. For such considerations, wave functions of the form given by eqs. (14.75)–(14.77) are especially suited. By substituting these functions into the projection equation (14.1) one obtains, as has already been mentioned, a coupled set of integral equations for the superposition amplitudes F^J_{MK} which depend on a set of variables $R, \beta, \Delta_1, \ldots, \Delta_s$.† These superposition amplitudes can be considered as new wave functions depending on a small number of collective variables, i. e., variables in the set $R, \beta, \Delta_1, \ldots, \Delta_s$ which assume continuous values. The collective behaviour of nuclear states can then be discussed in terms of these new wave functions [WI 76].

It is clear that via the projection equation (14.1) the coupling of the collective states discussed here to any kind of incoming and outgoing channel can be described. For this one has to add to the collective wave functions, as usual, cluster wave functions with boundary conditions appropriate to these channels.

15. Brief Discussion of Time-Dependent Problems

15.1. General Remarks

Even though the basically correct way to treat nuclear scattering and reaction processes is a time-dependent treatment, we have so far treated these processes as stationary problems. As was pointed out in chapter 1, the reason why we could do this is connected with the fact that the solutions of the time-dependent projection equation (2.1) can always be constructed by linearly superposing the solutions of the time-independent projection equation (2.3).

In this chapter, we shall discuss in some detail the connection between the stationary solutions of eq. (2.3) and the time-dependent solutions of eq. (2.1). In particular, we wish

† Some of the variables in the Δ_i set can be discrete. These discrete values can be used to indicate, e. g., internal excitations of the intrinsic wave functions.

to examine, from a time-dependent viewpoint, the physical significance of certain quantities appearing in a time-independent treatment as, for instance, the energy width of a resonance level.

In section 15.2, we first briefly explain why the scattering phase shift must, due to causality, increase rapidly in passing through a narrow resonance. After this, we discuss the connection between the level width and the lifetime of a compound resonance state. A fairly detailed discussion is then presented to show how one can compute, under certain circumstances, the level width approximately, with the nucleus ^{6}Li used as an example for this type of calculation. Finally, in section 15.3, we consider a specific example involving a time-dependent interaction, namely, the photodisintegration of ^{6}Li into tritons and ^{3}He.

15.2. Connection between the Lifetime of a Compound State and its Level Width

15.2a. Relationship between Phase Shift and Time Delay

We consider at first the time development of an incoming wave packet having a very small radial-momentum spread and, hence, a large spatial extension, which describes an incoming particle in a given orbital angular-momentum state. Since we are interested in an energy region where a sharp resonance level of the compound nucleus is located, we assume that the scattering energy is so well-defined that its kinetic-energy spread in the asymptotic region is much smaller than the energy width of the considered resonance. Furthermore, for simplicity, we shall consider the case where, besides the elastic channel, no other channels are open, and assume that the colliding particles have no charge and no internal spin.

As has already been stated several times, the solution of the time-dependent projection equation (2.1) with arbitrary boundary conditions can be represented by linear superpositions of the solutions of the time-independent projection equation (2.3). Therefore, we make for the radial part $\bar{F}_l(R, t)$ of the relation-motion function of the wave packet the following ansatz in the asymptotic region:

$$\bar{F}_l(R, t) \sim \int_0^\infty f(k)\, e^{ig(k) - i\omega(k)t} \left[e^{-i(kR - \frac{1}{2} l\pi)} - e^{2i\delta_l(k)}\, e^{i(kR - \frac{1}{2} l\pi)} \right] dk, \tag{15.1}$$

where f(k), g(k), and the phase shift $\delta_l(k)$ are all real functions of k. In eq. (15.1) the integrand, when multiplied by the corresponding angular-momentum function and the internal functions of the colliding particles, represents after antisymmetrization the asymptotic form of a solution of the projection equation (2.3) at an energy $E_R = \hbar\omega(k)$, with the parts containing the factors $\exp[-i(kR - \frac{1}{2} l\pi)]$ and $\exp[i(kR - \frac{1}{2} l\pi)]$ describing the behavior of the incoming and outgoing waves, respectively. The functions f(k) and g(k) have to be chosen in such a way that, for $t \ll 0$, $\bar{F}_l(R, t)$ represents an incoming wave packet with a very sharp energy; in other words, f(k) must possess a very sharp maximum at a value of $k = k_0$ and g(k) must be rather smooth.

To find the place where $\overline{F}_l(R, t)$ is a maximum, we use the condition that at this place the phase of the wave must be stationary for $k = k_0$. Thus, the position of the incoming wave packet is at

$$R_i = \left(\frac{dg}{dk}\right)_{k_0} - \left(\frac{d\omega}{dk}\right)_{k_0} t_i, \quad (t_i \ll 0) \tag{15.2}$$

and the position of the outgoing wave packet is at

$$R_0 = -\left(\frac{dg}{dk}\right)_{k_0} + \left(\frac{d\omega}{dk}\right)_{k_0} t_0 - 2\left(\frac{d\delta}{dk}\right)_{k_0} . \quad (t_0 \gg 0) \tag{15.3}$$

From eqs. (15.2) and (15.3), one then concludes that the time which the wave packet needs to go in from a radius R_i and then go out to a radius R_0 is given by

$$T = \frac{R_i + R_0 + 2\left(\frac{d\delta}{dk}\right)_{k_0}}{\left(\frac{d\omega}{dk}\right)_{k_0}} . \tag{15.4}$$

In the derivation of eq. (15.4), it is of course necessary to assume that both R_i and R_0 are much larger than the extension of the wave packet. Also, it is noted that, for a wave packet of well-defined energy, one has

$$\left(\frac{d\omega}{dk}\right)_{k_0} = \frac{1}{\hbar}\left(\frac{dE_R}{dk}\right)_{k_0} = \frac{\hbar k_0}{\mu} = v_R , \tag{15.5}$$

with v_R being the relative velocity of the colliding particles in the asymptotic region.

One sees from eq. (15.4) that

$$\Delta T = \frac{2}{v_R}\left(\frac{d\delta_l}{dk}\right)_{k_0} \tag{15.6}$$

is the reduction or protraction in the travelling time of the wave packet produced by the nuclear interaction. In the energy region around a narrow elastic-scattering compound resonance, the derivative of the phase shift $(d\delta_l/dk)_{k_0}$ must therefore be large and positive, because only under this condition can one obtain a large time delay required by the long lifetime of the compound system.

These considerations do not change essentially if, besides the elastic channel, other reaction channels are open. In this case, the phase shift $\delta_l(k)$ in eq. (15.1) becomes complex with a positive imaginary part which describes the absorption of the incident wave due to transitions to these reaction channels. With this modification taken into account, one can then proceed in exactly the same way as described above for the case of pure elastic-scattering, and obtain a similar conclusion.

From the above discussion we see already that there must exist an intimate relationship between the level width of a compound resonance and its lifetime. In the next subsection, we shall quantitatively formulate this relationship.

15.2b. Quantitative Relationship between the Level Width and the Lifetime of a Compound Nuclear State

For simplicity, we assume again that only the elastic channel is open. In this case, the stationary solution of eq. (2.3) in a state of total angular momentum j can then be written as (see subsection 9.2a)

$$\psi_j(E) = \psi_{Dj}(E) + \psi_{Cj}(E) = \psi_{Dj}(E) + a_k(E)\,\psi^k_{Cj} + \sum_{n \neq k} a_n(E)\,\psi^n_{Cj}, \tag{15.7}$$

where $\psi_{Dj}(E)$ describes the behaviour of the system in the elastic channel. The functions ψ^k_{Cj} and ψ^n_{Cj} represent an orthonormalized set of bound-structure wave functions which describe the quasi-bound states of the compound nucleus. As was discussed in subsection 9.2c, we can construct $\psi_{Dj}(E)$ by using the projection formalism such that it is orthogonal to ψ^k_{Cj}, the wave function of the particular quasi-bound state whose decay we wish to investigate. Also, the amplitude $a_k(E)$ in eq. (15.7) is given by (see eq. (9.37))

$$a_k(E) = -\frac{\langle \psi^k_{Cj} | H - E | \psi'^+_{Dj}(E)\rangle}{E^k_C - E - \Delta_j(E) - \frac{1}{2}\,i\Gamma_j(E)}, \tag{15.8}$$

which can be approximately written, in the energy region around the resonance, as

$$a_k(E) \approx -\frac{\langle \psi^k_{Cj} | H - E_r | \psi'^+_{Dj}(E_r)\rangle}{E_r - E - \frac{1}{2}\,i\Gamma_{jr}} \tag{15.9}$$

when the resonance is sharp and isolated. It should be mentioned that, because of the use of the projection formalism in the construction of ψ_{Dj}, the function $\psi'^+_{Dj}(E_r)$ is also orthogonal to ψ^k_{Cj}.

By superposing the stationary solutions $\psi_j(E)$ in an energy range of the order of Γ_{jr} around E_r, one can construct a time-dependent wave packet $\Psi_j(t)$ which is a solution of the time-dependent projection equation (2.1) and which has the property that, at $t = 0$, only the compound configuration ψ^k_{Cj} is present. That is, one can write

$$\Psi_j(t) = \int \lambda(E)\,\psi_j(E)\,e^{-iEt/\hbar}\,dE \tag{15.10}$$

with the property

$$\Psi_j(0) = \psi^k_{Cj}, \tag{15.11}$$

if one chooses judiciously a function $\lambda(E)$ which is different from zero in an energy region centered at E_r and having an extension larger than but of the order of Γ_{jr}.

The probability that, after a certain time t, the compound nucleus remains in its resonance configuration is given by

$$W(t) = |w(t)|^2, \tag{15.12}$$

with

$$w(t) = \langle \psi^k_{Cj} | \Psi_j(t)\rangle. \tag{15.13}$$

Using the properties of ψ_{Dj}, ψ_{Cj}^k, and ψ_{Cj}^n mentioned above, one obtains

$$w(t) = \int \lambda(E)\, a_k(E)\, e^{-iEt/\hbar}\, dE = -\int \frac{\lambda(E)\, e^{-iEt/\hbar} \langle \psi_{Cj}^k | H - E_r | \psi_{Dj}'^{+}(E_r) \rangle}{E_r - E - \frac{1}{2} i \Gamma_{jr}}\, dE. \tag{15.14}$$

We now compute w(t) at t large enough ($t \gg \hbar/\Gamma_{jr}$) such that all transient effects associated with the formation of the compound configuration ψ_{Cj}^k have died off. By carrying out a contour integration around the lower half of the complex plane, then one finds that the only appreciable contribution comes from an integration around the pole at $E = E_r - \frac{1}{2} i \Gamma_{jr}$ and the result is†

$$w(t) \sim e^{-iE_r t/\hbar}\, e^{-\frac{1}{2\hbar}\Gamma_{jr} t}. \tag{15.15}$$

Substituting the above equation into eq. (15.12) then yields, for t sufficiently large,

$$W(t) \sim e^{-\Gamma_{jr} t/\hbar}. \tag{15.16}$$

Equation (15.16) describes an exponential decay with a lifetime

$$\tau = \hbar/\Gamma_{jr}, \tag{15.17}$$

where Γ_{jr} has been given previously by eq. (9.24).

One can easily generalize the result represented by eq. (15.17) to the case where more than one channel is open. The only point one needs to note is that, in this case, the total energy width Γ_{jr} now consists of a sum of partial widths (see eq. (9.77)) belonging to the various open channels.

It is of course possible to obtain the result of eq. (15.17) also in the framework of other reaction theories as, for instance, the R-matrix theory (see, e.g., ref. [LA 58]). However, in contrast to these theories, Γ_{jr} and therefore τ are defined here explicitly in terms of the wave function of the nuclear system under consideration (see chapter 9) and the formula for Γ_{jr} is applicable in any system including that where the nuclei involved in both the incident and the decay channels are all composite particles.

15.2c. Calculation of the Level Width – ^{6}Li as an Example for a Decaying System

In a nuclear system which consists of not too many nucleons and where the excitation energy is not too large, Γ_{jr} can be calculated explicitly in rather good approximation. For heavier systems, such explicit calculations become quite difficult, but even here one can devise a consistent method to calculate certain decay properties, such as the α-decay width, approximately. In this subsection, we shall discuss this method and illustrate it with the example of ^{6}Li decaying into an α particle and a deuteron.

---

† For a detailed discussion, see ref. [FR 65].

For convenience in discussion, we write down again the expression for Γ_{jr} in the single-open-channel case, which was previously given in subsection 9.2a. This expression is

$$\Gamma_{jr} = \left[\frac{\Gamma_j(E)}{1 + \frac{d}{dE} \Delta_j(E)} \right]_{E = E_r} , \tag{15.18}$$

where

$$\Gamma_j(E) = 2\pi |\langle \psi_{Cj}^k | H - E | \psi_{Dj}^{\prime +}(E) \rangle|^2 , \tag{15.19}$$

$$\Delta_j(E) = \langle \psi_{Cj}^k | (H - E)\, G_{Dj}^{\prime P} (H - E) | \psi_{Cj}^k \rangle , \tag{15.20}$$

and the resonance energy E_r is defined by the equation

$$E_C^k - E_r - \Delta_j(E_r) = 0, \tag{15.21}$$

with E_C^k being the characteristic energy of the resonance level under consideration (see eq. (9.7)). Also, we define two quantities, the compound-nucleus radius R_{CN} and the classical turning radius R_T, which will appear frequently in the following discussion. The compound-nucleus radius R_{CN} is defined in such a way that, for $R < R_{CN}$, the effective interaction between the clusters is strong and over-all attractive, while the classical turning radius R_T is defined by the equation

$$\frac{\hbar^2}{2\mu} \frac{l(l+1)}{R_T^2} + \frac{z_1 z_2 e^2}{R_T} = E_R , \tag{15.22}$$

with E_R being the relative energy of the clusters in the decay channel. For resonance levels occurring well below the Coulomb-plus-centrifugal barrier, R_T will be appreciably larger than R_{CN}; for instance, in the case of long-living heavy α-decay nuclei, R_T is of the order of 30 fm which is more than three times the radius of a heavy nucleus.

We consider now the case where the decay channel is shielded from the compound region by a high Coulomb-plus-centrifugal barrier. In this case, the resonance level is narrow and the bound-state wave function ψ_{Cj}^k can be constructed in such a way that it describes very well the behavior of the resonance state in the compound region, with the consequence that the energy shift $\Delta_j(E)$ becomes very small. In fact, we shall assume that, up to a radius R_0 which is smaller than but close to R_T, ψ_{Cj}^k is an eigenfunction of the Hamiltonian operator H with eigenvalue $E_C^k = E_r$, i.e.,

$$H \psi_{Cj}^k = E_r \psi_{Cj}^k . \quad (R < R_0) \tag{15.23}$$

In an actual calculation, it is of course impractical to demand that the above equation be exactly satisfied even in a limited region of the configuration space with $R < R_0$. However, even with a function which satisfies eq. (15.23) only approximately, the result obtained for $\Gamma_j(E_r)$ will not be very different from that obtained with the exact solution of eq. (15.23). Therefore, for simplicity in discussion, we shall always adopt the assumption that in the above-specified region of the configuration space, eq. (15.23) is satisfied.

Under the above assumption, $(d\Delta/dE)_{E = E_r}$ is much smaller than 1 and one obtains the approximate relation

$$\Gamma_{jr} \approx \Gamma_j(E_r). \tag{15.24}$$

For the computation of $\Gamma_j(E_r)$, we separate the normalized function ψ_{Cj}^k into two parts, i.e.,

$$\psi_{Cj}^k = a\,\psi_a + b\,\psi_b, \tag{15.25}$$

with ψ_a and ψ_b properly normalized and orthogonal to each other. The function ψ_a is chosen to be large only in the region of the compound nucleus ($R < R_{CN}$), while the function ψ_b is defined to have the cluster structure of the decay channel and properly describe the motion of the clusters, as given by ψ_{Cj}^k, in the barrier region ($R_{CN} < R < R_T$). Also, we shall further construct the function $\psi_{Dj}^{\prime +}(E_r)$ to be orthogonal not only to ψ_{Cj}^k but also to ψ_a and ψ_b separately. By using the projection formalism discussed in chapter 8 and subsection 9.2c, this can always be done.

Using eqs. (15.18), (15.24), and (15.25), we obtain

$$\begin{aligned}\Gamma_{jr} &\approx 2\pi |\langle a\,\psi_a + b\,\psi_b | H - E_r | \psi_{Dj}^{\prime +}(E_r)\rangle|^2 \\ &= 2\pi | a \langle \psi_{Dj}^{\prime +}(E_r)|(H - E_r)\,\psi_a\rangle + b\langle \psi_{Dj}^{\prime +}(E_r)|(H - E_r)\,\psi_b\rangle|^2 .\end{aligned} \tag{15.26}$$

Now, because of the way of construction, $\psi_{Dj}^{\prime +}(E_r)$ does not penetrate deeply into the barrier region. Therefore, since ψ_a has an appreciable magnitude only in the region of the compound nucleus, the first term in eq. (15.26) is small and can be neglected in good approximation. Thus we can write

$$\Gamma_{jr} \approx 2\pi |b|^2 |\langle \psi_{Dj}^{\prime +}(E_r)|(H - E_r)\,\psi_b\rangle|^2 , \tag{15.27}$$

which, due to the choice of ψ_b described above, involves essentially a surface integral.

In eq. (15.27), the only unknown quantity is the amplitude b, because $\psi_{Dj}^{\prime +}(E_r)$ and ψ_b can be determined very simply. If one writes the radial part of the relative-motion function in ψ_b as $\chi_b(R)/R$, then, in the region $R_{CN} \leqslant R \leqslant R_T$, $\chi_b(R)$ can be taken in very good approximation as the decaying solution (bound-state boundary condition) of a two-cluster equation in which the only interaction term comes from the Coulomb potential. In other words, one can determine $\chi_b(R)$ in this particular region by solving the equation

$$H_R\,\chi_b(R) = (E_r - E_{Cl})\,\chi_b(R), \tag{15.28}$$

where E_{Cl} represents the sum of the internal energies of the clusters in the decay channel and H_R is given by

$$H_R = -\frac{\hbar^2}{2\mu}\left[\frac{d^2}{dR^2} - \frac{l(l+1)}{R^2}\right] + \frac{Z_1 Z_2 e^2}{R} . \tag{15.29}$$

For $R \lesssim R_{CN}$, ψ_b is not uniquely defined, since a precise way to divide ψ_{Cj}^k into ψ_a and ψ_b has not been specified. But this is not important because, as has just been pointed out, the integral in eq. (15.27) has its contribution coming from a surface part which is far outside of the compound region.

The radial part $f_D(R)/R$ of the relative-motion function in $\psi_{Dj}^{\prime +}(E_r)$ can be obtained in a similar manner. Here one solves eq. (15.28) for a scattering solution, and the result is $f_D(R) = C_D\,F_l(R)$ with $F_l(R)$ being the regular Coulomb function and C_D being a proportionality factor to be determined from the normalization condition of eq. (9.71).

For the calculation of $f_D(R)$, one should in principle include in H_R also a nonlocal potential which comes from the fact that, with the use of a projection formalism, the function ψ'^{+}_{Dj} is only defined in a reduced subspace of the elastic-channel Hilbert subspace (see subsection 9.2c). In practice, however, this does not need to be done, since this additional nonlocal potential is appreciably different from zero only for $R \lesssim R_{CN}$ and will therefore not significantly affect the behaviour of $f_D(R)$ in the region near R_T from which the integral in eq. (15.27) obtains its contribution.

It should be noted that the way in which one lets χ_b approach zero in the region $R > R_T$ affects the quantity $\langle \psi'^{+}_{Dj}(E_r)|(H-E_r)\,\psi_b\rangle$ in only a minor manner. This can be seen in the following way. Because of eq. (15.23), one can write

$$\langle \psi'^{+}_{Dj}(E_r)|(H-E_r)\,\psi_b\rangle = \iint_{R_0}^{\infty} \psi'^{+*}_{Dj}(E_r)\,(H-E_r)\,\psi_b\, d\mathbf{R}\, d\tau_i. \tag{15.30}$$

where $d\tau_i$ indicates an integration over cluster internal spatial coordinates and a summation over all spin and isospin coordinates. By observing that, for $R > R_0$, the antisymmetrization between the clusters can be neglected in good approximation, eq. (15.30) can be further reduced to

$$\langle \psi'^{+}_{Dj}(E_r)|(H-E_r)\,\psi_b\rangle \approx \int_{R_0}^{\infty} f_D^*\,(H_R - \bar{E}_r)\,\chi_b\, dR$$

$$= \int_{R_0}^{\infty} \chi_b\,(H_R - \bar{E}_r)\,f_D^*\, dR - \frac{\hbar^2}{2\mu}\left[f_D^*\,\frac{d\chi_b}{dR} - \chi_b\,\frac{df_D^*}{dR}\right]_{R=R_0}, \tag{15.31}$$

where $\bar{E}_r = E_r - E_{Cl}$, and we have assumed that the cluster internal functions and the angular part of the relative-motion function are properly normalized. Now, in the region where R is larger than R_0, $(H_R - \bar{E}_r)$ operating on f_D^* is equal to zero; therefore, we obtain

$$\langle \psi'^{+}_{Dj}(E_r)|(H-E_r)\,\psi_b\rangle \approx -\frac{\hbar^2}{2\mu}\left[f_D^*\,\frac{d\chi_b}{dR} - \chi_b\,\frac{df_D^*}{dR}\right]_{R=R_0}. \tag{15.32}$$

Thus we see that the form of χ_b for $R > R_T$ does not directly influence the value of the integral $\langle \psi'^{+}_{Dj}(E_r)|(H-E_r)\,\psi_b\rangle$. The only indirect influence it has is a rather minor one, which results from the fact that the function ψ_b must be normalized to one.

From this discussion it is seen that, besides the energy E_r, the only other quantity in Γ_{jr} of eq. (15.27) which depends on the compound-nucleus structure of the decaying state is the amplitude b.† Certainly, in heavy nuclei (e.g., α-decay nuclei), it would be very

† The quantity $|b|^2$ can be interpreted as the probability of finding the cluster structure of the decay channel in the surface region of the compound nucleus. But due to the fact that both this region and the function ψ_b are not precisely defined, we must emphasize that this interpretation is certainly a rather vague one.

difficult to calculate b by solving the many-particle Schrödinger equation in the same way as one does for a much lighter system. However, it is important to note that, especially in these nuclei, only nucleons in the Fermi surface can be readily excited. This is very useful, since it enables one to obtain reasonably accurate bound-state solutions by performing oscillator shell-model calculations with configuration mixing, where only the excitations of the nucleons in the outer subshells are taken into consideration.

We describe now an approximate way to calculate b. Here what one does is to first perform a translationally invariant shell-model calculation with configuration mixing in an oscillator well of width parameter β and obtain as well as possible a wave function $\overline{\psi}^k_{Cj}$ which describes the compound-nucleus structure of the decaying state. Then, one takes as b the following expression:

$$b = \langle \overline{\psi}^k_{Cj} | \overline{\psi}_b \rangle, \tag{15.33}$$

where $\overline{\psi}_b$ is an antisymmetrized two-cluster function of the type

$$\overline{\psi}_b = A \left\{ \tilde{\phi}(\tilde{A})\, \tilde{\phi}(\tilde{B})\, \frac{1}{R}\, \overline{\chi}_b(R)\, Y_{lm}(\theta, \varphi) \right\}, \tag{15.34}$$

with $\tilde{\phi}(\tilde{A})$ and $\tilde{\phi}(\tilde{B})$ being oscillator-type wave functions, characterized by width parameter β, describing the internal behaviour of the clusters in the decay channel, and the radial function $\overline{\chi}_b(R)$ adjusted such that b acquires a maximum absolute value.[†]

The above procedure yields a reasonable value for b, but one cannot simply use $\overline{\psi}_b$ as ψ_b in calculating Γ_{jr}. The reason is that $\overline{\chi}_b(R)$ does not have a correct behaviour in the region where R is larger than R_{CN}. But this can be remedied in a rather simple manner. What one needs is to define another radial function $\chi_b(R)$ which is the same as $\overline{\chi}_b(R)$ for $R < R_b$ ($R_b \approx R_{CN}$), but equal to the decaying solution of eq. (15.28) for $R > R_b$. This can be done by simply adjusting the connection radius R_b and the amplitude of the decaying solution such that the radial function $\chi_b(R)$ becomes continuous and has a continuous first derivative at $R = R_b$. It should be noted that by such a construction the resultant function $\psi^k_{Cj} = a\,\psi_a + b\,\psi_b$, which differs from the original function $\overline{\psi}^k_{Cj}$, is no longer normalized to one. But this is not expected to have any serious consequences because, especially in heavy nuclei, the contribution of $b\,\psi_b$ to the normalization of ψ^k_{Cj} is very small.

In the above procedure, it has also been assumed that the width parameters of the cluster internal functions $\tilde{\phi}(\tilde{A})$ and $\tilde{\phi}(\tilde{B})$ are both equal to that which characterizes $\overline{\psi}^k_{Cj}$. This is done not for any physical reason but merely for computational convenience. If one chooses to use width parameters which are more appropriate for the clusters in

† Because of the equivalence between oscillator shell-model and oscillator cluster-model wave functions discussed in chapter 3, the calculation of the overlapping integral in eq. (15.33) is not too difficult. In an actual calculation, one can simplify the procedure even further by simply taking $\overline{\chi}_b(R)$ in such a way that the resultant function $\overline{\psi}_b$ becomes a shell-model wave function of the lowest possible configuration in an oscillator well of width parameter β. This will introduce some underestimate into the magnitude of b, but this underestimate will, in general, not be a large one.

the decay channel, then the computation of the overlapping integral in eq. (15.33) will become much more difficult, although it is worth mentioning that this difficulty may be considerably alleviated if one uses the generator-coordinate technique described in subsection 5.2b.

Another point which should be mentioned is that, in calculating Γ_{jr} one should in general use the experimental value for E_r rather than the value determined from the shell-model calculation. This is important because, especially in the case where the resonance state lies rather close to the threshold of the decay channel, Γ_{jr} will be sensitively dependent upon the difference between the resonance energy and the threshold energy (see section 9.6).

The method to compute Γ_{jr} described above was tested in a model calculation [MA 74] in which the nucleus ^{6}Li was considered as a decaying system. In that calculation, a single-channel $d + \alpha$ resonating-group calculation was first performed and, by adjusting the strength of the central nucleon-nucleon potential, a narrow $l = 2$ resonance state deeply imbedded under the Coulomb-plus-centrifugal barrier was created. The level width of this state decaying into a deuteron and an α particle was then calculated by using eq. (15.27) and compared with that obtained from a single-level analysis of the resonating-group phase shifts. The result of this comparison showed that the value for Γ_{jr} calculated from eq. (15.27) is in fact fairly accurate, being only about 25 % smaller than the value obtained from the resonating-group calculation.

In this example, the narrow resonance is almost a pure potential-resonance (see subsection 9.2c). Therefore, the magnitude of b is very close to one. Also, because the resonance state occurs well below the Coulomb-plus-centrifugal barrier, the radial function $\chi_b(R)$ decreases rapidly in the barrier region; consequently, the replacement of $\overline{\chi}_b(R)$ by $\chi_b(R)$ does not appreciably change the normalization of the compound-nucleus wave function.

The method discussed here to calculate decay widths or lifetimes should be of interest in the study of particle decays of certain heavy nuclei where the compound-nucleus configuration of the decaying nucleus and the structure of the decay products are comparatively simple. Examples for such nuclei are the unstable α-decay heavy nuclei with long lifetimes ($\tau \gg 10^{-20}$ sec), for which a microscopic description was first given by *Mang* [MA 60a]. The advantage of the method described here over the work of Mang and others is that the approximations which go into the calculation of the lifetime are clearly defined and one knows at least in principle how to improve the calculation. For instance, one of the important improvements would be to refine the compound-nucleus wave function by adding bound-state type cluster wave functions with linear variational amplitudes to the shell-model wave function with configuration mixing, in order to achieve a better description of α-cluster formation on the nuclear surface.

15.3. Time-Dependent Projection Equation with Time-Dependent Interaction

So far in this chapter, we have considered time-dependent problems where the Hamiltonian involved is time-independent. Now we shall consider problems in which the Hamiltonian itself is time-dependent. In nuclear physics, one of the important problems of this type are the γ-absorption processes. As is well known, the common way to treat these processes is the time-dependent perturbation method. In this section, we shall therefore first derive, by starting from the time-dependent projection equation, the formula for the transition rate from an initial state to a continuous set of final states. Afterwards, we then discuss the application of this formula to a study of the photo-disintegration of ^{6}Li into a triton plus a ^{3}He particle.

To start the discussion, we write the Hamiltonian in the form

$$H = H_0 + V(t), \tag{15.35}$$

where H_0 represents the time-independent part. The quantity $V(t)$ is the time-dependent perturbation term which is responsible for the transition between initial and final states. In this section, we shall consider as final states only those in decay channels, i. e., final states with a continuous energy spectrum.

The time-dependent Schrödinger equation, written as a projection equation, has the form

$$\left\langle \delta \Psi \,|\, H_0 + V(t) + \frac{\hbar}{i}\frac{\partial}{\partial t} \,|\, \Psi \right\rangle = 0. \tag{15.36}$$

For Ψ we make the ansatz

$$\Psi = b_i(t)\, \psi_i\, e^{-iE_i t/\hbar} + \int b_{E'}(t)\, \psi_{E'}\, e^{-iE' t/\hbar}\, dE', \tag{15.37}$$

where the function $\psi_i e^{-iE_i t/\hbar}$ represents the initial stationary state and the functions $\psi_{E'} e^{-iE't/\hbar}$ represent the final stationary states which the system goes into as a result of the perturbing interaction $V(t)$. Both the functions ψ_i and $\psi_{E'}$ are orthonormalized eigenfunctions of H_0 in a chosen Hilbert subspace; that is, if one chooses a usually nonorthogonal set of basis functions ϕ_k to define this subspace, then ψ_i and $\psi_{E'}$ are orthonormalized solutions of the time-independent projection equation

$$\langle \delta \psi | H_0 - E | \psi \rangle = 0, \tag{15.38}$$

where

$$\psi = \mathrm{S}\, a_k\, \phi_k, \tag{15.39}$$

with a_k being linear discrete and continuous variational parameters and E being the energy eigenvalue. It should be noted that, in the ansatz of eq. (15.37), the eigenstates ψ_i with $E = E_i$ and $\psi_{E'}$ with $E = E'$ represent usually only a small part of the total number of eigenstates belonging to H_0. In other words, we are assuming that the transition from the initial to the final states occurs directly, without going through intermediate states of the system. This is certainly allowed, since $V(t)$ is usually weak enough to be considered only in first order, as in the case of electromagnetic transitions.

The quantities $b_i(t)$ and $b_{E'}(t)$ in eq. (15.37) are time-dependent amplitudes which will be varied to satisfy eq. (15.36). By proceeding in the usual manner, one finds easily that they are solutions of the following set of coupled equations:

$$\frac{\hbar}{i}\frac{db_i}{dt} + \langle \psi_i | V(t) | \psi_i \rangle\, b_i + \int \langle \psi_i | V(t) | \psi_{E''} \rangle\, e^{-i(E''-E_i)\,t/\hbar}\, b_{E''}\, dE'' = 0, \tag{15.40a}$$

$$\frac{\hbar}{i}\frac{db_{E'}}{dt} + \langle \psi_{E'} | V(t) | \psi_i \rangle \, e^{-i(E_i - E')t/\hbar} \, b_i$$

$$+ \int \langle \psi_{E'} | V(t) | \psi_{E''} \rangle \, e^{-i(E'' - E')t/\hbar} \, b_{E''} \, dE'' = 0. \tag{15.40b}$$

Starting from eqs. (15.40a) and (15.40b), one can then carry out a time-dependent perturbation calculation using the standard procedure.

Now, let us assume that $V(t)$ has a harmonic time dependence, i.e.,

$$V(t) = 2 V_0 \sin \omega t, \tag{15.41}$$

then, after some algebraic manipulation, we obtain for the transition rate W the following well-known expression (see, e.g., ref. [DA 65a]):

$$W = \frac{2\pi}{\hbar} |\langle \psi_f | V_0 | \psi_i \rangle|^2 \rho(E_f), \tag{15.42}$$

where ψ_f is the wave function of a typical final state which has a total energy E_f given by

$$E_f = E_i + \hbar\omega, \tag{15.43}$$

and $\rho(E_f)$ represents, at $E = E_f$, the density of states in the decay channel.

Equation (15.42) can be used to calculate the rate of disintegration of ^6Li into tritons and ^3He particles by γ-absorption. In this case, a first-order perturbation calculation is sufficient and the perturbing potential $V(t)$ is given by†

$$V(t) = -\frac{e}{Mc} \sum_i \mathbf{A}(\mathbf{r}_i, t) \cdot \mathbf{p}_i \, \frac{1 - \tau_{i3}}{2}, \tag{15.44}$$

where $\mathbf{A}$ is a vector potential satisfying the gauge condition

$$\nabla \cdot \mathbf{A} = 0, \tag{15.45}$$

and can be treated as a classical quantity. Now, if one chooses the photons with angular frequency ω to travel in the z-direction and polarize in the x-direction, then $\mathbf{A}$ assumes the form

$$\mathbf{A} = \hat{e}_x A_0 \sin \omega \left(\frac{z}{c} - t \right), \tag{15.46}$$

where $\hat{e}_x$ denotes a unit vector along the x-axis. Since, in the energy region under consideration, the wavelength of the photon is much larger than the nuclear radius, one can consider the vector potential to be constant over the nuclear volume and write it approximately as (electric-dipole approximation)

$$\mathbf{A} = -\hat{e}_x A_0 \sin \omega t. \tag{15.47}$$

† The interaction between the magnetic field and the intrinsic magnetic moments of the nucleons is neglected.

With this approximation, the perturbing potential V(t) then becomes

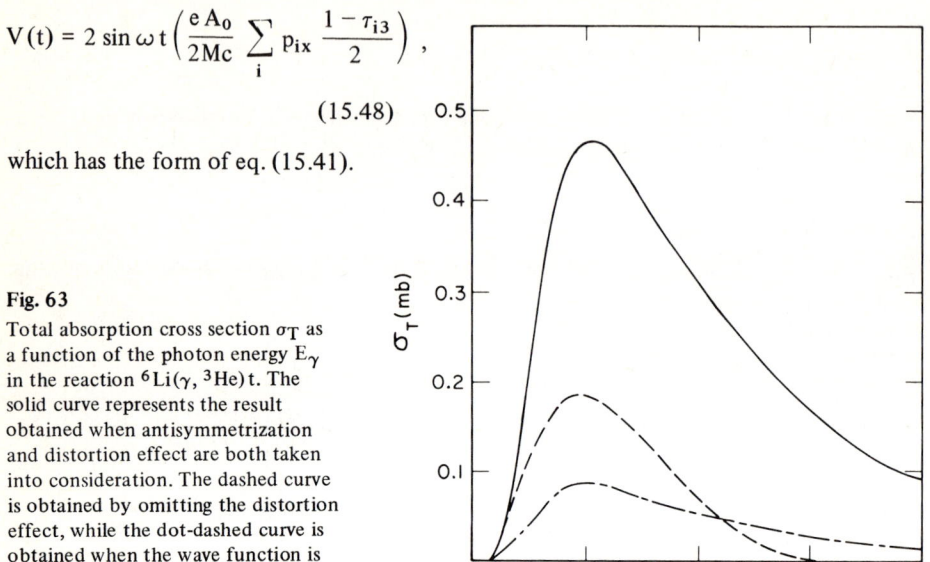

$$V(t) = 2 \sin \omega t \left(\frac{e A_0}{2Mc} \sum_i p_{ix} \frac{1 - \tau_{i3}}{2} \right), \tag{15.48}$$

which has the form of eq. (15.41).

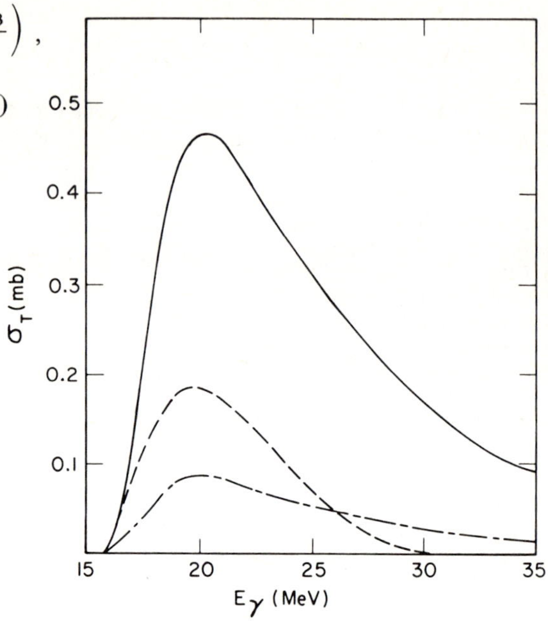

Fig. 63
Total absorption cross section σ_T as a function of the photon energy E_γ in the reaction $^6Li(\gamma, {}^3He)t$. The solid curve represents the result obtained when antisymmetrization and distortion effect are both taken into consideration. The dashed curve is obtained by omitting the distortion effect, while the dot-dashed curve is obtained when the wave function is unantisymmetrized. (Adapted from ref. [CL 74a].)

Numerical calculation of the absorption cross section for the process $^6Li(\gamma, {}^3He)t$ has been carried out by *Clement* and *Zahn* [CL 74a]. In that calculation, the ^6Li ground-state wave function given in subsection 5.3b was taken as ψ_i. Since this wave function describes a T = 0 ^3S configuration, the final-state wave function must be chosen to describe the scattering of tritons by ^3He in a T = 1 ^3P configuration, in order that the absorption can proceed by electric-dipole transition. In the calculation of *Clement* and *Zahn*, the scattering wave function obtained from a single-channel resonating-group $^3He + t$ calculation [TH 68] was used for ψ_f. This was a reasonable choice, since this single-channel calculation did yield scattering cross sections in fairly good agreement with experimental results.

The calculated total absorption cross section σ_T is shown as a function of the photon energy E_γ by the solid curve in fig. 63. But before we discuss the result of a comparison with experimental data, we wish to mention first the following interesting findings:

(a) In the final-state wave function ψ_f, the distortion of the relative-motion function between the ^3He and triton clusters by the nuclear interaction must be taken into consideration. If one omits this distortion effect, then the absorption cross section will have a much smaller magnitude, as is shown by the dashed curve in fig. 63.

(b) The $^6Li(\gamma, {}^3He)t$ calculation is a good example to show the influence of the Pauli principle. If the procedure of antisymmetrization is not carried out, then the absorption cross section (dot-dashed curve in fig. 63) will become smaller by a factor of about 6. On the other hand, it is noted from fig. 63 that, aside from this change in magnitude, the energy dependence of the absorption cross section is not appreciably influenced. This is in fact not entirely unexpected, but can be qualitatively explained in the following manner. The electric-dipole operator is a long-range operator, with the consequence that the transition matrix element obtains its contribution mainly from the surface region where the antisymmetrization has a comparatively minor effect on the character of the relative motion between the deuteron and the α clusters in the ground state of ^{6}Li. Instead, its main effect is to influence, roughly speaking, the degree of ^{3}He + t clustering which is present in the surface region. Indeed, as has been shown explicitly by Mader [MA 71], this degree of clustering is increased very much by the procedure of antisymmetrization if the ^{6}Li wave function is, as in the calculation of Clement and Zahn, given in the d + α cluster representation.

It is rather difficult to draw definite conclusions from a comparison between calculation and experiment, because various measurements yielded conflicting results. Thus, the absorption cross sections measured by *Sherman et al.* [SH 66, SH 68] are about a factor of four smaller than those inferred from the inverse-reaction measurements of *Blatt et al.* [BL 68] and of *Ventura et al.* [VE 71], and the ^{6}Li-electrodisintegration measurement of *Shin et al.* [SH 75]. The calculated result of *Clement* and *Zahn* agrees very well with the experimental result of *Sherman et al.* But if the results of the other measurements mentioned above should turn out to be the correct ones, as they appear to be, then it would indicate that the ^{6}Li wave function used in the calculation of *Clement* and *Zahn* does not adequately describe the phenomenon of ^{3}He + t clustering in the surface region. This would in fact not be too surprising, since this wave function was chosen to properly describe the feature of d + α clustering in ^{6}Li and the finding of a rather large average separation distance between the deuteron and α clusters might suggest that the degree of ^{3}He + t clustering in the surface region may have been quite severely underestimated. To see if this is indeed so, what one can do is to improve the ^{6}Li ground-state wave function by the addition of ^{3}He + t cluster wave functions with linear parameters into the variational calculation (see also the discussion in subsection 5.3e). This would make both the ^{6}Li calculation and the ^{6}Li (γ, ^{3}He) t absorption calculation somewhat more difficult, but, in view of the present experimental situation, we feel that such an undertaking is certainly a worthwhile one.

With the time-dependent projection equation, one can certainly also treat γ-absorption processes in a completely quantal manner by quantizing the electromagnetic field. In this case, one includes in the time-dependent Hamiltonian not only the γ-nucleon coupling term but also a term corresponding to the field energy, and the time-dependent variational amplitudes now refer to states of both the γ quanta and the nucleons. Then, by proceeding in a way similar to that described above, one obtains in first-order perturbation theory again eq. (15.42), with V_0 given in eq. (15.48) for the γ-absorption process.

In closing, we should mention that the purpose of this chapter is simply to show how time-dependent problems can also be treated by using the unified theory of this monograph. Of course, there are many other time-dependent problems which have not even been touched upon here, but it should be clear that these problems can certainly be similarly treated by starting from the time-dependent projection equation, as expressed by eq. (2.1).

16. Qualitative Considerations of Some Nuclear Problems

16.1. General Remarks

In the preceding chapters, we have discussed how one can formulate a microscopic theory to describe all low energy phenomena of the nucleus from a unified viewpoint. This theory is very flexible, because the basis wave functions which one needs for the description of specific nuclear phenomena can be selected in the most suitable manner. On the other hand, we have seen that, mainly because of the Pauli principle, the amount of computation necessary for a quantitative study of nuclear bound states and nuclear reactions increases very rapidly with increasing number of nucleons or clusters involved in the system. This is especially true if one uses in the calculation realistic two-nucleon potentials including a repulsive core, as was shown by some examples given in chapters 5 and 7. Therefore, up to the present, quantitative microscopic calculations have been carried out only in a rather small number of cases which involve essentially relatively light systems [TA 75].†

In spite of the above-mentioned limitation on quantitative calculations, one should still obtain valuable insight into the structure of nuclei and reaction mechanisms by using the physical concepts which have been developed in chapters 8–15. In this chapter, we shall therefore show how one can apply these physical concepts to a qualitative or semi-quantitative description of some nuclear problems where completely quantitative calculations are either very complicated or practically impossible. The situation may be compared with that in molecular physics, where quantum mechanics is believed to give a correct description of the interactions between the electronic structures of constituent atoms but quantitative calculations are very often extremely difficult. Nevertheless, great progress in understanding molecular structure has resulted from a more or less qualitative application of quantum mechanical ideas.

16.2. Coulomb-Energy Effects in Mirror Levels

The first problem which we shall consider concerns the Coulomb-energy effects in mirror nuclei. We choose this particular problem for investigation, because it will serve to show that even relatively simple arguments can give us considerable insight into the structure of specific nuclear levels.

Because of the charge independence of the nuclear force, the level spectra of mirror nuclei should be identical, provided that the influence of the Coulomb force is neglected. If, instead of neglecting the charge effect entirely, one now assumes the Coulomb energy to be the same in all states (ground and excited states) of the same nucleus, then the total spectrum of each nucleus will be shifted up by a constant amount equal to the Coulomb

† Some of the heavier systems which have been quantitatively treated in a microscopic manner are the scattering of α by ^{18}O [SU 75], ^{16}O by ^{16}O [FR 74, TO 75], and α by ^{40}Ca [FR 75].

energy of that nucleus. Hence, under this assumption, mirror levels of two mirror nuclei will all differ by the same energy difference and the corresponding level spacings in these nuclei should be exactly the same. Due to the long-range character of the Coulomb force, this assumption is rather well fulfilled. The small remaining differences in the energy spacing of mirror nuclei must therefore come from the different charge distributions in mirror levels. Since the charge distribution of a nucleus is directly related to the wave function which describes the state in question, one would expect that the cluster structure exhibited by the wave function will affect the Coulomb energy to a significant extent. By studying the Coulomb-energy behaviour from the energy spectra of mirror nuclei, one can thus hope to gain some information about the cluster structure of the levels [WI 61a]. We should emphasize of course that this possibility arises because the influence of the purely nuclear force is the same in corresponding mirror levels.†

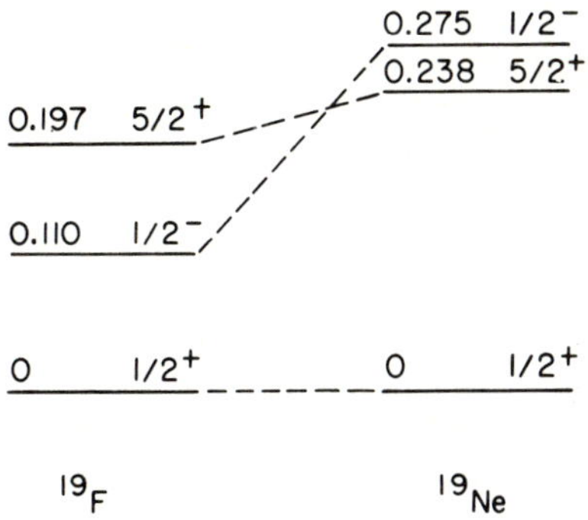

Fig. 64. Level diagram of ^{19}F and ^{19}Ne.

To see the Coulomb-energy effect in more detail, we consider as a specific example the three lowest energy levels (fig. 64) of the mirror pair ^{19}F and ^{19}Ne [AJ 72]. From detailed studies on the energy spectra, it was concluded [SH 60] that the cluster structure of the $\frac{1}{2}^+$ and $\frac{5}{2}^+$ levels of ^{19}F (^{19}Ne) is described by an unexcited ^{16}O cluster and a triton (^3He) cluster in relative motion of orbital angular momentum $l = 0$ and $l = 2$, respectively. These two levels are therefore members of a rotational band, and we expect that the Coulomb energies should be approximately the same in both of these levels. Experimentally, we indeed find that the excitation energies of the $\frac{5}{2}^+$ levels are nearly the same in both ^{19}F and ^{19}Ne. For the $\frac{1}{2}^-$ levels, however, we have a different situation. These levels are best described by an unexcited ^{15}N (^{15}O) cluster and an α cluster in relative oscillation of

† Exceptions occur if levels with the same quantum numbers, except the isobaric-spin quantum number, are located very close to each other. In this case, the Coulomb force can also appreciably affect the bound-state wave functions themselves by mixing states with different values of isobaric spin. For our considerations here, it is essential that the Coulomb force has only a negligible influence upon the structure of the mirror levels. In general, this condition is very well fulfilled, because levels with the same quantum numbers, except the isobaric-spin quantum number, are usually well separated in energy. For the examples which we consider here, this is certainly the case.

orbital angular momentum $l = 0$. To obtain the configuration of these states from that of the ground states, one moves a proton from the ^{16}O core in ^{19}F and a neutron from the ^{16}O core in ^{19}Ne. Together with the outside triton or ^{3}He cluster, an α cluster is then formed. This means that, in first approximation, there is a decrease of Coulomb energy for ^{19}F but not for ^{19}Ne. Therefore, it is expected that the $\frac{1}{2}^-$ level of ^{19}F has a smaller excitation energy than the corresponding mirror level in ^{19}Ne. Experimentally, this is indeed the case.

For a crude estimate of the difference ΔE_x in excitation energies of the $\frac{1}{2}^-$ mirror levels in ^{19}F and ^{19}Ne, we proceed as follows. Since ΔE_x arises only from Coulomb-energy differences, we may write it as

$$\Delta E_x \equiv E_x\left({}^{19}\mathrm{Ne};\tfrac{1}{2}^-\right) - E_x\left({}^{19}\mathrm{F};\tfrac{1}{2}^-\right) = \left[E_C\left({}^{19}\mathrm{Ne};\tfrac{1}{2}^-\right) - E_C\left({}^{19}\mathrm{Ne};\tfrac{1}{2}^+\right)\right] - \left[E_C\left({}^{19}\mathrm{F};\tfrac{1}{2}^-\right) - E_C\left({}^{19}\mathrm{F};\tfrac{1}{2}^+\right)\right], \tag{16.1}$$

where the Coulomb energy E_C of the indicated level is equal to the sum of the internal Coulomb energies of the constituent clusters plus their interaction Coulomb energy. For example, for the Coulomb energy of the $\frac{1}{2}^-$ level in ^{19}Ne, we have

$$E_C\left({}^{19}\mathrm{Ne};\tfrac{1}{2}^-\right) = E_C({}^{15}\mathrm{O}) + E_C(\alpha) + E_C(\alpha - {}^{15}\mathrm{O}). \tag{16.2}$$

To evaluate these Coulomb energies, we now crudely assume that the internal Coulomb energies of the clusters are equal to those of the corresponding free nuclei.† For the interaction Coulomb energies between the clusters in the various levels, we equally crudely assume that the charges of the ^{16}O, ^{15}O, and ^{15}N clusters in the different levels are distributed uniformly over a sphere of common radius R_0, and the charges of the ^{3}He and triton clusters, in their respective motions about the heavier clusters, are smeared out uniformly over a larger sphere of radius R_F. The interaction Coulomb energy between two clusters is then taken as the electrostatic energy between two concentric uniformly-charged spheres. With these assumptions, we obtain from eq. (16.1)

$$\Delta E_x \approx \Delta E_C({}^{15}\mathrm{O}, {}^{15}\mathrm{N}) - \Delta E_C({}^3\mathrm{He}, t) - \frac{9e^2}{R_F}\left(1 - \frac{1}{5}\frac{R_0^2}{R_F^2}\right), \tag{16.3}$$

where R_0 and R_F are considered as the radii of ^{16}O and ^{19}F, respectively. The value of R_0 is taken from ref. [ME 59], while the value of R_F is evaluated from the Coulomb-energy difference in the ground states of ^{19}F and ^{19}Ne. Using further the experimental values of the Coulomb-energy differences $\Delta E_C({}^{15}\mathrm{O}, {}^{15}\mathrm{N})$ and $\Delta E_C({}^3\mathrm{He}, t)$, we then obtain ΔE_x as 0.21 MeV; experimentally, the value of ΔE_x is 0.17 MeV.

For all mirror levels whose cluster structures are known, one can calculate in a similar way the excitation-energy differences ΔE_x. In table 3, we list a number of such examples. As is seen, the agreement between the calculated and experimental values is fairly good.

---

† This means that we neglect all antisymmetrization effects in the calculation of the various Coulomb energies. Because of the long-range character of the Coulomb force, this is approximately justified.

Table 3. Excitation-energy differences of mirror levels

Nuclear levels	Excitation energy (MeV)	Cluster structure[a)]	Cluster structure of ground state	ΔE_x (expt) (MeV)	ΔE_x (calc) (MeV)	Remarks
$^7Be\ (^7Li),\ \frac{1}{2}^-$	0.431 (0.478)	$^3He\ (^3H) + \alpha$, 2p	$^3He\ (^3H) + \alpha$, 2p	− 0.05	≈ 0	c)
$^7Be\ (^7Li),\ \frac{7}{2}^-$	4.55 (4.63)	$^3He\ (^3H) + \alpha$, 1f	$^3He\ (^3H) + \alpha$, 2p	− 0.08	− 0.06	b), c)
$^7Be\ (^7Li),\ \frac{5}{2}^-$	7.19 (7.48)	^6Li + p (n), 1p	$^3He\ (^3H) + \alpha$, 2p	− 0.29	− 0.36	
$^{11}C\ (^{11}B),\ \frac{1}{2}^-$	1.995 (2.124)	$^8Be + {}^3He\ (^3H)$, 2p	$^8Be + {}^3He\ (^3H)$, 2p	− 0.13	≈ 0	
$^{13}N\ (^{13}C),\ \frac{1}{2}^+$	2.366 (3.086)	^{12}C + p (n), 2s	^{12}C + p (n), 1p	− 0.72	− 0.60	d)
$^{19}Ne\ (^{19}F),\ \frac{1}{2}^-$	0.275 (0.110)	$^{15}O\ (^{15}N) + \alpha$, 5s	$^{16}O + {}^3He\ (^3H)$, 4s	+ 0.17	+ 0.21	
$^{19}Ne\ (^{19}F),\ \frac{5}{2}^+$	0.238 (0.197)	$^{16}O + {}^3He\ (^3H)$, 3d	$^{16}O + {}^3He\ (^3H)$, 4s	+ 0.04	≈ 0	c)

a) In this column, the internal structures and the modes of relative motion are listed. For instance, the $\frac{7}{2}^-$ level of ^7Be is described by a ^3He cluster and an α cluster in relative 1f-state motion.

b) For the calculation of ΔE_x, we have used the cluster-model wave functions described in ref. [PE 60].

c) Rotational level.

d) The average distance of the outside nucleon from the center of the nucleus in the $\frac{1}{2}^+$ state of $^{13}N\ (^{13}C)$ has been assumed to be the same as that in the corresponding $\frac{1}{2}^+$ state of $^{17}F\ (^{17}O)$; the latter distance can be calculated from their Coulomb-energy difference.

In particular, the sign of ΔE_x is always predicted correctly. For levels in the same rotational band as the ground state, the magnitude of ΔE_x is always smaller than 0.1 MeV. From our arguments presented above, this should indeed be expected.

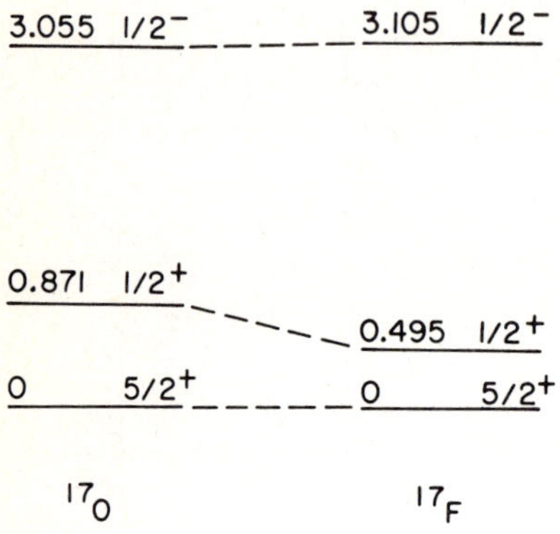

Fig. 65. Level diagram of ^{17}O and ^{17}F.

We should mention that one can also go the other way; that is, from the ΔE_x value of mirror levels one can also learn about the cluster structure of nuclear levels. For example, let us consider briefly the $\frac{1}{2}^-$ levels of the mirror pair ^{17}O and ^{17}F at excitation energies of 3.055 and 3.105 MeV [AJ 71], respectively (fig. 65). As is clear, the change of parity in going from the ground state to the $\frac{1}{2}^-$ state requires in the shell-model picture that one of the nucleons be raised to a higher shell. To obtain this state, one could try exciting the odd nucleon outside the closed 1p shell from the 1d shell to the 2p shell. This would yield the proper total angular momentum (spin and orbital angular momenta antiparallel to each other) and parity of the level. In the case of ^{17}O, it would be the odd neutron that would be moved further away from the core nucleus, and there would be no Coulomb-energy change. However, in the case of ^{17}F, it would be a proton which would be excited to the next higher shell, thus resulting in a relatively large decrease of Coulomb energy (about 0.5 MeV). By this reasoning, we would expect the excitation energy of the $\frac{1}{2}^-$ level in ^{17}F to be about 0.5 MeV smaller than the excitation energy of the ^{17}O mirror level. However, one does not find this effect experimentally; indeed, the excitation energy of the ^{17}F $\frac{1}{2}^-$ level is even a little larger than the excitation energy of the mirror level in ^{17}O.†

This shows that the structure described above is not the correct structure of these levels, and therefore one must look for another structure. One has to assume for the cluster structure of these mirror levels essentially an α-cluster excitation of the ^{16}O-cluster core. Although the lowest excitation of this kind in ^{16}O needs an energy of 6.06 MeV (see the discussion in section 3.7), one can account for the decrease of the excitation energy in

† This tendency is even more clearly expressed, if one considers the energy distances of the above mirror levels to the $\frac{1}{2}^+$ mirror levels in ^{17}O and ^{17}F rather than to the $\frac{5}{2}^+$ ground states.

^{17}O and ^{17}F by the fact that the odd nucleon can now be in the 1p shell, which is left partly empty by the ^{16}O excitation. The cluster structure of the $\frac{1}{2}^-$ ^{17}O level is therefore now that of a ^{13}C cluster in its ground state plus an α cluster. Similar consideration yields for the structure of the $\frac{1}{2}^-$ level in ^{17}F a ^{13}N cluster in its ground state plus an α cluster. There is a large spatial overlap of the clusters in both cases. However, in the case of ^{17}F, we are moving a proton to an inner shell and thereby increasing the Coulomb energy. The excitation energy of the $\frac{1}{2}^-$ level in ^{17}F should therefore be slightly larger than that of the $\frac{1}{2}^-$ level in ^{17}O. This is indeed indicated by the experimental data. Furthermore, it was found [GR 56] that this level in ^{17}O does not exhibit stripping in the (d, p) reaction, thus further supporting our assumption concerning the structure of this level which we inferred from the study of the Coulomb-energy behaviour.

One can obtain similar information about the cluster structure of nuclear levels by studying the Coulomb-energy behaviour in nuclear isobars with the isobaric-spin quantum number T not necessarily equal to $\frac{1}{2}$. For example, there exist low-lying T = 1 levels in ^6He, ^6Li, and ^6Be, and a simple study of the excitation-energy difference in this isobaric triad indicates that these levels should have an $\alpha + d^*$ cluster structure, with d^* being an excited deuteron cluster having S = 0 and T = 1.

A further point we wish to mention is that, as stated above, rotational levels in a rotational band of a given cluster structure have all approximately the same charge distribution and hence the same Coulomb energy. This means that corresponding rotational bands, i. e., bands with mirror cluster structures, will have the same relative spacing of the rotational levels in both nuclei; however, the distances of the rotational bands to the respective ground states of the mirror nuclei will not be the same if the ground states have a cluster structure different from that of the rotational bands, because different cluster structures in general have a different Coulomb energy. Experimentally, this has also been borne out [EV 65].

16.3. Reduced Widths and γ-Transition Probabilities

In the last section, we discussed how the influence of the Coulomb interaction on the energy spectra of nuclear levels can be used to obtain valuable information about the cluster structure of these levels. The disadvantage of this method is that it is restricted to mirror or analogue levels. Therefore, we shall discuss here two other methods which can also yield information on the cluster structure. The first one is connected with the reduced widths of resonance levels, while the other with the γ-transition probabilities between nuclear states [RO 60].

16.3a. Reduced Widths of Nuclear Levels

We start with a brief discussion of the meaning of the reduced width. For an isolated resonance (labelled as k) having a partial width Γ_{kb} for the decay into the channel b, one defines the reduced partial width γ_{kb}^2 by the equation [EI 58, PR 62, VO 59]

$$\Gamma_{kb} = 2\,k\,P\,\gamma_{kb}^2, \qquad (16.4)$$

where k is the relative wave number of the two particles in channel b. The quantity P is the penetrability; it is equal to 1 when the channel b involves a high energy uncharged particle with relative orbital angular momentum $l = 0$ with respect to the other particle. The reason why one commonly prefers to deal with γ_{kb}^2 rather than Γ_{kb} is because γ_{kb}^2 depends only on the properties of the internal or compound-nucleus region.†

For convenience, one further defines a dimensionless quantity

$$\theta_{kb}^2 = \gamma_{kb}^2 \left(\frac{3\hbar^2}{2\mu R_{CN}} \right)^{-1} \tag{16.5}$$

in terms of the Wigner limit $3\hbar^2/2\mu R_{CN}$, where R_{CN} is the radius of the compound nucleus.†† If the resonance state under consideration has predominantly the cluster structure of the decay channel, then θ_{kb}^2 will have a value rather close to 1. On the other hand, if other cluster structures have large amplitudes in this resonance state, then θ_{kb}^2 will be much less than unity. This is so, because the square root of the reduced width for the channel b has the expression

$$\gamma_{kb} = \left(\frac{\hbar^2}{2\mu R_{CN}} \right)^{1/2} \int_S \phi_b^* X_k \, dS, \tag{16.6}$$

where X_k is the characteristic wave function of the resonance state in the compound-nucleus region, ϕ_b is the surface function for the decay channel, and the integration is over a surface S which separates the compound-nucleus and the external regions in the configuration space. Therefore, by examining the value of θ_{kb}^2 deduced from the measured value of Γ_{kb}, one can learn about the cluster structure of this particular resonance state.

To obtain the reduced width from the experimentally observed width, one must first calculate the penetrability P. The usefulness of the reduced width for the investigation of the cluster structure of a nuclear level comes about because it generally varies greatly enough with the change in nuclear structure that moderate uncertainty in the calculation of the penetrability will not obscure the qualitative information carried by the reduced width.

The reduced width γ_{kb}^2 is a quantity introduced in the R-matrix theory, which is a formal theory where the many-particle Hamiltonian is not explicitly considered. It is not uniquely defined because, as can be seen from eq. (16.6), its definition depends on the position of the surface S surrounding the internal region. Another quantity which can be

---

† It should be noted that the γ^2 defined here is not the same as the γ^2 of chapter 9.

†† One can obtain an approximate expression of the Wigner limit by considering a high-energy incoming particle for which $P \approx 1$. Due to the uncertainty principle, the width Γ of the level is related to the time τ which the particle spends inside the target nucleus. Under the assumed condition, this particle will not be reflected back and forth within the nucleus and we have $\tau \approx R_{CN}/v$, with $v = \hbar k/\mu$ being the relative velocity. By now using eq. (16.4), we then obtain

$$\gamma^2 = \hbar^2/2\mu R_{CN}$$

which coincides, except for a factor of the order of unity, with the Wigner limit.

used as a measure of the degree of clustering contained in a compound state is the absolute square of the amplitude b defined in eq. (15.25). From the experimentally measured partial energy width, one can determine the magnitude of $|b|^2$ by using eq. (15.27), since the other quantity present in this equation depends essentially only on the behaviour of the cluster configuration in a region outside of the region of strong interaction between the two clusters. It should be noted, however, that $|b|^2$ is also not defined in a unique manner, because the division of ψ_{Cj}^k into ψ_a and ψ_b in the compound-nucleus region is not completely specified; from the discussion given in subsection 15.2c, it is seen that only $b\,\psi_b$ in the barrier region is precisely defined.

It is clear that γ_{kb}^2 and $|b|^2$ are closely related and either of these can be used to examine the cluster structure of resonance levels. However, since the reduced width γ_{kb}^2 is, at the present moment, a more familiar quantity, we shall use it exclusively in the following discussion.

To illustrate that cluster-structure information can be obtained by studying the reduced widths of resonance levels, we consider the case of ^{16}O. In table 4, a number of ^{16}O low-lying levels are listed, together with their cluster configurations assigned according to energetical arguments and information based on known values of spin and parity.† Also, we list in this table the values of the dimensionless reduced width θ_α^2 for the scattering of α particles by unexcited ^{12}C, which are obtained from eqs. (16.4) and

Table 4. Cluster structure and reduced widths of ^{16}O levels

E_x a) (MeV)	J^π	Cluster structure b)	θ_α^2 c)
0	0^+	$\alpha + {}^{12}C$, 3s	
6.05	0^+	$\alpha + {}^{12}C$, 4s	
6.13	3^-	$\alpha + {}^{12}C$, 2f	
6.92	2^+	$\alpha + {}^{12}C$, 3d	
7.12	1^-	$\alpha + {}^{12}C$, 3p	
8.87	2^-	$\alpha + {}^{12}C^*$, 3p	
9.60	1^-	$\alpha + {}^{12}C$, 4p	0.67
9.85	2^+	$\alpha + {}^{12}C^*$, 4s	0.002
10.34	4^+	$\alpha + {}^{12}C$, 2g	1.88
11.08	3^+	$\alpha + {}^{12}C^*$, 3d	
11.26	0^+	$\alpha + {}^{12}C$, 5s	0.76
11.52	2^+	$\alpha + {}^{12}C^*$, 3d	0.03
11.63	3^-	$\alpha + {}^{12}C$, 3f	0.73
12.44	1^-	$\alpha + {}^{12}C^*$, 4p	0.04
12.53	2^-	$\alpha + {}^{12}C^*$, 4p	

a) The $\alpha + {}^{12}C$ and the $p + {}^{15}N$ thresholds are at 7.16 and 12.13 MeV, respectively.

b) The internal structures and the modes of realtive motion are listed. The symbol $^{12}C^*$ refers to the first excited state of ^{12}C.

c) The quantity θ_α^2 is the reduced α-particle width in units of the Wigner limit.

† The assignment of cluster structure for many of these levels is discussed in subsection 3.6c.

(16.5) by using measured values of the level width [AJ 71]. For the five lowest levels there are no reduced-width data, since they occur below the $\alpha + {}^{12}C$ threshold of 7.16 MeV.

Based on the cluster-structure assignment, we can make the following predictions:

(a) The lowest of these levels which is energetically accessible by $\alpha + {}^{12}C$ scattering is the 8.87-MeV 2^- level. However, this level cannot be excited by the above scattering process because the $\alpha + {}^{12}C$ partial wave of $l = 2$ has positive parity. Therefore, its α-particle reduced width is zero.

(b) The 9.60-MeV 1^- level should have a large reduced width, since it has the structure of an α cluster plus an unexcited ${}^{12}C$ cluster.

(c) For the 9.85-MeV 2^+ level, the value of θ_α^2 should be much smaller than 1. This is due to the fact that this state is mainly described by the relative motion of an α cluster around an internally excited ${}^{12}C$ cluster.

(d) The 10.34-MeV 4^+ level is a rotational level in which an α cluster rotates around an unexcited ${}^{12}C$ cluster, as discussed in subsection 3.6c. Therefore, its reduced width should be of the order of the Wigner limit.

(e) The 11.08-MeV 3^+ level cannot be reached by $\alpha + {}^{12}C$ scattering, since the $l = 3$ partial wave has negative parity. Consequently, its reduced width is zero.

(f) In a similar way, the 11.26-MeV 0^+ and the 11.63-MeV 3^- states should have large reduced widths, while the 11.52-MeV 2^+ and the 12.44-MeV 1^- states should have small reduced widths because of their excited ${}^{12}C$ clusters.

By examining the empirical values of the dimensionless reduced width given in the last column of table 4, one sees that these predictions are very well confirmed. In this respect, it is also interesting to note that the proton reduced widths for the 12.44-MeV and 12.53-MeV states are only a few percents of the Wigner limit [AJ 59, HE 60]. These relatively small values are to be expected from the cluster structure of these states. The 12.53-MeV state has, in fact, been found [HO 55, KR 53] to decay into an α particle and a ${}^{12}C^*$ nucleus as well as decaying by γ and proton emission. This again is expected from the cluster structure assigned to this state.

The method discussed here is useful, only because the change in the reduced width from level to level is so marked that rather crude calculations of the penetrabilities can be tolerated. Thus, even though the values of the reduced width given in table 4 are subject to fairly large uncertainties, they still enable us to reach important qualitative conclusions.

16.3b. γ-Transition Probabilities

We discuss now the second method which can be used to obtain cluster-structure information. In this method, one considers the ratio

$$|M_\gamma|^2 = \frac{W_{exp}}{W_M}, \tag{16.7}$$

where W_{exp} is the experimentally determined rate of γ-transition between nuclear levels and W_M is the so-called Moszkowski single-particle estimate or unit [MO 66] given by

$$W_M(EL) = \frac{4.4(L+1)}{L[(2L+1)!!]^2}\left(\frac{3}{L+3}\right)^2\left(\frac{E_\gamma}{197}\right)^{2L+1} R^{2L} \times 10^{21}\ \text{sec}^{-1} \tag{16.8}$$

and

$$W_M(ML) = \frac{0.19(L+1)}{L[(2L+1)!!]^2}\left(\frac{3}{L+2}\right)^2\left(\mu_P L - \frac{L}{L+1}\right)^2\left(\frac{E_\gamma}{197}\right)^{2L+1} \times R^{2L-2} \times 10^{21}\ \text{sec}^{-1}, \tag{16.9}$$

with $R = 1.2\,A^{1/3}$ being the nuclear radius in units of fm, E_γ being the γ-ray energy in units of MeV, and $\mu_P = 2.79$ being the intrinsic magnetic moment of the proton in units of nuclear magneton.† If the value of $|M_\gamma|^2$ has a magnitude of the order of unity, this would indicate that the nuclear levels under consideration have relatively similar cluster structure. On the other hand, if $|M_\gamma|^2$ is much less than 1, then one would conclude that the cluster structures of these levels are rather different.

To demonstrate the usefulness of this method, we consider γ-transitions in ^{19}F [RO 60]. In fig. 66, the low-lying levels of this nucleus [AJ 72, PO 69], together with their cluster-structure assignment and the observed values of the branching ratio,†† are shown. For the three lowest levels, the cluster structures have already been discussed in section 16.2. The 1.554-MeV $\frac{3}{2}^+$ state is an $l = 2$ rotational level; it is the spin-orbit partner of the $\frac{5}{2}^+$ level at 0.197 MeV. The 1.346-MeV $\frac{5}{2}^-$ and 1.459-MeV $\frac{3}{2}^-$ states also form a spin-orbit doublet, with a cluster structure of an α cluster in 4d oscillation with respect to a ^{15}N cluster; they are the $l = 2$ rotational levels of a band with the band head being the 0.110-MeV $\frac{1}{2}^-$ state. The spin-orbit splitting of these levels is much smaller than that of the $\frac{5}{2}^+$ and $\frac{3}{2}^+$ levels, because the relative orbital angular momentum has to be shared among all 15 nucleons of the ^{15}N cluster and the intrinsic spins of the nucleons in ^{15}N are paired off to yield a total spin angular momentum of only $\frac{1}{2}$ [SH 60].

In table 5, we list the values of $|M_\gamma|^2$ for the various transitions indicated in fig. 66. The important features contained in this table are as follows:

(a) For transitions C_1, B_3, and A_4 where the initial and final states involved have rather similar cluster structure, the transition rates are of the order of the Moszkowski estimates.

(b) For transitions B_1, A_2, C_2, C_3, and A_5 where the initial and final states involved have quite different cluster structure, $|M_\gamma|^2$ is much smaller than 1.

(c) The transition A_1 is between two states of similar cluster structure. However, according to the cluster-structure assignments given in fig. 66, it involves both an orbital

† For a crude estimate, the statistical factor S in eqs. (77a) and (77b) of ref. [MO 66] is taken as 1. We use the *Moszkowski* instead of the *Weisskopf* [WE 51] estimate here, simply because the Moszkowski estimate for magnetic-multipole transitions is somewhat more reasonable.

†† Transitions between 1.554- and 1.346-MeV levels and between 0.197- and 0.110-MeV levels are not included in the analysis here, because for these transitions only rather unreliable upper limits for the branching ratio are known [CO 70, JO 54].

Table 5. γ-transition probabilities in ^{19}F

Transition	Multipolarity	W_{exp} (sec^{-1})	W_M (sec^{-1})	$\|M_\gamma\|^2$	Large change in cluster structure
A_1	M1	5.0×10^{12}	1.1×10^{14}	0.05	No
B_1	E1	1.1×10^{13}	2.3×10^{15}	4.8×10^{-3}	Yes
C_1	M1	2.1×10^{14}	7.2×10^{13}	2.9	No
A_2	E1	2.7×10^{12}	2.3×10^{15}	1.2×10^{-3}	Yes
B_2	M1	9.0×10^{12}	7.2×10^{13}	0.13	No
C_2	E1	1.3×10^{12}	1.5×10^{15}	8.7×10^{-4}	Yes
A_3	M2	$< 2.1 \times 10^{9}$	2.7×10^{9}	< 0.8	Yes
B_3	E2	2.0×10^{11}	1.1×10^{10}	18.2	No
C_3	E1	7.1×10^{9}	1.1×10^{15}	6.5×10^{-6}	Yes
A_4	E2	7.8×10^{6}	1.1×10^{6}	7.1	No
A_5	E1	1.2×10^{9}	1.0×10^{12}	1.2×10^{-3}	Yes

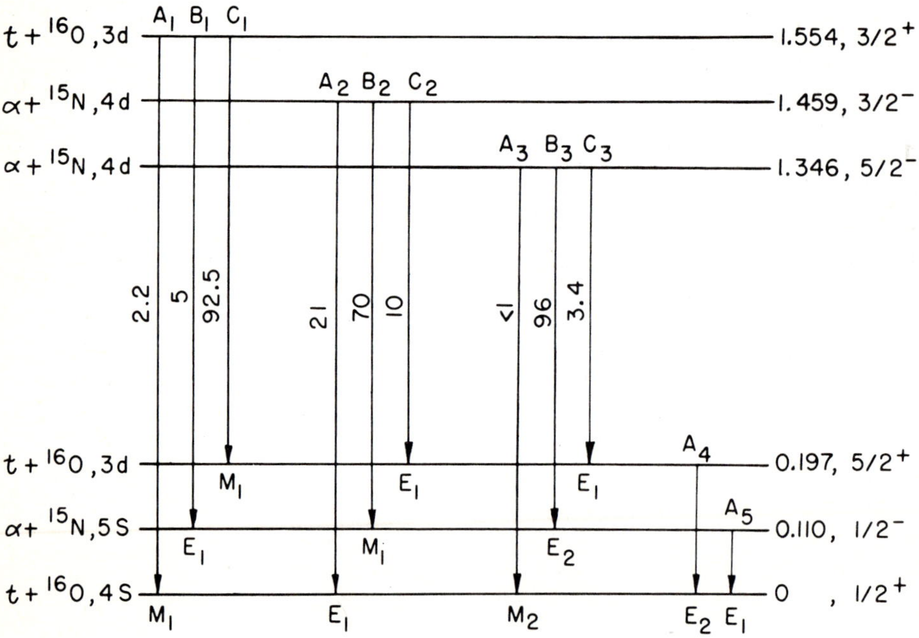

Fig. 66. γ-transitions among the low-lying states of ^{19}F. Numbers on the transition lines indicate branching ratios. Symbols $A_1, \ldots, A_5$ are used in the text for identifying transitions.

angular-momentum change of 2 units and a spin-flip. The magnetic-dipole transition matrix element which enters into the first-order calculation of the corresponding transition probability is, therefore, equal to zero. This means that the M1 component in this transition should be greatly hindered† and the resultant value of $|M_\gamma|^2$ should be relatively small [PO 69, WI 60]. By a similar argument, one can show that the M1 transition rate for B_2 should also be hindered with respect to the single-particle estimate (see, e.g., the discussion given in ref. [PO 69]). Even so, however, one sees from table 5 that the values of $|M_\gamma|^2$ for these two transitions are appreciably larger than those for the transitions mentioned above in (b).

(d) For the transition A_3, only an upper limit for the branching ratio is known [PO 69]. Therefore, no definitive statements concerning this particular transition can be made, and a more detailed measurement of this branching ratio should be desirable.

The important point responsible for the successful application of this method is that the value of $|M_\gamma|^2$ is sensitively dependent upon the difference in cluster structure between initial and final states. This is associated with the fact that electromagnetic multipole transitions probe mainly the nuclear surface region where different cluster structures exhibit especially different nature of spatial correlation, because there the influence of the Pauli principle is very much reduced, as has often been mentioned. The result of this is that our predictions work rather well, even though we have not taken into consideration the indistinguishability of the nucleons in our qualitative arguments.††

Finally, it should be mentioned that, in a similar way as one does here with γ-transition probabilities, the β-transition probabilities can also be used sometimes to test the cluster-structure assignment of nuclear levels (see, e.g., ref. [BA 63]).

16.4. Level Spectra of Neighbouring Nuclei

In section 3.6, we have discussed how one can qualitatively or even semi-quantitatively understand the level schemes of light nuclei in the low-excitation region where the energy levels have a relatively simple cluster structure. For example, as was mentioned in subsection 3.6c, the fact that the second 2^+ level in ^{16}O lies about 4 MeV above the second 0^+ level can be easily understood. In this particular example, the reason why it is possible to make such a rather definitive statement is that, in the low-excited positive-parity α-cluster states of ^{16}O, the outside α cluster is relatively far away from the unexcited or low-excited ^{12}C core. This means that, for an approximate determination of the energy difference between these two states, one can neglect the antisymmetrization between the outside α cluster and the ^{12}C core. Consequently, also because the nuclear forces are short-ranged, one can consider the excitation energy of the first 2^+ state for the ^{12}C cluster in these α-cluster states of ^{16}O as approximately equal to that for a free ^{12}C nucleus.

---

† The E2 component in the transition A_1 is not hindered.

†† Another example where the antisymmetrization between the clusters can be omitted in good approximation is given in section 6.2. There we have studied the E2 transition between the first excited 2^+ and the ground state of ^{20}Ne.

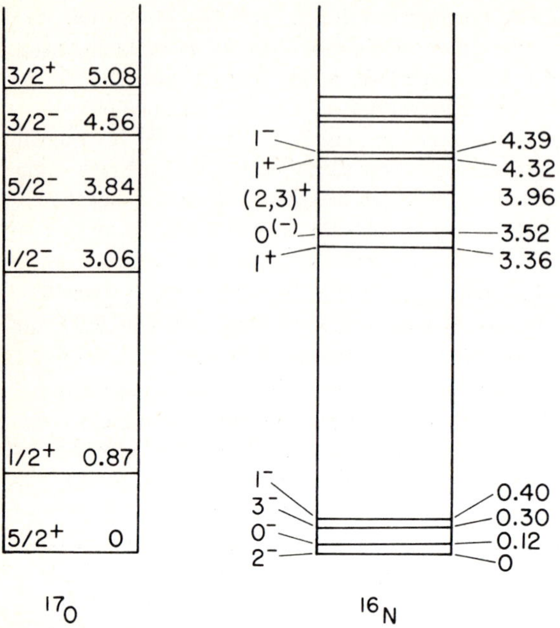

Fig. 67
Level spectra of ^{17}O and ^{16}N.

This method of using the information from one nucleus to discuss qualitatively or even semi-quantitatively the low-energy spectra of neighbouring nuclei is of course not limited to the case of the ^{16}O–^{12}C pair. As another example, we shall briefly discuss here how one can relate the low-energy spectrum of ^{16}N with that of the neighbouring nucleus ^{17}O [WI 59].

The level spectra of ^{17}O and ^{16}N [AJ 71] are shown in fig. 67. For the nucleus ^{17}O, the ground $\frac{5}{2}^+$ and the first excited $\frac{1}{2}^+$ states have a n + ^{16}O cluster structure, with the neutron moving around the ^{16}O core in 1d and 2s states, respectively.† The cluster structure for the lowest states in ^{16}N is similar; the only main difference is that the ^{16}O core with $J^\pi = 0^+$ is replaced by a ^{15}N core with $J^\pi = \frac{1}{2}^-$. This latter internal spin can now be coupled with the total angular momentum of the outside nucleon to yield ^{16}N-states with $J^\pi = 0^-, 1^-, 2^-$, and 3^-. As is seen from fig. 67, the four lowest states in ^{16}N do indeed have these particular values of spin and parity. To determine the energetical order of these states is, however, not so simple, because this order is critically influenced by an interplay between the noncentral and the central exchange parts of the nucleon-nucleon potential [WI 59].

† The order of the corresponding levels in ^{19}F (see fig. 64) is just the opposite. This comes from the fact that the spin-orbit coupling of the triton cluster with the ^{16}O core in the $l = 2$ state of ^{19}F is much weaker than that of the nucleon with the ^{16}O core in the $l = 2$ state of ^{17}O. The reason is that in ^{19}F the relative orbital angular momentum has to be distributed among three nucleons; this decreases the spin-orbit effect, because the intrinsic spins of the nucleons in the tritron cluster are partially paired off.

The excitation energy of the first excited state in ^{15}N is 5.27 MeV. Therefore, as in the ^{17}O case, one expects that the group of the four lowest levels in ^{16}N should be separated by an energy gap from the next group of higher levels. As is seen from fig. 67, this is indeed the case. The energy gap, being about 3 MeV, is appreciably smaller than the excitation energy of 5.27 MeV in ^{15}N mentioned above; this can be understood as due to rearrangement effects similar to those in ^{17}O and ^{17}F discussed near the end of section 16.2.

For the higher excited levels of ^{16}N, one cannot yet make a similar analysis at this moment, because the spins and parities of some of these levels have not yet been determined with certainty. In addition, we should mention that, in conducting a future analysis of this kind on these excited levels, one must take note of the fact that the structure of the excited states of the ^{15}N core and that of the excited states of the ^{16}O core are rather different from each other.

Another example of this type involves the nuclei ^{20}Ne and ^{19}F. Here again, one can show by using similar arguments that the low-lying positive-parity states of ^{20}Ne and the low-lying negative-parity states of ^{19}F have a close relationship with respect to each other [WI 59].

The above procedure to determined nuclear spectra can be considered as a generalization of a method of *de Shalit* [DE 53] where one connects, especially in the shell-model j-j coupling scheme, the low-energy level schemes of medium-heavy and heavy even-even nuclei with the low-energy level schemes of neighbouring odd-even nuclei.

16.5. Optical Resonances in Nuclear Reactions

In subsection 11.2d it was shown that, because of the indistinguishability of the nucleons, composite clusters can sometimes have a relatively long lifetime in even rather heavy nuclei. Together with the fact that the depth of the real part of the cluster-cluster interaction potential is usually quite large, one then expects that optical resonances† can frequently appear in scattering processes where these composite clusters are used as incident particles. Indeed, such resonances have often been observed and one of the simplest examples is the occurrence of $\alpha + \alpha$ resonances with $J^\pi = 0^+, 2^+$, and 4^+ in ^8Be. Here we wish to discuss qualitatively some more examples concerning this type of resonances. Also, one anticipates that optical or potential resonances should occur not only in the incoming channel but also sometimes in specific reaction channels. In this section, we shall therefore further discuss these latter resonances and present an illustrative example.

In the following subsections, we consider only examples in which the excitation of the compound nucleus is so large that compound levels overlap strongly and, consequently, the energy-averaging procedure discussed in section 11.2 does not need to be carried out.

† They are often called giant resonances. But the expression "optical resonances" is more appropriate, because it can refer also to rather narrow resonances.

16.5a. Optical Resonances in the Incoming Channel

We consider first the scattering of α particles by ^{13}C at medium bombarding energies of about 20 to 25 MeV. A rough estimate [WI 61], which we shall not describe here, shows that the lifetime τ for an α cluster in ^{13}C is approximately 2×10^{-20} sec.† On the other hand, the transit time τ_{tr} for such bombarding energies is about 3×10^{-22} sec.†† Therefore, since the lifetime is much longer than the transit time, the energy widths for the expected optical resonances are determined by the transit time (see the discussion given in section VII A of ref. [WI 66]) and are given approximately by

$$\Delta E = \hbar/\tau_{tr} \approx 2 \text{ MeV}. \tag{16.10}$$

The energy spacings between these optical resonances should be about a few MeV, as can be inferred from the energy separation of the $\alpha + {}^{12}C$ cluster states in ^{16}O. As yet, $\alpha + {}^{13}C$ elastic-scattering cross sections in this bombarding-energy region have not yet been measured; but from related experimental data to be discussed in the next subsection, we are able to conclude that these qualitative predictions are substantially correct.

As next example, we consider the scattering of ^{12}C by ^{12}C and of ^{16}O by ^{16}O, which have been first measured by *Almqvist et al.* [AL 60]. The remarkable feature of these measurements is that, in the neighbourhood of the corresponding Coulomb barriers, an optical-resonance structure is observed in the $^{12}C + {}^{12}C$ scattering but not in the $^{16}O + {}^{16}O$ scattering (see fig. 68). It is, therefore, of considerable interest to attempt an explanation of these findings from the viewpoint of the cluster representation.

Because ^{12}C clusters have unfilled outer shells, two ^{12}C clusters can overlap each other rather strongly in a scattering process. Hence, one expects a strong mutual interaction between these clusters. This can be expressed by using a deep optical well, which allows the occurrence of several optical resonances below and near the top of the Coulomb-plus-centrifugal barrier, in agreement with experimental observation. Also, as the excitation function for $^{12}C + {}^{12}C$ scattering in fig. 68 shows, optical resonances exist even up to several MeV above the Coulomb barrier. This indicates that most of these resonances must have relatively large spin values. Indeed, this indication has been supported by a detailed analysis of these resonances [AL 64, BO 65].

As was discussed in section 14.4, compound states of strongly collective nature can exist even in the region of high excitation. An example of this is the rotational band in ^{24}Mg, starting at a level with an excitation energy of 11.75 MeV. The existence of such states is connected with the fact that their coupling with other excitation modes, such as single-

† The lifetime τ is defined as the average length of time which the bombarding cluster stays inside a target nucleus without being resolved by making inelastic collisions with the target nucleons.

†† The transit time τ_{tr} is defined as the average length of time which the bombarding cluster spends in the target nucleus (provided it is not resolved) before leaking out without compound-nucleus formation. Equivalently, it may be defined as the average time required for the bombarding cluster to go across the target nucleus (transversal time), multiplied by the number of times it is reflected back at the nuclear surface.

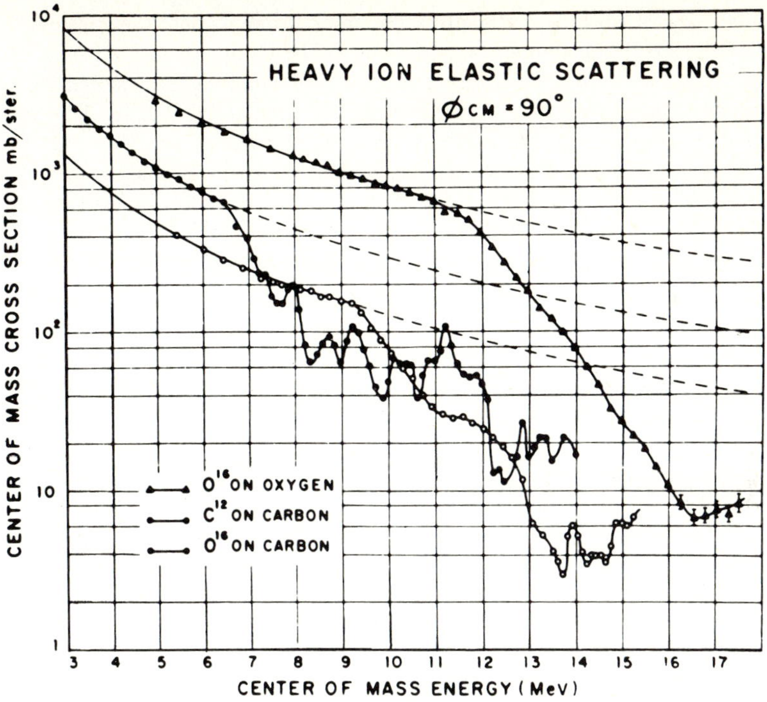

Fig. 68. Excitation functions for $^{16}O + ^{16}O$, $^{12}C + ^{12}C$ and $^{16}O + ^{12}C$ scattering. The dashed curves are the Mott-scattering predictions (Form ref. [BR 60a].)

particle excitation modes of approximately the same excitation energy, is very small. This is especially the case if the spins of these collective states are large because, as a result of centrifugal effects, few-nucleon excitation modes of even high excitation energy usually have rather small spin values. Therefore, it is understandable that the optical resonances mentioned above could have relatively small widths, as observed experimentally.

At energies near the top of the Coulomb barrier, the transit time τ_{tr} is large and the energy widths are determined by the lifetime τ of the ^{12}C clusters in their region of strong interaction. A crude estimate of this lifetime yields a value of $(2–5) \times 10^{-21}$ sec [WI 61], thus resulting in energy widths of the order of 100 keV. As the scattering energy becomes higher, the resonance structure should smear out and eventually disappear. This happens because of the decreasing transit time. At high energies, the ^{12}C clusters spend little time near each other ($\approx 5 \times 10^{-23}$ sec), since the l-dependent reflection coefficient at the nuclear surface approaches zero. By examining the behaviour of the scattering cross section shown in fig. 68 and especially the reaction cross sections reported in ref. [ÅL 60], one finds that the description given here is indeed qualitatively correct.

The situation in the case of $^{16}O + {}^{16}O$ scattering is quite different. Here the nucleons in each of the ^{16}O clusters fill up completely the 1p shell. Therefore, due to the Pauli principle, the mutual penetration in a scattering process of two internally unexcited ^{16}O clusters is appreciably smaller than that of two internally unexcited ^{12}C clusters. This has the consequence that, in contrast to the $^{12}C + {}^{12}C$ case, the real part of the $^{16}O + {}^{16}O$ effective potential for internally unexcited ^{16}O clusters becomes so weak that it can no longer allow the occurrence of optical resonance phenomena even at energies near the top of the $^{16}O + {}^{16}O$ Coulomb barrier.†

To understand more quantitatively the difference in the $^{16}O + {}^{16}O$ and $^{12}C + {}^{12}C$ effective potentials, one must perform resonating-group calculations for these systems. With the generator-coordinate and related techniques, these calculations are certainly feasible at the present time. In fact, single-channel $^{16}O + {}^{16}O$ resonating-group calculations have already been performed [FR 74, TO 75]††, but the more difficult $^{12}C + {}^{12}C$ calculation has not yet been attempted.

16.5b. Optical Resonances in Reaction Channels

Our previous considerations have dealt with examples of optical-resonance phenomena where the resonances are associated directly with the incoming channel of the nuclear reaction. Under certain circumstances, however, one should also expect optical-resonance phenomena to occur in a specific reaction channel even if resonances do not occur in the capture cross section or elastic-scattering cross section. This can be seen in the following way. Consider a nuclear reaction which is initiated, for instance, by the capture of a nucleon with an energy of several MeV in a medium-heavy or heavy nucleus.††† After the capture of the bombarding particle, the highly excited compound nucleus formed goes through many different excitation modes (described, for instance, by many different cluster and shell-model configurations) allowed by conservation laws. For example, suppose that the compound nucleus allows the formation of a two-cluster state of the general form

$$\psi = A' \{\hat{\phi}(\widetilde{A})\, \hat{\phi}(\widetilde{B})\, \chi(\mathbf{R}_A - \mathbf{R}_B)\}. \tag{16.11}$$

---

† This means, for example, that the ground state and the low-excited states of ^{32}S cannot be described by a cluster structure of two unexcited ^{16}O clusters in relative motion. In these states, at least one of the clusters must be an excited $^{16}O^*$ cluster. To have a configuration with two unexcited ^{16}O clusters in close proximity would require such strong correlations among the nucleons that, because of the Pauli principle, the gain in potential correlation energy in keeping the ^{16}O clusters unexcited would not be sufficient to compensate for the resultant large increase in relative kinetic energy between these clusters. In the oscillator shell-model description of the low-lying states in ^{32}S, this implies that some of the nucleons have to be lifted up to the 2p-1f and higher oscillator shells.

†† A careful study of these resonating-group results in terms of an $^{16}O + {}^{16}O$ effective potential has not yet been carried out.

††† The assumption of a bombarding energy of several MeV allows the nucleon to excite many compound states with different spin and parity. In addition, it means that we shall be concerned with an energy region in which usually no sharp resonances exist and many outgoing channels are open for the subsequent decay of the compound nucleus.

If the lifetime of such a cluster state is of the order of 10^{-21} sec or longer, then one would expect optical resonances to occur in the appropriate partial reaction cross section and these resonances can be fairly narrow provided that, in addition, the relative energy of the decay products is close to the top of their mutual Coulomb barrier.

For a somewhat more detailed discussion of the optical resonance behaviour in a reaction channel, we start with eqs. (12.8a) and (12.8b) of chapter 12. By omitting the feedback term and proceeding as described in section 12.2, we obtain for the reaction amplitude the expression

$$A_{j,12} \propto \left(\frac{v_1}{v_2}\right)^{1/2} \langle \widetilde{\psi}_{2j}(E) | H' - E | \psi_{1j} \rangle, \tag{16.12}$$

where $H' = H + W_R$, with W_R being a nonhermitian operator representing the influence of all other open channels and bound structures on the incoming channel and the reaction channel under consideration.† From eq. (16.12) one sees that, for an unequivocal identification of optical resonances in the reaction channel, the operator W_R must not be strongly energy-dependent in the excitation region of interest, because otherwise all resonances would result from a complicated superposition of optical effects in the reaction channel plus this energy dependence. In actual practice, however, this difficulty should not usually arise. Since the average kinetic energy between two clusters in the region of the compound nucleus is usually much larger than the bombarding energy of the captured particle,†† one generally expects that the wave functions of cluster configurations in the compound region should change rather slowly with changing incident energy. Consequently, it follows from eqs. (12.6) and (12.10) that W_R should in general be a smooth function of energy in the excitation region under consideration here.†††

With W_R, and consequently H', having a smooth energy dependence, the optical resonance behaviour is therefore originated in the wave function ψ_{1j} of the incoming channel and/or the wave function $\widetilde{\psi}_{2j}$ of the reaction channel. If the incoming channel is known to exhibit no optical resonance phenomena, then one can identify a true optical resonance behaviour in a specific reaction channel by simply examining the relevant experimental data on the partial reaction cross section. Otherwise, one will have

---

† In section 12.2, we have made a further approximation with the consequence that the operator H, instead of H′, appears in eq. (12.17).

†† For example, the average kinetic energy of two ^{12}C clusters in the region of strong interaction is about 50–100 MeV [WI 61].

††† Because of approximate selection rules, it can happen that, in the considered excitation region where compound levels overlap strongly, there appear relatively sharp resonances which are not optical resonances in either the incoming channel or the reaction channel under consideration. An example for this is the occurrence of analogue resonances in medium-heavy and heavy nuclei. In such cases, the analysis of the experimental situation would become quite complicated. In principle, one can of course take these resonances into account by explicitly including them in the formulation, as described briefly in section 12.2. Here we do not consider such cases simply because, in the example which we discuss afterwards, this type of complication does not arise.

to look at the behaviour of an energy-dependent quantity **R**, defined as the ratio of the partial reaction cross section to the total reaction cross section; this is necessary, because in such cases the optical effects we are concerned with will be superposed upon the variation with incident energy of the capture probability of the bombarding particle.

From the above considerations, we can now draw the following conclusions:
(a) In the presence of optical resonances in a reaction channel, the quantity **R** should exhibit similar features independent of how the compound nucleus is formed.
(b) The optical resonance behaviour in a reaction channel should be very similar to that in a corresponding elastic-scattering process involving the nuclei in this reaction channel, provided that Coulomb effects are properly corrected for.
(c) When optical resonances are present in a specific reaction channel, the angular distribution of the partial reaction cross section should change slowly with energy over an energy range of the order of the width of the optical resonance.

As a specific example, we consider the reactions $^{16}O(p, \alpha)^{13}N$ and $^{14}N(^3He, \alpha)^{13}N$ which have the same outgoing channel [GU 71]. In fig. 69 the integrated excitation functions for these reactions are shown. For the $(^3He, \alpha)$ curves, effects from the ^3He

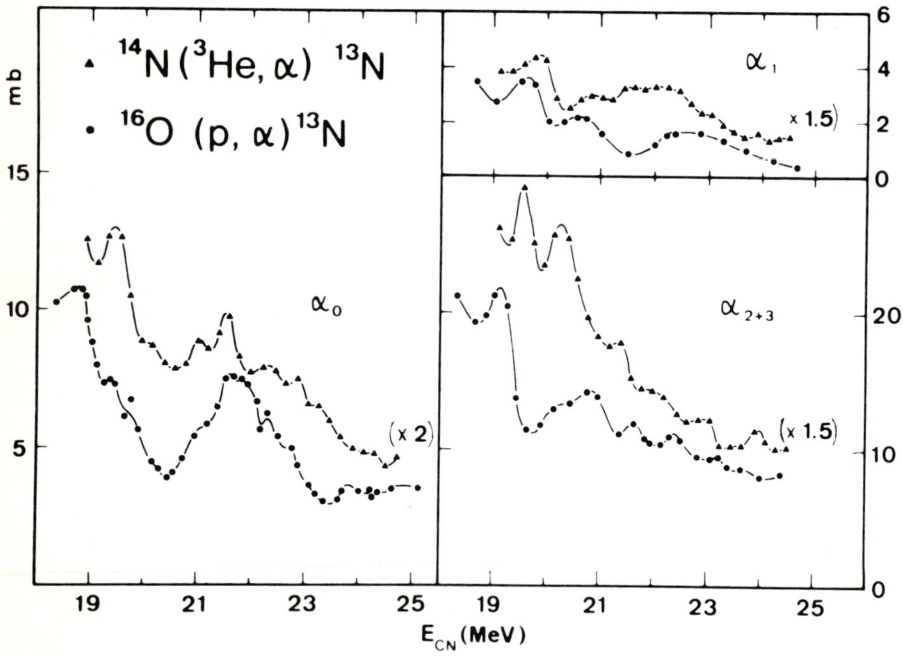

Fig. 69. Integrated excitation functions for the $^{16}O(p, \alpha)^{13}N$ and $^{14}N(^3He, \alpha)^{13}N$ reactions going to the ground and the three lowest excited states of ^{13}N. The $(^3He, \alpha)$ curves have been corrected for the ^3He Coulomb barrier. (From ref. [GU 71].)

Coulomb barrier have been appropriately corrected. Also, since the p + ^{16}O and ^{3}He + ^{14}N elastic-scattering cross sections show in the considered energy range a rather smooth energy dependence, these excitation functions can be simply used to study the resonance behaviour in the α + ^{13}N reaction channel.

From fig. 69 one sees that, for the transition leading to the ground state of ^{13}N, the two integrated excitation functions do have a very similar behaviour, in agreement with our conclusion given above. Also, as is seen from this figure, the energy width of the optical resonances in the α + ^{13}N reaction channel is of the order of 0.5 MeV; this again agrees reasonably well with our crude estimate given by eq. (16.10) for the scattering of α particles by ^{13}C.

For transitions leading to the three lowest excited states of ^{13}N, the similarity between the integrated excitation functions of the (p, α) reaction and those of the (^{3}He, α) reaction is less clear (see fig. 69). As has been explained by *Guazzoni et al.* [GU 71], this lack in close similarity could arise from the interference with a nonresonating part which may be different in these two reactions.

From our considerations given in this section and in chapters 11 and 12, it is clear that optical resonance phenomena should be generally observed in nuclear scattering and reactions. When these resonances occur at energies near the top of the Coulomb barrier, they should have relatively small width of the order of 100 keV. At higher energies, the transit time becomes quite small and these resonances should be rather broad.† Indeed, in the many examples discussed here and in chapters 11 and 12, we have always seen these general features.

† For these resonances not to become excessively broad, it is necessary that the effective potential between the clusters has a rather large slope near its edge.

17. Nuclear Fission

17.1. General Remarks

We now use the cluster viewpoint, in the framework of our unified theory, to discuss the problem of nuclear fission. In particular, we wish to understand the main features of the phenomenon of asymmetric fission occurring in many heavy nuclei [FA 62, FA 64]. As is well known, a partially successful macroscopic theory of nuclear fission has been available already for a long time, namely, that provided by the liquid-drop model (LDM) [BO 39, FR 39]. In this model one envisions a heavy nucleus, when excited for instance by the capture of a thermal neutron, as deforming until it splits into two parts. The key effect which lowers the energy of the intermediate deformed state (i. e., the deformation energy) enough to yield an observable fission rate for a nucleus of large atomic weight A_0 is the compensation of the increased surface energy by a decrease in the Coulomb energy. Using this picture of the fission process, one can then make some hydrodynamical calculations and predict a symmetric curve (shown schematically by the solid curve in fig. 70) for the mass distribution of the fission fragments, which has a peak at half the mass number A_0 of the compound nucleus.

For induced fission of some nuclei, the observed results are in general agreement with the symmetric curve. However, in contradiction to the prediction of this model, one obtains experimentally in many cases a curve with two peaks located symmetrically about $A_0/2$, known as the asymmetric fission curve (dashed curve in fig. 70).[†] This asymmetric distribution is observed in nuclei with neutron number $N_0 \geqslant 132$, both for spontaneous fission and for induced fission at low excitation energies. The question arises, therefore, as to how one is able to understand this asymmetric curve.

Since the peaks in the asymmetric cases correspond to one fission fragment with 82 neutrons or slightly more and the other with 50 neutrons or slightly more and since asymmetric fissions are observed only in nuclei with at least $50 + 82 = 132$ neutrons, it is natural to assert that they occur because of the stability of magic-number configurations with 50 and 82 neutrons [KO 50, MA 48, ME 50, WI 48]. Indeed, this assertion is further supported by the experimental fact that asymmetric fragments tend to come off with a lower mean excitation energy than symmetric fragments and the average total kinetic energy of asymmetric fission fragments is larger than that of symmetric fission fragments [FE 73][††]. Thus, the important question is: how can one incorporate these magic-number effects into the liquid-drop picture? For this it is clear that one has to make some microscopic considerations of the fission process.

---

† For binary fission, the mass distribution of fission fragments must necessarily be symmetric about $A_0/2$.

†† An exception occurs in the Fm-region ($Z_0 = 100$), which we shall discuss later in this chapter.

17.2. Substructure Effects in Fission Processes

The preceding discussion shows that one has to add essentially new features to the pure LDM consideration of nuclear fission in order to understand asymmetric fission. On the other hand, one must of course only refine the liquid-drop model but not invalidate it completely, because it does provide a very natural basis for the explanation of many important features of nuclear fission.† The refinement one seeks is to explicitly introduce the effects of cluster substructures, particularly those of closed-shell clusters,†† into the liquid-drop model [FA 62, FA 64]. In this way, one obtains a hybrid microscopic-macroscopic model which can better describe the many observed fission properties.

From the following considerations, we shall see that the reaction mechanism for fission is principally the same as that in lighter systems (see, e. g., the d + α scattering with specific distortion discussed in subsections 4.2b and 7.3c, and the optical resonances discussed in section 16.5). Because of the large number of nucleons involved in nuclei undergoing fission, the corresponding quantitative calculations become certainly much more complicated; however, it will be seen that, with a proper consideration of cluster correlations, one can already explain qualitatively many important features of low-energy fission in a unified manner even without detailed computations [GÖ 75].

To begin our discussion from the cluster viewpoint, we consider the case of induced binary fission where the target nucleus absorbs a slow neutron to form a fission-prone intermediate (compound) nucleus with Z_0 protons and N_0 neutrons ($N_0 \geqslant 132$).††† Then, because of the binding energy of the neutron, the intermediate nucleus is in effect excited to an energy of 6–8 MeV.†††† Now, since any wave function may be represented by a linear superposition of two-cluster wave functions, we may, for the consideration of the fission process as a time-dependent process, write the time-dependent wave function of the excited intermediate nucleus as

$$\Psi(\tilde{r}_1, \ldots, \tilde{r}_{A_0}; t) = \sum_h \sum_{ijk} a^h_{ijk}(t)\, \psi^h_{ijk}(\tilde{r}_1, \ldots, \tilde{r}_{A_0}), \tag{17.1}$$

with

$$\psi^h_{ijk}(\tilde{r}_1, \ldots, \tilde{r}_{A_0}) = A' \{\hat{\phi}_i(\widetilde{A}_h)\, \hat{\phi}_j(\widetilde{A}_l)\, \chi_k(\mathbf{R}_h - \mathbf{R}_l)\}. \tag{17.2}$$

In eq. (17.2), $\hat{\phi}_i(\widetilde{A}_h)$ represents the ith internal state of the heavy cluster with Z_h protons, N_h neutrons, and A_h nucleons, while $\hat{\phi}_j(\widetilde{A}_l)$ represents the jth internal state of the light

† For example, it explains the general decreasing trend of the spontaneous fission half-life as a function of the fissility parameter.

†† In this chapter, the phrase "closed-shell clusters" refers to clusters with just a few nucleons outside of the last closed proton or neutron shell.

††† For simplicity, we do not consider fission processes in which the compound nucleus breaks up into three or more fragments, although our qualitative considerations would not essentially change by the inclusion of such processes.

†††† In slow-neutron fission, the intermediate nucleus must contain an even number of neutrons, because in odd-N nuclei the missing pairing effect reduces the binding energy of the odd neutron below the threshold excitation for fission [BO 39].

cluster with Z_l protons, N_l neutrons, and $A_l = A_0 - A_h$ nucleons. The function $\chi_k(\mathbf{R}_h - \mathbf{R}_l)$ represents the kth state of their collective relative motion, with $\mathbf{R}_h$ and $\mathbf{R}_l$ being the centre-of-mass coordinates of the heavy and light clusters, respectively. The time dependence of Ψ is contained entirely in the amplitudes $a^h_{ijk}(t)$. Also, we should mention that the two-cluster states in eq. (17.1) are not all orthogonal. However, they do not overlap in the asymptotic region of large cluster separation; hence, there is no problem in defining various reaction cross sections (see subsection 7.2b). With the wave function Ψ written in the form of eq. (17.1), we may then think of the intermediate nucleus as passing simultaneously through many different kinds of cluster configurations, symmetric and asymmetric, as the coefficients $a^h_{ijk}(t)$ develop in time.

De-excitation of the excited intermediate nucleus may now occur in only very few ways. In addition to the fission modes, γ-emission and evaporation of a neutron are the only other possible decay modes.† The usefulness of the two-cluster expansion of eq. (17.1) lies in the fact that one can examine the behaviour of the specific terms which contribute to a particular de-excitation process. In fact, we shall show that at low energies only those terms representing asymmetric closed-shell cluster configurations (with some extra nucleons) can contribute significantly to de-excitation by fission.

For the following discussion, it is important to note one essential characteristic of closed-shell clusters. In chapter 6 it was shown that free clusters which are especially tightly bound, i. e., clusters with filled shells N = 50, N = 82, or Z = 50, etc., remain relatively tightly bound even when surrounded by many other nucleons.†† As will be seen below, this tight binding of closed-shell clusters plays a crucial role in low-energy asymmetric fission.

For a long time, it was not understood how a heavy nucleus ($N_0 \geqslant 132$) could have especially tightly bound substructures with N = 50 and N = 82 without a complete breakup of the N = 126 and other closed shells of the nucleus, which would more than offset the correlation energy gained by closing the shells of the substructures. But, as has often been pointed out, antisymmetrization (i. e., the Pauli principle) so strongly reduces the differences between different cluster structures that this difficulty is resolved.

To express this last point in more mathematical terms, let us consider the nucleus ^{236}U in the oscillator model and write the wave function as

$$\psi = A' \{\hat{\phi}(\widetilde{A}_h)\, \hat{\phi}(\widetilde{A}_l)\, \chi(\mathbf{R})\}. \tag{17.3}$$

---

† Both of these modes are long-lived however. The lifetimes τ_γ and τ_{evap} are of the order of magnitude 10^{-12} to 10^{-15} sec. The value of τ_{evap} is large, because one neutron must regain practically all the "heat" energy before it can escape again, which is very improbable. In terms of the expansion given by eq. (17.1), this means that neutron evaporation can occur only when the intermediate nucleus passes through a cluster configuration having $A_h = A_0 - 1$ and $A_l = N_l = 1$ with the heavy cluster unexcited or only very slightly excited depending upon the energy of the absorbed slow neutron.

†† It is perhaps more appropriate to talk about the correlation energy associated with a particular cluster configuration rather than the binding energy, because the binding energy of a cluster within a larger nucleus is somewhat artificially defined (see the discussion given in chapter 6).

For instance, $\hat{\phi}(\widetilde{A}_h)$ might represent a heavy cluster with $Z_h = 52$ and $N_h = 82$ in which the 50-proton and 82-neutron shells of the cluster are filled and the few odd nucleons lie in the next shells of the cluster, and $\hat{\phi}(\widetilde{A}_l)$ a light cluster with $Z_l = 40$ and $N_l = 62$ in which the 40-proton and 50-neutron shells are filled.† If we were to expand this oscillator cluster function into oscillator shell-model functions, then we would expect to find the single-particle shells with 82 protons and 126 neutrons of the intermediate nucleus almost filled and the remaining nucleons in some superposition of single-particle orbitals of the next shells such that the (negative) potential part of their correlation energy becomes especially large.†† Because of the restriction imposed by the Pauli principle, the strong correlations expressed by wave functions of the form (17.3) can very often have the effect that the associated (positive) kinetic part of the correlation energy is also rather large. In the shell-model representation, this means that some holes will be present in the lower oscillator shells, as has been discussed in more detail near the end of section 6.3.

We are now ready to discuss the role of closed-shell clusters in fission processes. The principal argument may be stated as follows. As the amplitudes $a^h_{ijk}(t)$ in eq. (17.1) vary in time, the intermediate (compound) nucleus passes through many cluster configurations of the same (within the limitation of the uncertainty principle) energy. Statistically, it is clear that it will spend most of its time in configurations for which the two clusters have nearly equal mass (more accurately, it will be found with a higher probability in such configurations). With a lesser probability, it will pass through configurations in which the masses of the two clusters are quite different. However, these various cluster configurations are not all equally fission-prone, because they are not all equally deformable.†††
A deformation of the intermediate nucleus connected with a specific cluster configuration may be simply regarded as measured by the average separation distance between the two ultimate clusters composing it at any instant. From the liquid-drop model we know that, the more these two clusters can separate, the lower and more penetrable will be the fission barrier holding them together and, hence, the greater will be the probability for the clusters to become free of each other. But the average separation distance between the clusters can be especially large, only if the energy in the relative motion of the clusters about each other is especially large. Because the intermediate nucleus has a virtually fixed total energy, this large relative energy can only be present in those configurations where the clusters have large (negative) correlation energies. As mentioned above, this happens when one or both clusters have closed-shell structure plus perhaps

† The number 40 is only a magic number in a shell model without strong spin-orbit coupling, but does seem to have some significance in fission processes [FA 64].

†† The nucleus ^{6}Li, which is small enough for the expansion to be carried out explicitly, may serve as an example. This particular case was treated in subsection 3.5c. There we have seen that, in the expansion into oscillator shell-model states, the proton and the neutron in the 1p shell (above the filled 1s shell) are coupled to an $L = 0, S = 1$ configuration so as to make the magnitude of their correlation energy especially large. A second example is ^{8}Be which has been repeatedly discussed in earlier chapters.

††† We emphasize that the deformation we have in mind here is not that of the constituent clusters, but that of the compound nucleus.

a few outside nucleons. Therefore, in low-energy fission, the intermediate nucleus ($N_0 \geqslant 132$) is particularly fission-prone only in those asymmetric cluster configurations for which the two ultimate clusters have unbroken closed shells.

The probability that the intermediate nucleus possesses the rather highly ordered closed-shell substructures leading to asymmetric fission is of roughly the same order of magnitude as the probability that one neutron regains all the excitation energy. Hence, in heavy nuclei, the asymmetric fission process and the neutron evaporation process generally compete with each other. Because of the presence of higher fission barriers when the intermediate nucleus is in its more probable symmetric configurations, the rate of low-energy symmetric fission is small and the intermediate nucleus must simply wait until it passes through one of these less probable configurations before it can readily decay.†

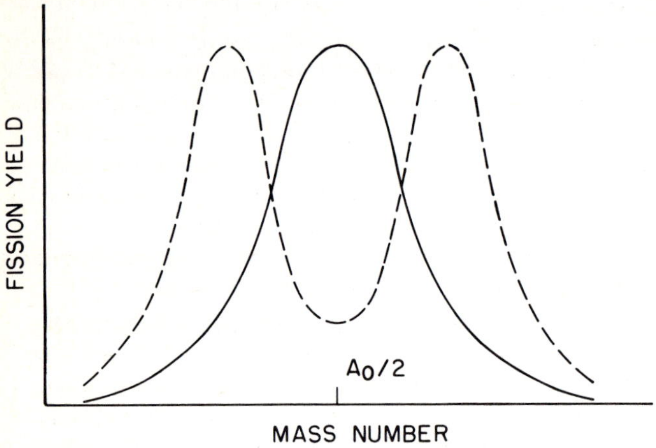

Fig. 70. Schematic mass distribution of fission fragments. The solid curve is for symmetric fission, while the dashed curve is for asymmetric fission.

We see from this discussion that the relative height and width of the fission barrier for different fission products plays a decisive role in determining the peak-to-valley (P/V) ratio in the two-humped mass distribution (see fig. 70). The sensitivity of the quantum mechanical tunnel effect to the shape of this barrier makes it immediately understandable why small differences in the fission barrier can produce the observed large differences in the decay probabilities of the fission products ($P/V \approx 100-1000$ for low-energy fission [VA 73]). Certainly, we repeat, the probability that the intermediate nucleus passes through a configuration in which the two clusters have almost the same number

† The compound state represented by the wave function of eq. (17.1) has a very well-defined energy; consequently, many different cluster configurations are present simultaneously. Therefore, we should understand our picturesque language above in the same manner by which one usually connects time-developing wave packets with stationary states.

of protons and neutrons is greater than that for an asymmetric closed-shell-cluster configuration. But, to obtain clusters of almost equal mass, we must break up the 50-proton and 82-neutron shells of the larger cluster.† Consequently, nearly symmetric cluster configurations have much less (negative) correlation energy and the resultant fission barrier for the separation of nearly symmetric clusters is usually so high that, at low excitation energies, symmetric fission will occur only rarely compared to asymmetric fission.

As the excitation energy of the intermediate nucleus increases, the influence of the difference in fission barriers for different fission products certainly decreases. This gives a natural explanation for the observed fact that, with increasing excitation energy, the relative probability for symmetric fission compared with asymmetric fission increases rather rapidly [VA 73]. On the other hand, it is clear from this excitation-energy dependence that spontaneous fission, which occurs when the nucleus is in its ground state, should exhibit cluster effects most strongly. Experimentally, this is indeed the case. Contributions are found to be restricted almost entirely to mass bands corresponding to unbroken closed-shell cluster configurations, and the symmetric mode is strongly suppressed.

It is also experimentally confirmed that, if the fissioning nucleus has less than 132 neutrons, the symmetric-fission mode prevails [FA 58, JE 58, JE 60]. This is expected, since it is now impossible for the heavy and light fragments to contain closed-shell structures simultaneously.

Our discussion indicates that the decision as to how the intermediate nucleus will fission at some given excitation energy is made at a comparatively early stage of the fission process, i.e., at relatively small deformations long before the scission deformations. Therefore, symmetric and asymmetric fission should be considered as two different processes [GÖ 75] in the same way as, for instance, the neutron evaporation process is considered to be different from the fission processes. In fact, at intermediate energies, both modes have been observed to show up distinctly in the fission of the nucleus ^{226}Ra by 13-MeV proton bombardment [KO 68] (see fig. 71).

With the cluster viewpoint discussed above, we have now shown the key effect for low-energy asymmetric fission. Using this viewpoint, we may proceed and briefly outline the remainder of the fission process. The part of the compound-nucleus wave function (eq. (17.1)) describing the saddle-point deformation before scission is rather simple; that is, mainly only the relatively few terms representing "cold" closed-shell clusters contribute to this part. At low excitation energies, there is little energy available at the saddle point for anything but deformation energy. Consequently, the clusters themselves must be rather "cold" [BO 56] and the collective kinetic energy is necessarily small. Because of this small kinetic energy, the collective motion is essentially rotational rather than vibrational. We should mention that, for heavy nuclei where asymmetric fission is

---

† The neutron-to-proton ratio in each cluster must be very nearly the same as that in the original compound nucleus. Otherwise, the symmetr-energy term in the mass formula will acquire a large value and the corresponding compound-nucleus configuration will be energetically unfavourable [FA 64].

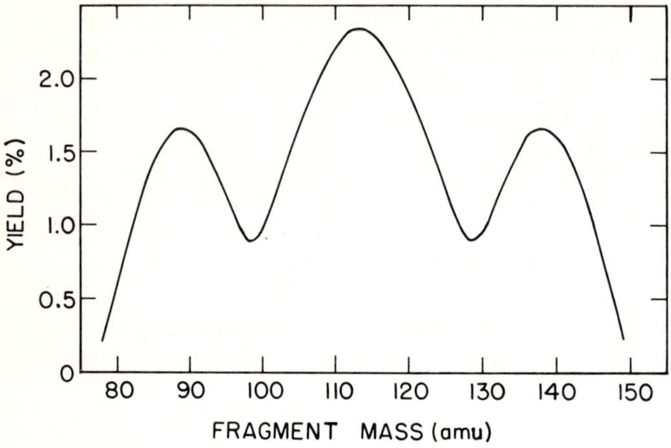

Fig. 71. Pre-neutron-emission fragment mass distribution in 13-MeV proton-induced fission of ^{226}Ra. (Adapted from ref. [KO 68].)

important, there exist generally not one but two saddle-point deformations. The existence of this second saddle point results in the appearance of fission isomers and intermediate structures in sub-threshold fission cross sections [VA 73]. In section 17.4, where we present a more quantitative discussion of the behaviour of the deformation energy, we shall discuss this further.

The final stage of the fission process is the descent from the second saddle point until the actual separation of the fragments is achieved at the scission point. During this relatively short time, extreme deformations occur which might be described by the classical model of a pear-shaped drop getting more and more elongated [VL 57, WH 59].

The "neck" between the two separating clusters is primarily formed from the more loosely bound nucleons outside the closed shells of the clusters. These outside nucleons are so distributed between the two clusters that the neutron-to-proton ratios of the clusters are as nearly as possible the same as the parent intermediate nucleus, i. e.,

$$N_h/Z_h \approx N_l/Z_l \approx N_0/Z_0 . \tag{17.4}$$

In this way, the symmetry energy is kept to a minimum. Immediately after the scission, these outside nucleons originally composing the neck are now distended, so that the fragments are strongly deformed [VA 73] and will carry out strong surface oscillations which should be accompanied by multipole radiation [TE 62]. These surface oscillations will be gradually transformed into unordered compound-nucleus excitation of the fission products, thereby certainly exciting the outer closed shells of the fragments. Most of this excitation energy will be lost in neutron emission, which occurs at this time to relieve the considerable neutron excess of the prompt-yield fragments.

We should point out that scission need not necessarily happen near the center of the neck joining the two unequal clusters [VA 73]. One can imagine scission taking place quite close, say, to the heavy cluster. In this case, the masses of the two fragments will be more nearly equal, but the original asymmetry of the saddle-point deformation will still show up in an asymmetry in excitation energy. That is, the almost undeformed large fragment will end up with much less excitation energy than the smaller one, which is now strongly distended by the loosely bound nucleons originally in the neck. Similarly, in the opposite case where the scission occurs close to the light cluster, one will find the light fragment to have relatively few loosely bound nucleons and hence little excitation, but the heavier fragment will now have a higher excitation energy. Thus, this means that scission near a tightly bound cluster is indicated by a minimum in the excitation energy of the corresponding fragment.

On the other hand, the fragments in symmetric fission are always quite far from closed shells. The many outside nucleons make these fragments much more deformable than closed-shell products. Hence, the energy in surface deformation at the moment of scission will be much greater for symmetric fragments than for closed-shell fragments, with the consequence that the products of symmetric fission should show a higher mean excitation energy than those of asymmetric fission [VA 73].

17.3. Mass Distribution of Fission Fragments

By taking explicit account of the especially fission-prone closed-shell cluster configurations described above, we now wish to show that the cluster viewpoint can be used to make more refined predictions about the general form of the mass-distribution curve of the fission products. For asymmetric fission, the mass distribution can be calculated approximately under the following assumptions [FA 62, FA 64]:

(a) An appreciable contribution comes only from compound-nucleus states with unbroken closed-shell cluster configurations.

(b) The fission probability is independent of the distribution of surplus nucleons, except that the neutron-to-proton ratios in the fission products should be approximately equal to N_0/Z_0. In the classical model [VL 57, WH 59], where the surplus nucleons are pictured to form the neck joining the two separating fragment cores, this assumption corresponds to equal scission probability anywhere along the neck.

(c) To both sides of the allowed mass bands defined by assumptions (a) and (b), the decreasing probability for asymmetric fission is represented by a Gaussian function. This is, of course, simply a mathematically convenient way to express the rapid decrease of the fission probability as soon as the energetically favoured cluster structures are broken up.

The main problem is the specification of the proper cluster states. Both neutron and proton shells have to be considered. The symmetry energy imposes severe restrictions on the possible combination of neutron and proton numbers. Any neutron-to-proton ratio differing appreciably from the value of the fissioning parent nucleus would mean an energetically unfavourable charge polarization of nuclear matter. For nuclei exhibiting asymmetric fission, the ratio N_0/Z_0 is generally around 1.56. The only combination of relevant magic numbers giving N/Z close to this value is N = 82 and Z = 50. Consequently, the heavy fragment should always be dominated by this structure. For the light fragment, the possibilities are N = 50 and Z = 40; but these neutron and proton numbers cannot be combined without deviating drastically from the ratio N_0/Z_0. Because of the high stability associated with N = 50, this shell will only very rarely be broken up in the light cluster.†

The above considerations may be summarized in the following prediction which has been experimentally verified [VA 73]. The mass distribution of the heavy fragment has its low-mass shoulder at $A_h = 134$ ($N_h = 82$, $Z_h = 52$) for all heavy nuclei undergoing asymmetric fission. Correspondingly, the high-mass shoulder of the light-fragment mass distribution is at $A_l = A_0 - 134$ and depends, therefore, on A_0. It is easy to see that, for this combination of $N_h = 82$ and $Z_h = 52$, the closed-shell structures of 82 neutrons and 50 protons are not broken up and, at the same time, the neutron-to-proton ratios in both the heavy and the light fragments differ very little from the neutron-to-proton ratio in the intermediate parent nucleus.†† Further, both fragments are even-even clusters, which are additionally energetically favoured as a result of the pairing interaction [GO 55].

The width of the mass peaks varies with A_0 and N_0. Its determination is somewhat difficult because, due to the restriction imposed by the neutron-to-proton ratio, a cluster which has 50 neutrons does not have an energetically favoured number of protons. At attempt to determine the width of these peaks may be made by using the information available from the measured excitation energies E^* of the fragments [GI 61, VA 73]. As has been pointed out earlier, the observed sharp minima in E^* at certain mass numbers are a direct indication of the relevant tightly bound cluster structures. In thermal-neutron induced fission of ^{235}U, for example, the most prominent mass number of the light fragment seems to be $A_l = 90$, which is very likely associated with a neutron number of 54 and a proton number of 36.

Figure 72 shows the result of an attempt to explain the mass distribution of the fission fragments in the case of thermal-neutron induced fission of ^{235}U. In this attempt, the assumptions (a), (b), and (c) were used, together with the adoption of $A_l = 90$ and $A_h = 134$ as favoured cluster numbers. Gaussian functions were fitted to the experimental points starting in each case 2 mass units to the inside of the shoulders fixed by the cluster numbers, in order to fulfill assumption (a). The height of the approximately trapezoidal peak is determined by normalization. The primary-yield curve (curve a) obtained in this manner has further been corrected for neutron emission using experimental values (curve b). As is seen from this figure, the agreement between curve b and the experimental points is indeed quite satisfactory.

More quantitative predictions [FA 64] are possible, if one uses in more detail the information contained in the measured excitation energies of the fission fragments.

† If N_0 and Z_0 are sufficiently large, there will also be some favouring not to break up the Z = 40 shell of the light cluster.

†† The measurement of the proton distribution in the fission products shows indeed that the N/Z ratios in the fragments are very nearly the same as the ratio N_0/Z_0 of the intermediate nucleus [HE 61, VA 73]. The is in agreement with the symmetry-energy argument presented here.

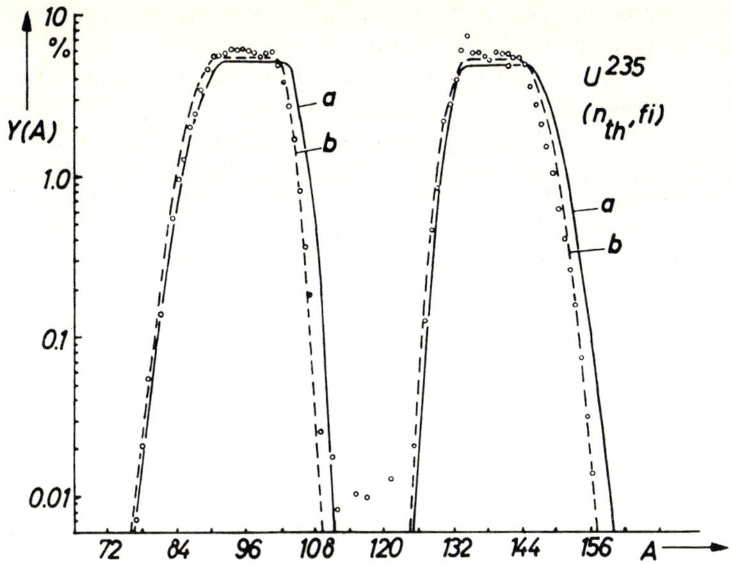

Fig. 72. Mass distribution for thermal-neutron induced fission of ^{235}U. Curve a: calculated prompt yield. Curve b: same as curve a, but corrected for neutron emission. Experimental data are represented by open circles. (From ref. [FA 62].)

It is also interesting to note that, for nuclei with $A_0 \gtrsim 264$ ($N_0 \gtrsim 164$ and $Z_0 \gtrsim 100$), the mass distribution is expected to become symmetric with a rather flat peak having a width $\Delta A \approx A_0 - 264$. Experimentally, there is indeed strong indication for a trend toward such a mass distribution when A_0 is increased toward this special number of 264 [BA 71a, JO 71, VA 73, UN 74].

17.4. Deformation Energy of Fissioning Nucleus

From the above considerations, we see that the application of the cluster viewpoint to the problem of fission can yield a qualitative understanding of some of the main features of the fission process. The important point in these considerations was concerned with the dependence of the deformation energy on the cluster substructure of the intermediate nucleus. In this section, we shall therefore study the nature of the deformation energy (potential-energy surface) in more quantitative terms and discuss ways in which the shell or substructure effects can be approximately taken into account.

17.4a. Dynamical Consideration of the Fission Process

Before we proceed to discuss the deformation energy, we shall first briefly indicate how one can, at least in principle, formulate the fission problem in the framework of our unified theory. Certainly, because of the very large number of nucleons involved in a fission process such as neutron-induced fission, it is very difficult to formulate a microscopic theory which is simple enough to yield quantitative predictions and yet describes adequately all the stages in this process including the capture of the neutron to form the intermediate nucleus. However, in the later stage of fission where the deformation is relatively large, the description of the behaviour of the intermediate nucleus does become simpler, because at this stage the available excitation energy is almost completely transformed into the deformation energy and the collective kinetic energy in the fission degree of freedom. In the following discussion concerning a microscopic formulation, we shall therefore concentrate on a description of this later stage in the fission process.

For simplicity, let us assume at this moment that the deformed intermediate-nucleus configuration is specified by a single deformation parameter α. Then, for the wave function describing the later stage of a fission process, we can make the Hill-Wheeler-ansatz [HI 53]

$$\psi(\tilde{r}_1, \ldots, \tilde{r}_{A_0}) = \int \phi(\tilde{r}_1, \ldots, \tilde{r}_{A_0}; \alpha)\, f(\alpha)\, d\alpha, \tag{17.5}$$

where $f(\alpha)$ is a variational function, and $\phi(\tilde{r}_1, \ldots, \tilde{r}_{A_0}; \alpha)$ or simply $\phi(\alpha)$ is, for example, the antisymmetrized ground-state wave function in a deformed potential well specified by this deformation parameter α.† If we substitute now this wave function into the projection equation (2.3), then we obtain

$$\int \langle \phi(\alpha) | H - E | \phi(\alpha') \rangle\, f(\alpha')\, d\alpha' = 0. \tag{17.6}$$

Because the intermediate nucleus consists of many nucleons and the deformation is relatively large, the overlapping integral $\langle \phi(\alpha) | H - E | \phi(\alpha') \rangle$ will, to a good approximation, have its largest magnitude at $\alpha' = \alpha$ and will go very rapidly to zero when α' deviates from α.†† For the evaluation of the integral in eq. (17.6), it is therefore useful to write

$$f(\alpha') \approx f(\alpha) + \frac{df(\alpha)}{d\alpha}(\alpha' - \alpha) + \frac{1}{2!}\frac{d^2 f(\alpha)}{d\alpha^2}(\alpha' - \alpha)^2. \tag{17.7}$$

By substituting eq. (17.7) into eq. (17.6), one then finds that, because of the argument given above, the term linear in $(\alpha' - \alpha)$ does not contribute. Further, it can be shown rather easily that, again as a consequence of the large number of nucleons involved, any overlapping term of the form $\langle \phi(\alpha) | O_i | \phi(\alpha') \rangle$, with O_i being a symmetric n-particle operator, can be approximately written as

$$\langle \phi(\alpha) | O_i | \phi(\alpha') \rangle \approx \langle \phi(\alpha) | O_i | \phi(\alpha) \rangle\, F(\alpha' - \alpha), \tag{17.8}$$

where $F(\alpha' - \alpha)$ is in a good approximation the same real function independent of the nature of the operator O_i, as long as the condition $n \ll A_0$ is satisfied [WI 76].

---

† For a wave function to describe all the stages in a fission process, one must also use excited-state wave functions in this deformed potential well. Consequently, the description will become very complicated, because many variational functions will then be required.

†† Due to the large number of nucleons $\langle \phi(\alpha) | H - E | \phi(\alpha') \rangle$ can be assumed to be real (see eq. (17.8)).

By assuming further that $\phi(\alpha)$ is normalized, we obtain, after the substitution of eq. (17.7) into eq. (17.6), the following equation:

$$\langle\phi(\alpha)|H|\phi(\alpha)\rangle\, f(\alpha) + \left[\frac{1}{2K(\alpha)}\int\langle\phi(\alpha)|H - E|\phi(\alpha')\rangle\,(\alpha'-\alpha)^2\, d\alpha'\right] \times \frac{d^2 f(\alpha)}{d\alpha^2} = E\, f(\alpha), \tag{17.9}$$

with

$$K(\alpha) = \int F(\alpha' - \alpha)\, d\alpha'. \tag{17.10}$$

If we now define the deformation energy $V_{def}(\alpha)$ as

$$V_{def}(\alpha) = E(\alpha) - E(0) \tag{17.11}$$

with

$$E(\alpha) = \langle\phi(\alpha)|H|\phi(\alpha)\rangle \tag{17.12}$$

and the inertial parameter $M(\alpha, E)$ as

$$M(\alpha, E) = -\hbar^2\left[\frac{1}{K(\alpha)}\int\langle\phi(\alpha)|H - E|\phi(\alpha')\rangle\,(\alpha'-\alpha)^2\, d\alpha'\right]^{-1}, \tag{17.13}$$

then eq. (17.9) becomes

$$-\frac{\hbar^2}{2M(\alpha, E)}\frac{d^2 f(\alpha)}{d\alpha^2} + V_{def}(\alpha)\, f(\alpha) = [E - E(0)]\, f(\alpha). \tag{17.14}$$

Equation (17.14) looks formally like a one-particle Schrödinger equation in one dimension, which contains an effective mass depending both on the total energy E and on the deformation of the intermediate nucleus.

Certainly, to describe the configuration of the deformed intermediate nucleus by only one parameter α is usually not adequate. Generally, one will have to use a set of deformation parameters or coordinates $\alpha_1, \alpha_2, \alpha_3, \ldots$, which will be collectively denoted by the symbol s_α. For example, if one wishes to consider an intermediate nucleus of axially symmetric shape with neck constriction, then one needs at least two deformation coordinates. In this more general case, the derivation of the Schrödinger equation for the variational function $f(s_\alpha)$ proceeds in the same way as in the one-parameter case. The result is that one obtains now a many-dimensional Schrödinger equation which contains a number of inertial parameters in the kinetic-energy term.

In some cases, it may be necessary to assume a more flexible wave function of the form

$$\psi(\tilde{r}_1, \ldots, \tilde{r}_{A_0}) = \int \phi(\tilde{r}_1, \ldots, \tilde{r}_{A_0}; s_\alpha)\, f(s_\alpha)\, ds_\alpha + \int \chi(\tilde{r}_1, \ldots, \tilde{r}_{A_0}; s_\gamma)\, g(s_\gamma)\, ds_\gamma, \tag{17.15}$$

where ϕ and χ are antisymmetrized ground-state wave functions in deformed potential wells specified by the sets of deformation coordinates s_α and s_γ, respectively. Then, by proceeding again from the projection equation (2.3), one obtains a set of coupled equations for the variational functions $f(s_\alpha)$ and $g(s_\gamma)$.†

Let us now go back for a discussion of the simpler equation (17.14). In this equation, the most difficult quantity to calculate is the inertial parameter $M(\alpha, E)$. However, in the case where the deformation is rather large and where the parameter α is chosen so as to represent the separation distance between the fission products when they are well separated, we can use the crude approximation

$$M(\alpha, E) = M_0/4, \tag{17.16}$$

where M_0 is the total mass of the intermediate nucleus. The reason is that, at large deformations, the mutual penetration of the clusters becomes relatively small and, consequently, the influence of the exchange terms between the nucleons in these clusters becomes less important. The result of this is that, under these conditions, $M(\alpha, E)$ should approach the reduced mass of the fission fragments. Thus, as long as the masses of these fragments do not differ too much from $M_0/2$, a value for $M(\alpha, E)$ given by eq. (17.16) would be a reasonable choice.††

With $M(\alpha, E)$ approximated as indicated by eq. (17.16), one can then solve eq. (17.14) to obtain the barrier penetration probability (see, e. g., ref. [CR 70]) and thereby the fission rate, if the deformation energy $V_{def}(\alpha)$ given by eqs. (17.11) and (17.12) is additionally computed. This indicates, therefore, that the computation of $V_{def}(\alpha)$ is the central problem in the study of nuclear fission. Of course, because of the many nucleons involved, it would be extremely difficult to calculate V_{def} microscopically by using two-nucleon potentials and one would usually have to rely on semi-empirical methods for its calculation (see refs. [MI 73, VA 73] and references contained therein). In the next two subsections, we shall describe briefly two such methods which have been commonly employed. From the results obtained, we shall see that these methods do yield the essential features of the potential-energy surface, which are necessary for an explanation of the many observed phenomena in a fission process.

17.4b. Calculation of the Deformation Energy — Strutinsky Prescription

The first method which we shall discuss was proposed by *Strutinsky* [ST 67]. In this method, the assumption is that the total energy $E(s_\alpha)$ of a heavy deformed nucleus (see eq. (17.12)), specified by a set of deformation parameters s_α, is given by the energy $E_{LDM}(s_\alpha)$ of the liquid-drop model plus a correction term arising from the shell structure of the nucleus. That is, one writes $E(s_\alpha)$ as

$$E(s_\alpha) = E_{LDM}(s_\alpha) + \Delta E(s_\alpha), \tag{17.17}$$

† In eq. (17.15) also cluster-terms can be introduced which describe the asymptotic decay-products of the fissioning nucleus.

†† Later, we shall mention another possibility to obtain $M(\alpha, E)$.

where the structure-correction term $\Delta E(s_\alpha)$ is composed of two parts, i.e.,

$$\Delta E(s_\alpha) = \sum_{n,p} [\Delta E_{sh}(s_\alpha) + \Delta E_{pr}(s_\alpha)] \tag{17.18}$$

with ΔE_{sh} and ΔE_{pr} being the shell-energy correction and the pairing-energy correction, respectively. It should be noted that in eq. (17.18) the symbol $\Sigma_{n,p}$ appears, which indicates that the corrections must be applied both for neutrons and protons.

Let us first discuss qualitatively the shell effect on the energy $E(s_\alpha)$. The influence of the shell structure, obtained for instance from a Nilsson potential with deformation parameters s_α, is related to the density of single-particle levels at the Fermi energy in the following manner. For a given well depth, the total binding energy of the nucleons is larger when fewer of them lie close to the Fermi surface, since this means that the other nucleons must lie deeper in the well. Therefore, the stability of the nucleus, and consequently its resistance against further deformation, is greater when the density of single-particle levels near the Fermi surface is smaller. This is nicely illustrated, for example, by the permanent deformation in the ground state exhibited by nuclei in the rare-earth and actinide region where neutron and proton numbers are between shells at sphericity. There the single-particle level density is large, thus forcing the nucleus to depart from sphericity in order to locate a region of greater stability, where a decreased level density will increase the binding energy.† In the above example, the deformations involved are rather small, but similar effects can also exist at larger deformations where shells and associated "magic" properties are found [MI 73, VA 73].

Following *Strutinsky*'s prescription [ST 67], one writes the shell-energy correction ΔE_{sh} as

$$\Delta E_{sh}(s_\alpha) = \sum_\nu 2\epsilon_\nu(s_\alpha)\, n_\nu - 2\int_{-\infty}^{\lambda_F} \epsilon\, \tilde{g}(\epsilon, s_\alpha)\, d\epsilon. \tag{17.19}$$

In the first term on the right side of eq. (17.19), ϵ_ν are single-particle energies of nucleon levels in a realistic shell-model potential and n_ν are the occupation numbers of these levels. In the second term, the function $\tilde{g}(\epsilon, s_\alpha)$ is the background level density which represents the slow and smooth variation of the density of levels over a wide energy range at deformation s_α; it is obtained by averaging the actual shell-model level density over a sufficiently large energy interval to wash out the shell effects. The quantity λ_F is the chemical potential or Fermi energy defined by the condition

$$2\int_{-\infty}^{\lambda_F} \tilde{g}(\epsilon, s_\alpha)\, d\epsilon = N_0 \text{ or } Z_0, \tag{17.20}$$

where the factor 2 in front of the integral reflects the fact that there are two nucleons (neutrons or protons) in every filled orbit. Strutinsky's prescription assumes that the

† For this argument, it is important to note that the Fermi energy is essentially independent of the deformation.

second term contains those parts of the sum of the single-particle energies already given by the LDM energy.

The pairing energy is also determined by the density of single-particle states close to the Fermi surface. The correction term ΔE_{pr} can be calculated by finding the best ground-state wave function of the Bardeen-Cooper-Schrieffer (BCS) type [BA 57] and comparing its energy with the sum of single-particle energies [NI 69]. As is explained by *Vandenbosch* and *Huizenga* [VA 73], this correction and the shell-energy correction reach their maximal and minimal values simultaneously, so that there is some compensation between these two corrections. The result is that the total deformation energy is somewhat less modulated than would be given by the shell-energy correction alone.

The deformation energy $V_{def}(s_\alpha)$ is given by

$$V_{def}(s_\alpha) = E(s_\alpha) - E(0), \tag{17.21}$$

where E(0) is the energy of the nucleus at sphericity. From $V_{def}(s_\alpha)$ one can further define a deformation-energy function (fission barrier) $\widetilde{V}_{def}(\beta)$ which is obtained by minimizing $V_{def}(s_\alpha)$ with respect to the parameters α_n $(n \neq 1)$ for a fixed value of $\beta = \alpha_1$. In other words, $\widetilde{V}_{def}(\beta)$ represents a cut through the many-dimensional potential-energy surface along the potential-energy-minimum path (fission path) with the energy projected on the β-axis.

A schematic representation of $\widetilde{V}_{def}(\beta)$ for a fissioning actinide nucleus having a neutron number $N_0 \approx 146$ is given in fig. 73. Because of the structure-correction term ΔE, the fission barrier is shown to have two humps at deformations β_2 and β_4, with a deep second minimum occurring in-between at deformation β_3. Such a double-humped curve is of great importance for the understanding of the fission process for nuclei with $N_0 \approx 146$ and will be further discussed later in this subsection.

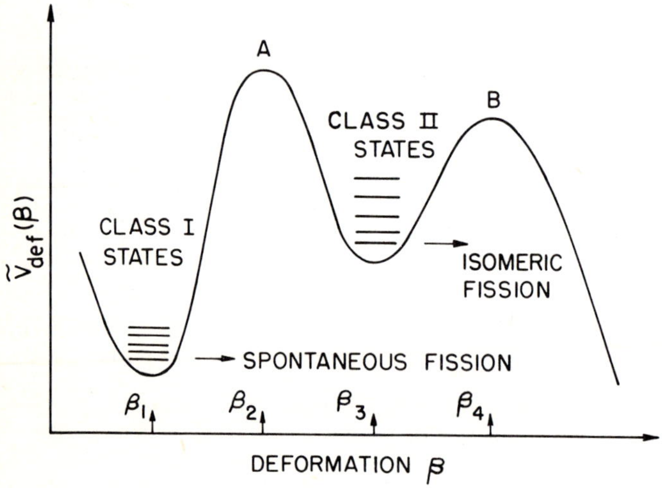

Fig. 73. A schematic plot of $\widetilde{V}_{def}(\beta)$ for an actinide nucleus having a neutron number $N_0 \approx 146$.

We shall now discuss qualitatively the physical reason for the occurrence of a double-humped fission barrier. The existence of a first minimum in $\widetilde{V}_{def}(\beta)$ at a nonzero β value ($\beta = \beta_1$) accounts for the observation that the ground state of a nucleus with $N_0 \approx 146$ is permanently deformed. This minimum occurs as a consequence of the formation of energetically favoured substructures by the nucleons outside of the closed 82-proton and 126-neutron shells, which leads to a prolate deformation of the nucleus. If the deformation is increased beyond β_1, then the deformation energy will increase until β reaches the value β_2. This increase is connected with the fact that some nucleons are excited out of the 82-proton and 126-neutron shells, thus breaking somewhat these favoured configurations. Upon further increasing the value of β, then one expects that the energetically favoured closed-shell substructures with N = 82, N = 50, and Z = 50 will come into play and the formation of these substructures will bring an energy gain and thereby a decrease in the deformation energy.† This decrease tends to be not too abrupt however, because there is a repulsive component in the effective interaction potential between these closed-shell substructures, representing the effect of the Pauli principle. The situation is in fact very similar to that in the $^{16}O + ^{16}O$ case discussed in subsection 16.5a; there we have seen that, because of the completely filled 1p-shells in the two unexcited ^{16}O clusters, the Pauli principle inhibits rather strongly the mutual penetration of these two clusters, in contrast to the situation existing in the case of $^{12}C + ^{12}C$ where the clusters involved have unfilled outer shells (see also the end of section 6.3).

At the second minimum of $\widetilde{V}_{def}(\beta)$ where $\beta = \beta_3$, the formation of the closed-shell substructures with N = 82 and Z = 50 is mainly completed. With a further increase in β, $\widetilde{V}_{def}$ is expected to increase again because these substructures themselves now become somewhat deformed which is energetically unfavoured†† and, in addition, no other strongly correlated substructures are formed. This increase in $\widetilde{V}_{def}$ will continue until β is equal to β_4. Beyond which, the decrease in Coulomb energy starts to become the dominant factor and $\widetilde{V}_{def}$ then decreases monotonically all the way to the scission point.

There are certain indications which support the qualitative explanation given above. These indications are:

(a) For the nucleus ^{236}U (N_0 = 144) which exhibits asymmetric fission, the calculations of *Möller* and *Nilsson* [MÖ 70] and of *Mustafa et al.* [MU 73] show that the nuclear shape is asymmetrically deformed already before the second saddle point. This is consistent with our contention that closed-shell configurations of N = 82 and Z = 50 are essentially formed at the second minimum of the deformation-energy curve. In fact, the calculations mentioned above further indicate that, if only reflection-symmetric deformations are allowed, the second barrier would be higher by more than 2 MeV.

---

† In the shell-model representation, the formation of these energetically favoured closed-shell substructures is related to the change, especially near the Fermi surface, of the density of single-particle levels as a function of β. By correlating the nucleons in these single-particle levels, energetically favoured substructures can be formed, depending upon the value of β.

†† Besides the additional increase in surface energy.

(b) For symmetric-fissioning nuclei ($A_0 \lesssim 214$), where the number of nucleons involved is small enough such that the substructures N = 82, N = 50, and Z = 50 cannot be approximately formed at the same time, the second minimum in $\widetilde{V}_{def}(\beta)$ is much less prominent [MÖ 70].

(c) In the case where $N_0 \gtrsim 164$ and $Z_0 \gtrsim 100$, one expects from the above argument that the intermediate nucleus will remain symmetrically deformed as β increases. This is so, because in both fragments the substructures N = 82 and Z = 50 can be formed with a large probability. Experimentally, it is indeed found that with A_0 increasing toward 264, there is a definite trend toward symmetric fission [BA 71a, JO 71, UN 74]. For example, in the spontaneous fission of ^{257}Fm, the peak-to-valley ratio in the mass distribution is only about 1.5 which is much smaller than the value found in the spontaneous fission of a lighter nucleus (e. g., $P/V \approx 750$ in ^{252}Cf [FL 72]).

In addition, it should be mentioned that the result obtained from the second method of calculating $V_{def}(s_\alpha)$, to be described in subsection 17.4c, shows in a rather direct was that our qualitative explanation is basically correct.

The potential-energy surface $V_{def}(s_\alpha)$ of ^{227}Ac, which exhibits both symmetric and asymmetric fission modes, must be rather complicated. It is expected to have two valleys separated by a potential barrier having a height of the order of a few MeV; in this way, one obtains two fission paths, one for symmetric deformations and the other for asymmetric deformations. In addition, it is important that these fission paths must start to divide before the second saddle point. All these features must be present; otherwise, the triple-humped mass-distribution curve plotted in fig. 71 would not show up.†

We now discuss briefly the consequences which follow from a deformation-energy curve of the type shown schematically in fig. 73. Because of the existence of a second minimum, one expects to find an excited state (class II state) at a few-MeV excitation, which has a much larger intrinsic deformation than the ground state (class I state). As usual, both the ground state and this excited state should give rise to rotational bands, commonly referred to as the ground-state band and the isomeric-state band, respectively. The isomeric-state band is expected to be characterized by a higher moment of inertia, since the states in this band are described by wave functions which have large magnitudes mainly in the second well. Experimentally, the existence of these class II states has indeed been verified by *Specht et al.* [SP 72] who have observed a well-formed rotational band with an unusually large moment of inertia in the nucleus ^{240}Pu.

It should be mentioned that similar states exist also in lighter nuclei. For instance, the first excited 0^+ state of ^{16}O and its associated rotational states are strongly deformed states, in contrast to the ground state of ^{16}O. Another example is the rotational band in ^{24}Mg, with its band head at 11.75 MeV excitation [MO 56]. Also, the low-lying $K = 0, 1^-$ levels in heavy even-even nuclei with $N_0 \geqslant 132$ [ST 54] are other examples of this type of state. The lightest nucleus in which such a 1^- state occurs is ^{218}Rn which has a neutron number equal to 132, indicating that these states are mainly produced by the formation of closed-shell substructures with N = 50 and 82, and thus are strongly deformed [FA 61].

† To adequately describe these two fission paths, one may need to use a wave function of the form given by eq. (17.15).

The existence of strongly deformed class II states in fissioning nuclei affords a very natural explanation of the phenomenon of isomeric fission. Isomeric fission was discovered in 1962 [PO 62] in an experiment where ^{238}U was bombarded by ^{22}Ne ions. In the subsequent reactions, the nucleus ^{242}Am was produced in an isomeric state with an excitation energy of 2.9 MeV, which showed a fission activity with a half-life of 14 msec, substantially shorter than the half-life of about 10^{12} years† for the spontaneous fission of ^{242}Am. Similar fission activities were later also observed in many other nuclei [VA 73]. In all instances, the half-life for isomeric fission was found to be much shorter than that for spontaneous fission in the same nucleus.

The phenomenon of isomeric fission is now explained by assuming that, during the reaction process, class II states can be weakly populated through certain decay modes of the hot intermediate nucleus which is excited above at least the first barrier (barrier A in fig. 73) of $\widetilde{V}_{def}(\beta)$. With this simple assumption, one can then qualitatively explain the following observation:

(a) The spontaneous fission half-life of the highly deformed class II state in the second well is always smaller than that of the ground state. This is so, because the isomeric state has a rather high excitation energy and because fission from this state occurs by tunneling through the second barrier (barrier B in fig. 73) only, instead of through the whole barrier as in the case of the ground state.

(b) γ-decay from the isomeric state (in the second well) to states in the first well is strongly inhibited by the very small spatial overlapping between the wave functions describing these two kinds of many-nucleon states.

It should also be mentioned that all isomeric fissions are expected to yield asymmetric mass distributions, because they occur only in heavy nuclei with $N_0 \approx 146$ where closed-shell substructures of $N = 82$, $N = 50$, and $Z = 50$ can be substantially formed.

17.4c. Calculation of the Deformation Energy—Cluster Prescription

In this subsection we discuss the second method, to be called the cluster prescription, which was proposed recently by *Schultheis et al.* [SC 70, SC 71a, SC 71b, SC 73a, SC 74a] to approximately calculate the potential-energy surface. This method has so far not been as widely employed in explicit calculations as the Strutinsky method described in the preceding subsection, but has lately been used to explain a number of interesting phenomena in fission [GÖ 75, GÖ 75a]. As compared with the Strutinsky method, the main advantage of this method is that it is computationally simpler and contains the useful feature that substructure effects are included in a rather direct manner.

In this method, one assumes again that the main part of the total energy $E(s_\alpha)$ is given by the liquid-drop energy; that is, one writes

$$E(s_\alpha) = E_{LDM}(s_\alpha) + \Delta E(s_\alpha), \tag{17.22}$$

† This is the fission half-life for ^{242}Am in its 48.6 keV long-lived excited state [CA 67a].

where $\Delta E(s_\alpha)$ is, as in eq. (17.17), a structure-correction term. For the calculation of this correction term, one makes use of an important finding, discussed in section 6.3, concerning the behaviour of cluster correlations when two large clusters penetrate each other. There it was shown that the motion of the nucleons in the higher shells of each cluster (i.e., the nucleons which are more or less localized in the surface region of the compound nucleus) is not influenced in first approximation by the presence of the nucleons in the other cluster. This has the consequence that the binding energies of the nucleons in these higher shells of the clusters should be approximately the same as those of the corresponding free particles. Therefore, by treating the deformed intermediate nucleus as described by a linear superposition of two-cluster configurations, one can approximately relate $\Delta E(s_\alpha)$ to the structure or shell corrections of the relevant spherical clusters and these latter corrections can be obtained by using the above-mentioned property of the nucleons in the higher shells of the individual clusters.

To see in more detail how one determines $\Delta E(s_\alpha)$, we write the wave function $\phi(\tilde{r}_1, \ldots, \tilde{r}_{A_0}; s_\alpha)$ describing the behaviour of the intermediate nucleus, specified by a set of deformation parameters s_α, as

$$\phi(\tilde{r}_1, \ldots, \tilde{r}_{A_0}; s_\alpha) = \sum_h \sum_{ijk} b^h_{ijk}(s_\alpha)\, \psi^h_{ijk}(\tilde{r}_1, \ldots, \tilde{r}_{A_0}), \tag{17.23}$$

where $\psi^h_{ijk}(\tilde{r}_1, \ldots, \tilde{r}_{A_0})$ is given by eq. (17.2). To utilize the finding mentioned above, we separate $\phi(\tilde{r}_1, \ldots, \tilde{r}_{A_0}; s_\alpha)$ or simply $\phi(s_\alpha)$ into two parts, i.e.,

$$\phi(s_\alpha) = \phi_I(s_\alpha) + \phi_{II}(s_\alpha), \tag{17.24}$$

with $\phi_I(s_\alpha)$ and $\phi_{II}(s_\alpha)$ describing the interior and the surface region of the intermediate nucleus, respectively. It is a rather simple matter to decide the type of cluster terms which should be included in ϕ_I and ϕ_{II}. With a two-cluster function given by eq. (17.2), one notes that the average separation distance between the clusters becomes large when the relative motion is highly excited. Thus, ϕ_{II} should contain those functions which are characterized by unexcited clusters ($i = 0, j = 0$) and consequently relative-motion functions of maximum allowable excitation ($k = k_{max}$ or K). Therefore, for $\phi_{II}(s_\alpha)$ we write

$$\phi_{II}(s_\alpha) = \sum_h b^h_{00K}(s_\alpha)\, \psi^h_{00K}. \tag{17.25}$$

The other part ϕ_I of the total wave function is associated with relative-motion functions of low excitation. This means that, in contrast to the clusters in ϕ_{II}, the clusters contained in ϕ_I are in general highly excited and therefore show hardly any shell-structure effect.

With $\phi(s_\alpha)$ separated into $\phi_I(s_\alpha)$ and $\phi_{II}(s_\alpha)$, the total energy $E(s_\alpha)$ becomes

$$\begin{aligned} E(s_\alpha) &= \langle \phi(s_\alpha) | H | \phi(s_\alpha) \rangle \\ &= \langle \phi_I | H | \phi_I \rangle + \langle \phi_I | H | \phi_{II} \rangle + \langle \phi_{II} | H | \phi_I \rangle + \langle \phi_{II} | H | \phi_{II} \rangle. \end{aligned} \tag{17.26}$$

Based on the above discussion, one easily sees that the first term $\langle \phi_I | H | \phi_I \rangle$ contributes only to the liquid-drop energy E_{LDM}. The mixed terms $\langle \phi_I | H | \phi_{II} \rangle$ and $\langle \phi_{II} | H | \phi_I \rangle$ are expected to be small and can similarly be included in the LDM, because ϕ_{I} and ϕ_{II} are many-nucleon wave functions which have large magnitudes in different parts of the configuration space (see the discussion given in subsection 17.4a) and therefore shell effects are essentially averaged out. Therefore, the structure correction $\Delta E(s_\alpha)$ of the intermediate nucleus is contained only in the term $\langle \phi_{II} | H | \phi_{II} \rangle$.

The way to compute $\langle \phi_{II} | H | \phi_{II} \rangle$ was discussed qualitatively in section 6.3. There it was shown that $\langle \phi_{II} | H | \phi_{II} \rangle$ can be divided into three kinds of terms: one from the higher single-nucleon states in the clusters, one from the lower single-nucleon states in the clusters, and a third containing small interaction terms between the nucleons in the higher and lower states. The second and third kinds of terms contribute only to the liquid-drop energy. Only the first kind of terms involving higher single-nucleon states contributes to the structure-correction term $\Delta E(s_\alpha)$. As was discussed there, these higher single-nucleon states are almost unaffected by the procedure of antisymmetrization between the clusters; consequently, the structure corrections of the clusters can be calculated as if they were free particles. Furthermore, since the mixed terms arising from the higher single-nucleon states in different clusters are small and show essentially no shell effects, we can therefore write the total shell or structure correction $\Delta E(s_\alpha)$ as

$$\Delta E(s_\alpha) = \sum_h |b^h_{00K}(s_\alpha)|^2 \, \epsilon^h_{Cl}, \tag{17.27}$$

where

$$\epsilon^h_{Cl} = \delta E(N_h, Z_h) + \delta E(N_l, Z_l), \tag{17.28}$$

with $\delta E(N, Z)$ being the shell or structure correction for an unexcited cluster with N neutrons and Z protons.

For $\delta E(N, Z)$ in eq. (17.28) one can use, for instance, the quantity $S(N, Z)$ of the *Myers-Swiatecki* mass formula [MY 66] or a similar quantity from other appropriate mass formulae such as that of *Seeger* and *Howard* [SE 75]. This yields then

$$\Delta E(s_\alpha) = \sum_h |b^h_{00K}(s_\alpha)|^2 \, [S(A_h) + S(A_l)]. \tag{17.29}$$

In the above equation, one notes that a single variable is used as the argument of the function S. This is possible, because the neutron and proton numbers are further related through eq. (17.4) which expresses the fact that no charge polarization in the intermediate nucleus should exist.

In actuality, one does not compute $\Delta E(s_\alpha)$ by directly using eq. (17.29), because to calculate $b^h_{00K}(s_\alpha)$ microscopically would obviously be a very difficult task. Thus, what *Schultheis et al.* did was to propose a prescription [SC 74a] which is simple to use but does contain the essential features of the above discussion. This prescription is as follows:

a) The rotationally symmetric nuclear surface, specified by a set of parameters s_α, is approximated as closely as possible by a large number N of spherical calottes with radii R_i.

(b) The contribution of a particular spherical cluster to the total shell or structure correction $\Delta E(s_\alpha)$ is plausibly assumed as proportional to the solid angle subtended by the corresponding calotte. That is, one writes

$$\Delta E(s_\alpha) = c \frac{\sum_{i=1}^{N} \frac{\Delta\sigma_i}{4\pi R_i^2} S(A_i)}{\sum_{i=1}^{N} \frac{\Delta\sigma_i}{4\pi R_i^2}}, \tag{17.30}$$

where c is a proportionality constant, $\Delta\sigma_i$ is the surface area of the ith calotte, and A_i is the nucleon number of the spherical cluster which is related to R_i by the well-known equation $R_i = r_0 A_i^{1/3}$.† For obvious reasons, the summation (or integration if $N \to \infty$) should cover only such surface elements for which R_i is positive (i.e., surface elements which concave toward the axis) and for which R_i is smaller than the radius of the nucleus at sphericity. This means that one makes the reasonable assumption of considering those surface elements not included in the summation as formed by highly excited clusters which have little shell correction.

(c) The total energy $E(s_\alpha)$ is minimized with respect to s_α. The resultant optimum value is then compared with the ground-state energy of the nucleus under consideration; in this way, one obtains the value of the proportionality constant c. It should be noted that the constant c is introduced mainly to compensate for the lack of complete theoretical justification for this prescription. Its value should come out to be relatively close to 1, because for a spherical nucleus the value of $\Delta E(s_\alpha)$ should obviously be equal to $S(A_0)$.

As an example of applying this prescription, we show the result of a simplified calculation of $\Delta E(s_\alpha)$ for the nucleus ^{236}U [SC 74a]. In this example, the surface of the nucleus is assumed to be formed by two spheres which overlap in a way according to the condition of volume conservation. The deformation parameters are the elongation parameter l and the mass-asymmetry parameter $\lambda = (R_1/R_2)^3$ (see fig. 74). Because of the choice of simple surface shapes, the correction term ΔE can be very easily calculated. An application of the procedure (c) then yields $c = 0.7$ which is reasonably close to 1. The result is shown in fig. 75, where the solid curve is obtained by imposing the restriction $\lambda = 1$ and the dashed curve is obtained by also allowing for mass asymmetry. From this figure, one notes the following interesting features:

(a) The ΔE curve has two minima between the spherical shape and the shape corresponding to touching fragments.

(b) The ground-state minimum is associated with the presence of the doubly magic $A = 208$ cluster ($Z = 82$, $N = 126$).

(c) The second minimum is due to the doubly magic $A = 132$ cluster ($Z = 50$, $N = 82$).

(d) At the barriers in the $\lambda = 1$ case, the doubly magic clusters are broken up.

† The quantity $S(A_i)$ in eq. (17.30) is understood to contain also a contribution from the complementary cluster with nucleon number equal to $(A_0 - A_i)$.

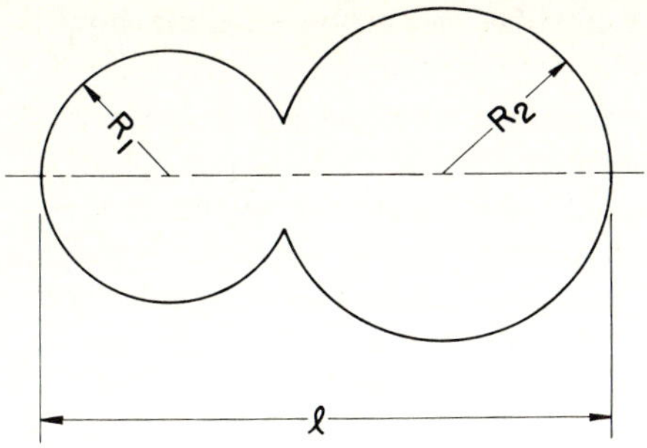

Fig. 74. Shape of nuclear surface in a two-sphere model.

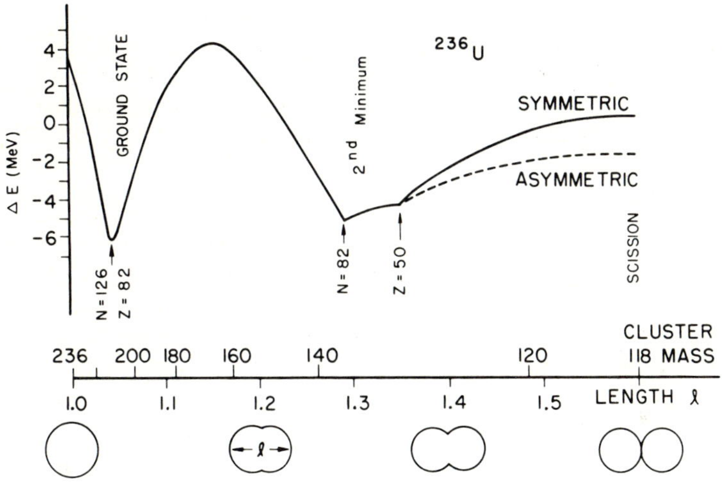

Fig. 75. Structure correction ΔE as a function of the elongation l for the nucleus ^{236}U. The solid curve is obtained with the restriction $\lambda = 1$. The N and Z values marked on the curve identify the kinds of spherical clusters which are present at the minima. The shape of the nucleus and the mass number of each cluster are also shown in the $\lambda = 1$ case. The dashed curve is obtained by using optimum values of λ which are found by minimizing the total energy E with respect of the mass-asymmetry parameter. (From ref. [SC 74].)

(e) On the path of minimum energy (fission path), the deformation is symmetric up to the second minimum.

(f) The second barrier is lowered by the inclusion of mass asymmetry.

(g) Between the second minimum and the scission point, the path of minimum energy corresponds to those asymmetric deformations which leave the magic cluster (A = 132) largely unbroken.

A more refined calculation with more shape parameters, which can allow, for instance, the presence of deformed clusters at scission, has also been performed [SC 75a]. The result shows that the features obtained from the simplified calculation described above remain essentially unchanged.

The Strutinsky prescription and the cluster prescription yield generally similar results for the potential-energy surface. This is mainly a consequence of the Pauli principle of course. The same structure effects are merely considered in different representations; the single-particle representation for the Strutinsky prescription and the cluster representation for the cluster prescription. We should point out, however, that the approach with the cluster representation does have the advantage of allowing a more physical interpretation and thereby providing more detailed information about the fission mechanism by tracing back complicated features of the deformation energy to the formation of substructures.

The cluster prescription is much simpler to use than the Strutinsky prescription. However, in this prescription the procedure to approximately calculate the coefficients $|b^h_{00K}(s_\alpha)|^2$ in eq. (17.29) is a rather crude one which should work best in the later stage of fission where the deformation is relatively large and the formation of clusters is evident. Therefore, another useful way to calculate the whole potential-energy surface is perhaps to adopt a hybrid procedure in which one employs the Strutinsky prescription at smaller deformations up to about the second minimum and then the cluster prescription at larger deformations.

17.4d. Discussion

With V_{def} calculated by using either of the prescriptions described in the above subsections, one can then use eq. (17.14), or a similar equation for the more general case involving a set of deformation coordinates s_α, to discuss at least qualitatively the various properties of fission from a dynamical point of view. But before we do this, we wish to make some remarks about eq. (17.14). In the derivation of this equation, we have used for $\phi(\alpha)$ wave functions which are neither translationally invariant nor eigenstates of the total angular momentum and the parity. Because of the large number of nucleons contained in the intermediate nucleus, the lack of translational invariance is not expected to result in any serious consequence. As for the deficiency in the wave function concerning the requirement of rotational and reflectional invariance, one can remedy it by employing angular momentum and parity eigenfunctions which are obtained from $\phi(\alpha)$ by adopting the projection technique described in chapter 14. If this is done, then one can examine with eq. (17.14) not only the gross features of the fission process but also the properties of the class II states mentioned in subsection 17.4b.

We start by considering the peak-to-valley (P/V) ratio in the fragment mass-yield distribution. For this is is necessary to know the inertial parameter $M(\alpha, E)$ near the second saddle point, in order to estimate the probability of barrier penetration. As was discussed in subsection 17.4a, this parameter can be taken as equal to $M_0/4$ in a crude calculation.† Now, if one makes further the rough approximation that symmetric and

† Another way to estimate $M(\alpha, E)$ near the second saddle point is by using experimental information on excitation-energy values of class II collective states.

asymmetric two-cluster configurations appear with equal probability, then one can calculate the ratio of symmetric and asymmetric fission rates by solving eq. (17.14) using, for instance, the WKB method. The result of such a calculation [FA 64] in the case of thermal-neutron induced fission of ^{235}U shows that, in order to explain the experimentally determined P/V ratio of about 600 [VA 73], the asymmetric fission barrier must be lower than the symmetric fission barrier by about 6 MeV or somewhat less. This is in fact qualitatively consistent with the potential-energy calculation of *Mustafa et al.* for ^{236}U [MU 73]. In that calculation it was found that the asymmetric barrier is favoured by about 2.3 MeV relative to the symmetric barrier and near scission the most probable asymmetric shape is preferred by about 7 MeV in potential energy relative to the symmetric shape.

From the above discussion, one might naively argue that if the excitation energy in ^{236}U is increased by about 2 MeV beyond the threshold for asymmetric fission, then the dip in the mass-distribution curve should approximately vanish because symmetric fission should now take place without tunneling. Experimentally, this is certainly not the case. Even for much higher excitation energies, a partially filled-up mass valley remains [VA 73]. The reason for this relatively slow decrease in the P/V ratio is that with increasing excitation energy the intermediate nucleus can fission (symmetrically and asymmetrically) in more and more configurations where the internal excitation energy is not completely transformed into the deformation energy and the collective kinetic energy in the fission degree of freedom. If one takes this effect into account by a statistical consideration, then one finds that, in agreement with experiment, the P/V ratio does decrease much slower with increasing excitation energy than what would follow from the naive argument given above.†

Next, we consider briefly the fission of the nucleus ^{227}Ac and the Fm-isotopes. In the case of low-energy ^{227}Ac fission which proceeds in two distinct fission valleys as was discussed in subsection 17.4b, a rough calculation shows that the function $f(s_\alpha)$ has oscillation amplitudes in directions perpendicular to the fission paths which are small, when measured in units of energy, compared to the corresponding valley depths. This must be so of course, in order that these oscillations do not practically overcome the potential barrier separating these two valleys. Also, because the separation of the fission valleys starts already at a relatively small deformation near the second minimum, the decision as to what the fission mode will be is made at a rather early stage. Therefore, together with the reason just given above, this explains why both the symmetric and the asymmetric modes are observed even when the intermediate nucleus ^{227}Ac is relatively highly excited (about 10–15 MeV above the symmetric and asymmetric fission barriers).

The symmetric and asymmetric modes in ^{227}Ac induced fission show some interesting differences. The mean excitation energy in symmetric fission products was found to be higher than that in asymmetric fission products, and the average total fragment kinetic energy was observed to be lower in symmetric fission than in asymmetric fission [KO 68].

---

† Certainly, the total fission probability increases rapidly with increasing excitation energy. This is seen, for instance, from the ratio of fission events to events of neutron evaporation as a function of the excitation energy [KO 74a, VA 73].

Both of these experimental findings are of course intimately related and can be explained by using the fact that, in comparison with the symmetric fission fragments, the asymmetric fission fragments are less deformable. The consequence of this is that, at scission, asymmetric fission fragments have smaller prolate deformation and are, therefore, closer to each other. This results in larger Coulomb repulsion and, thereby, larger average total kinetic energy for the asymmetric fragments.

We discuss now the Fm-isotopes which have $Z_0 = 100$. A study of these isotopes is particularly significant, because it gives strong support to the importance of substructure effects in symmetric and asymmetric fission. In the case of asymmetric fission, the heavy cluster contains closed-shell substructures with $N = 82$ and $Z = 50$, just as in lighter nuclei with $N_0 \geqslant 132$. For the symmetric fission of Fm-isotopes with $A_0 < 264$, the closed-shell substructure of $N = 82$ must of course be broken up. But now both clusters contain an energetically favoured $Z = 50$ substructure, which cannot happen in asymmetric fission because of the restriction expressed by eq. (17.4). Indeed, a recent calculation of the shell-energy [GÖ 75a] has shown that symmetric and asymmetric modes should have comparable weight in the low-energy fission of Fm-isotopes with $A_0 < 264$. Experimentally, one does find that the P/V ratios for the spontaneous fission of these Fm-isotopes are rather small compared with those of neighbouring lighter nuclei [UN 74].

That the symmetric fission mode in the Fm-isotopes is associated with closed-shell substructure formation is further supported by the experimental observation that, in the spontaneous fission of ^{257}Fm, the average total fragment kinetic energy is larger for symmetric than for asymmetric fission [BA 71a]. This is exactly opposite to what was found in lighter nuclei with $Z \leqslant 98$ (e.g., ^{254}Cf), and can be explained in the following manner. In the symmetric fission of Fm isotopes, closed-shell clusters with $Z = 50$ exist at least with a large probability in both fission fragments. Thus, these fragments are rather stiff against deviation from sphericity; consequently, the mean separation distance of the two fragments at the scission point is relatively small and, therefore, their mutual Coulomb energy, later transformed into the kinetic energy of the fragments, is relatively large. In the asymmetric fission of these isotopes, the situation is quite different. There the light fragment contains no energetically favoured closed-shell substructures and thus can have an appreciable prolate deformation at scission. The result of this is that the mutual Coulomb energy of the nascent fragments is smaller for asymmetric fission than that in the symmetric case, just as was observed in the experimental study of the nucleus ^{257}Fm.

By using similar arguments, one can easily predict that, for a nucleus with $N_0 \gtrsim 164$ and $Z_0 \gtrsim 100$, the symmetric fission mode should predominate and the pre-neutron mass-yield distribution should have a single peak centered at $A_0/2$ [FA 64]. Experimentally, this is yet to be seen.

In closing this chapter, we should mention that our main concern is to show how the fission problem can be described with the unified nuclear theory presented in this monograph. Clearly, there are many topics which have not even been touched upon. In fact, the only subject which has been dealt with at length is the effect of cluster formation. As our discussion shows, we do have at the present moment a qualitative understanding of this subject, but much work has yet to be done in order to acquire detailed quantitative knowledge concerning the importance of substructure effects in nuclear fission.

18. Conclusion

For every dynamical many-particle theory, one of the main requirements is that it must correctly describe the asymptotic behaviour of bound states and reaction products. This can be achieved in two different ways. The first way, which was developed for instance by *Feshbach* [FE 58, FE 62], is to employ a complete set of orthogonal wave functions in the total Hilbert space belonging to the many-particle system under consideration and derive from the many-particle Hamiltonian, through the use of projection operators, effective Hamiltonians and transition Hamiltonians which belong only to subspaces of the total Hilbert space but which contain the correct asymptotic boundary conditions for the various reaction products. With these Hamiltonians, one can then investigate the general properties of many-particle systems, such as the Breit-Wigner resonance formula, the direct-reaction theory, and so on. In this approach, a major drawback is that the projection operators, which one needs to formulate the various effective Hamiltonians, have generally such a complicated structure that it is usually very difficult to see a direct connection between the general formulae and the equations with which one obtains numerical results for comparison with experiment. Especially, this is true when one considers reactions involving composite particles.

In the second way, one proceeds as follows. Instead of manipulating the Hamiltonian of the total system to obtain correct boundary conditions, one introduces these boundary conditions directly into the ansatz for the total wave function ψ. To accomplish this, one separates ψ into a sum of usually non-orthogonal terms which are constructed to contain the desired boundary conditions for the problem under consideration. In other words, the various Hilbert subspaces into which the total Hilbert space is divided are now defined directly through these non-orthogonal terms which make up the total wave function ψ, in contrast to the first way where the subspaces are defined by the introduction of effective Hamiltonians. Thus, in this way, the asymptotic behaviour in the incoming channel and the various outgoing channels is described by the relative motions of the reaction partners in regions where the only interaction between them is the Coulomb interaction. As for the compound-nucleus region, it can be properly described by a superposition of bound-state-type functions, such as translationally invariant shell-model and Hill-Wheeler type wave functions, cluster-type wave functions, or a combination of all. We should mention that, by defining ψ in this way, one is committed to work in a truncated Hilbert space whose extension is determined by the requirement that all the open channels at a given total energy be properly taken into consideration. This is not a restriction however, because as long as the bound-state-type wave functions form in the compound region a complete set, then the exact solutions can certainly be obtained by solving the problem in this truncated region of the Hilbert space.

The main advantage of the second way is that the incoming and the outgoing channels are treated in a completely symmetrical manner. The coupled equations, which determine the linear variational functions and linear variational amplitudes contained in the ansatz for ψ, are obtained by writing the many-particle Schrödinger equation in the form of a projection equation. As has been shown in chapter 2, the solutions of these coupled equations are mutually orthogonal, if all degeneracies are removed. In addition, there exists the desirable feature that this orthogonality property is preserved and the Hamiltonian can be properly diagonalized, even when on physical grounds one truncates the Hilbert space further (e. g., by limiting the number of bound-state-type wave functions in the compound region) to obtain approximate solutions for the system. The only important point to note is that one must always use in ψ linear variational functions and amplitudes.

In this monograph, we have used the second approach to formulate a microscopic theory which treats nuclear bound-state, scattering, and reaction problems from a unified viewpoint. As has been described in detail, this theory is very flexible and can be used not only to discuss the general properties of many-particle systems but also to carry out explicit calculations. Also, it is quite general, in the sense that one can use this theory to study cases where the particles involved in the incoming and outgoing channels are arbitrary composite particles. In addition, we should point out that a close connection always exists between the formulae which describe the general features of the system and the equations which one uses for numerical calculations. Because of the properties mentioned in the preceding paragraph concerning the orthogonality of the wave functions and the diagonalization of the Hamiltonian, this close connection remains even when one performs only approximate calculations for the system.

The influence of the Pauli principle, expressed by the antisymmetrization of the wave function, has been examined in detail and shown to play a decisive role in all our considerations. One of its important consequences is that it reduces greatly the differences between apparently different structures when the nucleons are close to one another. From this it follows that, at relatively low excitation energies, the number of many-particle configurations which one needs for approximate calculations can usually be made so small that such calculations become quantitatively feasible. Also, this is the reason why seemingly conflicting nuclear models, such as collective and single-particle models, can be reconciled. Furthermore, it is responsible for the fact that the lifetimes of composite clusters in nuclear matter are sufficiently long, such that optical-model considerations can frequently be used to study scattering and reaction processes involving such clusters as incident particles.

Also because of the Pauli principle, there arises the interesting fact that even the same nuclear state may appear to exhibit different structures depending on the type of measurement which is made. For instance, in the ^{6}Li (p, ^{3}He) α reaction which has been carefully studied, we have shown that, at high bombarding energies, the ground state of ^{6}Li behaves like having a d + α cluster structure when the differential reaction cross section in the forward angular region is examined and a t + ^{3}He cluster structure when the differential reaction cross section in the backward angular region is considered. Indeed, this type of behaviour has been seen in many other examples and seems to be a general feature of nuclear states, which comes from the fact that nucleons in the interior of a nucleus are rather closely packed and, consequently, the effect of the Pauli principle is enhanced. Therefore, in a picturesque way one might say that nuclear states have frequently a chameleon-like character and it depends on the kind of experiment what structures these states will exhibit.

Another consequence of the Pauli principle, which is closely associated with the chameleon-like character mentioned above, is that reaction cross sections obtained from calculations employing totally antisymmetrized wave functions are generally quite different from those obtained by making the simplifying assumption that nucleons were distinguishable. An example for this is the photodisintegration of ^{6}Li into a triton and a ^{3}He particle discussed in chapter 15. There it was shown that in the case where one describes the ^{6}Li ground state by a d + α cluster structure, the procedure of antisymmetrization increases the absorption cross section by a factor of about 6.

On the other hand, if the clusters which participate in nuclear reactions do not strongly penetrate into each other, then the mutual antisymmetrization between the nucleons in different clusters becomes relatively unimportant and the clusters will behave very much like the corresponding free particles.

As a result of the above-mentioned influence of the Pauli principle, it is possible to describe the behaviour of the system in the compound-nucleus region and in the channel regions by using wave functions in completely different representations without having to be concerned about the problem of smooth transition of the nuclear configurations in the intervening surface region. For instance, one could use a superposition of translationally invariant shell-model and Hill-Wheeler type functions for the compound region and cluster wave functions for the channel regions. Especially in reactions involving heavier nuclei, such a mixed representation is usually convenient because one can then make use of extensively developed techniques in shell-model calculations and projected Hartree-Fock calculations.

In most of the numerical examples chosen for illustration, two-nucleon potentials of relatively simple form have been employed. In particular, the repulsive-core part of the nucleon-nucleon potential has usually not been explicitly considered and the saturation property has been taken into account only in a crude manner by mostly fixing the radii of the clusters involved. We should emphasize, however, that although the use of more refined forces will undoubtedly influence the numerical results to a certain extent, the general features of nuclear processes will be essentially unchanged. This must be so, since these general features are determined mainly by the fact that nucleons are fermions and nuclear forces are short-ranged, saturating, and on the average attractive.

Within the framework of the microscopic nuclear theory presented here, we have shown how the various subjects in nuclear physics, ranging from bound-state and scattering problems in light systems to heavy-ion reactions and fission processes, can be considered from a unified point of view. For all these subjects, we have attempted to discuss as clearly as possible the main underlying ideas. However, some of the considerations are necessarily brief and essentially qualitative, and much work has yet to be done in order to fully utilize the quantitative aspects of this theory.

Appendix A

Cluster Hamiltonians and Jacobi Coordinates

In this appendix, we describe first how the internal eigenstates of a cluster may be obtained for an internal Hamiltonian derived under the oscillator assumption. For definiteness, we consider an α cluster whose internal Hamiltonian is given by (see eq. (3.12))

$$H_\alpha = \sum_{j=1}^{3} \sum_{k=1}^{3} \left[\left(\delta_{jk} - \frac{1}{4} \right) \frac{1}{2M} \bar{\mathbf{p}}_j \cdot \bar{\mathbf{p}}_k + (\delta_{jk} + 1) \frac{1}{2} M \omega^2 \, \bar{\mathbf{r}}_j \cdot \bar{\mathbf{r}}_k \right], \qquad \text{(A1)}$$

where $\bar{\mathbf{p}}_j = -i\hbar \nabla_{\bar{\mathbf{r}}_j}$ and $\bar{\mathbf{r}}_j$ (j = 1, 2, 3) are cluster internal coordinates defined by eq. (3.8). To eliminate the cross terms in eq. (A1), we introduce the following Jacobi coordinates (see eq. (3.43)):

$$\begin{aligned}
\hat{\mathbf{r}}_1 &= \mathbf{r}_1 - \mathbf{r}_2 = \bar{\mathbf{r}}_1 - \bar{\mathbf{r}}_2, \\
\hat{\mathbf{r}}_2 &= \frac{1}{2}(\mathbf{r}_1 + \mathbf{r}_2) - \mathbf{r}_3 = \frac{1}{2}(\bar{\mathbf{r}}_1 + \bar{\mathbf{r}}_2) - \bar{\mathbf{r}}_3, \\
\hat{\mathbf{r}}_3 &= \frac{1}{3}(\mathbf{r}_1 + \mathbf{r}_2 + \mathbf{r}_3) - \mathbf{r}_4 = \frac{4}{3}(\bar{\mathbf{r}}_1 + \bar{\mathbf{r}}_2 + \bar{\mathbf{r}}_3), \\
\hat{\mathbf{r}}_4 &= \frac{1}{4}(\mathbf{r}_1 + \mathbf{r}_2 + \mathbf{r}_3 + \mathbf{r}_4) = \mathbf{R}_\alpha .
\end{aligned} \qquad \text{(A2)}$$

In terms of these Jacobi coordinates, the single-particle coordinates are given by

$$\begin{aligned}
\mathbf{r}_1 &= \frac{1}{2}\hat{\mathbf{r}}_1 + \frac{1}{3}\hat{\mathbf{r}}_2 + \frac{1}{4}\hat{\mathbf{r}}_3 + \hat{\mathbf{r}}_4 = \bar{\mathbf{r}}_1 + \mathbf{R}_\alpha, \\
\mathbf{r}_2 &= -\frac{1}{2}\hat{\mathbf{r}}_1 + \frac{1}{3}\hat{\mathbf{r}}_2 + \frac{1}{4}\hat{\mathbf{r}}_3 + \hat{\mathbf{r}}_4 = \bar{\mathbf{r}}_2 + \mathbf{R}_\alpha, \\
\mathbf{r}_3 &= -\frac{2}{3}\hat{\mathbf{r}}_2 + \frac{1}{4}\hat{\mathbf{r}}_3 + \hat{\mathbf{r}}_4 = \bar{\mathbf{r}}_3 + \mathbf{R}_\alpha, \\
\mathbf{r}_4 &= -\frac{3}{4}\hat{\mathbf{r}}_3 + \hat{\mathbf{r}}_4 = -(\bar{\mathbf{r}}_1 + \bar{\mathbf{r}}_2 + \bar{\mathbf{r}}_3) + \mathbf{R}_\alpha .
\end{aligned} \qquad \text{(A3)}$$

In practice, it is often more convenient to transform directly, by using eqs. (A2) and (A3), the original oscillator Hamiltonian expressed in single-particle coordinates as in eq. (3.5); thus, for completeness, we have included the transformation to the single-particle coordinates $\mathbf{r}_i$, as well as to the cluster coordinates $\bar{\mathbf{r}}_i$ and $\mathbf{R}_\alpha$.

The transformations between the various canonical momenta can also be easily derived. For example, by using eq. (A2), we find for the x-component of $\bar{\mathbf{p}}_1$ the following expression:

$$\bar{p}_{1x} = \frac{\hbar}{i} \frac{\partial}{\partial \bar{r}_{1x}} = \sum_{j=1}^{4} \frac{\partial \hat{r}_{jx}}{\partial \bar{r}_{1x}} \left(\frac{\hbar}{i} \frac{\partial}{\partial \hat{r}_{jx}} \right) = \sum_{j=1}^{4} \frac{\partial \hat{r}_{jx}}{\partial \bar{r}_{1x}} \hat{p}_{jx} = \hat{p}_{1x} + \frac{1}{2} \hat{p}_{2x} + \frac{4}{3} \hat{p}_{3x} . \qquad \text{(A4)}$$

Proceeding in a similar way, we then obtain

$$\bar{\mathbf{p}}_1 = \hat{\mathbf{p}}_1 + \frac{1}{2}\hat{\mathbf{p}}_2 + \frac{4}{3}\hat{\mathbf{p}}_3 = \mathbf{p}_1 - \mathbf{p}_4,$$
$$\bar{\mathbf{p}}_2 = -\hat{\mathbf{p}}_1 + \frac{1}{2}\hat{\mathbf{p}}_2 + \frac{4}{3}\hat{\mathbf{p}}_3 = \mathbf{p}_2 - \mathbf{p}_4,$$
$$\bar{\mathbf{p}}_3 = -\hat{\mathbf{p}}_2 + \frac{4}{3}\hat{\mathbf{p}}_3 = \mathbf{p}_3 - \mathbf{p}_4,$$
$$\mathbf{P}_\alpha = \hat{\mathbf{p}}_4 = \mathbf{p}_1 + \mathbf{p}_2 + \mathbf{p}_3 + \mathbf{p}_4, \tag{A5}$$

where $\mathbf{P}_\alpha$ is the canonical momentum to the center-of-mass coordinate $\mathbf{R}_\alpha$.

Substituting eqs. (A3) and (A5) into eq. (A1), we obtain the transformed internal Hamiltonian

$$H_\alpha = \frac{1}{2M}\left(\frac{\hat{\mathbf{p}}_1^2}{1/2} + \frac{\hat{\mathbf{p}}_2^2}{2/3} + \frac{\hat{\mathbf{p}}_3^2}{3/4}\right) + \frac{1}{2}M\omega^2\left(\frac{1}{2}\hat{\mathbf{r}}_1^2 + \frac{2}{3}\hat{\mathbf{r}}_2^2 + \frac{3}{4}\hat{\mathbf{r}}_3^2\right), \tag{A6}$$

which separates now into a sum of oscillator terms with identical frequency ω and varying mass $Mj/(j+1)$. The solution is thus a product of oscillator functions. For an unexcited α cluster, this solution is

$$\phi_0 = \exp\left[-\frac{a}{2}\left(\frac{1}{2}\hat{\mathbf{r}}_1^2 + \frac{2}{3}\hat{\mathbf{r}}_2^2 + \frac{3}{4}\hat{\mathbf{r}}_3^2\right)\right] = \exp\left(-\frac{a}{2}\sum_{i=1}^{4}\bar{\mathbf{r}}_i^2\right), \tag{A7}$$

where $\bar{\mathbf{r}}_4 = -(\bar{\mathbf{r}}_1 + \bar{\mathbf{r}}_2 + \bar{\mathbf{r}}_3)$ as in eq. (3.8) and $a = M\omega/\hbar$ is the width parameter. It should be noted that this wave function is the same as the spatial part of $\tilde{\phi}_0\,(\tilde{\alpha}_1)$ given by eq. (3.19).

As long as one remains with the oscillator assumption, the internal Hamiltonian of a cluster with any number of nucleons can always be brought to a separable oscillator form by employing Jacobi coordinates. This can be shown easily by using the method of mathematical induction. Suppose that the single-particle oscillator Hamiltonian for a cluster of A nucleons, given in eq. (3.1), has been reduced to a separable form by the following Jacobi coordinate transformation:

$$\hat{\mathbf{r}}_j = \frac{1}{j}\sum_{k=1}^{j}\mathbf{r}_k - \mathbf{r}_{j+1}, \quad (j = 1, 2, \ldots, A-1)$$
$$\hat{\mathbf{r}}_A = \frac{1}{A}\sum_{k=1}^{A}\mathbf{r}_k. \tag{A8}$$

Now consider the addition of another nucleon to the cluster. The total Hamiltonian for this new cluster of A + 1 nucleons is then

$$H = H_A + H_{cm}(A) + \frac{1}{2M}\mathbf{p}_{A+1}^2 + \frac{1}{2}M\omega^2\mathbf{r}_{A+1}^2, \tag{A9}$$

where H_A is the internal Hamiltonian of the A-nucleon cluster, given by

$$H_A = \sum_{j=1}^{A-1} \left[\frac{1}{2Mj/(j+1)} \hat{\mathbf{p}}_j^2 + \frac{1}{2} \frac{j}{j+1} M \omega^2 \hat{\mathbf{r}}_j^2 \right] \tag{A10}$$

and $H_{cm}(A)$ is the Hamiltonian for the center-of-mass motion of the A-nucleon cluster, given by

$$H_{cm}(A) = \frac{\hat{\mathbf{p}}_A^2}{2\,AM} + \frac{1}{2} AM\omega^2 \hat{\mathbf{r}}_A^2 . \tag{A11}$$

We now introduce Jacobi coordinates $\hat{\mathbf{r}}_j'$ for the new cluster, which are defined in terms of the Jacobi coordinates $\hat{\mathbf{r}}_j$ for the old cluster and the coordinate $\mathbf{r}_{A+1}$ by the equations

$$\begin{aligned} \hat{\mathbf{r}}_j' &= \hat{\mathbf{r}}_j, \quad (j = 1, \ldots, A-1) \\ \hat{\mathbf{r}}_A' &= \hat{\mathbf{r}}_A - \mathbf{r}_{A+1}, \\ \hat{\mathbf{r}}_{A+1}' &= \frac{1}{A+1} (A\hat{\mathbf{r}}_A + \mathbf{r}_{A+1}). \end{aligned} \tag{A12}$$

The corresponding canonical momenta $\hat{\mathbf{p}}_j'$ can be computed in a similar way as described by eq. (A4). The result is

$$\begin{aligned} \hat{\mathbf{p}}_j' &= \hat{\mathbf{p}}_j, \quad (j = 1, \ldots, A-1) \\ \hat{\mathbf{p}}_A' &= \frac{1}{A+1} \hat{\mathbf{p}}_A - \frac{A}{A+1} \mathbf{p}_{A+1}, \\ \hat{\mathbf{p}}_{A+1}' &= \hat{\mathbf{p}}_A + \mathbf{p}_{A+1}. \end{aligned} \tag{A13}$$

Using eqs. (A12) and (A13), we obtain then from eq. (A9)

$$H = H_{A+1} + H_{cm}(A+1) \tag{A14}$$

with

$$H_{A+1} = \sum_{j=1}^{A} \left[\frac{1}{2Mj/(j+1)} \hat{\mathbf{p}}_j'^2 + \frac{1}{2} \frac{j}{j+1} M\omega^2 \hat{\mathbf{r}}_j'^2 \right] \tag{A15}$$

and

$$H_{cm}(A+1) = \frac{\hat{\mathbf{p}}_{A+1}'^2}{2(A+1)M} + \frac{1}{2}(A+1) M\omega^2 \hat{\mathbf{r}}_{A+1}'^2 . \tag{A16}$$

This shows that if the internal Hamiltonian for a cluster of A nucleons can be made separable, so can that for a cluster of A + 1 nucleons. Since eq. (A10) has been demonstrated to be valid in the four-nucleon case (two- and three-nucleon cases are trivial), we have now shown by mathematical induction that eq. (A15) is valid for a cluster of any number of nucleons.

Next, we show that, for an A-nucleon system, the total orbital angular momentum expressed in Jacobi coordinates has the same form as that expressed in single-particle coordinates, i. e.,

$$\mathbf{L}_A = \sum_{j=1}^{A} (\hat{\mathbf{r}}_j \times \hat{\mathbf{p}}_j). \tag{A17}$$

To prove this, we again use the method of mathematical induction. That is, we assume this to be true for an A-nucleon system and then show that it is also true for an (A + 1)-nucleon system.

The total orbital angular momentum of an (A + 1)-nucleon system is

$$\mathbf{L}_{A+1} = \mathbf{L}_A + (\mathbf{r}_{A+1} \times \mathbf{p}_{A+1}), \tag{A18}$$

where the last term represents the orbital angular momentum of the additional nucleon. Using eqs. (A12), (A13), and (A17), one obtains easily

$$\mathbf{L}_{A+1} = \sum_{j=1}^{A+1} (\hat{\mathbf{r}}_j' \times \hat{\mathbf{p}}_j'), \tag{A19}$$

thus proving our assertion made above. In the case where there are N clusters, one can proceed in an analogous manner and obtain the result represented by eq. (3.49).

Appendix B

Designation of Oscillator States

In our discussions, we have made extensive use of oscillator wave functions. In fig. B1 (see p. 372), we show therefore the way by which the various subshells in an isotropic harmonic-oscillator well are labelled and the maximum number of neutrons or protons which can occupy these subshells.

Appendix C

Demonstration of the Projection Technique

To demonstrate the use of the projection formalism discussed in section 8.3 and subsection 9.2c, we describe in this appendix a simple resonance model [BE 69]. In this model, we consider the s-wave scattering of a neutron by a potential barrier of the form

$$V(r) = A\,\delta(r - a), \tag{C1}$$

with A = 207 MeV-fm and a = 3.14 fm. Because of the shielding effect of this barrier, there exists in this system a number of more or less sharp potential resonances. Here we shall consider only the energy region around the lowest resonance and calculate the phase shifts both exactly and in terms of a single-level Breit-Wigner resonance formula (see eq. (9.23)).

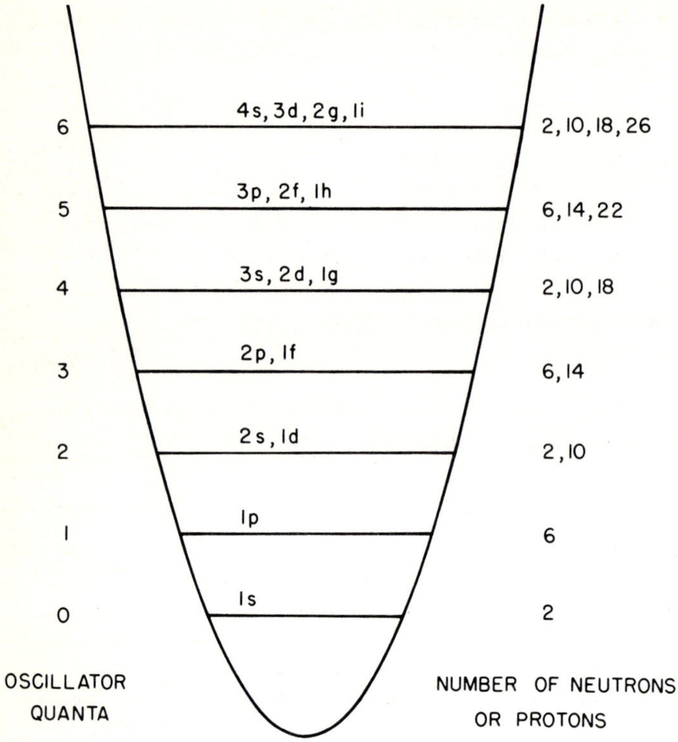

Fig. B1. Designation of subshells in an oscillator well.

The equation satisfied by the radial wave function u(r) is

$$H\,u(r) = E\,u(r), \tag{C2}$$

with

$$H = -\frac{\hbar^2}{2M}\frac{d^2}{dr^2} + A\,\delta\,(r-a). \tag{C3}$$

This equation can be solved easily to yield the exact s-wave phase shift δ_0. To obtain the phase shift by means of a single-level Breit-Wigner formula, we must first construct a normalized bound-state wave function u_C which adequately describes the behaviour of the neutron in the interior region ($r < a$). This wave function can be chosen as

$$\begin{aligned} u_C &= \left(\frac{2}{a}\right)^{1/2} \sin\frac{\pi r}{a}, \quad (r \leqslant a) \\ &= 0, \qquad\qquad\qquad (r > a) \end{aligned} \tag{C4}$$

which represents the lowest eigenstate in a potential well of the form

$$\begin{aligned} \widetilde{V}(r) &= 0, \quad (r \leqslant a) \\ &= \infty. \quad (r > a) \end{aligned} \tag{C5}$$

For this wave function, the energy expectation value

$$E_C = \langle u_C | H | u_C \rangle, \tag{C6}$$

where the brackets indicate only an integration over the radial variable, is equal to 20.7 MeV. Adopting this bound-state wave function, we can then introduce the projection operators

$$P = |u_C\rangle\langle u_C|, \tag{C7}$$

$$Q = 1 - |u_C\rangle\langle u_C|, \tag{C8}$$

and define an elastic-scattering subspace specified by the variation

$$\langle \delta\widetilde{u}_D | = \langle \delta u | Q. \tag{C9}$$

As discussed in subsection 9.2c, we must now solve the equation

$$\langle \delta\widetilde{u}_D | H - E_D^\beta | \widetilde{u}_D^\beta \rangle = 0, \tag{C10}$$

in order to obtain the resolvent operator $\widetilde{G}_D^+$ and the scattering function $\widetilde{u}_D^+ = \widetilde{u}_D^\beta (E_D^\beta = E)$ for the computation of the potential-scattering phase shift $\widetilde{\delta}(E)$, the shift function $\widetilde{\Delta}(E)$, and the width function $\widetilde{\Gamma}(E)$. Because of the presence of the operator Q, $\widetilde{u}_D^\beta(r)$ satisfies an integrodifferential equation instead of a differential equation as for $u(r)$ and, consequently, shows no resonance behaviour in the energy region around E_C.

Upon making the linear approximation described in subsection 9.2a, we obtain the following values for the resonance parameters:

$$\begin{aligned} E_r &= 0.94\ E_C, \\ \Gamma_r &= 0.011\ E_C, \\ \Delta(E_r) &= 0.06\ E_C. \end{aligned} \tag{C11}$$

The result for the potential-scattering phase shift $\widetilde{\delta}(E)$ is shown by the dashed curve in fig. C1. Here it is seen that, as expected, it varies rather slowly with energy in the energy-region under consideration. Also, in this figure, we have shown by the solid curve the total s-wave phase shift δ_0 calculated exactly. The curve for the phase shift obtained from the Breit-Wigner formula is not shown, since it coincides so closely with the solid curve that a clear distinction cannot be made.

With a different but reasonable choice for u_C, we find that an appreciable change can occur in E_C, $\widetilde{\Delta}(E)$, and $\widetilde{\Gamma}(E)$ but, as discussed in subsection 9.2b, not in δ_0, E_r, and Γ_r.

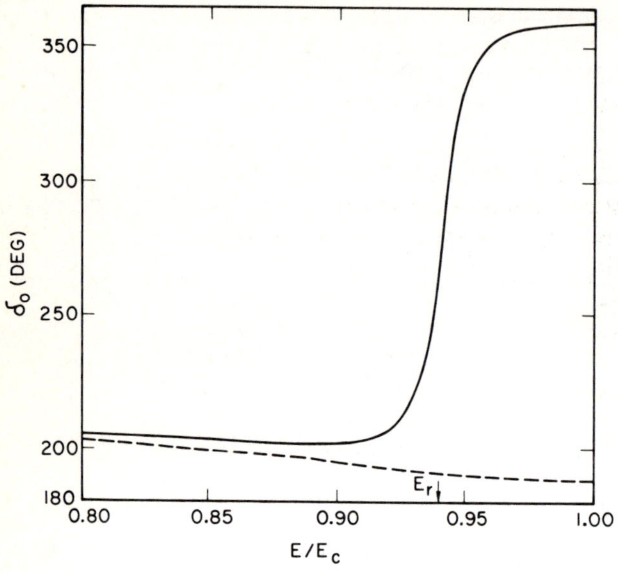

Fig. C1
s-wave phase shift as a function of energy. The solid curve represents the total phase shift δ_0, while the dashed curve represents the potential-scattering phase shift $\tilde{\delta}$. (Adapted from ref. [BE 69].)

Appendix D

Connection with Conventional Direct-Reaction Theory

In this appendix we briefly discuss, by means of the ^{6}Li(p, ^{3}He) α example, the connection [HU 72a] between the direct-reaction theory presented in chapter 12 and the conventional direct-reaction theory commonly used in phenomenological analyses of experimental data. As will be seen, one obtains the latter theory from the former by essentially making approximations with respect to the procedure of antisymmetrization.

At relatively high energies where the conventional direct-reaction theory is usually employed, it was discussed in chapter 12 that one may appropriately adopt the weak-coupling approximation. Thus, as a starting point for our discussion, we use the reaction amplitude given by eq. (12.17), i.e.,

$$A_{j,12} \propto (v_1/v_2)^{1/2}\, M_{12}\,, \tag{D1}$$

where the matrix element M_{12} has the form

$$M_{12} = \langle \tilde{\psi}_{2j}(E) | H - E | \psi_{1j}(E) \rangle \tag{D2}$$

with

$$H = \sum_{i=1}^{7} \frac{1}{2M} p_i^2 + \sum_{i<j=1}^{7} V_{ij} - T_C . \tag{D3}$$

As in section 12.4, we shall consider the wave function $\tilde{\psi}_{2j}$ to be totally antisymmetric but the wave function ψ_{1j} as unantisymmetrized.

The target nucleus ^{6}Li in the initial channel (channel 1) will be considered in the d + α cluster representation. Because of our choice in not antisymmetrizing ψ_{1j}, we can label the nucleons which form the various clusters in this channel. Thus, in the following discussion, we shall consider nucleons 1 to 4 as forming the α cluster, nucleons 5 and 6 as forming the deuteron cluster, and nucleon 7 as the incoming proton.

By introducing the p + ^{6}Li reduced mass μ_1 and relative momentum $\hat{\mathbf{p}}_1$, and the d + α reduced mass $\mu_{d\alpha}$ and relative momentum $\hat{\mathbf{p}}_{d\alpha}$, we can write the total Hamiltonian H as

$$H = H_d + H_\alpha + \frac{\hat{\mathbf{p}}_{d\alpha}^2}{2\mu_{d\alpha}} + \frac{\hat{\mathbf{p}}_1^2}{2\mu_1} + V_{d\alpha} + V_{p\alpha} + V_{pd}, \tag{D4}$$

where H_d and H_α are the internal Hamiltonians of the deuteron and α clusters, respectively. The potential term $V_{d\alpha}$ represents a sum of two-nucleon potentials between nucleons in the deuteron cluster and nucleons in the α cluster, i.e.,

$$V_{d\alpha} = \sum_{i=1}^{4} \sum_{j=5}^{6} V_{ij}, \tag{D5}$$

and the potential terms $V_{p\alpha}$ and V_{pd} are defined in a similar manner.

The ^{3}He nucleus in the final channel (channel 2) will be considered as formed by a proton and a deuteron cluster. In the simplest conventional direct-reaction theory, what one does is then to omit the antisymmetrization in $\widetilde{\psi}_{2j}$ with regard to nucleons in different clusters. By carrying out now a summation over all spin and isospin coordinates and an integration over the deuteron-cluster and α-cluster internal spatial coordinates, one obtains, by omitting the index j for simplicity, the expression

$$M_{12} = \langle \chi_2(\mathbf{R}_{pd}) F_2(\mathbf{R}_2) | \overline{H}_{d\alpha} + \overline{V}_{p\alpha} + \overline{V}_{pd} + \frac{1}{2\mu_1}\hat{\mathbf{p}}_1^2 + E_d + E_\alpha - E | \chi_1(\mathbf{R}_{d\alpha}) F_1(\mathbf{R}_1)\rangle, \tag{D6}$$

where the various spatial coordinates are defined as

$$\mathbf{R}_1 = \mathbf{r}_7 - \mathbf{R}_{Li}, \qquad \mathbf{R}_2 = \mathbf{R}_h - \mathbf{R}_\alpha,$$
$$\mathbf{R}_{d\alpha} = \mathbf{R}_d - \mathbf{R}_\alpha, \qquad \mathbf{R}_{pd} = \mathbf{r}_7 - \mathbf{R}_d, \tag{D7}$$

with $\mathbf{R}_d$, $\mathbf{R}_h$, $\mathbf{R}_\alpha$, and $\mathbf{R}_{Li}$ being the center-of-mass coordinates of the deuteron, ^{3}He, α, and ^{6}Li clusters, respectively. The functions χ_1 and χ_2 describe, respectively, the internal relative motions in the ^{6}Li and ^{3}He clusters, while the functions F_1 and F_2 describe, respectively, the p + ^{6}Li and ^{3}He + α relative motions in channel 1 and channel 2. The quantity $\overline{H}_{d\alpha}$ is the effective Hamiltonian operator for the d + α relative motion in the ^{6}Li cluster, and E_d and E_α are the internal energies of the deuteron and the α clusters. Finally, the quantities $\overline{V}_{p\alpha}$ and $\overline{V}_{pd}$ represent the effective interactions between the relevant clusters; they are obtained from $V_{p\alpha}$ and V_{pd} by summing and integration over appropriate cluster internal coordinates. For the derivation of eq. (D6), it is important that the deuteron and α clusters in channel 2 are not distorted as compared with the corresponding clusters in channel 1. In other words, one must assume that the deuteron-cluster and α-cluster internal wave functions in channels 1 and 2 are exactly the same.

By using the relation

$$E = E_d + E_\alpha + E_{d\alpha} + \frac{\hbar^2}{2\mu_1} k_1^2, \tag{D8}$$

where $E_{d\alpha}$ is the negative of the $d + \alpha$ separation energy in ^{6}Li and $\mathbf{k}_1$ is the $p + {}^6$Li relative wave vector in the asymptotic region, and the equation

$$\overline{H}_{d\alpha}\, \chi_1 = E_{d\alpha}\, \chi_1, \tag{D9}$$

we can further reduce M_{12} to the form

$$M_{12} = \langle \chi_2(\mathbf{R}_{pd})\, F_2(\mathbf{R}_2) | T_1 + \overline{V}_{p\alpha} + \overline{V}_{pd} - \frac{\hbar^2}{2\mu_1} k_1^2 | \chi_1(\mathbf{R}_{d\alpha})\, F_1(\mathbf{R}_1)\rangle \tag{D10}$$

with

$$T_1 = \frac{1}{2\mu_1} \hat{\mathbf{p}}_1^2. \tag{D11}$$

As has been mentioned in section 12.2, one commonly introduces into a direct-reaction calculation a local, complex, energy-dependent optical potential $\widetilde{V}_1$ which describes, to a good approximation, the scattering and absorption of clusters in channel 1, and chooses F_1 in such a way that the equation

$$(T_1 + \widetilde{V}_1)\, F_1 = \frac{\hbar^2}{2\mu_1} k_1^2\, F_1 \tag{D12}$$

is satisfied. Substituting eq. (D12) into eq. (D10), we then obtain

$$M_{12} = \langle \chi_2 F_2 | \overline{V}_{p\alpha} + \overline{V}_{pd} - \widetilde{V}_1 | \chi_1 F_1 \rangle. \tag{D13}$$

As is noted, this is the matrix element used in a conventional distorted-wave Born-approximation calculation, which describes a deuteron-pickup process in a definite (j, π) state. For the construction of the function F_2, one similarly introduces an optical potential $\widetilde{V}_2$ which describes the scattering and absorption of clusters in channel 2. Also, in such a direct-reaction calculation, the interactions $\overline{V}_{p\alpha}$ and $\overline{V}_{pd}$ are chosen such that the experimental results in the $p + \alpha$ and $p + d$ systems are reasonably reproduced.†

By using optical potentials $\widetilde{V}_1$ and $\widetilde{V}_2$ which yield good agreement with experimental elastic-scattering data, the influence of the Pauli principle can be phenomenologically taken into account in an approximate manner. However, it should be noted that, even at relatively high energies, one could only use eq. (D13) to describe approximately the ^{6}Li(p, ^{3}He)α differential reaction cross section in the forward angular region. As has been explained in section 12.4, it will be necessary to include another process in which

† For example, one could use the $p + \alpha$ empirical phase shifts and the proton separation energy in ^{3}He for this purpose.

the incoming proton is exchanged with one of the protons in the α cluster, if a reasonable description of the differential reaction cross section in the backward angular region is also desired. Within the framework of the conventional direct-reaction theory, this can be accomplished by further considering the α cluster as consisting of a proton and a triton cluster and introducing a function χ_3 to describe the p + t relative motion. For the reaction amplitude, one then uses not only the amplitude corresponding to the deuteron-pickup process as described by eq. (D13), but also an amplitude corresponding to a process involving the exchange of protons as mentioned above, which in the language of section 12.4 is called a single-particle exchange process.

The above discussion shows that, if one divides the p + ^{6}Li system into more and more clusters in a conventional direct-reaction calculation, then the Pauli principle will eventually be fully taken into account and the result will approach that of the antisymmetrized direct-reaction calculation described in chapter 12.

References

[AB 64] *Abramowitz, M.* and *I. A. Stegun,* "Handbook of Mathematical Functions" (National Bureau of Standards, 1964) (*9.6*)

[AF 68] *Afnan, I. R.* and *Y. C. Tang,* Phys. Rev. **175** (1968) 1337 (*5.3c*)

[AJ 59] *Ajzenberg-Selove, F.* and *T. Lauritsen,* Nucl. Phys. **11** (1959) 1 (*16.3a*)

[AJ 66] *Ajzenberg-Selove, F.* and *T. Lauritsen,* Nucl. Phys. 78 (1966) 1 (*5.3b, 5.3c, 6.2, 7.3b, 13.2a, 16.2*)

[AJ 68] *Ajzenberg-Selove, F.* and *T. Lauritsen,* Nucl. Phys. **A114** (1968) 1 (*5.4*)

[AJ 70] *Ajzenberg-Selove, F.* and *T. Lauritsen,* Nucl. Phys. **A152** (1970) 1 (*9.2c*)

[AJ 71] *Ajzenberg-Selove, F.*, Nucl. Phys. **A166** (1971) 1 (*11.3b, 16.2, 16.3a, 16.4*)

[AJ 72] *Ajzenberg-Selove, F.*, Nucl. Phys. **A190** (1972) 1 (*16.2, 16.3b*)

[AJ 74] *Ajzenberg-Selove, F.*, Nucl. Phys. **A227** (1974) 1 (*10.2b*)

[AL 56] *Alder, K., A. Bohr, T. Huus, B. Mottelson,* and *A. Winther,* Rev. Mod. Phys. **28** (1956) 432 (*14.3*)

[AL 57] *Allred, J. C., D. K. Froman, A. M. Hudson,* and *L. Rosen,* Phys. Rev. **82** (1957) 786 (*7.3c*)

[AL 60] *Almqvist, E., D. A. Bromley,* and *J. A. Kuehner,* Phys. Rev. Lett. **4** (1960) 365, 515 (*16.5a*)

[AL 64] *Almqvist, E., J. A. Kuehner, D. McPherson,* and *E. W. Vogt,* Phys. Rev. **136** (1964) B84 (*16.5a*)

[AL 66] *Ali, S.* and *A. R. Bodmer,* Nucl. Phys. **80** (1966) 99 (*11.4b*)

[AN 70] *Anni, R.* and *L. Taffara,* Rivista del Nuovo Cimento **2** (1970) 1 (*13.1*)

[AR 70] *Arima, A., V. Gillet,* and *J. Ginnochio,* Phys. Rev. Lett. **25** (1970) 1043 (*14.4*)

[AU 70] *Austern, N.*, "Direct Nuclear Reaction Theories" (Wiley, New York, 1970) (*12.1*)

[BA 52] *Barschall, H. H.*, Phys. Rev. **86** (1952) 431 (*11.2d*)

[BA 57] *Bardeen, J., L. N. Cooper,* and *J. R. Schrieffer,* Phys. Rev. **108** (1957) 1175 (*17.4b*)

[BA 63] *Bartis, F. J.*, Phys. Rev. **132** (1963) 1763 (*16.3b*)

[BA 67] *Bacher, A. D.*, Ph. D. Thesis, California Institute of Technology (1967) (*7.3d*)

[BA 67a] *Bayman, B. F.*, in "Proceedings of the International School of Physics «Enrico Fermi»", course 40 (1967) 380 (*10.3a*)

[BA 70] *Barrett, B. R., R. G. L. Hewitt*, and *R. J. McCarthy*, Phys. Rev. **C2** (1970) 1199 (*14.8b*)

[BA 71] *Barrett, B. R., R. G. L. Hewitt*, and *R. J. McCarthy*, Phys. Rev. **C3** (1971) 1137 (*14.8b*)

[BA 71a] *Balagna, J. R., G. P. Ford, D. C. Hoffman*, and *J. D. Knight*, Phys. Rev. Lett. **26** (1971) 145 (*17.3, 17.4b, 17.4d*)

[BA 73] *Banerjee, B., H. J. Mang*, and *P. Ring*, Nucl. Phys. **A215** (1973) 366 (*14.7*)

[BA 73a] *Baz, A. I.*, and *M. V. Zhukov*, Sov. Journal of Nucl. Phys. **16** (1973) 31, 529 (*7.2c*)

[BA 75] *Baur, G., T. Udagawa*, and *H. H. Wolter*, Annual Report 1975 KFA-IKP 10/76 (*13.2*)

[BE 68] *Benöhr, H. C.*, Ph. D. Thesis, University of Tübingen (1968) (*9.2b, 9.2d*)

[BE 69] *Benöhr, H. C.* and *K. Wildermuth*, Nucl. Phys. **A128** (1969) 1 (*4.3, 9.2b, 9.2d, C*)

[BE 69a] *Bentz, H. A.*, Z. Naturforsch. **24a** (1969) 858 (*5.3a*)

[BE 71] *Bethe, H. A.*, Ann. Rev. Nucl. Sci. **21** (1971) 93 (*1.1, 5.3d*)

[BL 68] *Blatt, S. L., A. M. Young, S. C. Ling, K. J. Moon*, and *C. D. Porterfield*, Phys. Rev. **176** (1968) 1147 (*15.3*)

[BO 39] *Bohr, N.* and *J. A. Wheeler*, Phys. Rev. **56** (1939) 426 (*17.1, 17.2*)

[BO 52] *Bohr, A.*, Kgl. Danske Videnskab. Selskab., Mat.-fys. Medd. **26** (1952) No. 14 (*14.2, 14.5*)

[BO 53] *Bohr, A*, and *B. R. Mottelson*, Kgl. Danske Videnskab. Selskab., Mat.-fys. Medd. **27** (1953) No. 16 (*14.2, 14.4, 14.7*)

[BO 54] *Bohr, A.*, Ejnar Munksgaards Forlag (1954) 24 (*14.4*)

[BO 56] *Bohr, A.*, in "Proceedings of the 1st International Conference on Peaceful Uses of Atomic Energy" (United Nations, New York, 1956) vol. 2, p. 151 (*17.2*)

[BO 65] *Bondorf, J. P.* and *R. B. Leachman*, Kgl. Danske Videnskab. Selskab., Mat.-fys. Medd. **34** (1965) 10 (*16.5a*)

[BO 67] *Bouton, M., P. van Leuven, H. Depuydt*, and *L. Schotsmans*, Nucl. Phys. **A100** (1967) 90 (*5.3c*)

[BO 67a] *Bouton, M., M. C. Bouton*, and *P. van Leuven*, Nucl. Phys. **A100** (1967) 105 (*5.3a, 5.3c*)

[BO 72] *Boykin, W. R., S. D. Baker*, and *D. M. Hardy*, Nucl. Phys. **A195** (1972) 241 (*7.3a*)

[BO 72a] *Bohlen, H., N. Marquardt, W. von Oertzen*, and *Ph. Gorodetzky*, Nucl. Phys. **A179** (1972) 504 (*13.3, 13.4*)

[BR 56] *Breit, G.* and *M. E. Ebel*, Phys. Rev. **103** (1956) 679 (*13.1*)

[BR 56a] *Breit, G.* and *M. E. Ebel*, Phys. Rev. **104** (1956) 1030 (*13.1*)

[BR 57] *Brink, D. M.*, Nucl. Phys. **4** (1957) 215 (*3.3, 3.5*)

[BR 57a] *Brussel, M. K.* and *J. H. Williams*, Phys. Rev. **106** (1957) 286 (*12.3a*)

[BR 58] *Brueckner, K.*, "The Many-Body Problem" (Methuen, London, 1958) (*5.3c*)

[BR 59] *Brown, G. E.*, Rev. Mod. Phys. **31** (1959) 893 (*10.3b*)

[BR 60] *Breit, G., M. H. Hull, Jr., K. E. Lassila*, and *K. D. Pyatt, Jr.*, Phys. Rev. **120** (1960) 2227 (*1.1*)

[BR 60a] *Bromley, D. A., J. A. Kuehner*, and *E. Almqvist*, in "Proceedings of the International Conference on Nuclear Structure", Kingston, Canada (University of Toronto Press, Toronto, 1960) p. 255 (*16.5a*)

[BR 62] *Breit, G., M. H. Hull, Jr., K. E. Lassila*, and *K. D. Pyatt, Jr.*, Phys. Rev. **128** (1962) 826 (*1.1*)

[BR 64] *Brown, G. E.*, "Unified Theory of Nuclear Models" (North-Holland, Amsterdam, 1964) (*3.6c*)

[BR 66] *Brink, D. M.*, in "Proceedings of the International School of Physics «Enrico Fermi»", course 36 (1966) 247 (*5.2b*)

[BR 68] *Brink, D. M.* and *A. Weiguny*, Nucl. Phys. **A120** (1968) 59 (*5.2b*)

[BR 68a] *Brown, R. E.* and *Y. C. Tang*, Phys. Rev. **176** (1968) 1235 (*7.3a, 11.4a, 11.4b*)

[BR 71] *Brown, R. E.* and *Y. C. Tang*, Nucl. Phys. **A170** (1971) 225 (*5.3d, 7.3b, 11.2c, 11.3a, 11.3b, 11.4a*)

[BR 73] *Brayshaw, D. D.*, Phys. Rev. **C7** (1973) 1731 (*1.1*)

[BR 74] *Brown, R. E., F. S. Chwieroth, Y. C. Tang*, and *D. R. Thompson*, Nucl. Phys. **A230** (1974) 189 (*11.4b*)

[BR 74a] *Brown, R. E., W. S. Chien*, and *Y. C. Tang*, in "Proceedings of the International Conference on Few-Body Problems in Nuclear and Particle Physics", Quebec, Canada (1974) (*11.4b*)

[BU 72] *Bumiller, F. A., F. R. Buskirk, J. N. Dyer*, and *W. A. Monson*, Phys. Rev. **C5** (1972) 391 (*5.3a*)

[CA 67] *Cameron, J. M.*, Technical Report P-80, UCLA (1967) (*11.3b*)

[CA 67a] *Caldwell, J. T., S. C. Fultz, C. D. Bowman*, and *R. W. Hoff*, Phys. Rev. **155** (1967) 1309 (*17.4b*)

[CH 71] *Chuang, L. S.*, Nucl. Phys. **A174** (1971) 399 (*7.3a*)

[CH 73] *Chwieroth, F. S.*, Ph. D. Thesis, University of Minnesota (1973) (*5.2a*)

[CH 73a] *Chwieroth, F. S.*, University of Minnesota Report C00-1764-180 (1973) (*7.2b*)

[CH 73b] *Chwieroth, F. S., R. E. Brown, Y. C. Tang*, and *D. R. Thompson*, Phys. Rev. **C8** (1973) 938 (*7.3f, 11.4b, 12.3a*)

[CH 73c] *Chien, W. S.*, Ph. D. Thesis, University of Minnesota (1973) (*7.3g*)

[CH 74] *Chwieroth, F. S., Y. C. Tang*, and *D. R. Thompson*, Phys. Rev. **C9** (1974) 56 (*12.3*)

[CH 74a] *Chwieroth, F. S., Y. C. Tang*, and *D. R. Thompson*, Phys. Rev. **C10** (1974) 406 (*12.3a, 12.4b*)

[CH 75] *Chien, W. S.* and *R. E. Brown*, Phys. Rev. **C10** (1975) 1767 (*11.4b*)

[CH 75a] *Chwieroth, F. S., Y. C. Tang*, and *D. R. Thompson*, unpublished results (1975) (*12.4a*)

[CL 74] *Clement, D., E. J. Kanellopoulos*, and *K. Wildermuth*, Phys. Lett. **52B** (1974) 309 (*13.3*)

[CL 74a] *Clement, D.* and *W. Zahn*, Phys. Lett. **48B** (1974) 183 (*15.3*)

[CL 75] *Clement, D.*, Ph. D. Thesis, University of Tübingen (1975) (*13.2a*)

[CL 75a] *Clement, D., E. J. Kanellopoulos*, and *K. Wildermuth*, Phys. Lett. **55B** (1975) 19 (*13.2a*)

[CL 75b] *Clement, D.*, private communication (1975) (*13.2a*)

[CO 59] *Corelli, J. C., E. Bleuler*, and *D. J. Tendam*, Phys. Rev. **116** (1959) 1184 (*7.3e*)

[CO 65] *Cohen, S.* and *D. Kurath*, Nucl. Phys. **73** (1965) 1 (*5.4, 8.2*)

[CO 70] *Cooper, J. C.* and *B. Crasemann*, Phys. Rev. **C2** (1970) 451 (*16.3b*)

[CR 70] *Cramer, J. D.* and *J. R. Nix*, Phys. Rev. **C2** (1970) 1048 (*17.4a*)

[DA 65] *Darriulat, P., G. Igo, H. G. Pugh*, and *H. D. Holmgren*, Phys. Rev. **137** (1965) B315 (*11.3a*)

[DA 65a] *Davydov, A. S.*, "Quantum Mechanics" (Pergamon, Oxford, 1965) (*15.3*)

[DA 67] *Darriulat, P., D. Garreta, A. Tarrats*, and *J. Arvieux*, Nucl. Phys. **A94** (1967) 653 (*7.3c*)

[DA 70] *Davies, W. G., C. Broude*, and *J. S. Forster*, Bull. Am. Phys. Soc. **15** (1970) 543 (*14.8b*)

[DA 71] *Danos, M.* and *V. Gillet*, Phys. Lett. **34B** (1971) 24 (*14.4*)

[DE 53] *De Shalit, A.*, Nucl. Phys. **7** (1953) 225 (*16.4*)

[DE 58] *De Shalit, A.* and *V. F. Weisskopf*, Ann. Phys. (N. Y.) **5** (1958) 282 (*6.3*)

[DE 63] *De Shalit, A.* and *I. Talmi*, "Nuclear Shell Theory" (Academic, New York, 1963) (*8.2*)

[DE 72] *De Takacsy, N.*, Phys. Rev. **C5** (1972) 1883 (*5.2b*)

[DR 63] *Drell, S.* and *K. Huang*, Phys. Rev. **91** (1963) 1527 (*1.1*)

[ED 60] *Edmonds, A. R.*, "Angular Momentum in Quantum Mechanics" (Princeton University Press, Princeton, New Jersey, 1960) (*14.2, 14.3, 14.4, 14.5*)

[EH 59] *Ehrenberg, H. F., R. Hofstadter, U. Meyer-Berkhout, D. G. Ravenhall,* and *S. E. Sobottka*, Phys. Rev. **113** (1959) 666 (*3.7*)

[EI 58] *Eisenbund, L.* and *E. P. Wigner*, "Nuclear Structure" (Princeton University Press, Princeton, New Jersey, 1958) (*16.3a*)

[EI 66] *Eikemeier, H.* and *H. H. Hackenbroich*, Z. Phys. **195** (1966) 412 (*5.3c*)

[EI 69] *Eigenbrod, F.*, Z. Phys. **228** (1969) 337 (*5.3a, 5.3c*)

[EI 71] *Eikemeier, H.* and *H. H. Hackenbroich*, Nucl. Phys. **A169** (1971) 407 (*7.3f, 7.3h*)

[EI 72] *Eisenberg, J. M.* and *W. Greiner*, "Microscopic Theory of the Nucleus" (North-Holland, Amsterdam, 1972) (*13.1*)

[EL 55] *Elliot, J. P.* and *T. H. R. Skyrme*, Proc. Roy. Soc. (London) **A232** (1955) 561 (*3.3, 3.4*)

[EL 56] *Elliot, J. P.* and *T. H. R. Skyrme*, Nuovo Cimento 10 (1956) 4, 164 (*3,4*)

[EL 57] *Elliot, J. P.* and *B. H. Flowers*, Proc. Roy. Soc. (London) **A242** (1957) 57 (*3.6c*)

[EL 61] *Elton, L. R. B.*, "Nuclear Sizes" (Oxford University Press, London, 1961) (*5.3a*)

[EN 67] *Endt, P.*, in "Nuclear Structure", edited by *A. Hossain, H. A. Rashid,* and *M. Islam* (North-Holland, Amsterdam, 1967) p. 58 (*10.3a*)

[EV 65] *Everling, F.*, private communication (1965) (*16.2*)

[FA 58] *Fairhall, A. W., R. C. Jensen,* and *E. F. Neuzil*, in "Proceedings of the 2nd International Conference on Peaceful Uses of Atomic Energy" (United Nations, Geneva, 1958) vol. 15, p. 452 (*17.2*)

[FA 61] *Faissner, H.* and *K. Wildermuth*, Naturwissenschaften **48** (1961) 400 (*17.4b*)

[FA 62] *Faissner, H.* and *K. Wildermuth*, Phys. Lett. **2** (1962) 212 (*17.1, 17.2, 17.3*)

[FA 64] *Faissner, H.* and *K. Wildermuth*, Nucl. Phys. **58** (1964) 177 (*17.1, 17.2, 17.3, 17.4d*)

[FA 76] *Faessler, A., K. R. Sandhya Devi, F. Grümmer, K. W. Schmid,* and *R. R. Hilton*, Nucl. Phys. **A256** (1976) 106 (*14.7, 14.8c*)

[FE 37] *Feenberg, E.*, Phys. Rev. **51** (1947) 597 (*14.2*)

[FE 58] *Feshbach, H.*, Ann. Phys. (N. Y.) **5** (1958) 357 (*8.3, 11.1, 18*)

[FE 62] *Feshbach, H.*, Ann. Phys. (N. Y.) **19** (1962) 287 (*8.3, 11.1, 18*)

[FE 70] *Federsel, H., E. J. Kanellopoulos, W. Sünkel,* and *K. Wildermuth*, Phys. Lett. **33B** (1970) 140 (*4.3, 7.3b*)

[FE 71] *Fetscher, W., E. Seibt, Ch. Weddingen,* and *E. J. Kanellopoulos*, Phys. Lett. **35B** (1971) 31 (*11.3a*)

[FE 73] *Ferguson, R. L., F. Plasil, F. Pleasonton, S. C. Burnett,* and *H. W. Schmitt*, Phys. Rev. **C7** (1973) 2510 (*17.1*)

[FE 74] *Feshbach, H.*, Rev. Mod. Phys. **46** (1974) 1 (*10.2a*)

[FE 74a] *Federsel, H.*, Ph. D. Thesis, University of Tübingen (1974) (*10.2a, 10.2b*)

[FL 72] *Flynn, K. F., B. Srinivasan, O. K. Manuel,* and *L. E. Glendenin*, Phys. Rev. **C6** (1972) 2211 (*17.4b*)

[FO 49] *Fox, L.* and *E. T. Goodwin*, Proc. Cambridge Phil. Soc. **45** (1949) 373 (*7.3b*)

[FO 64] *Fox, J. D., C. F. Moore,* and *D. Robson*, Phys. Rev. Lett. **12** (1964) 198 (*10.3a*)

[FR 39] *Frenkel, J. A.*, Phys. Rev. **55** (1939) 987 (*17.1*)

[FR 55] *Fröberg, C. E.*, Rev. Mod. Phys. **27** (1955) 399 (*7.3b*)

[FR 55a] *Friedman, F. L.* and *V. F. Weisskopf*, "Niels Bohr and the Development of Physics" (Pergamon, London, 1955) (*11.2b*)

[FR 65] *Franz, W.*, Z. Phys. **184** (1965) 181 (*15.2b*)

[FR 72] *Frisbee, P. E.*, Ph. D. Thesis, University of Maryland (1972) (*13.4*)

[FR 74] *Friedrich, H.*, Nucl. Phys. A224 (1974) 537 (*16.1, 16.5a*)

[FR 75] *Friedrich, H.* and *K. Langanke*, to be published (1975) (*16.1*)

[FU 75] *Furber, R. D.*, Ph. D. Thesis, University of Minnesota (1975) (*5.5, 7.3a, 11.3a*)

[GI 61] *Gibson, W. M., T. D. Thomas*, and *G. L. Miller*, Phys. Rev. Lett. 7 (1961) 65 (*17.3*)

[GI 63] *Giamati, C. C., V. A. Madsen*, and *R. M. Thaler*, Phys. Rev. Lett. **11** (1963) 163 (*11.4b*)

[GI 64] *Giamati, C. C.* and *R. M. Thaler*, Nucl. Phys. **59** (1964) 159 (*7.3a*)

[GI 73] *Girand, B., J. C. Hocquenghem*, and *A. Lumbroso*, Phys. Rev. **C7** (1973) 2274 (*5.2b*)

[GO 55] *Goeppert-Mayer, M.* and *J. H. D. Jensen*, "Elementary Theory of Nuclear Shell Structure" (Wiley, New York, 1955) (*6.3, 17.3*)

[GO 58] *Gomes, L. C., J. D. Walecka*, and *V. F. Weisskopf*, Ann. Phys. (N. Y.) **3** (1958) 241 (*6.3*)

[GO 73] *Goeke, K., H. Müther*, and *A. Faessler*, Nucl. Phys. **A201** (1973) 49 (*14.8b*)

[GÖ 75] *Gönnenwein, F., H. Schultheis, R. Schultheis*, and *K. Wildermuth*, to be published (1975) (*17.2, 17.4c*)

[GÖ 75a] *Gönnenwein, F., H. Schultheis, R. Schultheis*, and *K. Wildermuth*, Phys. Lett. **57B** (1975) 313 (*17.4c, 17.4d*)

[GR 56] *Green, T. S.* and *R. Middleton*, Proc. Phys. Soc. (London) **A69** (1956) 28 (*16.2*)

[GR 68] *Greenlees, G. W., G. J. Pyle*, and *Y. C. Tang*, Phys. Rev. **171** (1968) 1115 (*7.2a, 11.2d*)

[GR 71] *Greenlees, G. W.* and *Y. C. Tang*, Phys. Lett. **34B** (1971) 359 (*11.4b*)

[GR 72] *Greenlees, G. W., W. Makofske, Y. C. Tang*, and *D. R. Thompson*, Phys. Rev. **C6** (1972) 2057 (*11.3b*)

[GU 71] *Guazzoni, P., S. Micheletti, M. Pignanelli, F. Gentilin*, and *F. Pellegrini*, Phys. Rev. **C4** (1971) 1086 (*16.5b*)

[HA 66] *Hackenbroich, H. H., H. G. Wahsweiler*, and *K. Wildermuth*, Z. Naturforsch. 21a (1966) 870 (*6.1*)

[HA 67] *Hackenbroich, H. H.*, Forschungsbericht K67–93 (Frankfurt, ZAED, 1967) (*5.3c*)

[HA 69] *Hackenbroich, H. H.*, in "Proceedings of the International Conference on Clustering Phenomena in Nuclei", Bochum (IAEA, Vienna, 1969) p. 129 (*7.2c, 7.3g*)

[HA 70] *Halbert, E. C., J. B. McGrory, B. H. Wildenthal*, and *S. P. Pandya*, in "Advances in Nuclear Physics", edited by *M. Baranger* and *E. Vogt* (Plenum, New York, 1970) vol. 4 (*8.2*)

[HA 71] *Hackenbroich, H. H.* and *P. Heiss*, Z. Phys. **242** (1971) 352 (*7.3h*)

[HA 72] *Hardy, D. M., R. J. Spiger, S. D. Baker, Y. S. Chen*, and *T. A. Tombrello*, Nucl. Phys. **A195** (1972) 250 (*7.3a*)

[HA 72a] *Hackenbroich, H. H., P. Heiss*, and *H. Stöwe*, in "Few-Particle Problems in the Nuclear Interaction", edited by *I. Šlaus et al.* (North-Holland, Amsterdam, 1972) (*7.3i*)

[HA 74] *Hackenbroich, H. H., P. Heiss*, and *L. C. Niem*, Nucl. Phys. **A221** (1974) 461 (*12.2*)

[HE 32] *Heisenberg, W.*, Z. Phys. 77 (1932) 1 (*1.1*)

[HE 56] *Heydenburg, N. P.* and *G. M. Temmer*, Phys. Rev. **104** (1956) 123 (*7.3g*)

[HE 60] *Hebbard, D. F.*, Nucl. Phys. **15** (1960) 289 (*16.3a*)

[HE 61] *Herrmann, G.*, Habilitationsschrift, Universität Mainz (1961) (*17.3*)

[HE 66] *Herndon, R. C.* and *Y. C. Tang*, Methods in Computational Physics **6** (1966) 153 (*5.3b*)

[HE 69] *Heiss, P.* and *H. H. Hackenbroich*, Phys. Lett. **30B** (1969) 373 (*7.3f*)

[HE 71] *Heiss, P.* and *H. H. Hackenbroich*, Nucl. Phys. **A162** (1971) 530 (*3.7c*)

[HE 72] *Heiss, P.* and *H. H. Hackenbroich*, Nucl. Phys. **A182** (1972) 522 (*7.3h*)

[HE 72a] *Heiss, P.* and *H. H. Hackenbroich,* Z. Phys. **251** (1972) 168 (*7.3i*)

[HE 73] *Heiss, P.* and *H. H. Hackenbroich,* Nucl. Phys. **A202** (1973) 353 (*7.3i*)

[HI 53] *Hill, D. A.* and *J. A. Wheeler,* Phys. Rev. **89** (1953) 1102 (*4.3, 7.2c, 14.7, 14.9, 17.4a*)

[HO 55] *Hornyak, W.* and *R. Sherr,* Phys. Rev. **100** (1955) 1409 (*16.3a*)

[HO 57] *Hofstadter, R.,* Ann. Rev. Nucl. Sci. **7** (1957) 231 (*7.3e*)

[HO 61] *Holmgren, H. D.* and *L. M. Cameron,* in "Proceedings of the Rutherford Jubilee Conference", Manchester (1961) (*3.6a*)

[HO 61a] *Holmgren, H. D.* and *E. A. Wolicki,* in "Proceedings of the Rutherford Jubilee Conference", Manchester (1961) (*3.6a*)

[HO 66] *Hoop, B., Jr.* and *H. H. Barschall,* Nucl. Phys. **83** (1966) 65 (*9.4b*)

[HO 68] *Horiuchi, H.* and *K. Ikeda,* Progr. Theoret. Phys. **40** (1968) 277 (*14.2, 14.3, 14.8a*)

[HO 70] *Horiuchi, H.,* Progr. Theoret. Phys. **43** (1970) 375 (*5.2b*)

[HO 71] *Hodgson, P. E.,* "Nuclear Reactions and Nuclear Structure" (Clarendon Press, Oxford, 1971) (*11.1*)

[HO 72] *Horiuchi, H.,* Progr. Theoret. Phys. **47** (1972) 1058 (*5.2b*)

[HU 62] *Hull, M. H., Jr., K. E. Lassila, H. M. Ruppel, F. A. McDonald,* and *G. Breit,* Phys. Rev. **128** (1962) 830 (*1.1*)

[HU 68] *Hutzelmeyer, H.,* Ph. D. Thesis, Florida State University (1968) (*5.3c*)

[HU 69] *Hutzelmeyer, H., P. Kramer,* and *T. H. Seligman,* in "Proceedings of the International Conference on Clustering Phenomena in Nuclei", Bochum (IAEA, Vienna, 1969) p. 327 (*5.4*)

[HU 70] *Hutzelmeyer, H.* and *H. H. Hackenbroich,* Z. Phys. **232** (1970) 356 (*5.3c*)

[HU 72] *Hub, R., D. Clement,* and *K. Wildermuth,* Z. Phys. **252** (1972) 324 (*12.3, 12.3b, 12.4b*)

[HU 72a] *Hub, R.,* Ph. D. Thesis, University of Tübingen (1972) (*C*)

[IV 32] *Ivanenko, D.,* Nature **129** (1932) 798 (*1.1*)

[IN 39] *Inglis, D. R.,* Phys. Rev. **56** (1939) 1175 (*3.6a*)

[IW 57] *Iwamoto, F.* and *M. Yamada,* Progr. Theoret. Phys. **17** (1957) 543 (*5.3c*)

[JA 55] *Jastrow, R.,* Phys. Rev. **98** (1955) 1479 (*3.7, 5.3c*)

[JA 66] *Janssens, T., R. Hofstadter, E. B. Hughes,* and *M. R. Yearian,* Phys. Rev. **142** (1966) B922 (*5.3c*)

[JA 69] *Jacobs, H., K. Wildermuth,* and *E. Wurster,* Phys. Lett. **29B** (1969) 455 (*4.2b, 7.3c*)

[JA 70] *Jacobs, C. G.* and *R. E. Brown,* Phys. Rev. **C1** (1970) 1615 (*11.3a*)

[JE 58] *Jensen, R. C.* and *A. W. Fairhall,* Phys. Rev. **109** (1958) 942 (*17.2*)

[JE 60] *Jensen, R. C.* and *A. W. Fairhall,* Phys. Rev. **118** (1960) 771 (*17.2*)

[JE 70] *Jenkin, J. G., W. D. Harrison,* and *R. E. Brown,* Phys. Rev. **C1** (1970) 1622 (*7.3d*)

[JO 54] *Jones, G. A., W. R. Phillips, C. M. P. Johnson,* and *D. H. Wilkinson,* Phys. Rev. **96** (1954) 547 (*16.3b*)

[JO 71] *John, W., E. K. Hulet, R. W. Lougheed,* and *J. J. Wesolowski,* Phys. Rev. Lett. **27** (1971) 45 (*17.3, 17.4b*)

[KA 59] *Kanellopoulos, Th.* and *K. Wildermuth,* Nucl. Phys. **14** (1959) 349 (*3.5b, 3.6c*)

[KA 74] *Kankowsky, R.* and *D. Fick,* to be published (1974) (*7.3i*)

[KA 74a] *Kamimura, M.* and *T. Matsuse,* Progr. Theoret. Phys. **51** (1974) 438 (*7.3e*)

[KE 59] *Kerman, A. K.* in "Nuclear Reactions", edited by *P. M. Endt* and *M. Demeur* (North-Holland, Amsterdam, 1959) vol. I, p. 427 (*14.4*)

[KI 72] *King, T. R.* and *R. Symthe,* Nucl. Phys. **A183** (1972) 657 (*12.3a*)

[KO 50] *Kowarski, L.,* Phys. Rev. **78** (1950) 477 (*17.1*)

[KO 68] *Konecny, E.* and *H. W. Schmitt*, Phys. Rev. **172** (1968) 1213 (*17.2, 17.4d*)

[KO 74] *Koepke, J. A., R. E. Brown, Y. C. Tang,* and *D. R. Thompson*, Phys. Rev. **C9** (1974) 823 (*5.3d, 5.5, 7.3a*)

[KO 74a] *Konecny, E., H. J. Specht,* and *J. Weber*, in "Proceedings of the 3rd IAEA Symposium on Physics and Chemistry of Fission" (IAEA, Vienna, 1974) vol. II, p. 3 (*17.4d*)

[KO 75] *Korennoy, V. P., V. I. Kukulin, V. G. Neudatchin,* and *Yu. F. Smirnov*, to be published (1975) (*11.4b*)

[KR 53] *Kraus, A. A., Jr., A. P. French, W. A. Fowler,* and *C. C. Lauritsen*, Phys. Rev. **89** (1953) 299 (*16.3a*)

[KR 66] *Kramer, P.* and *M. Moshinsky*, Nucl. Phys. **82** (1966) 241 (*5.4*)

[KR 69] *Kramer, P.* and *T. H. Seligman*, Nucl. Phys. **A136** (1969) 545 (*5.2*)

[KR 69a] *Kramer, P.*, in "Proceedings of the International Conference on Clustering Phenomena in Nuclei", Bochum (IAEA, Vienna, 1969) p. 163 (*14.2*)

[KR 73] *Kramer, P., K. Wildermuth,* and *J. Beam*, Part. and Nucl. **5** (1973) 145 (*14.2, 14.4*)

[KU 55] *Kunz, W.*, Phys. Rev. **97** (1955) 456 (*7.3f*)

[KU 67] *Kutschera, W., D. Pelte,* and *G. Schrieder*, Nucl. Phys. **A111** (1967) 529 (*14.8b*)

[KU 68] *Kuo, T. T. S.* and *G. E. Brown*, Nucl. Phys. **A114** (1968) 241 (*8.2*)

[LA 58] *Lane, A. M.* and *R. G. Thomas*, Rev. Mod. Phys. **30** (1958) 257 (*15.2b*)

[LA 66] *Lauritsen, T.* and *F. Ajzenberg-Selove*, Nucl. Phys. **78** (1966) 1 (*3.6a, 3.6b*)

[LI 71] *Li, G. C., I. Sick, R. R. Whitney,* and *M. R. Yearian*, Nucl. Phys. **A162** (1971) 583 (*5.3a, 5.3c*)

[MA 48] *Mayer, M. G.*, Phys. Rev. **74** (1948) 235 (*17.1*)

[MA 54] *Magnus, W.* and *F. Oberhettinger*, "Formulas and Theorems for the Functions of Mathematical Physics" (Chelsea Publishing Co., New York, 1954) (*5.2a*)

[MA 60] *Massey, H. W.*, in "Nuclear Forces and the Few-Nucleon Problem", edited by *T. C. Griffith* and *E. A. Power* (Pergamon, London, 1960) vol. II, p. 345 (*4.3*)

[MA 60a] *Mang, H. J.*, Phys. Rev. **119** (1960) 1069 (*15.2c*)

[MA 65] *Marion, J. B.*, Phys. Lett. **14** (1965) 315 (*10.1*)

[MA 69] *Mahaux, C.* and *H. A. Weidenmüller*, "Shell-Model Approach to Nuclear Reactions" (North-Holland, Amsterdam, 1969) (*1.2, 9.1, 10.3a, 10.3b, 10.3c*)

[MA 71] *Mader, R.*, Diplomarbeit, University of Tübingen (1971) (*15.3*)

[MA 74] *Mack, C.*, Ph. D. Thesis, University of Tübingen (1974) (*15.2c*)

[MA 75] *Matsuse, T., K. Kamimura,* and *Y. Fukushima*, Progr. Theoret. Phys. **53** (1975) 706 (*7.3e, 13.4*)

[MC 64] *McDonald, D. G., W. Haeberli,* and *L. W. Morrow*, Phys. Rev. **133** (1964) B1178 (*7.3h*)

[MC 67] *McIntyre, L. C.* and *W. Haeberli*, Nucl. Phys. **A91** (1967) 382 (*7.3c*)

[MC 73] *McGrory, J. B.*, Phys. Rev. **C8** (1973) 693 (*8.2*)

[ME 50] *Meitner, L.*, Nature **165** (1950) 561 (*17.1*)

[ME 59] *Meyer-Berkhout, H., K. W. Ford,* and *A. E. S. Green*, Ann. Phys. (N. Y.) 8 (1959) 119 (*16.2*)

[ME 68] *Meyerhof, W. E.* and *T. A. Tombrello*, Nucl. Phys. **A109** (1968) 1 (*7.3g*)

[MI 58] *Miller, P. D.* and *G. C. Phillips*, Phys. Rev. **112** (1958) 2048 (*7.3a*)

[MI 73] *Michaudon, A.*, in "Advances in Nuclear Physics", edited by *M. Baranger* and *E. Vogt* (Plenum Press, New York, 1973) vol. 6, p. 1 (*17.4a, 17.4b*)

[MO 56] *Morinaga, H.*, Phys. Rev. **101** (1956) 254 (*14.4, 17.4b*)

[MO 57] *Moszkowski, S. A.*, in "Encyclopedia of Physics" (Springer-Verlag, Berlin, 1957) vol. 39, p. 464 (*3.3, 14.5*)

[MO 60] *Mottelson, B. R.* and *J. G. Valatin*, Phys. Rev. Lett. **5** (1960) 511 (*14.7*)

[MO 66] *Moszkowski, S. A.*, in "Alpha, Beta, and Gamma-Ray Spectroscopy", edited by *K. Siegbahn* (North-Holland, Amsterdam, 1966) vol. 2, p. 863 (*16.3b*)

[MO 67] *Moldauer, P. A.*, Phys. Rev. **157** (1967) 907 (*10.3b*)

[MO 69] *Morrow, L. W.* and *W. Haeberli*, Nucl. Phys. **A126** (1969) 225 (*7.3h*)

[MÖ 70] *Möller, P.* and *S. G. Nilsson*, Phys. Lett. **31B** (1970) 283 (*17.4b*)

[MU 73] *Mustafa, M. G., U. Mosel*, and *H. W. Schmitt*, Phys. Rev. **C7** (1973) 1519 (*17.4b, 17.4d*)

[MY 66] *Myers, W. D.* and *W. J. Swiatecki*, Nucl. Phys. **81** (1966) 1; University of California Lawrence Radiation Laboratory Report UCRL 17070 (1966) (*17.4c*)

[NE 69] *Neuhausen, R.*, Z. Phys. **220** (1969) 456 (*5.3c*)

[NE 69a] *Neudatchin, V. G.* and *Yu. F. Smirnov*, Progr. in Nucl. Phys. **10** (1969) 273 (*5.6*)

[NI 58] *Nilson, R., W. K. Jentschke, G. R. Briggs, R. O. Kerman*, and *J. N. Snyder*, Phys. Rev. **109** (1958) 850 (*7.3g, 11.4b*)

[NI 69] *Nilsson, S. G., C. F. Tsang, A. Sobiczewski, Z. Szymanski, S. Wycech, C. Gustafson, I. L. Lamm, P. Möller*, and *B. Nilsson*, Nucl. Phys. **A131** (1969) 1 (*17.4b*)

[NI 71] *Niem, L. C., P. Heiss*, and *H. H. Hackenbroich*, Z. Phys. **244** (1971) 346 (*3.4, 7.3g*)

[NO 71] *Norbeck, E., J. L. Gadeken*, and *F. D. Ingram*, Phys. Rev. **C3** (1971) 2073 (*10.2b*)

[NU 69] *Nuttale, J.*, Ann. Phys. (N. Y.) **52** (1969) 428 (*7.2c*)

[OE 72] *Oeschler, H., H. Schröter, H. Fuchs, L. Baum, G. Gaul, H. Lüdecke, R. Santo*, and *R. Stock*, Phys. Rev. Lett. **28** (1972) 694 (*11.3c*)

[OH 64] *Ohlsen, G. G.* and *P. G. Young*, Phys. Rev. **136** (1964) 1632 (*7.3c*)

[OK 66] *Okai, S.* and *S. C. Park*, Phys. Rev. **145** (1966) 787 (*7.3a*)

[OL 72] *Olsen, D. K., T. Udagawa, T. Tamura*, and *R. E. Brown*, Phys. Rev. Lett. **29** (1972) 1178 (*12.2*)

[OL 73] *Olsen, D. K., T. Udagawa, T. Tamura*, and *R. E. Brown*, Phys. Rev. **C8** (1973) 609 (*12.2*)

[OL 75] *Olsen, D. K., T. Udagawa*, and *R. E. Brown*, Phys. Rev. **C11** (1975) 1557 (*12.2*)

[OS 76] *Osterfeld, F.* and *H. H. Wolter*, Phys. Lett. **60B** (1976) 253 (*12.2*)

[PA 75] *Partridge, R. A., R. E. Brown, Y. C. Tang*, and *D. R. Thomspon*, in "Proceedings of the 2nd International Conference on Clustering Phenomena in Nuclei", College Park, Maryland (ERDA, Washington, D. C., 1975) p. 225 (*11.4b*)

[PE 56] *Perring, J. K.* and *T. H. R. Skyrme*, Proc. Phys. Soc. (London) **69** (1956) 600 (*14.2*)

[PE 57] *Peierls, R. E.* and *J. Yoccoz*, Proc. Phys. Soc. (London) **A70** (1957) 381 (*14.2, 14.3, 14.4, 14.5*)

[PE 60] *Pearlstein, L. D., Y. C. Tang*, and *K. Wildermuth*, Nucl. Phys. **18** (1960) 23 (*3.6d, 5.2a, 16.2*)

[PE 60a] *Pearlstein, L. D., Y. C. Tang*, and *K. Wildermuth*, Phys. Rev. **120** (1960) 224 (*5.2a, 7.3f*)

[PE 69] *Petitjean, C., L. Brown*, and *R. G. Seyler*, Nucl. Phys. **A129** (1969) 209 (*9.5b*)

[PH 60] *Phillips, G. C.* and *T. A. Tombrello*, Nucl. Phys. **19** (1960) 555 (*3.6a*)

[PH 61] *Phillips, G. C.* and *T. A. Tombrello*, Phys. Rev. **122** (1961) 224 (*3.6a*)

[PL 72] *Plattner, G. R., A. D. Bacher*, and *H. E. Conzett*, Phys. Rev. **C5** (1972) 1158 (*7.3f*)

[PO 62] *Polikanov, S. M., V. A. Druin, V. A. Karnaukhov, V. L. Mikheev, A. A. Pleve, N. K. Skobelev, V. G. Subbotin, G. M. Ter-Akop'yan*, and *V. A. Fomichev*, Soviet Phys. JETP **15** (1962) 1016 (*17.4a*)

[PO 69] *Poletti, A. R., J. A. Becker*, and *R. E. McDonald*, Phys. Rev. **182** (1969) 1054 (*16.3b*)

[PR 62] *Preston, M. A.*, "Physics of the Nucleus" (Addison Wesley, Reading, 1962) (*16.3a*)

[RE 69] *Reichstein, I.* and *Y. C. Tang*, Nucl. Phys. **A139** (1969) 144 (*5.3d*)

[RE 70] *Reichstein, I.* and *Y. C. Tang*, Nucl. Phys. **A158** (1970) 529 (*7.3f, 11.3b*)

[RE 75] *Remaud, B.* and *W. Laskar*, in "Proceedings of the 2nd International Conference on Clustering Phenomena in Nuclei", College Park, Maryland (ERDA, Washington, D. C., 1975) p. 232 (*13.2a*)

[RO 56] *Robertson, H. H.*, Proc. Cambridge Phil. Soc. **52** (1956) 538 (*7.3b*)

[RO 60] *Roth, B.* and *K. Wildermuth*, Nucl. Phys. **20** (1960) 10 (*16.3, 16.3b*)

[RO 67] *Rodberg, L. S.* and *R. M. Thaler*, "Introduction to the Quantum Theory of Scattering" (Academic, New York, 1967) (*8.2*)

[SC 54] *Schiff, L. I.*, Phys. Rev. **96** (1954) 765 (*5.3c*)

[SC 61] *Schwartz, C.*, Ann. Phys. (N. Y.) **16** (1961) 36 (*7.2c*)

[SC 62] *Schmid, E. W.*, Nucl. Phys. **32** (1962) 82 (*5.3b*)

[SC 63] *Schmid, E. W., Y. C. Tang*, and *R. C. Herndon*, Nucl. Phys. **42** (1963) 95 (*3.7c, 5.3b*)

[SC 63a] *Schmid, E. W., Y. C. Tang*, and *K. Wildermuth*, Phys. Lett. **7** (1963) 263 (*5.3b*)

[SC 69] *Schwandt, P., B. W. Ridley, S. Hayakawa, L. Put*, and *J. J. Kraushaar*, Phys. Lett. **30B** (1969) 30 (*11.3a*)

[SC 70] *Schultheis, H., R. Schultheis*, and *G. Süssman*, Nucl. Phys. **A144** (1970) 545 (*17.4c*)

[SC 71a] *Schultheis, H.* and *R. Schultheis*, Phys. Lett. **34B** (1971) 245 (*17.4c*)

[SC 71b] *Schultheis, H.* and *R. Schultheis*, Phys. Lett. **37B** (1971) 467 (*17.4c*)

[SC 72] *Schmelzbach, P. A., W. Grüebler, V. König*, and *P. Marmier*, Nucl. Phys. **A184** (1972) 193 (*7.3c*)

[SC 72a] *Schmid, E. W.*, in "Proceedings of the 11th International Universitätswochen für Kernphysik", Schladming, Austria (1972) (*7.3c*)

[SC 72b] *Schranner, R.*, Diplomarbeit, University of Tübingen (1972) (*9.4b*)

[SC 73] *Schwager, J.*, Nuovo Cimento **18A** (1973) 787 (*7.3c*)

[SC 73a] *Schultheis, H.* and *R. Schultheis*, Nucl. Phys. **A215** (1973) 329 (*17.4c*)

[SC 74] *Schenk, K., M. Mörike, G. Staudt, P. Turek*, and *D. Clement*, Phys. Lett. **52B** (1974) 36 (*12.3, 12.3b*)

[SC 74a] *Schultheis, H., R. Schultheis*, and *K. Wildermuth*, Phys. Lett. **53B** (1974) 325 (*17.4c*)

[SC 75] *Schranner, R.*, Ph. D. Thesis, University of Tübingen (1975) (*12.2*)

[SC 75a] *Schultheis, H.* and *R. Schultheis*, private communication (1975) (*17.4c*)

[SE 60] *Seagrave, J. D.*, in "Nuclear Forces and the Few Nucleon Problem", edited by *T. C. Griffith* and *E. A. Power* (Pergamon, London, 1960) vol. II, p. 583 (*7.3i*)

[SE 75] *Seeger, P. A.* and *W. M. Howard*, Nucl. Phys. **A238** (1975) 491 (*17.4c*)

[SH 60] *Sheline, R. K.* and *K. Wildermuth*, Nucl. Phys. **21** (1960) 196 (*3.7, 16.2, 16.3b*)

[SH 66] *Sherman, N. K., J. R. Stewart*, and *R. C. Morrison*, Phys. Rev. Lett. **17** (1966) 31 (*15.3*)

[SH 68] *Sherman, N. K., J. E. E. Baglin*, and *R. O. Owens*, Phys. Rev. **169** (1968) 771 (*15.3*)

[SH 75] *Shin, Y. M., D. M. Skopik*, and *J. J. Murphy*, Phys. Lett. **55B** (1975) 297 (*15.3*)

[SI 70] *Sick, I.* and *J. S. McCarthy*, Nucl. Phys. **A150** (1970) 631 (*5.3a*)

[SP 60] *Spencer, R. G., G. C. Phillips*, and *T. E. Young*, Nucl. Phys. **21** (1960) 310 (*3.6a*)

[SP 67] *Spiger, R. J.* and *T. A. Tombrello*, Phys. Rev. **163** (1967) 964 (*3.6a, 7.3a, 9.5b*)

[SP 67a] *Spiger, R. J.*, Ph. D. Thesis, California Institute of Technology (1967) (*7.3a*)

[SP 72] *Specht, H. J., J. Weber, E. Konecny*, and *D. Heunemann*, Phys. Lett. **B41** (1972) 43 (*17.4b*)

[ST 54] *Stephens, Jr., J., F. Asaro*, and *I. Perlman*, Phys. Rev. **96** (1954) 1568 (*17.4b*)

[ST 67] *Strutinsky, V. M.*, Nucl. Phys. **A95** (1967) 420 (*17.4b*)

[ST 71] *Stöwe, H., H. H. Hackenbroich*, and *H. Hutzelmeyer*, Z. Phys. **247** (1971) 95 (*5.3c*)

[SU 67] *Suelzle, L. R., M. R. Yearian*, and *H. Crannell*, Phys. Rev. **162** (1967) 992 (*5.3a, 5.3c*)

[SU 68] *Sünkel, W.* and *K. Wildermuth*, Phys. Lett. **26B** (1968) 655 (*7.3d*)

[SU 68a] *Sünkel, W.*, Ph. D. Thesis, University of Tübingen (1968) (*7.3d*)

[SU 72] *Sünkel, W.* and *K. Wildermuth*, Phys. Lett. **41B** (1972) 439 (*5.2b, 7.3e, 11.3c, 13.4, 14.8a*)

[SU 75] *Sünkel, W.* and *K. Wildermuth*, in "Proceedings of the 2nd International Conference on Clustering Phenomena in Nuclei" College Park, Maryland (ERDA, Washington, D. C., 1975) p. 156 (*16.1*)

[SU 76] *Sünkel, W.*, private communication (*7.3e*)

[SW 67] *Swan, P.*, Phys. Rev. Lett. **19** (1967) 245 (*11.4b*)

[TA 61] *Tang, Y. C., K. Wildermuth*, and *L. D. Pearlstein*, Phys. Rev. **123** (1961) 548 (*3.6a*)

[TA 62] *Tang, Y. C., K. Wildermuth*, and *L. D. Pearlstein*, Nucl. Phys. **32** (1962) 504 (*3.5c, 5.3e*)

[TA 63] *Tang, Y. C., E. W. Schmid*, and *K. Wildermuth*, Phys. Rev. **131** (1963) 2631 (*7.3a, 13.4*)

[TA 65] *Tang, Y. C., E. W. Schmid*, and *R. C. Herndon*, Nucl. Phys. **65** (1965) 203 (*5.3b*)

[TA 69] *Tang, Y. C.*, in "Proceedings of the International Conference on Clustering Phenomena in Nuclei", Bochum (IAEA, Vienna, 1969) p. 109 (*7.3a, 11.4b*)

[TA 71] *Tang, Y. C.* and *R. E. Brown*, Phys. Rev. **C4** (1971) 1979 (*11.3a, 11.3b, 11.4b*)

[TA 72] *Tabakin, F.*, Nucl. Phys. **A182** (1972) 497 (*5.2b*)

[TA 75] *Tang, Y. C.* and *D. R. Thompson*, in "Proceedings of the 2nd International Conference on Clustering Phenomena in Nuclei", College Park, Maryland (ERDA, Washington, D. C., 1975) p. 119 (*16.1*)

[TE 62] *Temmer, G. M.*, private communication (1962) (*17.2*)

[TE 73] *Teufel, G.*, Ph. D. Thesis, University of Tübingen (1973) (*9.5b*)

[TH 67] *Thompson, D. R.* and *Y. C. Tang*, Phys. Rev. **159** (1967) 806 (*7.3b, 7.3d*)

[TH 68] *Thompson, D. R.* and *Y. C. Tang*, Nucl. Phys. **A106** (1968) 591 (*7.3d, 15.3*)

[TH 69] *Thompson, D. R.* and *Y. C. Tang*, Phys. Rev. **179** (1969) 971 (*5.4*)

[TH 69a] *Thompson, D. R., I. Reichstein, W. McClure*, and *Y. C. Tang*, Phys. Rev. **185** (1969) 1351 (*7.3b*)

[TH 71] *Thompson, D. R.* and *Y. C. Tang*, Phys. Rev. **C4** (1971) 306 (*7.2a, 7.3e, 11.3a, 11.3b, 11.4b*)

[TH 72a] *Thompson, D. R., Y. C. Tang*, and *R. E. Brown*, Phys. Rev. **C5** (1972) 1939 (*11.2c, 11.3a*)

[TH 73] *Thompson, D. R.* and *Y. C. Tang*, Phys. Rev. **C8** (1973) 1649 (*4.2b, 5.3d, 5.4, 7.3c, 11.4b, 13.2a*)

[TH 73a] *Thompson, D. R., Y. C. Tang, J. A. Koepke*, and *R. E. Brown*, Nucl. Phys. **A201** (1973) 301 (*5.3d, 7.3d*)

[TH 74] *Thompson, D. R., Y. C. Tang*, and *F. S. Chwieroth*, Phys. Rev. **C10** (1974) 987 (*7.3c, 7.3g*)

[TH 74a] *Thompson, D. R.* and *Y. C. Tang*, Phys. Rev. **C10** (1974) 406 (*7.3c*)

[TH 75] *Thompson, D. R.* and *Y. C. Tang*, Phys. Rev. **C11** (1975) 1473; Phys. Rev. **C12** (1975) 1432 (*11.3b*)

[TO 61] *Tobocman, W.*, "Theory of Direct Nuclear Reactions "(Oxford University Press, London, 1961) (*12.1*)

[TO 63] *Tombrello, T. A.* and *L. S. Senhouse*, Phys. Rev. **129** (1963) 2252 (*7.3g, 11.4b*)

[TO 66] *Tombrello, T. A.*, Phys. Rev. **143** (1966) 772 (*7.3h*)

[TO 75] *Tohsaki, A., F. Tanabe*, and *R. Tamagaki*, Progr. Theoret. Phys. **53** (1975) 1022 (*16.1, 16.5a*)

[UN 74] *Unik, J. P., J. E. Gindler, E. L. Glendenin, K. F. Flynn, A. Gorski*, and *R. K. Sjoblom*, in "Proceedings of the 3rd IAEA Symposium on Physics and Chemistry of Fission" (IAEA, Vienna, 1974) vol. II, p. 19 (*17.3, 17.4b, 17.4d*)

[VA 69] *Van Oers, W. T. H.* and *J. M. Cameron*, Phys. Rev. **184** (1969) 1061 (*11.3b*)

[VA 73] *Vandenbosch, R.* and *J. R. Huizenga*, "Nuclear Fission" (Academic, New York, 1973) (*17.2, 17.3, 17.4a, 17.4b*)

[VE 63] *Verhaar, H. J.*, Nucl. Phys. **45** (1963) 129 (*14.2, 14.3, 14.5*)

[VE 63a] *Verhaar, H. J.*, Nucl. Phys. **41** (1963) 53 (*14.3, 14.5*)

[VE 64] *Verhaar, H. J.*, Nucl. Phys. **54** (1964) 641 (*14.3, 14.4, 14.5*)

[VE 71] *Ventura, E., C. C. Chang*, and *W. E. Meyerhof*, Nucl. Phys. **173** (1971) 1 (*15.3*)

[VL 57] *Vladimirskii, V. V.*, Soviet Physics JETP **5** (1957) 673 (*17.2*)

[VO 59] *Vogt, E.*, in "Nuclear Reactions", edited by *P. M. Endt* and *M. Demeur* (North-Holland, Amsterdam, 1959) vol. I, p. 215 (*16.3a*)

[VO 65] *Volkov, A. B.*, Nucl. Phys. **74** (1965) 33 (*5.3c*)

[VO 70] *Von Oertzen, W.*, Nucl. Phys. **A148** (1970) 529 (*13.1, 13.3, 13.4*)

[VO 74] *Votta, L. G., P. G. Roos, N. S. Chant*, and *R. Woody*, III, Phys. Rev. **C10** (1974) 520 (*12.3a, 13.4*)

[WA 68] *Wakefield, B.* and *B. E. F. Macefield*, Nucl. Phys. **A114** (1968) 561 (*16.2*)

[WE 51] *Weisskopf, V. F.*, Phys. Rev. **83** (1951) 1073 (*16.3b*)

[WE 64] *Werner, H.* and *J. Zimmerer*, Paris Conference on Nuclear Physics (1964) p. 241 (*11.4b*)

[WH 37] *Wheeler J. A.*, Phys. Rev. **52** (1937) 1083, 1107 (*4.3, 14.2*)

[WH 59] *Whetstone, Jr., S. L.*, Phys. Rev. **114** (1959) 581 (*17.2*)

[WI 48] *Wick, G. C.*, Phys. Rev. **76** (1948) 181 (*17.1*)

[WI 54] *Wild, W.* and *K. Wildermuth*, Z. Naturforsch. **9A** (1954) 299 (*5.3d*)

[WI 58] *Wildermuth, K.* and *Th. Kanellopoulos*, Nucl. Phys. **7** (1958) 150 (*3.3, 3.5a*)

[WI 59] *Wildermuth, K.* and *Th. Kanellopoulos*, CERN-Report 59-23 (1959) (*3.5b, 16.4*)

[WI 59a] *Wildermuth, K.* and *Th. Kanellopoulos*, Nucl. Phys. **9** (1958/59) 449 (*3.7b, 14.2, 14.4*)

[WI 60] *Wilkinson, D. H.*, in "Nuclear Spectroscopy", edited by *F. Ajzenberg-Selove* (Academic, New York, 1960) part B, p. 852 (*16.3b*)

[WI 61] *Wildermuth, K.* and *R. L. Carovillano*, Nucl. Phys. **28** (1961) 663 (*11.2d, 16.5a, 16.5b*)

[WI 61a] *Wildermuth, K.* and *Y. C. Tang*, Phys. Rev. Lett. **6** (1961) 17 (*16.2*)

[WI 62] *Wildermuth, K.*, "Lectures on the Cluster Model of the Nucleus", Technical Report No. 281, University of Maryland (1962) (*3.6b, 5.3d*)

[WI 63] *Willey, R. S.*, Nucl. Phys. **40** (1963) 529 (*5.3c*)

[WI 65] *Wittern, H.*, Nucl. Phys. **62** (1965) 628 (*11.2d*)

[WI 66] *Wildermuth, K.* and *W. McClure*, "Cluster Representation of Nuclei" (Springer-Verlag, Berlin, 1966) (*3.7d, 6.3, 7.3f, 16.5a*)

[WI 69] *Wildermuth, K.*, in "Proceedings of the International Conference on Clustering Phenomena in Nuclei", Bochum (IAEA, Vienna, 1969) p.3 (*4.3, 7.2c*)

[WI 72] *Wildermuth, K.*, in "The Structure of Nuclei", Trieste Lectures, (IAEA, Vienna 1972) p. 117 (*2*)

[WI 76] *Wildermuth, K.*, to be published (*14.9, 17.4a*)

[WU 62] *Wu, T.* and *T. Ohmura*, "Quantum Theory of Scattering" (Prentice Hall, London, 1962) (*1.1, 7.1, 7.2c, 8.4*)

[WU 67] *Wurster, E. J.*, Ph. D. Thesis, University of Tübingen (1967) (*7.3f*)

[YO 57] *Yoccoz, J.*, Proc. Phys. Soc. (London) **A70** (1957) 388 (*14.2, 14.3, 14.4, 14.5*)

[YU 72] *Yukawa, Y.*, Phys. Lett. **38B** (1972) 1 (*5.2b*)

[ZA 71] *Zaikin, D.*, Nucl. Phys. **A170** (1971) 584 (*5.2b*)

[ZA 71a] *Zakhariev, B. N. et al.*, Nuov Cim. **6A** (1971) 151 (*7.2c*)

Subject Index